PSYCHONEUROENDOCRINOLOGY RESEARCH TRENDS

PSYCHONEUROENDOCRINOLOGY RESEARCH TRENDS

MARTINA T. CZERBSKA
EDITOR

Nova Biomedical Books
New York

For permission to use material from this book please contact us:
Telephone 631-231-7269; Fax 631-231-8175
Web Site: http://www.novapublishers.com

LIBRARY OF CONGRESS CATALOGING-IN-PUBLICATION DATA

Psychoneuroendocrinology research trends / Martina T. Czerbska (editor).
p. ; cm.
Includes bibliographical references and index.
ISBN-13: 978-1-60021-665-7 (hardcover)
ISBN-10: 1-60021-665-X (hardcover)
1. Psychoneuroendocrinology. I. Czerbska, Martina T.
[DNLM: 1. Neuropsychology. 2. Hormones--physiology. 3. Mood Disorders--physiopathology. 4. Neuroendocrinology. WL 103.5 P927 2007]
QP356.45.P793 2007
612.8--dc22

2007013308

Published by Nova Science Publishers, Inc. ✦ New York

CONTENTS

PREFACE

Psychoneuroendocrinology is the clinical study of hormone fluctuations and their relationship to human behavior. It may be viewed from the perspective of psychiatry, where in certain mood disorders, there are associated neuroendocrine or hormonal changes. It may also be viewed from the perspective of endocrinology, where certain endocrine disorders can be associated with psychiatric illness. It is the blend of psychiatry and endocrinology. This new book presents the latest research advances in the field.

Expert Commentary - A vast amount of clinical and preclinical data support the concept that impaired glucocorticoid receptor (GR) signaling is a key mechanism in the pathogenesis of bipolar disorder (BD). The effects of antidepressants on GR signaling associated with the observed normalization of behavior have given further support to the Glucocorticoid Receptor Hypothesis of depression. Recent research uncovers that GR signaling is multicomponent, interacting with other cellular signaling pathways at various levels, implicating that elucidation of the GR signaling in BD is yet to be determined.

Intracellular GR signaling and glucocorticoid (GCs) sensitivity depend on several processes, e.g the interaction of GR with other cytoplasmic and nuclear proteins, such as the transcription factors AP-1, NF-kB, STAT-5, heat shock proteins (HSP90, HSP70) and an array of other chaperones, corepressors, coactivators, as well as with molecules residing to the inner surface of the plasma membrane (G protein complex components). The activation/inhibition of GR by stress-induced mitogen activated protein kinases (MAPKs), e.g JNK, p38, the discovery of the GRβ variant, as well as the recent data revealing cell-specific expression of multiple translational GR isoforms, add to the complexity of our understanding of GC-GR signaling.

In spite of the several chemical reactions and processes taking place at the molecular level, resulting in the overall GC action, much of the research effort so far in BD has been focused mainly upon determining the number and /or function of GR molecules. Clinical studies assessing the complex interactions of GR signaling with other cell signaling cascades in BD at the molecular level are only sparse and warrant further investigation. Existing data are limited in demonstrating the integrity of GRα and GRβ gene structure. Moreover, elucidation of the contributing roles of JNK/AP-1 and NF-kB signaling pathways in GR signaling cascade, have indicated that the derangements in GR and AP-1 mediated signaling (reduced GR-DNA binding and AP-1-DNA binding) observed in BD patients were normalized under effective medication. Taken together, a "localized GR resistance" has been implicated as integral to the underlying aetiopathogenetic mechanisms in bipolar disorder and the potential of GR signaling as molecular target in therapeutics. However, a further full understanding of the Glucocorticoid Receptor Hypothesis of depression from a molecular to a clinical level, must await the characterization of the interaction(s) of GR with the many other intracellular molecules and /or its interplay with other signaling pathways in BD patients, at different states of the disease, untreated or under different drug regimes.

In addition to the nuclear genomic effects induced by GC-GR complex and some rapid, non-genomic effects via plasma membrane GRs, GCs have shown substantial effects on mitochondrial function. Several lines of evidence indicate the involvement of GCs on energy transduction processes in mitochondria, associated with the existence of mitochondrial GR in brain cells. Such data implicate that a mitochondrial genome-GR interaction may be an important regulator of cellular energy metabolism and energy-dependent physiological processes in brain. It is of note, that a mitochondrial GR has been associated with apoptosis and GC-sensitivity in peripheral blood cells. These observations, together with the recent findings that BD patients are characterized by mitochondrial dysfunction, energy transduction alterations and increased apoptosis in brain cells, suggest that a novel pathway for the action of GC and GR in mitochondria of brain cells may play a key role in BD aetiopathogenesis and associated abnormalities. The delineation of such novel pathways and their integration to the behavioral level and clinical expression of BD should be a challenge of future research studies.

In conclusion, the rapid advances in molecular biology techniques, our knowledge of the human genome and the cross-talk between the nuclear and mitochondrial genome, the increasing understanding of cell signaling cascades, have revealed the complexity of GC-GR signal transduction. In view of above, the scrutiny of GR signaling in BD aetiopathogenesis and treatment is of high scientific interest and remains a continuous scientific challenging issue.

Chapter I - This chapter reviews the findings on differences between the sexes in cognition, emotion and brain functional organization. The role played by the organizational and activational actions of sex steroids, and the impact of these differences on brain function and pharmacological response to drugs are described in a laboratory rat model, with emphasis on electrical brain activity as a tool to assess the functional expression of sex differences in the brain. Brain oscillations are produced by neuronal assemblies that fire in synchrony reflecting basic mechanisms of brain function. Functional coupling between the oscillations from different brain regions appears to play a major role in neural communication and cognitive integration. Recent findings show substantial differences in oscillations and in functional coupling among brain regions between male and female rats, as well as between men and women. These differences are related with brain function, they are modulated by organizational and activational actions of sex steroids and may be related with sexually dimorphic aspects of cognition, emotion and pharmacological response to anxiolytic drugs. Benzodiazepines (BZ) involve allosteric modulation of $GABA_A$ receptors. Furthermore, sex steroids, specially progesterone and its metabolites exert modulatory effects on the $GABA_A$ receptor complex. However, although, BZ effects have been studied in groups of only men and of men and women, little attention has been paid to the study of the dimorphic effects of

BZ on brain function despite known sex differences in mood prevalence and anxiety disorders, in limbic circuits, and in frontal cortex progesterone receptors.

Present evidence calls attention to the need for further research on sex differences in brain organization to increase our understanding of psychopathology and drug prescription from a gender perspective.

Chapter II - Steroids produced in the periphery have long been of interest to reproductive biologists. The emergence of neuroendocrinology as a dominant subfield of behavioral studies appeared to demote steroids to a lesser status than the exciting neuropeptides that clearly interacted with many regions of the brain. Seminal findings in recent years, however, have propelled the study of steroids back onto center stage. In this chapter the authors identify seminal findings and describe why they are revolutionary. These new findings have demonstrated steroids have the capacity to modify the structure and function of brain regions both related to and unrelated to reproduction. The consequences on behavior are likely to be profound. This chapter focuses on one steroid, testosterone, because it occupies a strategic position in the metabolic cascade of the sex steroids. It is a position that can influence the actions of androgens and estrogens, as well as the adrenal precursors of testosterone and the recently identified neurosteroids, on the brain and behavior. The authors believe future research will reveal testosterone to have behavioral effects extending well beyond those related to reproduction. Still, the massive literature on testosterone – sociosexual behavior provides a solid background on which to begin that work.

Chapter III - Description and evaluation of new biomarkers in psychoneuroendocrinology is a constantly evolving field. Salivary biomarkers have received special attention since they are readily accessible and easily obtained. Salivary alpha-amylase has been proposed to be a sensitive biomarker for stress-related changes in the body, and a growing body of research is accumulating showing the validity and reliability of this parameter. This chapter attempts to describe salivary alpha-amylase as an emerging biomarker for stress and provides an overview of the current literature on stress-related alterations in salivary alpha-amylase. It is critically discussed how salivary alpha-amylase might reflect changes of the autonomic nervous system. New approaches in the measurement of salivary alpha-amylase and potential sources of confounding measurement results are identified. Finally, current and future fields of the application of salivary alpha-amylase measurement are outlined.

Chapter IV - The effect of estrogen on brain function and behavior has been well established in both humans and other animals. Although the changes in hormone levels that occur as part of the normal menstrual cycle in women influence neurochemistry, the changes themselves have not been systematically considered as variables of interest. As a redress, current literature is reviewed and it is proposed that the degree of natural estrogen flux is itself an individual difference variable worthy of study. An illustration of how such a model is possible is presented based on what is known about brain function focusing on speculation that serotonin receptors of subtype 1A could be a plausible vehicle for the effects of estrogen flux to occur. The literature regarding serotonin function and borderline personality disorder among women is reviewed and how the proposed model might account for various discrepancies in the research on borderline personality disorder and serotonin system reactivity is considered. As a whole, the theoretical model presented in which estrogen changes themselves can aggravate borderline personality disorder symptomology is shown to have preliminary support from several lines of research.

Chapter V - Despite recent legislative action to limit production and use, the abuse of methamphetamine (MA) has continued to increase. MA is a highly addictive stimulant that is both easy and cost effective to manufacture. As a result, the illicit use of MA is a growing problem in the United States. Immediate physical effects of MA use such as hypertension, hyperthermia, tachycardia, tachypnea, anxiety and paranoia are well documented, but the longer term physical and mental adverse effects of MA use are not as well defined. The importance of identifying the mental and physical effects of long-term MA use is essential for improved treatment. This chapter reviews the results of recent MA research in the areas of quality of life (QOL), depression and cognitive impairment. It also considers recent evidence that MA users exhibit atypical structural and functional brain characteristics that may underlie the suite of adverse cognitive and mental health characteristics associated with this drug of abuse. Finally, the potential additive and synergistic impact of users' sex and HIV status with MA use on physical and mental health disorders are considered.

Chapter VI - In major primates and humans hunger, satiety, food intake, food preference and aversion are regulated by a vast array of neurotransmitters, neuropeptides and peripheral peptides. The secretory patterns of most of them have been investigated in Eating Disorders (ED), including Anorexia Nervosa (AN), Bulimia Nervosa (BN) and Binge Eating Disorder (BED) and found to be mostly impaired. Alterations of the orexigenic neuropeptides, including neuropeptide Y, aguti-related peptide, opioid peptides, galanin, vasopressin, and of anorexigenic neuropeptides, including α-melanocyte stimulating hormone, brain derived neurotrophic factor, corticotropin-releasing hormone, thyreotropin-releasing hormone, neurotensin, somatostatin, and oxytocin, have been observed in AN and BN. Peripheral peptides stimulating or inhibiting hunger and satiety include ghrelin, leptin, insulin, cholecystokinin, peptide YY, bombesin, pancreatic polypeptide, gastrin-releasing peptide, neuromedin B, vasoactive intestinal peptide, gastrin, resistin and adiponectin. Their secretory patterns have also been found mostly impaired in ED. These central and peripheral peptides interfere also with the development of psychological aspects whose alterations occur in ED. In fact, they modulate mood, anxiety, aggressiveness, hedonic and rewarding patterns, and cognitive processes, particularly learning and memory. This suggests that neuropeptides and peripheral peptides specifically regulating hunger satiety and food intake may interfere not only in the eating pathology of AN and BN, but also in some of their psychological disorders.

Chapter VII - A significant literature exists on the role of psychosocial factors in cancer initiation and progression, and effects of psychosocial interventions on eventual survival, but research investigating the effects of psychosocial interventions on psychoneuroendocrine and psychoneuroimmune outcomes in cancer patients is rare. There is some evidence that stress-reduction interventions may affect cortisol secretion profiles and aspects of cellular immunity, but the clinical significance of any observed effects is not known. Questions that require additional investigation concern: 1) the significance of various endocrine and immune outcome measures for predicting disease outcome in cancer patients (i.e. disease recurrence and/or survival); 2) the optimal timing of psychosocial interventions to affect biological outcomes (i.e. pre- or post-surgery, chemotherapy); 3) the type and stage of cancers that are potentially most responsive to psychosocial interventions (e.g. early vs. late stage, tumour type), and; 4) consideration of other factors that may be mediating any biological changes seen as a result of psychosocial interventions (i.e. health behaviours). The research in these areas will be reviewed and fruitful directions for future research outlined.

Chapter VIII - Aspects of the hypothalamic-pituitary-adrenal (HPA) stress response system are programmed by experiential factors in rats during the early postnatal period. Important developmental events occur later on, during adolescence, in frontal brain regions that are critical for modulating HPA activity in the adult. However, it is currently unknown to what extent these alterations are sensitive to environmental conditions. In particular, the mesocorticolimbic dopamine system undergoes major restructuring in frontal regions during adolescence, including regions critically involved in regulating HPA output. The overall purpose of the work presented in this chapter is to begin evaluating the potential for environmental conditions to program adolescent development of the stress response system toward a context-specific optimum. In order to manipulate environmental conditions during this period, we have developed a novel, ecologically-relevant adolescent stressor paradigm involving repeated presentation of cat odour (periadolescent predator odour; PPO). This model was used in three separate objectives. In Objective 1 we examined long-term behavioural outcomes in the adult, in comparison with the commonly used maternal separation (MS) paradigm and a sham-separated control group. The purpose of Objective 2 was to examine adolescent corticosterone responses to the cat odour stressor, and in Objective 3 we investigated alterations in adult levels of D1 and D2 dopamine receptors in stress-responsive medial prefrontal cortex (mPFC) subregions. In this study, significant adolescent cort responses to cat odour were measurable in females but not males (Objective 2), and this corresponded with long-term alterations in the behavioural phenotype of PPO-exposed females but not males. Specifically, adult females who had been exposed to PPO exhibited a phenotype characterized by increases in measures of generalized anxiety (Objective 1). In addition, levels of D2 receptors were lower in mPFC subregions of both male and female adults who had been exposed to PPO. These findings support the contention that adolescence is a sensitive period for stress response programming, and a need for extensive future work is indicated.

Chapter IX - Early life stressors have long been known to have powerful immediate and lasting biobehavioral consequences. Though much has been learned about the neural and endocrine substrates of such effects, increasing evidence suggests that elements of the immune system may also play a substantial role. Neonatal exposure to lipopolysacchride (LPS) has been found to affect later behavioral and endocrine endpoints in rats and mice in ways that parallel the effects of more-traditional early experiences. In guinea pigs—a rodent model for studies of filial attachment and separation—proinflammatory factors appear to contribute to the behavioral reaction of pups during isolation in a novel environment. Evidence for this assertion includes findings that: (1) exposure of guinea pig pups to LPS produces the same constellation of passive behavioral effects as does protracted isolation from the mother in a novel environment; (2) isolation in a novel environment increases proinflammatory cytokine expression and core body temperature; and, (3) administration of anti-inflammatory agents attenuate passive responses during isolation. Further, it appears that corticotropin-releasing factor (CRF) may activate the immune system during isolation since administration of this peptide increases the passive responses in the same way as does prolonged isolation or injection of LPS, and anti-inflammatory treatment attenuates the behavioral effect of CRF. These findings will be reviewed, and their implications for the understanding of early-experience effects and the development of psychopathology will be discussed.

Chapter X - Early life stressors have long been known to have powerful immediate and lasting biobehavioral consequences. Though much has been learned about the neural and endocrine substrates of such effects, increasing evidence suggests that elements of the immune system may also play a substantial role. Neonatal exposure to lipopolysacchride (LPS) has been found to affect later behavioral and endocrine endpoints in rats and mice in ways that parallel the effects of more-traditional early experiences. In guinea pigs—a rodent model for studies of filial attachment and separation—proinflammatory factors appear to contribute to the behavioral reaction of pups during isolation in a novel environment. Evidence for this assertion includes findings that: (1) exposure of guinea pig pups to LPS produces the same constellation of passive behavioral effects as does protracted isolation from the mother in a novel environment; (2) isolation in a novel environment increases proinflammatory cytokine expression and core body temperature; and, (3) administration of anti-inflammatory agents attenuate passive responses during isolation. Further, it appears that corticotropin-releasing factor (CRF) may activate the immune system during isolation since administration of this peptide increases the passive responses in the same way as does prolonged isolation or injection of LPS, and anti-inflammatory treatment attenuates the behavioral effect of CRF. These findings will be reviewed, and their implications for the understanding of early-experience effects and the development of psychopathology will be discussed.

Chapter XI - The influence of interleukin-1 (IL-1) deficiency on aggressive behavior, social investigation, and endocrine status was studied in IL-1α/β deficient (IL-1-KO) male mice derived from the BALB/cA strain. In social investigation tests adult IL-1-KO males spent more time on chasing and sniffing and other form of interactions with juvenile intruders in comparison with wild type males. In contrast to very peaceful wild type males, which did not fight in intra-strain pair-wise test conducted 3-5 days after interactions with juveniles, IL-1-KO males showed fights in each test. Encounter in inter-strain pair-wise tests IL-1 KO and BALB/cA males experienced in contacts with juveniles also demonstrated higher aggressiveness in IL-1 deficient males. The difference in aggressiveness between the males of these strains was correlated with differences in androgen secretion. Concentration of testosterone in feces collected during 5 days of social isolation, including period of the short-term contacts with juvenile males, was higher in IL-1-KO males than in BALB/cA males. Males of both strains showed the similar concentrations and similar decline of fecal corticosterone after contacts with juveniles. Therefore, the reciprocal relationships between immunocompetence and aggressiveness predicted by life history trade-off could be established through the primary changes of the certain immunoregulatory functions such as cytokines.

Chapter XII - The glucocorticoid receptor (GR) has for long been considered in the aetiopathogenesis of bipolar disorder (BD). Reference is made on the GR gene and protein structure, its splicing variants, on important aspects of GR-mediated signaling events, such as the mode of action by gene activation through its glucocorticoid response elements GREs (transactivation effect) or via its interference with the cognate response elements of other transcription factors i.e. AP-1 and NF-kB (transrepression effect). The interaction of GR with other signaling molecules such as the mitogen activated protein kinases or the G-protein generated signals are presented. Recent advances on clinical and preclinical data delineating the role of GR signaling in BD as well as the effect of antidepressants on GR function are reviewed. Data regarding the presence of GR in the mitochondria of brain cells as well as the association of mitochondrial dysfunction with BD are briefly discussed.

Chapter XIII - The aim of this research project was to study the relationships between different coping styles adopted by subordinate mice with experience of chronic defeat and the neuroendocrine response to situations of social stress. Also, this study analyses whether different coping strategies for chronic and acute social stress are related to differences in emotional reactivity levels prior to stress. OF1 mice were subjected to different behavioral tests (Open field test, Plus-maze test and Boissier's test) in order to establish emotional reactivity profiles. Subsequently, they were socially stressed by repeated experiences of defeat for 23 days in a sensorial contact model. The authors results indicate a relationship between the coping style adopted by subordinate subjects in response to chronic social stress and the physiological response to stress. Subjects with a passive coping strategy for chronic social stress showed higher levels of corticosterone than those which adopted an active coping strategy. The data also indicate that a passive strategy for coping with chronic or acute social stress is related to greater emotional reactivity, and that the strategies adopted in response to acute social stress do not determine the behavior the subject will adopt in response to chronic social stress. The results in general may be of interest in the study of individual differences in susceptibility to illnesses related to social stress in humans, such as depression or anxiety.

Chapter XIV – The authors currently know that androgens act in the brain during its development, affecting both its structure and its neural function. In other words, in addition to the activating effects of adolescence, these hormones also have diverse organizational effects that, since they mold the central nervous system (CNS), influence both cognitive processes and the subject's behavior. Thus, a wide variety of data gathered over recent decades show that gonadal hormones affect the cognitive abilities of both men and women. Furthermore, recent studies carried out with normal populations instead of selected groups with hormonal abnormalities, have provided evidence that associates endogen hormone levels with behavior.

This chapter reviews existing research into both the relationship between androgen levels and diverse cognitive abilities, and the relationship between androgen levels and social behavior during childhood. First of all, it presents an overview of the principal androgens and their ontogenetic development, their main action mechanisms and their influence in the sexual differentiation of the brain and behavior. Next, it reviews the existing literature on the relationship between androgen levels and diverse cognitive abilities, paying special attention to social cognition (theory of mind, etc.) in children, before reviewing those research projects focusing on the relationship between androgens and social behavior in children, paying special attention to behaviors of dominance and aggression. Finally, the principal conclusions are discussed from an evolutionist or ultimate causes perspective.

Chapter XV - Stress is a universally known and well studied phenomenon that can be regarded as both necessary and potentially detrimental for a person's mental and physical well-being. Most mental disorders can be both triggered and worsened by stressful experiences. An important role of stress and its consequences has been assumed for a variety of mental afflictions ranging from depression to adjustment disorders. Accordingly, stress and the physiological stress response including changes of the hypothalamus-pituitary-adrenal (HPA) axis have been investigated extensively in mental health patients. In this course, stress hormone levels have been assessed in urine and blood and suppression and stimulation tests have been employed with important but often conflicting results. The assessment of salivary cortisol and the awakening cortisol response (ACR) has provided a new method to detect even subtle changes of the HPA axis by a reliable and non-invasive method, but the complexity of the system and the contradictory data imply that much about the role of stress and the HPA axis in mental disorders is so far not sufficiently understood. The present article gives an overview of the current literature concerning the HPA axis in mental disorders with a focus on depression and on the ACR. New results are presented on the ACR in psychotherapy inpatients. These preliminary results suggest that the ACR is a promising marker for evaluating the influence of therapeutic interventions on the HPA axis in psychiatric patients.

Chapter XVI - Patients with borderline personality disorder (BPD) frequently report early, multiple, and chronic adverse or even traumatic experiences like repeated sexual or physical abuse or emotional neglect. Early life stress has been suggested to be an important risk factor in the development of BPD, although this may not be the case in all patients. However, traumatization is not a specific risk factor for BPD. It has been shown that women with a history of childhood sexual or physical abuse are also more likely to exhibit anxiety or mood disorders and of cause posttraumatic stress disorder (PTSD). Noteworthy, PTSD and MDD are also common in BPD.

In patients suffering from PTSD and MDD alterations of the hypothalamic-pituitary-adrenal (HPA) axis were observed repeatedly. While several studies found decreased cortisol suppression after ingestion of 1mg dexamethasone (DEX) in MDD, PTSD was characterized by an enhanced suppression after a low dose (0.5mg) of DEX. In borderline personality disorder some authors reported low rates of cortisol non-suppressors after 1mg dexamethasone while others found high rates of non-suppressors in such patients. Up to now only few studies used the low dose DST in BPD. Findings indicate an impact of comorbid PTSD symptoms and in part of depressive symptoms on HPA axis feedback regulation.

In sum, the current literature does not suggest a clear pattern of HPA axis feedback dysregulation in BPD patients. Trauma-related and depressive symptoms seem to interact with regard to their effects on HPA axis regulation or might even cancel out each other by opposite effects. Further studies should continue to evaluate such interactions in the relationships of early life stress, BPD, PTSD, and MDD when investigating HPA axis function.

Chapter XVII - Functional somatic disorders are syndromes for which no clear or consistent organic pathology can be found. Since functional somatic disorders are associated with early life and chronic stress, the question raises how stress is linked to experiencing complaints. Psychoneuroendocrinological approaches combine psychological and biomedical views and provide a complex but promising new strategy to unravel somatization.

In this chapter, the authors will first introduce "the big three" functional somatic disorders and discuss the role of stress in these disorders. Next, the authors will give an overview of current theories and studies about psychoneuroendocrinological alterations in functional somatic disorders, with an emphasis on the two most important stress axes in the human body: the autonomic nervous system (ANS) and the hypothalamic-pituitary-adrenal-axis (HPA-axis).

The majority of alterations in ANS function found in functional somatic disorders consists of a decreased parasympathetic and/or an increased sympathetic activation, nonetheless, there are some dissonant findings. A critical approach will be taken to evaluate these findings.

Assessment of the HPA-axis usually demonstrates a lower cortisol output in patients with functional somatic disorders compared to healthy controls. However, a lot of studies examining the HPA-axis were not able to detect any differences, and few studies detect even an elevated cortisol output. The authors will make an attempt to show the most consistent evidence out of the heterogeneous methods and results reported.

Because of the heterogeneity in both ANS and HPA-axis findings in functional somatic disorders, the authors will provide some factors that may explain the discrepancies.

Finally, the authors discuss lacunas in our current knowledge about the role of the ANS and HPA-axis in somatization. Recommendations for future research include use of standardized research methods and performance of longitudinal cohort studies.

In: Psychoneuroendocrinology Research Trends
Editor: Martina T. Czerbska, pp. 1-2

ISBN: 978-1-60021-665-7

Expert Commentary A

GLUCOCORTICOID RECEPTOR SIGNALING IN BIPOLAR DISORDER: RESEARCH FRONTIERS

P. Moutsatsou*

Department of Biological Chemistry, Medical School, University of Athens, Goudi 11527, Athens, Greece.

ABSTRACT

A vast amount of clinical and preclinical data support the concept that impaired glucocorticoid receptor (GR) signaling is a key mechanism in the pathogenesis of bipolar disorder (BD). The effects of antidepressants on GR signaling associated with the observed normalization of behavior have given further support to the Glucocorticoid Receptor Hypothesis of depression. Recent research uncovers that GR signaling is multicomponent, interacting with other cellular signaling pathways at various levels, implicating that elucidation of the GR signaling in BD is yet to be determined.

Intracellular GR signaling and glucocorticoid (GCs) sensitivity depend on several processes, e.g the interaction of GR with other cytoplasmic and nuclear proteins, such as the transcription factors AP-1, NF-kB, STAT-5, heat shock proteins (HSP90, HSP70) and an array of other chaperones, corepressors, coactivators, as well as with molecules residing to the inner surface of the plasma membrane (G protein complex components). The activation/inhibition of GR by stress-induced mitogen activated protein kinases (MAPKs), e.g JNK, p38, the discovery of the GRβ variant, as well as the recent data revealing cell-specific expression of multiple translational GR isoforms, add to the complexity of our understanding of GC-GR signaling.

* Correspondence concerning this article should be addressed to P. Moutsatsou, Department of Biological Chemistry, Medical School, University of Athens, 75 Mikras Asias street, Goudi 11527, Athens, Greece. E-mail: pmoutsatsou@med.uoa.gr; Tel/Fax: 00302107462682.

In spite of the several chemical reactions and processes taking place at the molecular level, resulting in the overall GC action, much of the research effort so far in BD has been focused mainly upon determining the number and /or function of GR molecules. Clinical studies assessing the complex interactions of GR signaling with other cell signaling cascades in BD at the molecular level are only sparse and warrant further investigation. Existing data are limited in demonstrating the integrity of GRα and GRβ gene structure. Moreover, elucidation of the contributing roles of JNK/AP-1 and NF-kB signaling pathways in GR signaling cascade, have indicated that the derangements in GR and AP-1 mediated signaling (reduced GR-DNA binding and AP-1-DNA binding) observed in BD patients were normalized under effective medication. Taken together, a "localized GR resistance" has been implicated as integral to the underlying aetiopathogenetic mechanisms in bipolar disorder and the potential of GR signaling as molecular target in therapeutics. However, a further full understanding of the Glucocorticoid Receptor Hypothesis of depression from a molecular to a clinical level, must await the characterization of the interaction(s) of GR with the many other intracellular molecules and /or its interplay with other signaling pathways in BD patients, at different states of the disease, untreated or under different drug regimes.

In addition to the nuclear genomic effects induced by GC-GR complex and some rapid, non-genomic effects via plasma membrane GRs, GCs have shown substantial effects on mitochondrial function. Several lines of evidence indicate the involvement of GCs on energy transduction processes in mitochondria, associated with the existence of mitochondrial GR in brain cells. Such data implicate that a mitochondrial genome-GR interaction may be an important regulator of cellular energy metabolism and energy-dependent physiological processes in brain. It is of note, that a mitochondrial GR has been associated with apoptosis and GC-sensitivity in peripheral blood cells. These observations, together with the recent findings that BD patients are characterized by mitochondrial dysfunction, energy transduction alterations and increased apoptosis in brain cells, suggest that a novel pathway for the action of GC and GR in mitochondria of brain cells may play a key role in BD aetiopathogenesis and associated abnormalities. The delineation of such novel pathways and their integration to the behavioral level and clinical expression of BD should be a challenge of future research studies.

In conclusion, the rapid advances in molecular biology techniques, our knowledge of the human genome and the cross-talk between the nuclear and mitochondrial genome, the increasing understanding of cell signaling cascades, have revealed the complexity of GC-GR signal transduction. In view of above, the scrutiny of GR signaling in BD aetiopathogenesis and treatment is of high scientific interest and remains a continuous scientific challenging issue.

In: Psychoneuroendocrinology Research Trends
Editor: Martina T. Czerbska, pp. 3-5
ISBN: 978-1-60021-665-7

Expert Commentary B

Future Prospects for the Distinction of Depression Subtypes

Thomas Huber
Department of Clinical Psychiatry and Psychotherapy,
Medical School Hannover, Hannover, Germany

The ultimate reason why clinicians and researchers put a diagnosis to any mental affliction is the aim to draw therapeutic conclusions from this. In the case of depressive disorders, a number of different treatments have emerged including antidepressive drugs, different psychotherapeutic techniques, sleep deprivation, light therapy, electroconvulsive therapy, vagus nerve stimulation, stereotactic brain surgery, mood stabilizing drugs and recently medications originally invented as antipsychotics. The multitude of these approaches underscores two grave problems in clinical psychiatry: firstly the actual causes of depressive disorders are too complex to be pinpointed and can thus not be specifically aimed at therapeutically. And secondly, the rates of treatment resistance are still disconcertingly high. It is therefore still impossible to predict which patient will profit from which regimen. As a consequence, treating depressive disorders remains largely a matter of trial and error with patients often undergoing several treatment strategies before remission is achieved. It is as yet even impossible to predict whether a depressed patient will profit more from psychotherapy or pharmacotherapy or a combination of both.

The logical conclusion from this dilemma is to assume there are different subtypes of depressive disorders responding differently to the treatment strategies highlighted. Multiple attempts have been made to distinguish between such subtypes on the basis of an assumed cause, of symptom clusters or of pathophysiological alterations.

The most well known depressive entity distinguished by an assumed cause is adjustment disorder with depressed mood. Both the Diagnostic and Statistical Manual of Mental Disorders (DSM-IV) and the International Classification of Diseases (ICD-10) imply this disorder to be the direct response to an identifiable stressor; however, in clinical practice the onset of symptoms within three months of the onset of the stressor is often not met in spite of a convincing connection between both thus leading to a diagnosis of a depressive disorder.

Additionally, in many clear cut major depressive episodes, stressful life events can be identified and the decision whether to diagnose an adjustment disorder or a depressive disorder can sometimes seem rather arbitrary. Even more disturbing in this concept is the fact that adjustment disorders can be successfully treated with antidepressive agents and that depressive disorders without any identifiable connection to life events respond well to psychotherapy. The same appears true for the distinction between the classical "reactive" and "endogeneous" depression.

The current diagnostic systems of ICD-10 and DSM-IV largely refrain from any causal implications in the diagnosis of depressive disorders and base the diagnosis solely on the absence of an identifiable cause and the pattern of depressive symptoms. The abandoned entity of "endogenous depression" - a form of depression assumed to be largely independent of life events – is more or less found in the subtype "with melancholic features" of DSM-IV and "with somatic syndrome" of ICD-10. Similarly, the subtype of "depressive episode with atypical features" has been created on the basis of the symptom pattern. However, this distinction still appears unable to provide a reliable prediction of treatment response, although promising studies have suggested certain treatment modalities to be more likely advantageous in each of them.

When assuming the available treatment regimens for depression not to work on their cause and not on a specific subtype distinguished by its symptom pattern, a third option comes into view: since all treatment regimens have been proven to result in pathophysiological changes, possibly these changes are the most promising criteria for the distinction between therapeutically meaningful depression subtypes. One of the first targets in the attempts to identify such criteria has been the brain receptor function on the basis that different antidepressive drugs tackle different receptor systems. However, the hope to discern a serotonin from a more dopamine form of depression has not been fulfilled.

Functional and structural brain imaging studies have provided new insight into alterations of the brains of depressed subjects and the changes accompanying clinical improvement. Differing results suggest that there is more than one possible pattern of these alterations. Future research might be able to show distinguished subtypes of depression characterized by the pattern of brain function changes.

Another possibility could be endocrinological changes seen in depressive disorders. The hypothalamus-pituitary-adrenal axis has been investigated extensively and provided varying and often contradictory results as to the alterations associated with depression and its remittance. With respect to the physiological rise in salivary cortisol levels following awakening (awakening cortisol response, ACR), there appears to exist an exaggerated response in some depressed subjects but a blunted awakening cortisol rise in others. Tentative early results suggest depressed psychiatric inpatients as well as chronically depressed outpatients to show an exaggerated ACR whereas depressed psychotherapy inpatients as well as subjects suffering from burn out exhibit a blunted ACR. Regarding subjects undergoing inpatient psychotherapy, remittance of depressive symptoms appears not be linked with a simple "normalization" of altered ACR, but seems to follow a pattern reflecting the psychotherapeutic process. Preliminary results hint to the possibility that the ACR might have predictive quality as to which subjects will profit from the psychotherapy program investigated and when they are ready to move into the next treatment phase. However, these results have to be interpreted with extreme caution due to small sample sizes and lack of replication as yet.

It might be a fact that different subtypes of depression exist distinguished by symptom pattern, causes and pathophysiological changes including endocrinological and brain function alteration, but that none of these alleged subtypes actually predicts response to different treatments. In fact, convincing models have been suggested assuming multiple causes to lead to an alteration of a complex system of affective regulation and to converge in a final common pathway of depression. Any treatment tackling this dysregulation could be beneficial even though very different parts of the alleged system are targeted. In order to be able to predict treatment response more accurately, more must be known about the alterations in this alleged affective system, which includes the limbic system, multiple influences from different brain regions, changes of receptor pattern and function, of neuroplastic factors as well as endocrinological changes.

Research on the question whether new techniques may be able to differentiate subtypes of depression responding differently to different treatment strategies is still scarce. However, studies on brain functioning and structure and on the awakening cortisol response are promising with respect to a new distinction of depressive disorders which may be able to allocate patients to the right therapy more accurately in the future.

In: Psychoneuroendocrinology Research Trends
Editor: Martina T. Czerbska, pp. 7-72

ISBN: 978-1-60021-665-7

Chapter I

Sex-Steroid Dimorphic Effects on Functional Brain Organization: Differences in Cognition, Emotion and Anxiolysis

María Corsi-Cabrera[1], *Yolanda del Río-Portilla*[2] *and Zeidy Muñoz-Torres*[1]
[1]Sleep Laboratory, Facultad de Psicología, Universidad Nacional Autónoma de México;
[2]Sleep Laboratory, and Department of Psychophysiology, Facultad de Psicología, Universidad Nacional Autónoma de México.

Abstract

This chapter reviews the findings on differences between the sexes in cognition, emotion and brain functional organization. The role played by the organizational and activational actions of sex steroids, and the impact of these differences on brain function and pharmacological response to drugs are described in a laboratory rat model, with emphasis on electrical brain activity as a tool to assess the functional expression of sex differences in the brain. Brain oscillations are produced by neuronal assemblies that fire in synchrony reflecting basic mechanisms of brain function. Functional coupling between the oscillations from different brain regions appears to play a major role in neural communication and cognitive integration. Recent findings show substantial differences in oscillations and in functional coupling among brain regions between male and female rats, as well as between men and women. These differences are related with brain function, they are modulated by organizational and activational actions of sex steroids and may be related with sexually dimorphic aspects of cognition, emotion and pharmacological response to anxiolytic drugs. Benzodiazepines (BZ) involve allosteric modulation of $GABA_A$ receptors. Furthermore, sex steroids, specially progesterone and its metabolites exert modulatory effects on the $GABA_A$ receptor complex. However, although, BZ effects have been studied in groups of only men and of men and women, little attention has been paid to the study of the dimorphic effects of BZ on brain function despite known sex differences in mood prevalence and anxiety disorders, in limbic circuits, and in frontal cortex progesterone receptors.

Present evidence calls attention to the need for further research on sex differences in brain organization to increase our understanding of psychopathology and drug prescription from a gender perspective.

Keywords: Sex differences, cognition, emotion, EEG, coherence, brain anatomy, brain function, sex steroids

Introduction

Men and women are not identical copies of the human being template. Although they share most of the species characteristics, each sex as a group differs in many aspects that extend well beyond differences in reproductive physiology and behavior. However, except perhaps for endocrinology and cultural studies, in most spheres of science including behavioral and biological fields, research has mainly been conducted on groups of only males or of males and females mixed in the same group.

Almost 150 years have passed since Darwin´s observations on sex differences. However, it has not been until relatively recently that differences between men and women in behavior, cognition and emotion have been recognized and systematically investigated by scientific research in psychology, neuroscience and other related fields. Our understanding of sex differences has grown in the past few decades, first by evidence from post-mortem studies and now by more recently developed techniques such as neuroimaging and molecular biology that have shown extensive sexual dimorphism of the brain at structural, ultrastructural and biochemical levels, not only in regions dealing with reproduction control, but in areas widely distributed all over the brain. Differences in brain function have been disclosed by techniques assessing brain electrical activity such as EEG, EMG and ERP, and metabolic activations such as PET and fMRI and have revealed that structural brain differences are expressed as differences in functional organization which in turn impact cognition, emotion and behavior.

The identification of brain receptors for sex hormones and the growing evidence of the influence of sex hormones on brain structure and function demonstrating that sexual dimorphism depends to a large extent on the hormonal milieu during development and in adulthood, have reopened the question of the origin of brain sex differences. Although causal effects cannot be directly inferred, the degree of parallelism between the brain systems known to control behavior that show structural and functional sexual dimorphism is striking and allows to suggest interacting links.

Sex differences, when recognized, have been taken for granted or explained solely on the basis of cultural and environmental influence. Multiple factors appear to be responsible for sex differences. Plasticity is one of the main attributes of the brain, which is continuously transforming through learning and experience. Its remarkably plasticity offers a broader perspective on possible explanations of sex differences. However, the many instances of cultural influence do not preclude the important neurobiological substrates. Sociocultural influence that sculpts human behavior is acting on a pre-wired brain that has been developing for millions of years by adaptation for survival, as a species, and by personal experience in a life-time of each individual in an endless spiral of dynamic interactions.

The recognition of sex differences has nothing to do with legal, labor, social and interpersonal inequality or unjust/discrimination. Research has demonstrated that sex differences have an impact on developmental and neurodegenerative risks, psychiatric and affective disorders, neurological illness and response to drugs. Thus, sex differences are essentially important from a theoretical point of view that would allow us a deeper understanding of brain and behavior, and from a practical point of view to improve treatment approaches and educational programs.

Evidence in the human being is restricted to clinical cases and to data obtained in non-invasive studies. The understanding of hormonal, biochemical, genetic and neurophysiological mechanisms underlying sex differences requires experimental manipulations that have forced us to turn to animal models, which in turn impose inherent constraining limitations to extrapolate conclusions to the human being. It cannot be safely assumed that sex differences observed in different species are under a single organizing principle, nevertheless, animal models have made valuable contributions to the understanding of mechanisms underlying sex differences.

The aim of this chapter is to review evidence on sex differences in the human being and in animal studies that may help to explain the underlying biological mechanisms. The chapter is divided into five main sections. The first section begins with a summary of relevant evidence from human studies showing sex differences in emotion and cognition and explores pertinent data on structural and functional organization of the brain that may help to understand these differences. The second section presents data obtained in animal models for possible explanations of sexual dimorphism based on the role of sex steroids in establishing and maintaining sex differences. The third part is focused on findings of our research group using electrophysiological measures to assess the functional expression of brain sexual dimorphism in a rat model developed at the laboratory. Here we demonstrate that structural differences are translated into brain function and are influenced by organizational and activational actions of sex steroids. The fourth section outlines research demonstrating that behavioral and brain function are modified by benzodiazepine anxiolytics, and that these functions depend on sex steroid actions during development and in adulthood. The final section, after having glimpsed at some general principles from animal studies, describes research findings suggesting that brain response to anxiolytics is also sexually dimorphic in humans. In conclusion, a deeper understanding of sex differences is a pending assignment for psychology and neuroscience.

Further research is needed on sex differences in brain organization for the understanding, from a gender perspective, of social interaction, psychopathology, educational programs, therapeutic approaches, and drug prescription.

Behavioral and Brain Differences between Men and Women

Sex Differences in Cognition and Emotion

Convergent evidence from clinical and neuropsychological studies shows that men and women differ in emotional perception and experience, and in some cognitive abilities and strategies used for task solution.

Although men and women show similar overall level of intelligence (IQ), normal distribution of several cognitive abilities differs between men and women populations. Two of the cognitive abilities that show consistent sex differences are verbal speed and fluency, which favor women, and target directed motor tasks and spatial manipulations requiring mental rotation and shape transformation, which favor men (for review see 223 and 181; 211, 222, 139, 207, 37, 286). Despite some controversial results, there is evidence suggesting that women also perform better in several tests such as perceptual speed, visual memory, verbal reasoning, arithmetic calculation and fine motor skills, whereas men are better at mathematical reasoning and geometric principles and concepts [181].

Numerous studies on mental disorders have revealed that women suffer more often from anxiety disorders and depressive illness than men, and that psychoneuroendocrine response to stress and to challenging situations is also different for each sex [268]. Women are believed to experience and express most emotions more intensely and more frequently than men [84]. Women show better performance in both, speed and accuracy in recognizing facial expression of emotions [217], especially facial expressions of happiness, than men [45]. They also rely to a greater degree on emotional content when processing information [38], and identify better non-verbal auditory and body emotional cues [133], are more susceptible to interference from incongruent word valence and emotional prosodic messages, and integrate language and prosodic information faster than men [301]. Subjective emotional response to music also differ between women and men; women attribute overall greater positive feelings to music [5] and identify the feeling of being happy induced by music with greater number of positive emotions such as being cheerful, satisfied, inspired and pleased, while men identify happiness only by being comfortable and tranquil [95].

Menstrual Cyclic Changes in Cognition and Emotion

In addition to differences between men and women, steroid production during the menstrual cycle has been implicated in central nervous disorders such as hyperkinetic movement disorders [213] and epilepsy [299].

Many women report that they experience changes in intellectual efficiency and mood, temporally linked to the menstrual cycle and this is supported by experimental findings. Cognitive abilities associated with gender differences oscillate as a function of the menstrual cycle, whereas, contradictory results have been reported for other cognitive tasks. Abilities favoring women, such as motor coordination, verbal fluency and spatial memory tests show improvement during the periovulatory or early luteal phases when estrogen and progesterone levels are high and decline during the premenstrual period, whereas performance on spatial skills tasks improves and verbal and manual skills deteriorate [39,134]. Spatial attention and laterality of language also change through the cyclic variations in the hormonal milieu; the well-known right ear (left hemisphere) advantage for language is more evident in menstrual than in midluteal phase [140].

Women report a subjective feeling of difficulty to initiate activities and to confront challenging situations or problems, enhanced distractibility and lack of concentration in the premenstrual phase, suggesting transient changes in the executive functions of the frontal cortex, probably related with an influence of ovarian steroids on this part of the cortex; however, only a few studies have investigated frontal lobe functions during the different phases of the menstrual cycle. Worse performance during the premenstrual phase has been reported on random number generation correlated with the self-rated severity of premenstrual symptoms [40], and in the Stroop´s color-word naming test [209], whereas better performance has been found in this test during the follicular phase [176]. In an experiment with simultaneous recording of EEG activity during performance in the Wisconsin Card Sorting Test (WCST) that requires specific prefrontal functions, we found better performance in the early luteal phase when progesterone levels are at its highest and during the menstrual phase when hormone levels are at its lowest associated with decreased alpha power, and worst performance during the late luteal phase associated with decreased power in fast oscillations [329]. In a PET study, using the WCST, the subjects presented a tendency to perform best during hormone (progesterone or estrogen) administration associated with activation of the areas involved in the performance of the task, such as dorsolateral prefrontal cortex (DLPC), inferior parietal lobule and the posterior region of the inferior temporal lobule, in contrast to the attenuated activation pattern seen in the same areas with Lupron, a synthetic nonapeptide used to suppress gonadal steroid secretion [23].

Changes in mood and affective symptoms such as anxiety, depression and irritability have been also observed during the menstrual cycle [176,293]. Increased levels of autonomic and psychological arousal: skin conductance, temperature, heart rate and blood pressure, as well as urinary excretion of adrenaline and noradrenaline are highest in the luteal phase. Aggression and hostility, depression, irritability and anxiety, self-reports of stressful events, self-preoccupation and general stress level are also higher during this phase. Self-reported anxiety levels along the menstrual cycle are correlated with specific EEG oscillations; higher anxiety levels are associated with increased alpha activity, whereas low anxiety levels correlate with low theta activity and coherent activity, especially within the right hemisphere [328]. The increase in alpha oscillations with the level of anxiety is consistent with that observed after benzodiazepine anxiolytic administration [182] and withdrawal [148], and stressful situations [105,241,276], and the low theta activity and lower inter- and intrahemispheric coherent activity associated with low anxiety is in agreement with results reported in rats after benzodiazepine administration (see below; 348).

Sex Differences in Brain Morphology

Sex differences are not restricted to cognition and emotion; postmortem and, more recently, neuroimaging studies with magnetic resonance images (MRI) have shown sex differences in structural and ultrastructural brain organization in the human brain in neural systems involved in cognition and emotion. Structural brain differences confer a different substrate for functional organization and may underlie behavioral sex differences.

The male and female brain differs not only in brain size and weight corrected for body and/or head size, but also in specific brain structures not directly related to reproductive behavior. Men have greater intracranial volume [190,3,92] and larger lobes than women [239], while the frequency of sulcal and giral convolutions in frontal and parietal regions is less in men than women [210]. The right Heschl gyrus, but not the left, is sexually dimorphic [184]. Men also possess larger volumes of cerebellum [284,92] and parietal association cortex [67] both related with spatial ability and motor coordination, whereas women show overall larger volumes of cortex [111] and caudate [92]. The amygdala [110-111], hypothalamus and frontomedial area [111] related with emotional processing are larger in men, while other emotion-related areas such as hippocampus [92] orbitofrontal cortex [127] and medial paralimbic region are larger in women [111].

Independently of sex differences, left and right sides of the brain are not equal; there is a well-recognized overall leftward posterior and rightward anterior asymmetry or brain torque with significant asymmetries in frontal temporal and occipital lobes [113]. This brain asymmetry increases in the antero-posterior direction starting from the central region [190]. The leftward asymmetry is evident in the planum temporale, a triangle in the superior temporal gyrus coincident with part of the Wernicke area, cytoarchitectonically corresponding to the secondary auditory cortex [108,15,311,113], Heschl gyrus [113] and in the left Silvian fissure [97]. The parietal operculum [97] and the left temporo-parietal region [273] are also larger on the left side. The right central sulcus is larger and deeper, probably indicating increased connectivity between motor and somatosensory cortices in the left side, facilitating fine movements [67]. The left cingulate gyrus is also larger in the left side [3], whereas the entorhinal cortex [153] amygdala [92] and the temporo-parieto-occipital area have larger volume on the right side [273], as does the lateral ventricle [83].

Not only brain size or volume is sexually dimorphic, brain asymmetries are also different in men and women. Male brains are more asymmetric than female brains; meanwhile, women have lower degree of asymmetry between left and right cortical thickness [70]. Leftward asymmetries are larger in men in planum temporale [113,311], superior temporal gyrus [190], Heschl gyrus [113,190], deeper central sulcus [7], overall temporal and parietal [273] and inferior parietal lobule [99], thalamus and posterior cingulate [190]. However, some studies reveal contradictory results, and report that only women show planum temporale leftward asymmetry [184].

The sexual dimorphisms extend from gross morphological/anatomical differences to differences in brain tissue composition; the difference between men and women being greater for white matter proportion as a whole than for gray matter in women [3]. The proportion of gray to white matter (G/W) is also different in men and women. The overall G/W matter ratio is larger in women than in men [3,239,129,92] with total intracranial volume corrected [129], especially in dorsolateral prefrontal cortex and superior temporal gyrus [302]. Conversely, men have greater white matter proportion in mesial temporal lobe, entorhinal and perirhinal cortex and anterior cerebellar lobes [113]. The sex difference in G/W matter proportion is reversed in heteromodal association areas of inferior parietal lobule where men show larger gray matter volumes than females [100]. However, despite the overall greater white matter tissue in men, there are several studies showing that commissural tracts interconnecting both hemispheres are larger in women. Cross-sectional area of the corpus callosum proportional to cerebral volume [229] and to forebrain size [334] is larger in women, particularly in the splenium [78,68] also in rhesus monkeys [98], the anterior isthmus [361,103] and posterior section [170,103,296]. The anterior commisure is also larger in women and the presence of the massa intermedia is more frequent in women than in men [4]. However, other studies have found that the corpus callosum is longer [336] and larger [79] in men for their brain size especially in the genu [78], or alternatively, no differences between the sexes are observed [26].

Without considering sex variability, tissue composition is also asymmetric; cingulum bundles tractography show greater left asymmetry except in the posterior portion [112], while gray matter ratio is greater in the left insula [3]. Parasagital projections from anterior body and anterior third of the corpus callosum are more diffuse on the right side [210]. The overall proportion of gray matter is larger on the left side in men [129]. Males are more asymmetric with larger left white matter in frontal and temporal perisylvian region [273] and in temporal stem and optic radiation [190], while women have a larger proportion of gray matter in left superior temporal gyrus, Heschl gyrus, planum temporale, inferior frontal, cingulate gyrus and central sulci margins [113] and in the right parietal, than males [239]. Women also show gray matter concentrations in parahippocampal gyri and in the banks of cingulate and calcarine sulci [113]. Hemispheric asymmetry is correlated with corpus callosum size in males but not in females, suggesting a sex-dependent decrease in interhemispheric connectivity with increasing hemisphere asymmetries [77].

The difference in G/W matter composition confers a different brain organization in men and women, with men possessing higher intrahemispheric long-range interconnectivity than women, while women possess greater interhemispheric connectivity and more neurons favoring local diffuse networks.

Sex Differences in Functional Organization of the Brain

Although it is not possible to establish direct causal effects, structural brain differences confer sexually dimorphic attributes that are translated into functional brain differences as demonstrated by the major neuroimaging techniques available for the study of human brain function, electroencephalography (EEG), magnetoencephalography (MEG), positron emission tomography (PET) and functional magnetic resonance imaging (fMRI). Although there are some discrepancies among findings possibly due to differences in methodology, stimuli, tasks and recording techniques, neuroimaging evidence converges to show sex differences in functional organization of the brain at rest and during cognitive engagement, consistent with evidence from behavioral and clinical results. Three main patterns of sex differences in brain functional organization emerge, the male brain seems be more asymmetrically lateralized than that of women, women activate wider networks than men while men favor long-range interconnected areas, and, finally, men and women use different neuronal systems for the same cognitive task differing in antero-posterior cortical representation of functions within the left and the right hemisphere.

Regarding lateralization of cognitive functions, such as verbal and spatial processes, lower hemispheric specialization has been found in women as opposed to men in behavioral studies (for a review 223) using either visual [199] or auditory stimuli [222,37]. The strategies used for solution of hemispheric specialized tasks are also different. Men use self-centered reference for spatial position (are more field-independent), while women use external cues as a reference for spatial location (are more field-dependent) [222,207] or serial approaches [274,278].

EEG power asymmetry is higher in men than in women during verbal and spatial processing [19,60,286,151]. The leftward asymmetry in men and more diffuse and symmetrical pattern in women has been confirmed by metabolic (fMRI) for phonological [50,313] and semantic tasks [292,16,351], and during passive listening to narrative tests [266]. Other studies, have found no lateralization during hearing a story read aloud [126], and no sex differences in a verbal fluency task [304]. In a PET study, producing past tense forms, men showed left-lateralized activation while women recruited bilateral perisylvian cortex [160].

In relation to metabolic brain activation during spatial task solution, there is less consensus regarding hemispheric lateralization [164,342,107,126,121,30], probably due to the large variety of spatial tasks tested, each demanding different brain operations. Nevertheless, results indicate sex differences in engaged brain regions, probably related with the strategies used for their solution; men rely on a gestalt strategy for mental rotation tasks while women use a serial one more frequently; accordingly, a stronger activation in visuo-spatial brain areas, such as the parietal region, has been reported for men and in serial processing regions, such as the frontal cortex (BA 44/45) for women [358,342].

Brain responses to emotional perception and subjective experience are also less lateralized in women than in men. Metabolic results show predominant participation of bilateral networks in women and more lateralized to the right in men during recognition of facial expressions [177,352,46,306] and remembered unpleasant [46] or arousing pictures [44,300]. Bilateral EEG activation in women has also been documented for musically induced emotions [305,5], as well as bilateral uncoupled alpha oscillations in women for negative emotions, while in men the uncoupled network involves midline and posterior regions in the right hemisphere [94].

There is also evidence on functional sex differences, in overall cerebral blood flow [128], in the extent of activated networks and in the functional relationship between brain areas, which jointly point to a brain sexual dimorphism much greater than only hemispheric specialization; for example, in women, during rest, the left amygdala was associated with greater functional connectivity in regions such as subgenual prefrontal cortex and hypothalamus, while in men, the right amygdala was associated with greater functional connectivity with sensorimotor cortex, striatum and pulvinar [178]. Clinical outcome after brain lesions for some verbal and nonverbal processes shows that the basic speech and motor praxis functions are more diffusely organized within the left hemisphere in men, probably related with predominance of long-range connectivity, while they are more focally (anterior) localized in women [180], probably related with diffuse local networks provided by higher G/W matter ratio.

Brain oscillations recorded as field potentials or in EEG scalp recordings come from neuronal assemblies firing in synchrony. Usually spectral power and integrated amplitude estimates have been used to assess changes in EEG activity related to cognitive engagement. These analyses give information about the general level of energy in a particular band or frequency of oscillation. Even though changes in EEG power of specific oscillations or in metabolic activity may in part reveal the engagement of certain brain areas in particular operations, the sole activation of local brain regions during cognitive processing does not convey important information on functional relationships among them. The functional coupling or coherent activity among brain regions has become a relevant element in the explanation of the binding of spatially disperse operations over several brain regions into unified information necessary for cognitive functions [80,208,324]. Enhanced synchronization between neuronal assemblies has been shown to be involved in the efficacy of processing simultaneous common inputs [344]. Synchronized activity over widespread brain areas conforming networks of temporally related activation underlies cognitive complex functions such as language [264], consciousness [162] and distorted cognition during dreaming [261].

Coherence and cross-correlation analyses of the EEG are considered appropriate to study the functional relationship between two sites of the brain because they emphasize temporal coupling of simultaneous ongoing activity between them. Both analyses, the first one in the frequency domain and the latter in the time domain, provide mathematical values about the linear relationship between two brain regions [312], indicating the degree of functional coupling between them. Studies using both, power and coherence approaches, have found that changes in either of these variables can occur independently [19] since changes in power depend on the mass of synchronized neuronal assembly, while changes in coherent activity depend on the temporal patterning between two neuronal assemblies. Some studies have found that coherent activity is more sensitive than EEG power measurements to differences in cerebral organization and to ongoing regional interactions.

Men and women differ in their patterns of inter and intra-hemispheric relationships revealing a different functional organization during photic stimulation [237] and at rest, as well as during cognitive performance and emotion. Interhemispheric functional coupling is higher in women at rest as well as during cognitive activity [19,96,279,60,11], whereas intrahemispheric correlation [60], as well as long-range correlations [238] are higher in men, implying a different cortical organization in each sex. The higher interhemispheric correlation in women indicates higher interhemispheric cooperativity and lower level of hemispheric differentiation, in agreement with lower hemispheric specialization during verbal and spatial processing found in women as opposed to men and with larger commissural tracts in women, whereas the higher intrahemispheric correlation in men agrees with the higher proportion of white over gray matter within the same hemisphere reported in anatomical studies mentioned above, and with greater modulation of metabolic activity within each hemisphere in men when naming concrete nouns [116] and with clinical outcome (Kimura,

1987). A certain pattern of functional relationship may be beneficial for women and disadvantageous for men, for example, failure in information processing in men, but not in women, is associated with higher interhemispheric functional coupling, whereas successful trials are associated with lower interhemispheric coupling during the solution of verbal and mixed tasks [55].

A study carried out in our laboratory [94] showed that not only the local activation but also the functional relationship among brain regions is involved in pleasant and unpleasant emotions induced by music. Results revealed that the subjective experience of pleasant emotions was sustained by a larger network of coherent alpha activity in women, linking all frontal regions with the left anterior temporal and posterior parieto-temporal association areas, which suggests a greater interaction between the anterior limbic system and frontal region and posterior temporo-parietal association areas, probably mediating greater emotional and perceptual integration in women. These brain areas have shown metabolic [32,185,93] and electric activation [305,5] during positive emotion. The larger network mediating emotion in women is consistent with a higher number and more widespread metabolic and electric activation of brain regions [124,46,38], while men show more a focal and subcortical activation [306,177].

Brain Functional Organization as Intrinsic Individual Characteristic

Several lines of evidence have demonstrated that the network of activation and the pattern of functional relationships among cortical areas are not dependent on environmental demands or specific cognitive tasks only, but also reflect intrinsic characteristics in special populations, for example cognitive abilities, gender, and psychopathology, since they are maintained during rest.

Base line EEG activity is correlated with successful performance in specific cognitive tasks and abilities. EEG activation and spatial performance [102], and, specifically, right hemispheric activation at rest and better spatial problem solving have been reported for high spatial ability males, as well as high left hemispheric activation for low spatial ability, while women showed no marked relationship between EEG lateralization and cognitive abilities [281]. We found that brain functional coupling interacts with spatial ability and gender. Significant opposite correlations were observed between spatial ability assessed by the Differential Aptitude Test and interhemispheric functional coupling in the resting EEG: the higher the spatial ability, the lower the interhemispheric functional coupling in men, and vice versa in women, suggesting a different cortical organization for spatial ability in each sex [56]. Moreover, the EEG pattern of women with extremely high spatial ability, which is not a typical feminine characteristic, is closer to the male pattern, whereas the EEG pattern of males with extremely low spatial ability, which is infrequent in males, is closer to that of women, suggesting that that women with high spatial ability have a masculine type while low spatial ability men have a feminine type of functional organization during rest [9,53].

Stability of Brain Functional Organization

The pattern of intercortical functional coupling during rest is a stable characteristic of each individual since it is maintained in wakefulness and sleep [20], and for long periods of time: in men along two [137] and three sessions [238], and over longer time intervals [347,187], and in women over eleven sessions within a one-month period [62] and for 9 sessions over a 9-month period [54]. High multiple correlation coefficients were obtained along many sessions for inter- as well as for intrahemispheric coherent activity in each woman indicating that the pattern of functional relationship of ongoing background oscillations is quite constant during resting conditions when no specific information processing is required, especially between central and posterior homologous regions of left and right hemisphere with eyes closed. Interhemispheric coherent activity was more stable across sessions than spectral power and intrahemispheric coherent activity, which is consistent with greater variability in power than in coherence reported by others [86].

Neuronal Substrate of Sex Differences in the Brain

In conclusion, there are sex differences evident in cognition and emotion, in brain structure and in brain functioning. Increasing evidence has shown that the brain is a target for sex steroids, and that gonadal hormones influence the central nervous system providing neural foundations that may explain behavioral, emotional and cognitive sex differences. Most of these results come from animal models, given that experimental hormonal manipulations are precluded in the human being, and thus great care is needed when trying to extrapolate conclusions to the human being.

Sex Steroid Receptors in the Brain

Sex steroids effects on the central nervous system are both genomic, by increasing the transcription of specific genes after binding to intracellular receptors, and non-genomic affecting neurotransmission, by acting directly on neuronal membrane receptors, as in the case of progestins [215,307], or indirectly, by interacting, as in the case of estrogen, with second messenger systems, with transcription factors [220,248,354] and rapid signaling pathways [257,345]. Thus, sex steroids may affect neuronal structure and connectivity by genomic actions and may alter functional connectivity by modulating neuronal excitability by non-genomic actions.

Sex steroid receptors are widely expressed in the brain, both in neurons and in glia, not only in areas involved in reproduction, but also in areas not directly related to reproductive behavior, where they can influence diverse cellular events and brain function. The nature of these receptors and the possible mechanisms of action of sex steroids that determine sexual dimorphism in cognition and mood have been extensively studied and reviewed [220,245].

Progestin receptors (PR) are known to exist in two forms, a large molecular form B and a smaller form A [174]. PR have been described in hypothalamic nuclei, septum, striatum, hippocampus and amygdala [132,319], in substantia nigra [319,25], suprachiasmatic nucleus [196] and in cerebral cortex [319], especially at the frontal cortex where they show the highest concentration, are more abundant in adult females [214], and are present approximately since postnatal day 7 [173,321]. PR are also expressed in astrocytes and oligodendrocytes where progesterone (P) promotes myelin synthesis [168].

PR expression in glia [168], hypothalamus and other brain regions, such as the cerebellum [122] is primed by estrogen treatment [230], but not in the cerebral cortex [36,230,122] nor in hippocampus and amygdala [230].

Estrogen receptors (ER) are localized in neurons, either in the cell nucleus where they affect the expression of target genes, or in the cytoplasm, outside of the nuclei including glia and neuronal processes, dendrites and presynaptic terminals, where they can influence diverse cellular events and modulate neurotransmission.

ER have been localized in adult male and female rats, as well as in other species, in hypothalamic nuclei involved in neural control of gonadotropin release (anteroventral periventricular and arquate nuclei), in circuits controlling hormonal copulatory behavior (hypothalamic medial preoptic and ventromedial nuclei, lateral septal nuclei, medial and cortical nuclei of the amygdala, amygdalohippocampal area, bed nucleus of the stria terminalis), and in rodent [205] and human [196] suprachiasmatic nucleus, involved in the control of circadian rhythms. They are also expressed in widespread areas throughout the brain related to emotion, motivation, memory and cognition where they may influence information processing. ER expression in rodents has been detected in neurons all over the cerebral cortex [323,346,191] including the areas mediating complex and association functions, such as primate prefrontal [354] and human dorsolateral prefrontal cortex and parietal region [262]; in visceral information processing, such as insular cortex [297]; in sensory relay centers such as thalamic ventral group [323] and trigeminal brainstem complex and ganglion [21,323]; in networks controlling motor behavior such as cerebellum [149] and substantia nigra [280], and in dopaminergic circuits involved in reward and motivation [66]. In addition, they have been found in limbic regions, especially in hippocampus, in the Subiculum, hippocampal formation and Ammon´s horn [323,35] and amygdala [149,346] and in systems engaged in cardiopulmonary regulation such as parabrachialis nucleus [318], nucleus of the solitary tract [323], dorsal motor nucleus of the vagus and hypoglossal nucleus [303] and in mouse dorsal raphe nucleus participating in serotonergic circuits involved in affective tone [314].

Finally, they have also been located in several systems related to modulation of global brain activation, such as, thalamic intralaminar nuclei [323], human histaminergic system in tuberomamillary nucleus [156], locus coeruleus [13,367], and vertical and horizontal nuclei of the diagonal band [346]; more specifically, in cholinergic neurons in medial septum, vertical and horizontal limbs of the diagonal band, substantia innominata and nucleus basalis projecting to cerebral cortex and hippocampus in rodents [315] and in human nucleus basalis of Meynert [155]. ER distribution in rodents is similar to that found in monkeys [123] and in some instances in humans. In sum, ER are strategically localized to influence emotion, motivation, memory, cognition and brain global activation.

The recognition of two isoforms of estrogen receptors, ER-alpha and ER-beta, with distinct brain distribution may in the future provide new insight regarding the action of estrogen on brain function. Both isoforms are co-expressed in many brain regions for example the cerebral cortex and hippocampus [191] but have distinct distribution in other brain areas and cellular compartments [320]. ER-beta is the predominant isoform in the cerebral cortex [72,320,251,193], in the hippocampus [251], in locus coeruleus [367], and in the human histaminergic system [156]. ER-beta in the cerebral cortex is found not only in cells, but also in neuronal processes [367] and in apical dendrites in pyramidal cell layers in CA1 and CA3 hippocampal subfields [12]. Whereas ER-alpha is expressed to a higher degree than ER-beta in hippocampal interneurons [250], hypothalamus and amygdala [252], cardiopulmonary controlling regions [303], in cholinergic forebrain system [315] and human nucleus basalis of Meynert [155] and hippocampal subfields and dorsolateral prefrontal cortex [262].

Androgen receptors (AR) have been found in rat hypothalamus, in sensory relays (vestibular nuclei, choclear nuclei, medial geniculate nucleus and nucleus of the lateral leminiscus [323]), in all regions of the rat cerebral cortex [192] and cingulate gyrus [243].

It has been established that steroid hormones may produce either activational or organizational actions. The former represent short transient actions in already organized systems, while the latter are characterized by long-lasting permanent effects on developing systems during the perinatal period exerting important influence on the sexual differentiation of the mammalian brain, which involves structural, ultrastructural, biochemical and behavioral sex differences [220].

Organizational Actions of Sex Steroids on the Brain

Animal research has provided experimental results which have helped to understand the mechanisms underlying the sex differenciation processes and sex steroid actions on brain development, although there are important species differences that have to be considered when drawing conclusions. The process of male sexual differentiation is a complex phenomenon involving events that start very early in development by the secretion of androgens by the testes. In male rats, this dynamic process includes masculinization, the development of male-type behaviors and of other behaviors, not necessarily directly related with reproductive phenomena, and defeminization, the loss of the ability to express female-type behaviors, such as the suppression of gonadotropin secretion cyclicity and the inhibition of expression of the lordosis behavior (female sexual behavior) in adulthood [350].

The major biologically active metabolites of androgens in peripheral tissues, such as the prostate and seminal vesicles, are 5-alpha reduced androgens like 5alpha-dihydrotestosterone (DHT); however, brain tissue also converts androgens through the aromatization of testosterone (T) to 17beta-estradiol (E) or of androstenedione to estrone. The masculinizing effects of testosterone on the brain during the perinatal critical period are mediated mostly, but not exclusively, by the aromatization of the molecule in the brain to estradiol, since there is also evidence of non-estrogenic primary effects on the brain [212], while the defeminization process is probably mediated by ER-beta [197].

In relation to sexual differentiation of the female, alpha-fetoprotein prevents circulating estrogens from accessing the brain, thus the absence of gonadal steroid hormones reaching the central nervous system during early development produce typical female characteristics. Prenatal or postnatal administration of aromatizable androgens to females during the critical period of sexual differentiation alters their reproductive physiology: females show no ovarian or gonadotropin cyclicity with delayed or absent vaginal opening, suppression of lordotic behavior and a behavioral repertoire similar to that of males, including higher incidence of mounting [226,212,255]. Perinatal testosterone treatment also affects induced maternal behavior of virgin rats and pup-oriented behaviors [167]. These manipulations become less effective if performed a few days after birth.

There are certain behaviors, not directly related to reproduction, that show sex differences and that are also modified by exposure to testosterone treatment during the critical period of sexual differentiation. For example, male rats [188] and primates [24] show higher frequency of rough play than females, and advantage over females in spatial ability [310]. Both, spatial ability [291] and play behavior are modified by sex hormones. Estrogen implants into the amygdala masculinize social play in females, and girls born with congenital adrenal hyperplasia treated at birth also show male-like patterns of play, while progesterone exposure [224] and antiandrogenic treatment during the neonatal period of the rat reduces play-fighting in male rats [115].

Although the presence of sex steroid receptors is not conclusive of causal effects, sexual dimorphism is present in brain regions where sex steroid receptors are concentrated. Hypothalamic regions involved in the regulation of reproductive behaviors show morphological sex differences. In rodents, the preoptical area and the ventromedial nucleus show sex differences in the volume of cell nuclei and synaptic organization [275]. The sexually dimorphic nucleus of the medial preoptic area of the rat is several times larger and contains more neuronal cells in males than in females [114]. On the contrary, the density of cells in the anteroventral periventricular nucleus of the preoptic area is greater in female than in male rats on day 21 of gestation [337]. These sex differences can be affected by hormones during the critical period of sexual differentiation. Accordingly, early castration of the male rat (immediately after birth), administration of estrogen antagonists or aromatase inhibitors interfere with the sexual differentiation process of the anatomical brain differences that characterize this sex [114], while testosterone treatment induces brain anatomy with male characteristics in females [135].

As mentioned, the influence of sex hormones is not restricted to endocrine control, it extends to non-reproductive areas of the brain involved in memory processes, mood and emotion, such as the hippocampus, the cerebral cortex and hemispheric interconnections. The cerebral cortex is known to be sexually dimorphic in the rat, both cortical thickness and neuron number, as well as dendritic arborization and spine density, especially in the monocular and binocular region of the rat primary visual cortex, is larger in the male than in the female rat [309-310]. The pattern of cortical asymmetry is also sexually dimorphic, the male right cerebral cortex is thicker and contains a larger number of neurons than the left one, especially in cortical areas 10, 3 and 17, while in females it is thicker on the left side and the laterality is less defined with only some areas showing significant differences at all ages [76]. The male-female pattern is consistent with higher ER concentration in the left cerebral cortex in males and in the right side in females at postnatal day 2-3 suggesting a growth-inhibiting effect of estrogen [298]. An asymmetry pattern is also observed in hippocampus with males showing a thicker hippocampus on the right and females on the left side [76]. The corpus callosum, the main commissural tract interconnecting the hemispheres, also displays sex differences. The female splenium has overall more axons, especially unmyelinated axons [170], whereas the male splenium contains more myelinated axons, resulting in equal total size for both sexes [179]. The cellular level is also modified by sex hormones. Estradiol increases the expression of glutamine synthetase, the enzyme that inactivates GABA transforming it into glutamine [234]

ER expression in both cerebral cortex and hippocampus has been found throughout the late prenatal and early postnatal period, and shows a transient increase which suggests specific timetables. ER in the cerebral cortex is expressed as early as embryonic day 16 [228] with transient increases in newborn rats [142] and during the first postnatal week [258]. They have been found to be highest at postnatal day 2 in the right female cortex and in the left male cortex [298]. In the hippocampus, ER alpha and ER beta as well as aromatase activity are expressed prenatally in both sexes with higher levels in males [157], and the concentration increases from postnatal day 0 to postnatal day 4, when they peak [246]. The ER-alpha isoform seems to play an important role during early development; it is higher than ER-beta in cerebral cortex in neonatal than in adult rats, whereas ER-beta levels are similar in neonatal and adult rats [125], specifically at least in the primary auditory cortex [142]. Although both isoforms are present in the hippocampus, ER-alpha are there since postnatal day 3, reaching higher levels at postnatal day 10 [331], and their distribution in hippocampal subfields is more widespread than that of ER-beta during early development at postnatal days 3, 10 and 14 [260].

AR receptors are also present during the postnatal period of sexual differentiation in the primary visual cortex and in cingulate and frontal cortices, though at a lower level in the latter, and density is higher in males than in females on postnatal day 10 [243]. AR receptors may be involved in connections interlinking cortical regions, since they are specifically present in cortical pyramidal cells showing retrograde labeling from other cortical areas especially within the same hemisphere and to a lesser degree from the other hemisphere [192].

Accordingly, early postnatal T treatment of females alters sex differences in both cortex and hippocampus, increasing cortical thickness to resemble a male-like pattern [75], specially the granule cell layers [291], and the size of pyramidal cells in CA1 and CA3 hippocampal subfields [154]. Removal of the testes or ovaries at birth reverses the cortical pattern of male-right, female-left laterality, with females showing thicker right cortex similar to the male pattern, and males displaying thicker left cortex in frontal and somatosensory areas [75]. The size of the splenium is also increased in females by T treatment, whereas neonatal castration does not alter the male pattern [242].

Thus, most sex differences in the brain are dependent on organizational actions of sex steroids, though not solely, which occur during the critical period of brain sexual differentiation in the rat. The functional expression of these structural sex differences in the brain and its dependence on sex steroid actions may be revealed by electrical activity recording. We have developed a model in our laboratory to study sex differences and sex steroid influence on brain function by using quantitative analyses of electric field potentials in the rat.

SEX DIFFERENCES IN ELECTRIC FIELD POTENTIALS IN THE RAT

Electric field potential characteristics are well known in the rat. Among alert rats, two main patterns of field potentials (EEG activity) are recorded at the cortical surface; low voltage irregular fast oscillations between 10-80 Hz, corresponding to beta and gamma bands in humans, and theta rhythm or slow rhythmic activity in the range of 3-4 to 10-12 Hz. Simultaneous recordings of cellular activity, field potentials and behavior have demonstrated that these oscillatory frequencies originate in specific neuronal systems and their corresponding behavior patterns. Low voltage, irregular, fast activity, occurring during arousal and paradoxical sleep involves an overall increase in single spike discharges of thalamic intralaminar cells, with concomitant depolarization of cortical neurons and the cessation of oscillatory activity in thalamo-cortical projecting neurons [335]. Theta rhythm accompanies alert immobility, voluntary movements and paradoxical sleep and is generated by the activity of the so-called hippocampal theta cells [43]. Thus, two main systems participate in the generation of the field potential characteristics found in the arousal profile of rodents, the thalamo-cortical system for beta/gamma and, the septo-hippocampal system for theta, both under the control of global influences coming from several ascending activating systems in the brainstem, posterior hypothalamus and basal forebrain [335,285,338]. As mentioned previously, cerebral cortex and hippocampus express sex steroid receptors and are sexually dimorphic, so that it might be thought that the electrical activity would also be. Indeed, both, theta and beta bands are sexually dimorphic, extending the notion of behavioral and structural sex differences in the brain to include differences in functional brain organization in the case of each sex. In two independent experiments, with different groups of intact, adult rats, (100 days old), quantitative analysis of spontaneous field potentials from the parietal cortex has revealed three main sex differences [165-166]: 1) in inter-parietal asymmetry 2) in delta, theta and beta activity, and 3) in inter-parietal functional coupling.

Hemispheric Asymmetry

Males show marked parietal functional asymmetry (power asymmetry) with higher activation of the right compared to the left parietal in all bands except for delta (Figure 1), whereas this feature is absent in females. We have maintained the usual human nomenclature for human EEG bands for the sake of clarity. The greater hemispheric asymmetry found in males is consistent with anatomical and behavioral asymmetries and asymmetries relating to sexual interactions in the rat. Accordingly, when comparing the right cortex to the left in males, it is found to be thicker, and more neurons are located in specific brain areas, such as the occipital and parietal cortices [75,310], and express more ER than among females [298].

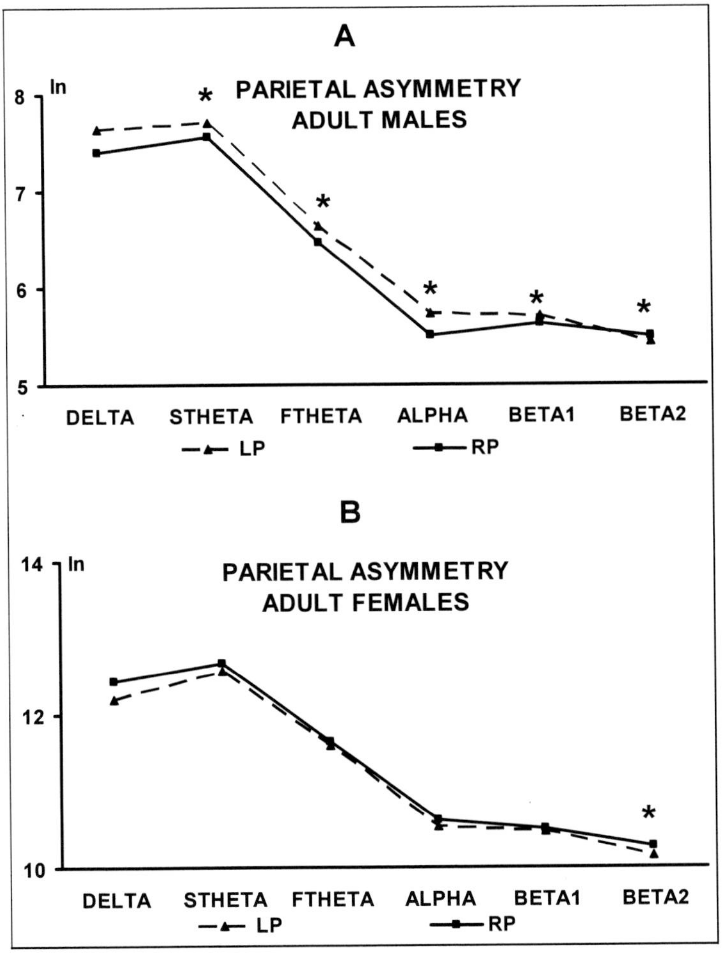

Figure 1. Parietal asymmetry in electric field potentials in intact adult male and female rats. Mean (n = 10 for all groups) of absolute power, log transformed, for delta (0.5-3.5 Hz), slow theta (4-7.5 Hz), fast theta (8-9.5 Hz), alpha (10-12 Hz), beta1 (12.5-17.5 Hz) and beta2 (18-25 Hz) bands. Human nomenclature for bands has been kept for sake of clarity. Interhemispheric differences (indicated by asterisks) were significant only in males for all bands except for delta. Same captions are used in subsequent figures. Data from 166.

Hippocampal/Cortical Arousal

Males show higher absolute power and proportional contribution to total power (relative power) of slow and fast theta than females, whereas females display a higher proportional contribution of delta and beta activity (Figure 2A). This profile suggests preponderance of hippocampal arousal mechanisms in males, and of thalamo-cortical mechanisms in females.

Theta rhythm in rats, when recorded on the cortical surface of the parietal cortex, overlying the hippocampal complex, is most probably of hippocampal origin and spread by volume conduction to the cortex [31,265,349]. Theta activity in rodents represents the main output of the limbic system [43] and it has been considered to reflect hippocampal arousal, as it increases during reaction to novelty [236], in situations which induce emotional arousal such as sex and fear [295,145], and when exposed to psychological stress, for example, fear-induced, when conditioned to expect an electroshock [158]. A higher hippocampal output in males accords with evidence from other research fields showing that the hippocampus (review: 357, 223, 35) and cholinergic neurons of the medial septum diagonal band projecting to the hippocampus, involved in theta generation [200], are sensitive to sex steroids during the sex differentiation period, with larger hippocampal volume [76,291,159] and pyramidal soma size, as found in males when compared to females [154], and larger excitatory postsynaptic potential amplitudes in hippocampal CA1 cells taken from male slices [326].

Fast oscillations can be regarded as the expression of thalamo-cortical system activation, under the activating influence of brainstem [335] and basal forebrain [285,338]. Functional systems involved in cortical slow and fast activity such as the cortex itself [323,346,191], intralaminar thalamic nuclei [323], locus coerules [13,367] and basal forebrain cholinergic system (review 22, 346, 317) are also sensitive to sex steroids during the sex differentiation period, as previously described.

Inter-Parietal Functional Coupling

Males also manifest a greater correlation between left and right oscillations or inter-parietal functional coupling in delta, and in slow and fast theta bands, indicating greater functional coupling between right and left parietals (Figure 2B)

A high correlation or coherent activity between two recording sites implies a higher degree of similar electrical activity, the greater the similarity between the two signals compared, the greater the correlation between them and vice versa, so that correlation reflects a shared neural activity. It is known that coherent activity between two brain regions depends on simultaneous synchronization of the two neuron assemblies involved. When two areas of the brain exhibit similar field potentials, it may be inferred that they are functionally related, whereas if each of the sites compared is being differentially activated, field potential activity must therefore be different and the correlation lower. A high correlation between two simultaneous field potentials indicates that field potential oscillations are synchronized with each other. Therefore, a high correlation is indicative of enhanced functional relationships, whereas a low correlation is evidence of functional uncoupling during particular cognitive processes or physiological states. Although correlation reflects the linear functional relationship of ongoing electrical activity between two brain areas, it does not reveal the specific causes responsible for this. It may depend in part on reciprocal intra-cortical long and short connections, and on common afferent inputs from a third system including those from sub-cortical generators [42,343], thus, the level of correlation may be the result of one of these factors or of all of them. Nevertheless, higher correlations are indicative of greater cooperation and functional interactions, whereas low correlation reflects low cooperation and greater local functional differentiation.

Thus, the higher inter-parietal functional coupling in males indicates a greater functional relationship between the left and right parietal. The corpus callosum is the main commissural tract connecting both hemispheres, and there are ultra-structural sex differences in the splenium of the corpus callosum in the rat. Males, despite overall similar callosal size [242], have larger myelinated axons and more abundant thick axons than females [170,179], thus providing for faster inter-hemispheric transfer of information and probably facilitating inter-parietal integration.

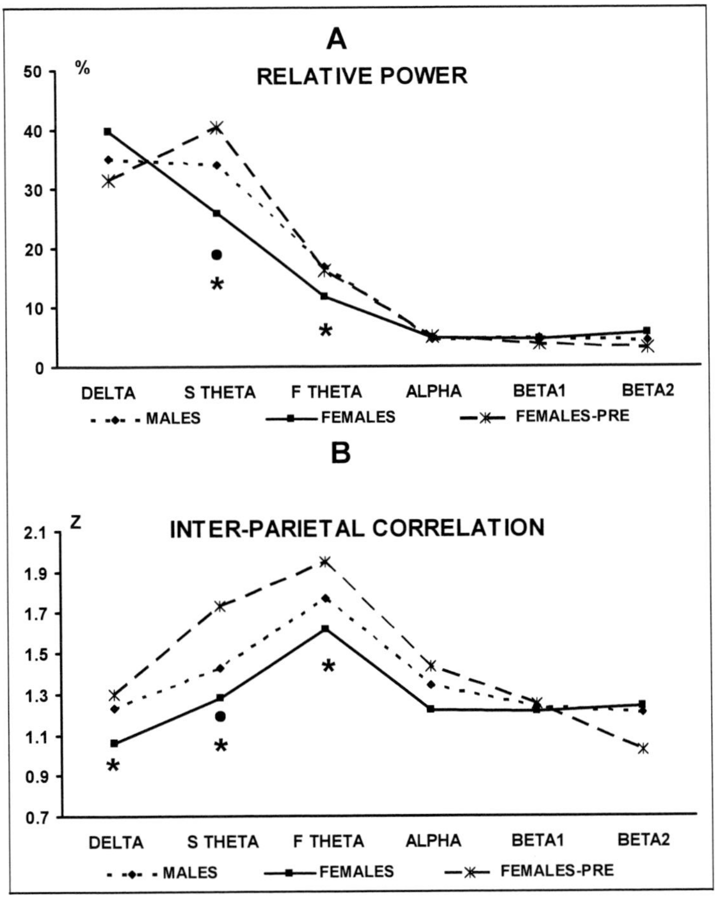

Figure 2. Sex differences in electric field potentials and inter-parietal functional coupling in adult intact males and females, and in prenatally testosterone propionate-treated females (FEMALES-PRE). Mean of relative power (absolute power of each band/total power x 100) in A and inter-parietal correlation, transformed to Fisher's Z scores in B. Significant differences between intact male and female rats are indicated by asterisks and between intact and prenatally treated females by dots. Data from 166.

Organizational Actions of Sex Steroids on Electric Field Potentials

Prenatal Organizational Actions

Sex differences in hemispheric power asymmetry, theta activity and inter-parietal functional relationship depend on the prenatal organizing effects of androgens during the critical prenatal period of brain sexual differentiation, since these are eliminated by prenatal testosterone treatment [166]. Exposure of female fetuses to a daily injection of 2 mg of testosterone propionate (TP) applied to the mother on days 14 to19 of gestation eliminates sex differences in brain oscillations found in untreated intact adult animals. Thus, in TP treated females, inter-hemispheric power asymmetry disappears, theta absolute and relative power (Figure 2A), and inter-parietal functional coupling (Figure 2B) when recorded in adulthood at 100 days of age is similar to that found in males [166].

The suppression of sex differences is mostly due to a masculinizing effect on females, since both electrical parameters were significantly higher among prenatally virilized females than among non-treated females (Figure 2A and 2B), whereas testosterone-treated males had similar overall EEG, to that found in untreated males. Female virilization was confirmed by the lack of vaginal opening, masculine ano-genital distance and body weight.

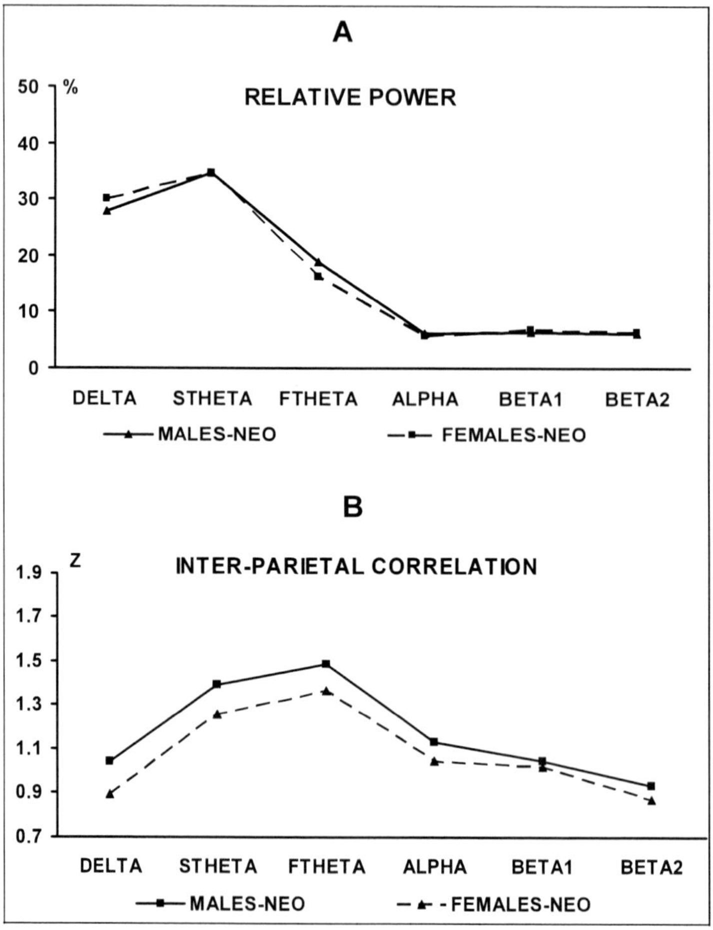

Figure 3. Sex differences in electric field potentials and inter-parietal functional coupling are abolished by neonatal gonadectomy in males (MALES-NEO) and neonatal testosterone propionate-treatment in females (FEMALES-NEO). Data from 63.

The masculinization of the theta rhythm caused by the prenatal exposure of females to testosterone, revealing higher hippocampal activity, concurs with the findings of other authors who have reported a higher number of AR in CA1 cells in males and in TP treated females [365], with the virilizing effects of female exposure to perinatal androgen on the hippocampus volume and pyramidal soma size [76,295,220,357,154], and with the greater hippocampal susceptibility among females to environmental influences during development, for example, when they were reared in isolation during the critical period of sex differentiation [74,171] and lower neuronal survival in cultures from females [147].

The masculinization of inter-parietal functional coupling, which increased to masculine levels when females were exposed to testosterone concurs with the increase in the size and number of myelinated axons in the female splenium, following testosterone treatment during the critical period of brain differentiation [242] and with the presence of AR in cortical pyramidal cells with callosal connections [192].

In conclusion, prenatal virilization of females alters the field potential oscillations by abolishing hemispheric asymmetry, increasing the strength of the theta generator and reducing the proportion of cortical to hippocampal activation, and by increasing the temporal functional coupling in delta, and slow and fast theta. All these findings point to the masculinization of the functional organization of the brain.

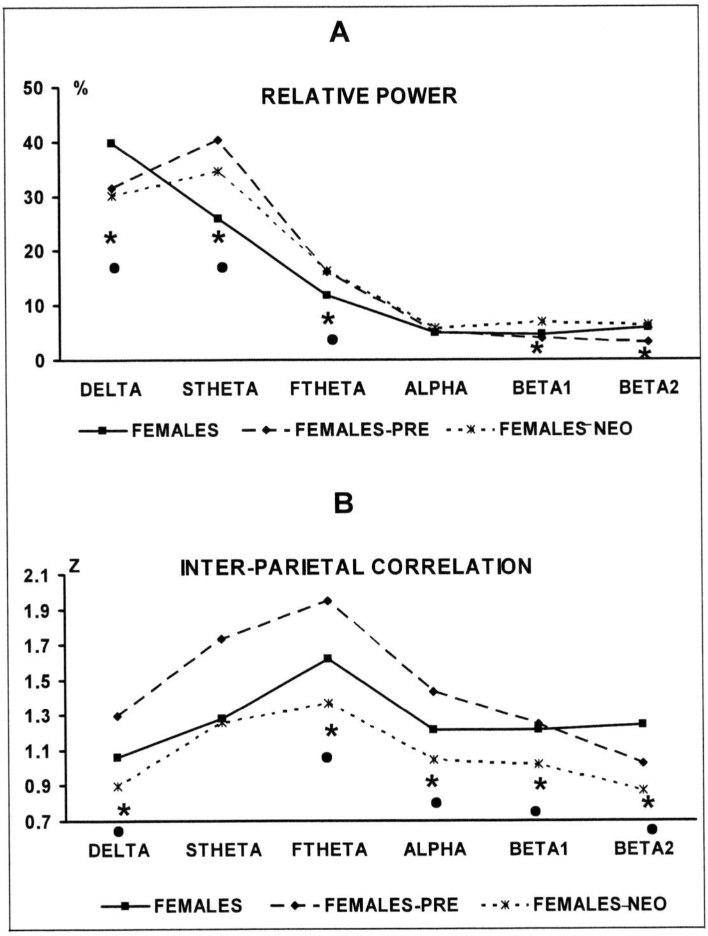

Figure 4. Virilizing effect of prenatal and neonatal testosterone propionate treatment in females on electric field potentials and inter-parietal functional coupling. Significant differences between intact and prenatally treated females (FEMALES-PRE) are indicated by asterisks and between intact and neonatally treated females (FEMALES-NEO) by dots. Data from 166 and 63.

Given that a single injection of a relatively low dose of testosterone administered around the first five days of life was still able to produce clear reproductive alterations in adulthood [226,212,255], and virilization in the case of other non-reproductive behaviors [18], the following study analyzed whether postnatal treatment of female rats with androgens is as effective as prenatal treatment for causing masculinization of field potential oscillations [63].

Neonatal Organizational Actions

Both, neonatal administration of TP (120 μg at day 4 after delivery) to females and castration of males on day 4 after delivery eliminate sex differences in terms of the proportional contribution to total power of slow and fast theta, delta and beta1 (Figure 3A), and in inter-parietal functional coupling of delta, and slow and fast theta; although female inter-parietal functional coupling is still lesser than that of males, the sex differences are not significant (Figure 3B). Both parameters are similar in both neonatally virilized females and in neonatally castrated males when recorded at 100 days of age.

The comparison among untreated females, females that received TP prenatally (in utero) and neonatally (day 4) showed that neonatal virilization of electrical activity in the female partially contributes to the lack of sex differences; neonatally treated females showed higher slow and fast theta and lower delta relative power than untreated females (Figure 4A). However, neonatal administration of testosterone was no longer effective in producing significantly higher inter-parietal functional correlation (Figure 4B), as happened in the case of prenatal treatment, which restricted the masculinizing effect of TP on inter-parietal functional coupling to the prenatal period of sex differentiation [63].

These findings agree with evidence showing that the effects on sexual differentiation produced by androgens in females are more effective if administered prenatally. Thus a neonatal influence caused by sex steroids on female electric field potentials is also observed, though more partial than the prenatal influence, a finding which is understandable given that the expression of estrogen receptors during development is not synchronous in all brain regions, providing different region-specific critical periods for different brain areas [247,198].

The previous experiments explored the role of sexual differentiation as they affect field potential oscillations in females; in males, however, little was known about this subject. Therefore, the following initial experiments analyzed whether neonatal castration, following puberty alters the electrical activity pattern of males, when compared to castrated (GNX) males [63].

Male neonatal castration results in the feminization of some features of electrical activity. It induced a lower proportion of delta and a higher proportion of beta1 and beta2 frequencies, but failed to modify theta activity (Figure 5A), indicating that the absence of sex steroids during this period in males affects cortical activation inducing a feminine-like pattern, but not hippocampal activation. Neonatal cortical male susceptibility is consistent with the neonatal estradiol content [6]) and expression of the gene encoding GAP-43, a protein implicated in neurite outgrowth and motility in the cerebral cortex, which is greater in males than in females [316], and feminization of anatomical characteristics of the posterior cortex [75], which occurs with similar treatments. These data support evidence indicating the vulnerability of the mechanisms underlying cortical activation in males during neonatal development as a result of environmental influences, which show that the cerebral cortex of males is differently affected by rearing in an impoverished or an enriched environment [76] and with the greater susceptibility of males to frontal cortex lesions [333], in comparison with a more robust hippocampus which is resistant to these same influences [169,75,76,171].

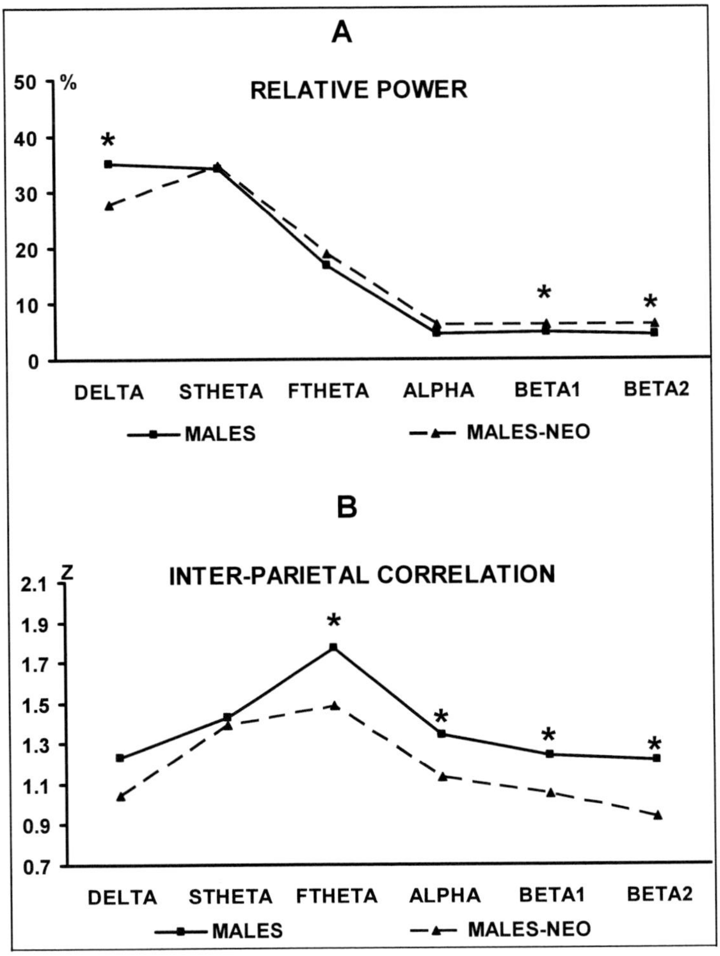

Figure 5. Feminization of electric field potentials and inter-parietal functional coupling in neonatally castrated males. Significant differences between intact males and neonatally castrated males (MALES-NEO) are indicated by asterisks. Data from 63.

Neonatal castration of males also resulted in lower inter-parietal functional coupling in fast theta, alpha, beta1 and beta2 bands forming a feminine-like pattern, when compared with intact adult males (Figure 5B), which is consistent with the idea that AR expression is involved in cortico-cortical connections interlinking cortical regions [192].

In conclusion, these experiments demonstrate an important organizational influence on the part of sex steroids during the perinatal period, which affect the adult pattern of brain electrical activity of male and female rats, such that administration of testosterone to females transforms it into a male pattern, and the lack of the hormone in males transforms it into a female pattern. These results extend the existing ample knowledge concerning anatomical sexual dimorphism and the masculinizing effects of testosterone during the perinatal period on the hippocampus, cerebral cortex and corpus callosum, to functional organization of the brain and they provide evidence to indicate the functional consequences of anatomical differences and the way these are manifested in adulthood.

Activational Actions of Sex Steroids on Electric Field Potentials

The activational influence of the female sex hormone appears not to be restricted to the behavior under direct endocrine control, but extends to overall behavior. It is well established that steroids secreted during the estrous cycle are involved in cognitive functions [221,357]. The estrous cycle seemingly plays a role in spatial learning in a hippocampal version of the Morris water maze, both of which improve during estrous, compared to performance during proestrous [355] and are deteriorated by E plus P treatment in ovariectomized adult females, but not by either E or P treatment alone [48]. Estrogens modify synaptogenesis in the hippocampus; they rapidly increase the number of dendritic spines and synapses in CA1 of the rat hippocampus [221], modify the survival and proliferation of cells in the hippocampus, enhancing the production of new cells in the dentate gyrus (for review 104), enhance cholinergic function [22] and long-term potentiation in CA1 [52], which is greatest during the afternoon of proestrous [356], probably related to an increase in ER-alpha expression in dendritic spines in this region during the same estrous phase [290].

Some studies have evaluated the effect of the estrous cycle on the excitability and the electrical activity of the brain [340,362]. The threshold and pattern of electroshock seizures in rats is modulated by the estrous cycle, being higher during diestrous, lower during proestrous and lowest during estrous [362]. Estrogen and progesterone exert modulating effects on neuronal activity and brain excitability. The arousal threshold for electrical stimulation of the brain and the electrical post-reaction to vaginal-cervix stimulation are both increased by estrogens and decreased by androgens and progesterone.

Progesterone has inhibitory effects on single unit responses in the cortex, thalamus and hypothalamus [204], and on the electrical arousal threshold [175], and causes anticonvulsant effects [299,204]. Progestins are also involved in the synaptic downregulation in the CA1 region during the estrous cycle of the rat [123], and are able to modulate hippocampal electrophysiological responses [203].

Estrogens, on the other hand, increase excitability in the CA1 pyramidal neurons of the hippocampus [341,363,47], have antidepressant effects [249] and increase convulsiveness [204].

As would be expected from the above evidence, field potentials also oscillate as a function of the estrous cycle in rats and during the menstrual cycle in women, indicating that sex steroids not only have organizational effects on cortical electrical activity but also activational influence. Inter-parietal coupling is significantly higher during the peri-ovulatory phase of the estrous cycle [57], and between the left and right frontal cortex during ovulation in women [330].

The next series of experiments was carried out in rats in order to elucidate the activational role of sex steroids on the functional organization of the brain and they demonstrated that the activational effects of sex steroids are necessary in order to maintain sex differences in the cortical electrical activity of the adult rat and they also revealed important changes induced by their influence [73].

Gonadectomy (GNX) was performed following puberty, at 80 days of age and electrical activity was recorded at 100 days of age to avoid the organizational actions of sex steroids, also evident during peri-pubertal periods of development [10] that affect sexually dimorphic, non-reproductive behaviors [17,271,325,333], as well as in the brain where they decrease cortical volume and the number of neurons, activities which are halted by removing the ovaries [172].

Results demonstrated that the lack of gonadal steroids following puberty affects the right over left parietal activation, eliminating the parietal asymmetry manifested in intact rats and even reversing the pattern of asymmetry in both sexes, with a greater contribution of alpha and beta2 to total power in the left parietal instead of in the right parietal, as reported for adult controls.

The pattern of right parietal asymmetry was re-established in both sexes by increasing the absolute power of the entire spectrum in the right compared to the left parietal in males, as well as in females, after three days of daily hormonal administration of either beta estradiol (E) or progesterone (P) to GNX females and either TP or DHT to GNX males, at dose levels sufficient for inducing estrous or for adequately restoring sexual activity.

The lack of gonadal steroids following puberty also reversed the sex difference in delta and theta activity which is typically found when comparing intact males and females. GNX females showed a decreased proportional contribution of delta and increased fast theta activity with a tendency towards masculinity (Figure 6A). Sex differences in inter-parietal functional coupling are also eliminated in GNX rats due to a slight decrease, evident in males, as although it remains lower in females, the differences are no longer significant (Figure 6B).

Hormonal treatment re-established the typical sex differences; males treated with TP showed an increase in theta power, which once again differentiated them from females, whereas DHT had no effect on theta power. These results taken together suggest that the effect may not be due to the reduction of testosterone to DHT, but is more probably due to a direct effect caused by testosterone or to the aromatization of testosterone to become estrogens. The activational effect of T for increasing theta power is consistent with the organizing effect of this hormone. Prenatal treatment with TP during the critical period of sex differentiation in the brain induces higher theta power in adult females as previously mentioned.

TP treatment of GNX males had a partial effect on inter-parietal functional coupling, increasing it in slow and fast theta, though the increase was not enough to re-establish the significant differences between intact males and females. The mild effect of TP may have several explanations. It could be that doses sufficient for restoring sexual activity are not sufficient to restore inter-parietal functional coupling or that it requires longer exposure to the activational influence of TP. It is known that the effect of sex steroids may depend on specific metabolites or may even be secondary, i.e., mediated by another hormone or neuro-modulator, and thus more time may be required for the change to become evident [220].

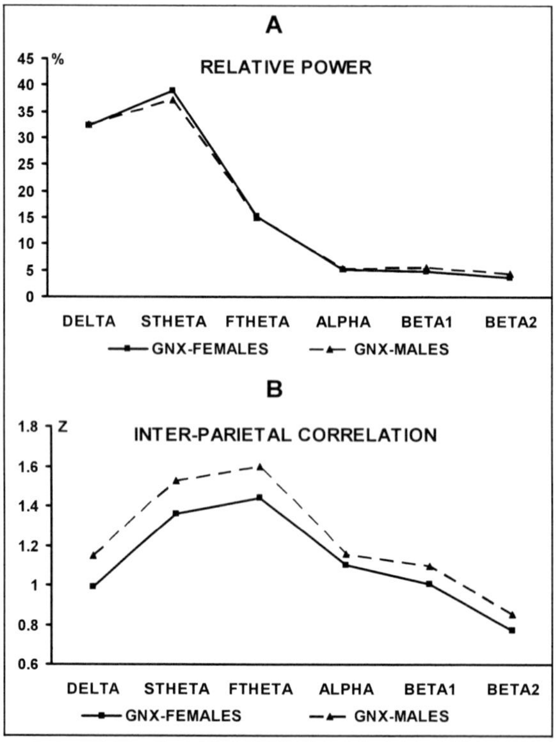

Figure 6. Sex differences in electric field potentials and inter-parietal functional coupling are eliminated by gonadectomy after puberty. Data from 73.

Estrogen treatment in GNX females accentuates a female-like pattern, decreasing theta and increasing beta in terms of their proportional contribution to total power (Figure 7A), probably due to an up-regulation of ER after prolonged GNX [2,231].

Inter-parietal functional coupling was not significantly modified by E treatment in females (Figure 7B). The lack of effect of female steroids on inter-parietal functional coupling among females contrasts with the increase which was observed during the estrous cycle [57], suggesting that the effects of the estrous cycle are likely to be due to the combined action of estrogen and progesterone, whereas isolated action on the part of each steroid may be ineffective, as happens in the case of other E/P interactions, i.e. the number of progesterone receptors in the brain increases and decreases as a function of estrogen priming [220] and the acquisition of the spatial Morris water-maze is differentially affected by each steroid or its combination [48].

Given the possibility of an estrogenic effect in males, the action of estrogen on GNX males, organized as males was investigated (del Río-Portilla, Fernández-Guasti, and Corsi-Cabrera unpublished results). Estrogen treatment decreased fast theta and increased delta and slow theta relative power (Figure 7C) and decreased inter-parietal functional coupling of delta, fast theta, alpha and beta1, forming a feminine-like pattern (Figure 7D).

The differential estrogenic effects which induce a female-like pattern in males and accentuate the female-like pattern among females may be due to E effects over a normal neonatal feminized or masculinized brain tissue. This explanation needs to be proven with further research, by administrating E to masculinized females, during the critical period of sex differentiation, and to castrated males during the same period.

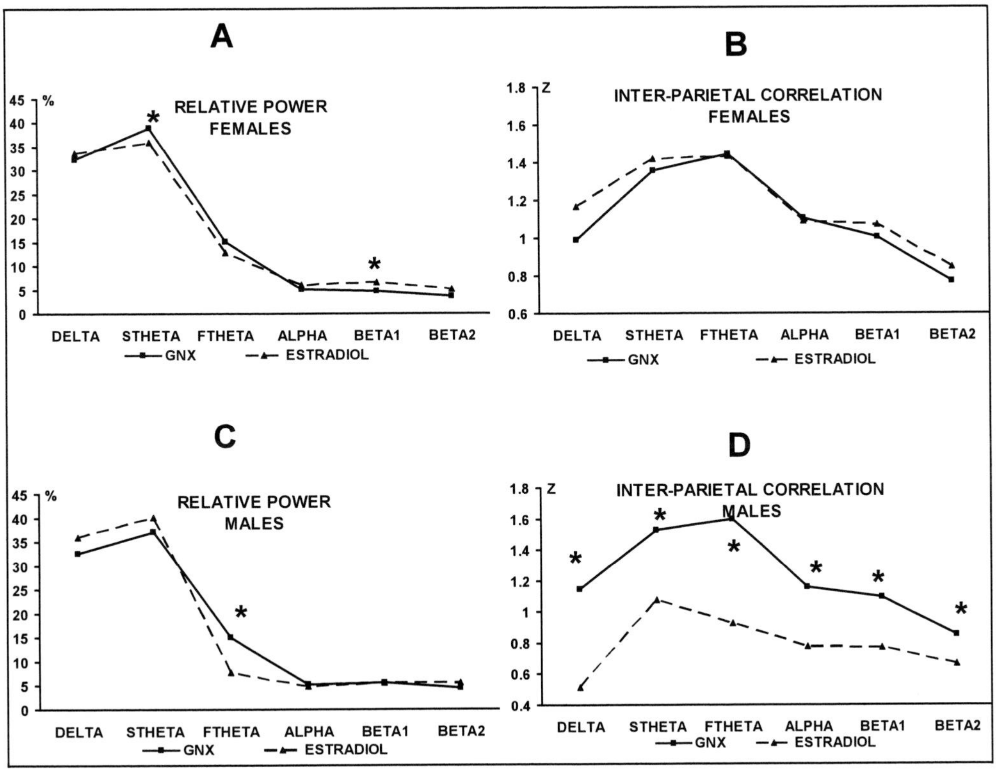

Figure 7. Activational effects of estrogen treatment in female and male rats gonadectomized (GNX) after puberty on electric field potentials and inter-parietal functional coupling. Asterisks indicate significant differences between GNX rats with and without estrogen treatment. Data from 73 and unpublished results.

As previously mentioned, progesterone affects the electrical excitability of the central nervous system and has anesthetic, hypnogenic and anticonvulsant effects in both males and females, thus the activational action of P was also investigated in GNX males, organized as males and in GNX females, organized as females [91]. The administration of P to males produced an increase in the power of frequencies from 10 to 25 Hz, reaching the same level as in females (Figure 8 A and B). This finding indicates a common effect of progesterone on fast oscillations in males and females independent of a feminine or a masculine brain.

Organizational-Activational Interaction of Sex Steroids on Electric Field Potentials

Given that the response to progesterone was similar in GNX males and females following puberty and thus organized as either male or female respectively, the question remained as to whether P would have the same effect on neonatally virilized females and castrated males [91].

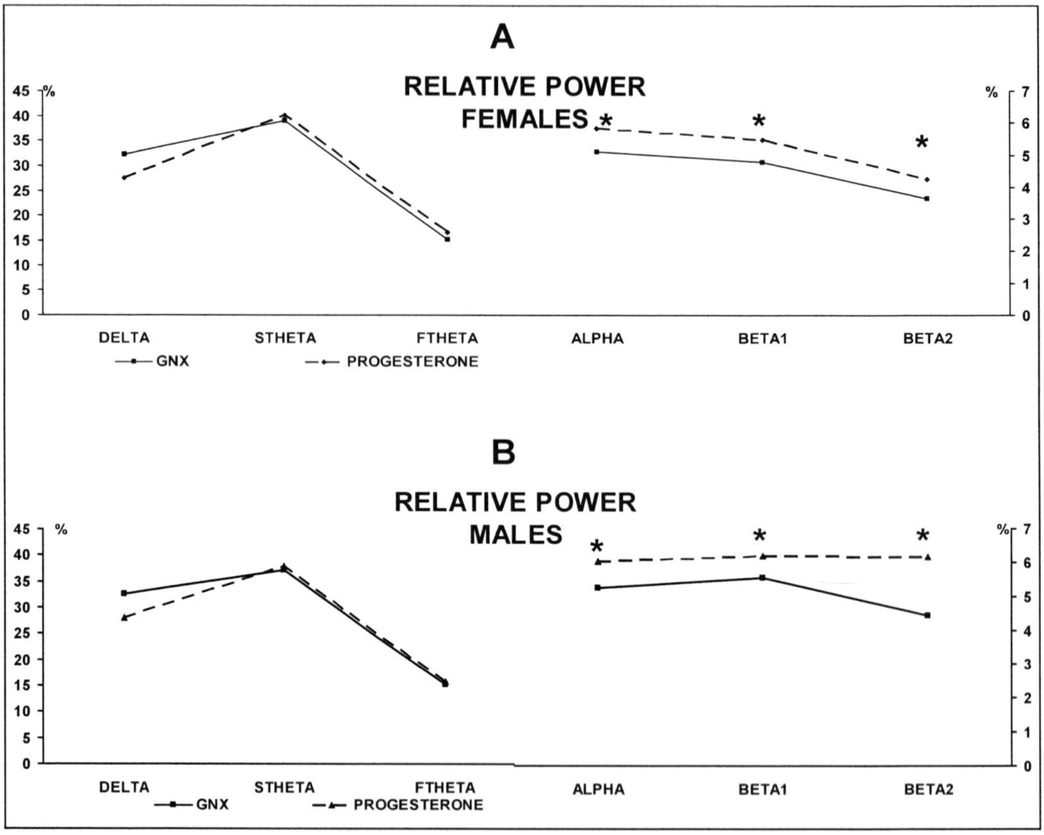

Figure 8. Activational effects of progesterone treatment in female and male rats gonadectomized (GNX) after puberty on electric field potentials. Asterisks indicate significant differences between GNX rats with and without progesterone treatment. Data from 73 and 91.

Interestingly, the actions of progesterone on power do not depend upon the sexual differentiation process which occurs neonatally. The increase in alpha, beta1 and beta2 power induced by progesterone was similar between neonatal castrated males and males castrated in adulthood and between neonatal virilized females and females ovariectomized in adulthood. The lack of effect caused by manipulating the availability of neonatal estrogen is consistent with the finding that PR expression in the cerebral cortex in neonatal mice is not dependent on ER-alpha expression [353], as it is in hypothalamus.

In conclusion, the activational effects of sex steroids are expressed in the electrical activity of the brain. The activational action of sex steroids, in addition to their organizing effects are necessary for greater inter-parietal coupling and greater theta activity, as well as for right over left parietal asymmetry in males. The activational effects of testosterone in males on theta power are probably exerted through testosterone or its aromatized metabolites. For sex differences in inter-parietal coupling in males to manifest, it is probably necessary for them to be exposed to sex steroids for longer, just as females require prolonged exposure to the combined action of E and P. E accentuated the feminine-like pattern in females and induced a partially feminine profile in males. The effect of progesterone on increasing fast frequencies does not depend upon the sexual differentiation process that occurs neonatally.

Dimorphic Effect of Benzodiazepine on Brain Function

Stress-related responses are sexually dimorphic and dependent on the hormonal status of rats as well as that of human beings. The stress response in rats is different for each sex depending on the kind of anxiety test, for example, in open-field tests, female rats ambulate more and defecate less than males [163], two responses which are modified by castration and hormone replacement therapy [17]; social interaction is modulated in adult, male rats, depending on whether the environment is familiar or unfamiliar but this is not the case in females, and the differential response to environment displayed by intact males disappears among pre-pubertal castrated males [271]. Therefore, the rat is an appropriate model to study the effect of anxiolytics on brain functional organization, showing how this is related to sex and sex steroid effects. Having established that sexual dimorphism in the brain is expressed in the sex differences in electrical activity dependent upon organizational and activational actions of sex steroids, electric field potentials can be used as a tool for investigating sex differences in brain response to anxiolytic drugs, as well as organizational and activational effects of sex steroids on the brain's response to drugs.

A wide range of neuroactive effects caused by progesterone and its alpha-ring-reduced metabolites; allopregnanolone (3alpha- hydroxy-5alpha-pregnane-20-one) and pregnanolone (3alpha-hydroxy-5beta-pregnan-20-one) have been known for some time, including anesthetic [308,130,232], hypnogenic [201], anticonvulsant and anxiolytic effects, similar to those elicited by benzodiazepines (BZ) and barbiturates [28,360,267,87,90,272].

These effects are to be expected if we consider that the main inhibitory neurotransmitter receptor of the brain, the gamma-aminobutiric acid (GABA) type A receptor complex ($GABA_A$) mediates the effects of several classes of clinically important drugs, particularly some anesthetics and axiolytic drugs, such as benzodiazepines, and that some synthetic and endogenous steroids have demonstrated that they act as allosteric modulators of the GABA receptor complex. The $GABA_A$ receptor is an oligomeric complex consisting of several subunits with independent binding sites for GABA, BZ and barbiturates, thus BZ exert their pharmacological effects through an allosteric modulatory site within the $GABA_A$ receptor [131,339] modulating GABA-ergic activity by affecting ligand binding and thus increasing the frequency of opening of the GABA-dependent chloride channels [14].

It is also well known that progesterone and its alpha reduced metabolites, allopregnanolone and pregnanolone have benzodiazepine-like effects, by interacting with the membrane receptors coupling primarily to the $GABA_A$ receptor subtype [216,219,364,294], where they exert a direct neuro-modulatory effect at specific recognition sites [106,189,216,233,235,267,256,263,272]. They can modulate both, the inhibitory actions of GABA and of BZ by potentiating the effects of GABA, enhancing receptor-mediated chloride ion uptake, and by increasing the binding of muscimol, a GABA agonist, and flunitrazepam, a BZ [106,216,263,256,189,267]. The anesthetic effect in rats is potentiated by pregnanolone [240] and sensitivity to triazolam is increased by progesterone, administered 2.5 hours prior to a challenge dose of triazolam in women [195]. Similarly, it has been reported that flumazenil, picrotoxin and bicuculline prevent the anxiolytic-type effects of progesterone and allopregnanolone [29,88,282]. Some of these metabolites are synthesized de novo from cholesterol in the brain during stressful situations and are called neurosteroids [152].

Although initially there was controversy in defining the role played by gender and the sexual differentiation process in the anxiolytic-type actions of progesterone or its metabolites, -Rodríguez Sierra et al. [288] reported that males were insensitive to the anxiolytic-type actions of this steroid- whereas this has now been demonstrated. Progestines and allopregnanolone have clear anxiolytic actions among both sexes [90]. Regarding males, several authors [359,360,225,287,90] have found that various progestines induce anxiolytic-type responses. In females, the anxiolytic-type actions of various progestines is even better documented [28,29,87,90,218,267,27].

Given that the stress-related response is sexually dimorphic, that endogenous steroids interact with GABA/BZ receptors and that estrogen during fetal life exerts permanent organizational actions on the limbic system, brain response to BZ anxiolytics in adulthood may depend on sex and on the sexual differentiation process. Since the electrical activity response to BZ is known, theta activity and inter-parietal functional coupling are sexually dimorphic and also dependent on the organizational actions of sex steroids, field potential recording may be used to investigate the effects of BZ on brain functional organization as a function of sex and the organizational and activational actions of sex steroids.

Dimorphic Effect of BZ on Electric Field Potentials

The EEG response to BZ, where the sex of the rats is not taken into consideration, is well described. An anxiolytic dose of diazepam (DZ), a well-known BZ with anxiolytic effects, induces an increase in absolute power of frequencies above 20 Hz [183] and in relative power of frequencies between 13 and 30 Hs [158].

Thus, in our first experiment we investigated the effects of DZ on the cortical waking electric field potentials of GNX male and female adult rats following puberty, and thus without the activational effect of gonadal steroids, but having being exposed to the normal organizational actions of sex steroids [348]. The anti-anxiety effect of DZ was experimentally corroborated, using the elevated plus-maze, according to procedures previously described [259].

DZ induced the well known increase in fast frequencies, regardless of sex. One main finding of this experiment, that had not been previously reported, is that DZ modified inter-parietal functional coupling, in both males and females, enhancing it in delta and decreasing it in slow and fast theta and beta1 bands. These results indicate that inter-parietal temporal coupling may also be related to the anxiolytic effects of diazepam. The effect of an axiolityc dose of DZ on cortical fast oscillations and on inter-hemispheric functional coupling suggests that DZ has an effect on the cortex and on hippocampus, which are rich in $GABA_A$ receptors [332,143], generates theta activity and participates in the regulation of anxiety [295,158,43].

An EEG pattern of increased beta has traditionally been associated with an activated state [335,117]; however, benzodiazepines have consistently been related to a lower level of activation. Thus, the EEG pattern of higher beta following diazepam has been termed pharmacological dissociation [49] and has been considered as a paradoxical effect or a special type of sedation. However, it has been demonstrated that in order to get larger amplitudes in field potentials, synchronized activity of a larger population of individual coherent sources is required [244]. GABA receptors in the cerebral cortex are abundant [332,143], and GABA-mediated inhibition has been proposed as a candidate for generating synchronization among neuronal populations [343,80,208,81]; accordingly, beta increase is the result of synchronization of a larger population of neurons under GABAergic inhibition. On the basis of the increase in fast frequencies produced by diazepam, it might be suggested that the anxiolytic effects of this compound were at least partly related to an increase in the absolute power of these frequencies. The role of GABAergic interneurons on the cortico-thalamic network, involved in the generation of slow and fast oscillations [335], on cholinergic basal forebrain neurons participating in the modulation of fast oscillations [285,338] and on septo-hippocampal system involved in theta generation [43] is well described. However, local interneuron networks are less well understood and introduce confounding effects which will only be made clear by further research. Nevertheless, these important results show sex differences concerning the effect of DZ on brain function.

The response to DZ was clearly sexually dimorphic. The increase in absolute power of fast frequencies was significantly lower in females compared to males; besides this, DZ decreased slow and fast theta absolute power (Figure 9A and B) and inter-parietal functional coupling of fast frequencies exclusively in GNX females, demonstrating a higher susceptibility to DZ among females in terms of effects on hippocampal activity and of the uncoupling effect of DZ (Figure 9C and D).

The greater effect of DZ on fast oscillations in males may be explained by the greater sensitivity of males to anxiolytics [327,141,219], whereas, the lesser effect of DZ on fast frequencies in females, despite clear anxiolytic-type effects may be due to the lower sensitivity of GABA receptors in them [85], together with a lower inter-parietal correlation, also decreased by diazepam, both resulting in a lowered synchronization of fast oscillations of neuronal assemblies. Additionally, the milder response of beta in gonadectomized females may be related to the lack of estradiol (see below).

The sexually dimorphic electrical response to DZ indicates that activation/inhibition mechanisms, responsible for synchronization of cortical neuronal assemblies, as reflected in power, are more susceptible to DZ among males, while the attenuation of hippocampal synchronization is more sensitive among females. As already mentioned, cerebral cortex in males and hippocampus among females are respectively more susceptible to environmental and developmental influences; thus this dimorphic response to DZ is not surprising.

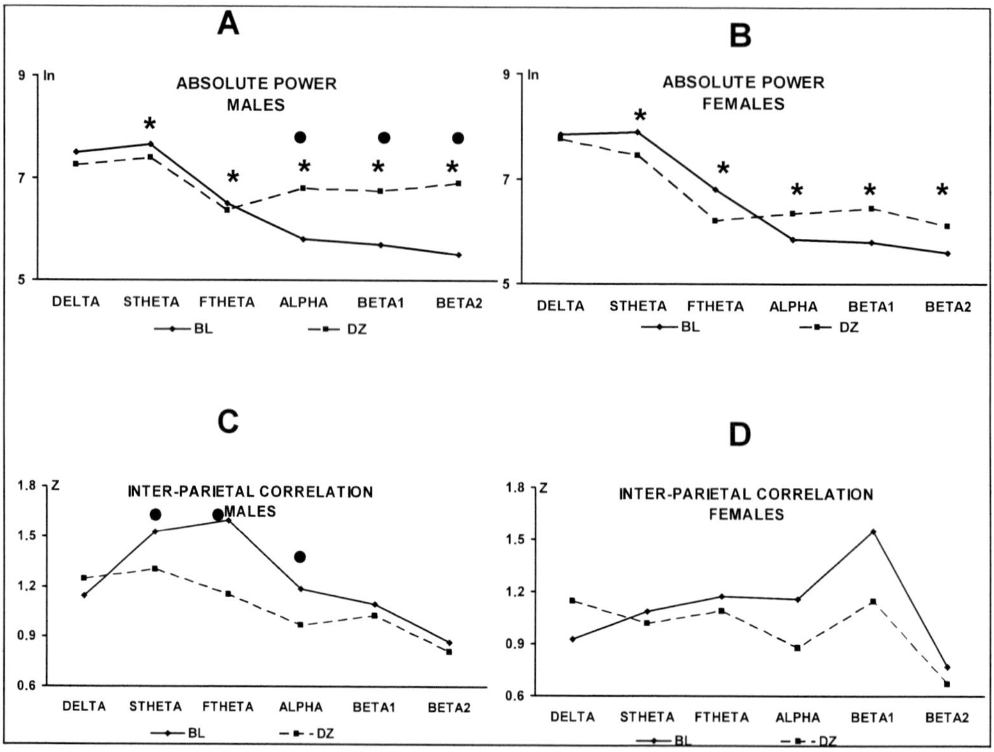

Figure 9. Sexually dimorphic effect of diazepam on electric field potentials and inter-parietal functional coupling in intact adult rats. Significant differences between vehicle baseline (BL) and diazepam (DZ) are indicated by asterisks, and between male and females by dots. Data from 348.

Organizational Actions of Sex Steroids on BZ Effects on Electric Field Potentials

Since theta activity and inter-parietal functional coupling are sexually dimorphic and dependent on organizational actions of sex steroids, in order to explore whether the sexually dimorphic effect of DZ on theta power, fast frequencies and inter-parietal functional coupling is dependent on the neonatal sexual differentiation process, DZ was administered in adulthood to neonatally virilized females and neonatally castrated males [91].

The increase in fast oscillations following DZ is observed to the same extent among neonatally castrated males as among adult castrated males, indicating that it is not dependent on neonatal organization in males; whereas, in neonatally virilized females the increase in alpha and beta1 power is greater than in GNX adult females and similar to that of males, producing a male-type effect, indicating that the effect of DZ which increases fast oscillations in females is dependent on the neonatal masculinizing effects of sex steroids.

The stronger effect of DZ on inter-parietal functional coupling in females, gonadectomized after puberty is no longer evident among neonatally virilized females, which show a similar response to that of neonatally castrated males. The indistinguishable male-type effect of DZ on alpha and beta power and on inter-parietal functional coupling observed in neonatally virilized females indicates that the feminine-type response is dependent on the organizational influence of gonadal steroids.

Activational Actions of Sex Steroids on BZ Effects on Electric Field Potentials

The levels of circulating progesterone influence the stress response in male and female rats. Shock-punished responses increase in both, adult females and neonatally castrated adult males following progesterone or estrogen injections [288]. Ovariectomized rats or those in metestrus, diestrus or early proestrus showed comparably higher levels of anxiety during the experiment, than females in late proestrus or estrous [87]. Furthermore, plasma levels of allopregnanolone and pregnanolone following an oral dose of progesterone correlate directly with plasma levels of progesterone and correlate inversely with tension and anxiety in women [101].

Electric field potentials and functional coupling between left and right parietal electrical activity is modulated by the estrous cycle [57] and by the activational actions of sex steroids as described previously. These data pointed out the importance of sex steroids for modulating behavioural and brain anxiolytic-type responses.

Therefore, in the following experiment [348], the interaction between DZ and sex steroids on brain function was investigated among adult rats normally organized as males and females respectively.

The effect of DZ on cortical activity in females depends on the activational effects of estrogen and progesterone. E treatment eliminated the decrease in absolute power of slow and fast theta and induced an increase in fast frequencies, similar to that observed in males, where they rendered the female response similar to the male response (Figure 10A). These findings concur with results showing that estradiol-filled canulae implanted into the cholinergic neurons of the medial septum diagonal band, which projects to the hippocampus and known to be involved in theta activity generation, induce higher spine density in GNX females [200].

Where females were treated with P, this was found not to modify the feminine pattern of spectral power induced by DZ; the same pattern was maintained as among female controls (Figure 10B), however, P treatment did additionally extend the inter-parietal functional uncoupling to alpha and beta2 bands (Figure 10D) and also introduced a left over right asymmetry, which was absent in GNX females, increasing alpha, beta1 and beta2 in the right parietal and theta in the left parietal, as if the right cortex and the left hippocampus had been affected. In summary, the effect of DZ is stronger in P-treated females, than the response shown by males or by females devoid of P or treated with E, inducing greater inter-parietal decoupling of cortical (beta) and hippocampal (theta) processes, lesser hippocampal activation, and stronger effects on the right parietal.

The exogenous administration of androgens, TP or DHT to GNX males following puberty does not modify the effects of DZ on the male profile of electric field potentials nor in inter-parietal functional coupling, indicating that the effect is not dependent on the activational effects of testosterone or its aromatized metabolites.

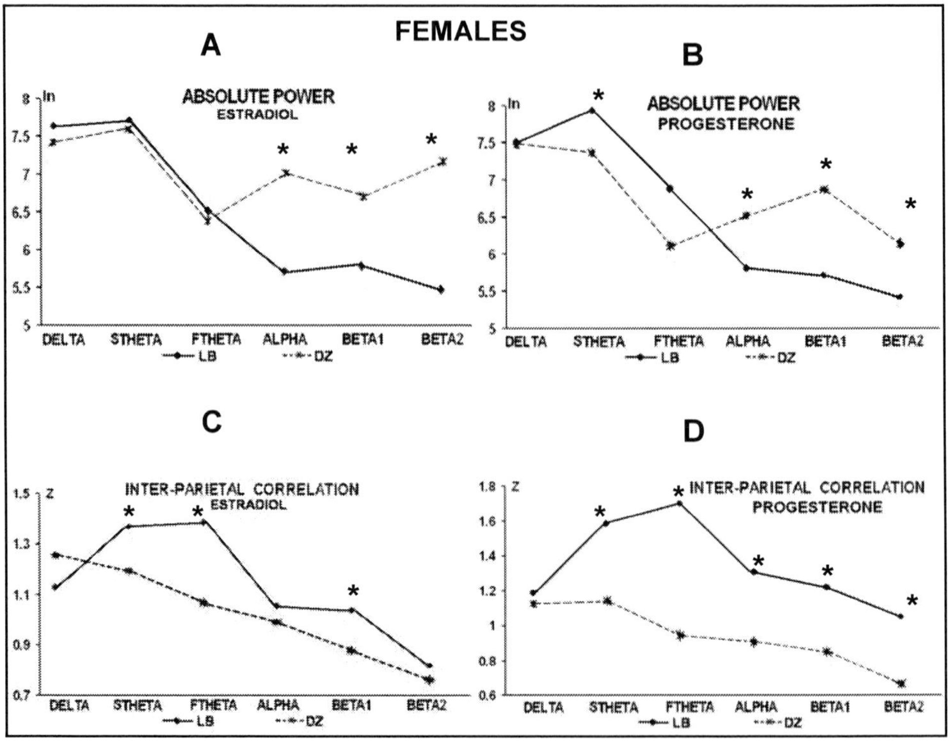

Figure 10. Activational effects of estrogen and progesterone treatment in adult gonadectomized females (GNX) on brain response to diazepam. Significant differences between vehicle baseline (BL) and diazepam (DZ) are indicated by asterisks. Data from 348.

DZ administration to GNX males following puberty when they had previously been treated with E (del-Río-Portilla and Fernández-Guasti, unpublished results), produced a further increase in absolute power of alpha, beta1 and beta2 oscillations and a decrease in slow theta activity in the female-type response, but did not modify the DZ effect on inter-parietal coupling (Figure 11 A and B). P treatment applied to GNX males has an additive effect on the increase in fast oscillations, induced by DZ and induced a female-type response to DZ, decreasing fast theta activity (Figure 11 C) and inter-parietal correlation (Figure 11 D).

The additive effect of DZ in P-treated males is not surprising since de GABA/BZ receptor is modulated by progesterone metabolites [138,216,267] and may be related to the increase in affinity for BZ after P administration [27] and to greater sensitivity to anxiolytics and greater sensitivity of progesterone receptors in males than in females [327,141,219].

It is worth mentioning that both progesterone and diazepam produce similar effects on the EEG, for example, an increased absolute power in high frequency bands. On the basis of the similarities between progesterone and some of its metabolites, such as allopregnanolone and the benzodiazepines or barbiturates as they affect the sleep architecture of the EEG, it has been proposed that progestines may influence sleep through the stimulation of the $GABA_A$ receptors [202]. Indeed, this research group has proposed that all the agonistic modulators of

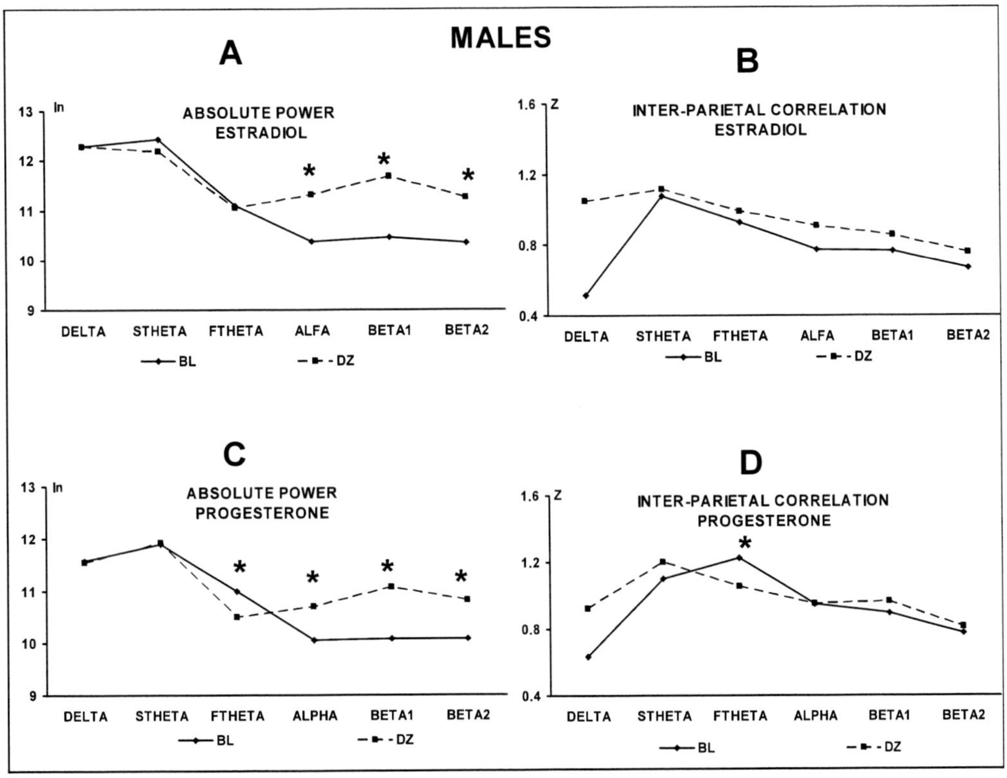

Figure 11. Activational effects of estrogen, progesterone and testosterone treatment in adult gonadectomized males (GNX) on brain response to diazepam. Significant differences between vehicle baseline (BL) and diazepam (DZ) are indicated by asterisks. Data from 348 and unpublished results.

the $GABA_A$ receptor (most of them with anxiolytic-type activity) induce a general enhancement of fast-frequency EEG signals. Present data agree with this observation and indicate that a similar effect is observed at doses much lower than those required to modify the sleep EEG pattern. Further evidence upholding the idea that progestines share a mechanism for action with benzodiazepines and which can change EEG in a similar way, comes from the study of antagonists. Thus, pre-treatment with picrotoxin markedly attenuated most of the effects of progesterone on sleep [201].

In conclusion, these data demonstrate that the effect of DZ on brain function among adult rats is not only sexually dimorphic but also dependent upon the activational effects of gonadal steroids. Among adult rats, gonadectomized after puberty, the increase in the power of frequencies from 10 to 32 Hz is significantly greater in males than in females. Following E administration, the effect on females becomes similar to the effect it has on males. Finally, DZ and more strongly DZ plus P treatment decreases inter-parietal functional coupling, indicating an effect on coherent activity between left and right parietals, particularly among females, caused by both compounds.

Organizational-Activational Interaction of Sex Steroids on BZ Effects on Electric Field Potentials

It is known that the sexual differentiation process has important effects on the anxiolytic-type actions of both, progestines and benzodiazepines during adulthood. Neonatally virilized females are extremely sensitive to the anxiolytic actions of diazepam [89] and insensitive to the anxiety-type effects of progesterone [288] and allopregnanolone [90].

The combined administration of DZ and P rendered interesting results among neonatally virilized females and castrated males in terms of brain function [91]. The increase induced by DZ plus P in alpha, beta1 and beta2 power among neonatally virilized females is as great as that observed in the case of DZ alone and as great as that shown by GNX males following puberty in a male-type pattern. The same can be said for neonatally castrated males; the increase in alpha beta1 and beta2 power is of the same magnitude as that found with DZ alone and similar to the effect observed among neonatally virilized females and among GNX males after puberty. The lack of sex differences introduced by progesterone treatment on relative power in males and females, neonatally treated in order to alter the sex differentiation processes, indicate that the effect of progesterone on DZ is independent of neonatal sexual differentiation.

Inter-parietal functional coupling decreased in slow theta, alpha and beta1 and beta2 bands only among GNX females following puberty and was not altered either among males or among virilized females or among castrated males indicating that the combined action of P and DZ in females is dependent on the prenatal organizational actions of sex steroids, rendering their brain response more susceptible to the uncoupling effects of DZ, whereas neonatal sex steroid actions are no longer effective for altering it. Inter-parietal functional coupling of fast theta, in contrast, was decreased among male and female neonatally treated rats to a similar extent as among GNX females following puberty, indicating that in males it is dependent of postnatal sexual differentiation.

In conclusion, the present series of experiments demonstrated that males are more sensitive than females to the combined effects of DZ plus P on fast frequency power, whereas females are more sensitive to the uncoupling effects of DZ, particularly when DZ and P are administered in combination. DZ effects, in contrast to those of P, are modified to a greater extent by the neonatal sexual differentiation.

Sexual Dimorphic Effect of Benzodiazepines on the Human Being

Limbic system functions involved in emotion, stress, anxiety and mood can be influenced by circulating sex steroids in adult life in a sexually dimorphic way; however, despite the ample evidence demonstrating BZ-steroid interactions and sex differences in the brain, BZ effects have been little explored among women, and sex has not been considered as a variable. It seems reasonable to expect that sex differences in brain organization would modify the effect of BZ on brain function in a way similar to that in the rat. To the best of our knowledge there are only a few basic, clinical studies conducted on sex differences concerning BZ effects on humans, apart from those concerning pharmacokinetics and pharmacodynamics [366,119,120]. One study reports that DZ (10 mg) impaired psychomotor and sensory functions [253], and affected reaction time and time estimation among women, more so than among men [322].

The changes in EEG activity induced by BZ in humans are very similar to those observed in the rat; alpha activity is decreased, while beta is increased in the waking EEG; some studies also report an increase in delta activity in the waking EEG depending on the type and dose of anxiolytics [182,51,8]; the time course of the increase in beta activity is directly related to plasma drug concentrations [117,118].

Given that the quantitative analysis of EEG makes use of a direct and objective method for determining the effects of psychotropic drugs on the human brain [146] and that the EEG changes induced by BZ in the human being are well known, in the subsequent series of experiments, sex differences in the brain response to BZ during rest [289] and during the performance of a sustained, attention demanding task (Muñoz-Torres et al., unpublished results) were investigated. EEG activity was recorded for this purpose in two double-blind placebo experiments, and spectral power and spectral inter- and intra-hemispheric functional coupling or coherent activity were assessed during resting conditions with eyes open in one experiment, and in the other where a one-second pre- stimulus was analyzed, during the sustained attention task performance.

During resting conditions with eyes open, both men and women, as a single group, showed the well-documented decrease in alpha1 and increase in beta1 and beta2 power. Functional coupling among cortical brain regions was also modified (Figure 12); DZ induced a selective, functional uncoupling between frontal and central regions in the theta band and between parietal and occipital regions in the alpha2 band of the right hemisphere. The loss of functional coupling between parieto-occipital and fronto-central regions may be involved in the impairment of cognitive and motor functions [150]. Additionally, interhemispheric functional coupling in the case of beta frequencies and intrahemispheric functional coupling between frontal regions was enhanced bilaterally in the case of theta and between right frontal regions in the case of delta. It is known that the frontal lobes are implicated in mood regulation, especially the right side [69].The increase in functional coupling among frontal regions indicates greater integration, which may be involved in the positive effect of DZ on anxiety and is consistent with the idea that inter- and intra-hemispheric pathways may function to synchronize activity across brain regions and balance the activation between them [206]. DZ effect on funtional coupling was stronger on the right hemisphere, which is known to be involved in novelty and exploration of the environment [109,161,269].

When men and women were analyzed separately, women but not men showed a decrease in theta and alpha2 power. Although great caution must be taken when making interspecies comparisons, the similarity in the selective decrease in theta activity, found in both, the case of female rats as well as among women is striking. Theta power decreased among women and not among men, as well as among gonadectomized female rats without hormonal treatment and among gonadectomized females receiving progesterone and not among males [348]. In contrast, inter-hemispheric correlation was more affected in the case of men, who showed a decrease in the inter-hemispheric correlation of alpha1. These results suggest that

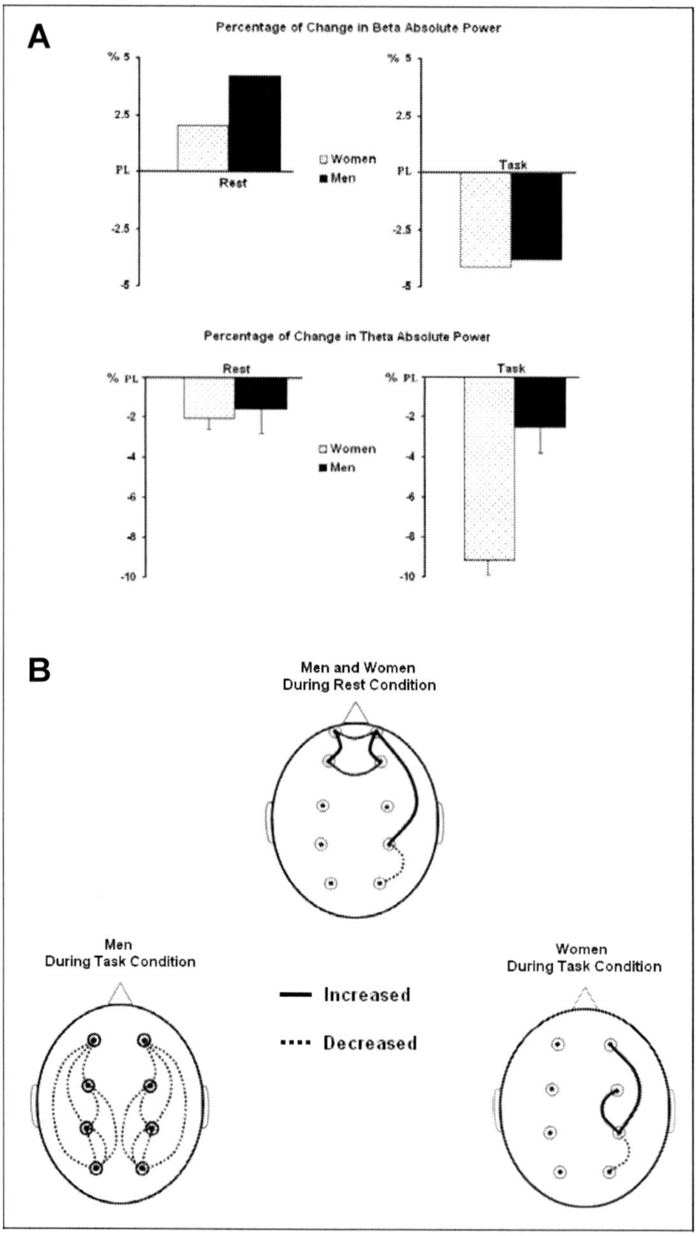

Figure 12. Sexually dimorphic effect of diazepam in absolute power in men and women during rest and during an attention demanding task. In A, percentage of change in Beta (Top) and Theta (Down) absolute power in comparison to placebo (PL). Asterisks indicate significant differences between PL and Diazepam (DZ). In B, schematic representation of the head illustrating pairs of electrodes showing significant changes in functional coupling under DZ. Solid lines indicate increased, and doted lines decreased functional coupling. Data from 289 and unpublished results.

theta rhythm is more sensitive to DZ in women than it is in men, whereas in fact the effect of DZ on inter-hemispheric functional coupling is stronger in men than in women. The milder effect of DZ on inter-hemispheric functional coupling in women may be related to sex differences in brain organization and anatomy. Under normal conditions, women have larger corpus callosum [78,68,70,361] and anterior commissure [4] and greater inter-hemispheric EEG correlation and coherence, than men [19,96,56,60,277]. The higher left and right hemisphere inter-connectivity may result in more robust inter-hemispheric functional coupling among women than among men, as it happens after sleep deprivation; inter-hemispheric correlation is decreased among men, whereas among women the effect of sleep deprivation on inter-hemispheric correlation is attenuated [58,59,61].

DZ effects on several cognitive tasks, such as reaction time, memory and attention are well documented (for review 41). Prefrontal cortex plays a central role in selective attention [270] and sex steroids modulation on prefrontal cortex function has been recently reported. Estrogen receptors are expressed in the cerebral cortex [323,149,346,191,354,262] and progesterone receptors are abundant especially in female frontal region [214]. Cortical activation and arousal result from a complex interplay where global influences coming from brainstem, thalamo-cortical networks, posterior hypothalamus and basal forebrain modulate activation/inhibition synaptic activity by GABAergic, cholinergic, noradrenergic and glutamatergic neurotransmitter systems. These brain regions express estrogen receptors [13,367,323,156,155,346,262,315] and are influenced by sex steroids. Estrogen has been reported to modulate the production of Ach in basal forebrain neurons, a system that projects to the cerebral cortex and hippocampus [315] and is thus involved in selective attention. Sex steroids also regulate the activity of GABAergic interneurons; ER in the cortex, basal forebrain and hippocampal formation appear to be almost exclusively in interneurons where they extensively colocalized with inhibitory GABAergic interneurons and in the cerebral cortex [33]. High estrogen states modulate cortical activation; C-Fos expression in DLPC cortical pyramids is increased by E treatment in primates [354], as well as norepinephrine, dopamine and serotonin turnover in the medial prefrontal cortex following exposure to a novel environment in male GNX rats [136]. Estrogen reduce cortical GABA levels by modulating the expression of inhibitory neurostransmitters in the cortex [34], and progesterone modulates $GABA_A$ receptor function [216,219,364,294]. Estrogen also regulate the expression of excitatory NMDA receptors and increases NMDA binding in female hippocampus [290] and regulates noradrenergic pre- and postsynaptic transmission in the cerebral cortex [144]; GNX in adult males induces a large transient decrease in tyrosine hydroxylase [194], and influences adrenergic receptor expression increasing tyrosine hydroxylase gen transcription and norepinphrine turnover in GNX female rats [254]. However, there are only few studies assessing DZ effects on executive functions reporting contradictory results [64,65,71,227,283,1]. Despite evidence demonstrating sex differences in the distribution and sensitivity of $GABA_A$ receptor on the frontal cortex [85], GABA levels [82] and sexual dimorphism in brain function, studies on cortical EEG activity during task performance exploring sex differences under DZ effects are lacking.

The analysis of EEG activity under the effect of 5 mg of DZ during performance of the sustained attention demanding task in healthy men and women revealed, that in contrast to the well known increase in power of fast frequencies during rest, beta power was decreased, inter-hemispheric coherent activity of fast frequencies was increased while in 9 and 10 Hz it was decreased, probably reflecting an effort to maintain attention [58,59,61].

The dimorphic effect of DZ on theta activity in women was maintained during task performance in accordance to the data obtained during rest, only woman showed decreased theta and slow alpha power. The DZ selective effect on the right intrahemispheric functional coupling during rest, was observed during task performance in women only, whereas in men intra-hemispheric coherent activity was globally decreased.

We conducted the first stage of a behavioral study in males in order to clarify the inhibitory influence of DZ on functions attributed to frontal lobes especially voluntary attention, decision making and the capability to inhibit inadequate motor responses. Under the influence of 10 mg of DZ reaction time increased as expected, especially for incongruent stimuli demanding a decision of motor response and when target appeared on the left hemisphere. We hypothesized that the dose employed is not enough to alter the facility of the right hemisphere, but it does affect the left less specialized hemisphere to shift spatial attention. This evidence suggests impairment in frontal cortex executive skills under DZ-enhanced GABAergic inhibition. DZ is probably modifying the synaptic inhibitory/excitatory ratio required for successful performance in frontal lobes tasks.

In sum, as hypothesized, DZ had a differential effect on men and women's waking EEG activity and a disrupting effect on functional coupling among cortical regions. A single oral dose of 5 mg of DZ, regardless of sex, increased the functional coupling of high frequencies between hemispheres, increased functional coupling selectively within frontal regions and decreased functional coupling between anterior and posterior regions in the case of slow frequencies. Theta relative power was more affected in women, whereas inter- and intrahemispheric functional coupling was more affected in men. The sexually dimorphic effect of DZ on EEG activity may be related to sex differences in brain organization and to the activational effects of the female gonadal steroids which are different to those found in men. These results may have clinically relevant implications for therapeutic strategies.

CONCLUSION

In this chapter we presented behavioral as well as brain structural and functional evidence of sexual dimorphism in human and non-human species. Irrespective of the impossibility to present an exhaustive description and explanation of each result, this review surveyed a large spectrum of experimental data on sex steroid actions during development and in adulthood, which provide evidence that brain anatomy and function are modulated by sex and sex steroid influences.

As was repeatedly shown in this survey, the early emergence of gonadal steroids and sex steroid receptors in the brain during perinatal development implies that sex differences in the brain are not only determined by genetic sex but also by sex steroid influence. What needs to be kept in mind, however, is that multiple factors appear to be responsible for sex differences in the brain and behavior, pre-wiring can be modulated by activational actions of sex steroids as well as by experience and interaction with the environment, and that, although sex differences in the brain and in non-reproductive behavior seem to be a universal phenomenon among almost all studied mammals, nature did not develop a unique model of sexual dimorphism; important interspecies differences may have developed under the pressure of biological adaptation.

A central argument in this chapter is that behavioral sex differences may be the expression of functional brain differences, which in turn depend on structural dimorphism. Although the findings reviewed in this chapter generally support a biological substrate for sex differences, the functional significance of structural differences and the mechanisms linking structure and function are still waiting to be clarified.

A series of experiments are revised in which brain electrical activity was used to assess sexual dimorphism in brain function. Quantitative analysis of ongoing brain electrical activity can be looked at as a reliable tool to assess the functional status of the brain and, thus, to probe the functional expression of sex differences, especially in humans where it is not so easily done with other invasive or costly techniques. The many instances of organizational and activational actions of sex steroids on brain activity and functional coupling are important sources of data to link brain structural and functional sex dimorphism, and add to the growing evidence that sex differences in cognition and emotion are subjected to direct organizational and activational influence exerted by sex steroids on the brain. There are, however, a number of limitations that should be considered in future research. The mechanisms underlying electric field potentials and functional relationships among brain regions are not fully understood yet. Correlation reflects the linear functional relationship of ongoing electrical activity between two brain areas, but it does not uncover nonlinear relationships, nor the specific causes responsible for the level of functional coupling. Ongoing electrical activity is a multifactorial product which includes intrinsic characteristics of the neural elements - for example, type of neuronal population-, and intrinsic membrane properties and specific organization of the neural network – for example, inhibitory and excitatory elements of local circuits, backward and forward feedback mediated by interneurons-. It also depends, at least partially, on reciprocal intracortical long and short connections, and on common afferent inputs from a third system including those from subcortical generators. There is also a clear need to obtain data from a larger coverage of brain locations that would provide for topographic mapping.

Although in the last decades many studies have addressed the question of sex differences, a major question that has only received marginal attention so far is sex differences in the brain response to anxiolytics. Only few studies have investigated this issue, even though experimental evidence shows that sex differences are probably more common than we are usually aware of. Research in this field in humans is at its very beginning; therefore, the number of unanswered questions in this field of research exceeds the number of solved problems and should be the focus of future research.

Although the above instances of recent research can at best provide preliminary evidence of possible sex differences in the brain response to anxiolytics, they give rise to the hope that future work will open new perspectives in medical and pharmacological thinking, providing for better treatment and dosage approaches from a gender perspective.

Acknowledgments

We thank Isabel Pérez-Montfort and Caroline Carslake for careful reading and correction of the English version of the manuscript. Research by the present authors was in part supported by grants from the Universidad Nacional Autónoma de México (DGAPA-UNAM) and the Consejo Nacional de Ciencia y Tecnología (CONACyT). Zeidy Muñoz-Torres is recipient of a scholarship by the Consejo Nacional de Ciencia y Tecnología (CONACYT).

REFERENCES

[1] Acheson, A., Reynolds, B., Richards, J. B., & de Wit, H. (2006). Diazepam impairs behavioral inhibition but not delay discounting or risk taking in healthy adults. *Experimental and clinical psychopharmacology, 14*, 190-198.

[2] Agarwal, V. R., Sinton, C. M., Liang, C., Fisher, C., German, D. C., & Simpson, E. R. (2000). Upregulation of estrogen receptors in the forebrain of aromatase knockout (ArKO) mice. *Molecular and cellular endocrinology, 162*, 9-16.

[3] Allen, J. S., Damasio, H., Grabowski, T. J., Bruss, J., & Zhang, W. (2003). Sexual dimorphism and asymmetries in the gray-white composition of the human cerebrum. *Neuroimage, 18*, 880-894.

[4] Allen, L. S., & Gorski, R. A. (1991). Sexual dimorphism on the anterior commissure and the massa intermedia of the human brain. *The Journal of comparative neurology, 312*, 97-104.

[5] Altenmüller, E. O., Shurman, K., Lim, V. K., & Parlitz, D. (2002). Hits to the left, flops to the right: different emotions during listening to music are reflected in cortical lateralisation patterns. *Neuropsychologia, 40*, 2242-2256.

[6] Amateau, S. K., Alt, J. J., Stamps, C. L., & McCarthy, M. M. (2004). Brain estradiol content in newborn rats: sex differences, regional heterogeneity, and possible de novo synthesis by the female telencephalon. *Endocrinology, 145*; 2906-2917.

[7] Amunts, K., Jancke, L., Mohlberg, H., Steinmetz, H., & Zilles, K. (2000). Interhemispheric asymmetry of the human motor cortex related to handedness and gender. *Neuropsychologia, 38*, 304-312.

[8] Ansseau, M., Doumont, A., Cerfontaine, J. L., Mantanus, H., Rousseau, J. C., & Timist-Berthier, M. (1984). Self-reports of anxiety level and EEG changes after a single dose of benzodiazepines. *Neuropsychopharmacology, 12*, 255-259.

[9] Arce, C., Ramos, J., Guevara, M. A., & Corsi-Cabrera, M. (1995). Effect of spatial ability and sex on EEG power in high school students. *International Journal Psychophysiology, 20*, 11-20.

[10] Arnold, A. P., & Breedlove, S. M. (1985). Organizational and activational effect of sex steroids on brain and behavior: a reanalysis. *Hormones and Behavior, 19*, 469-499.

[11] Azari, N. P., Pettigrew, K. D., Pietrini, P., Murphy, D. G., Horwitz, B., & Schapiro, M. B. (1995). Sex differences in patterns of hemispheric cerebral metabolism: a multiple regression/discriminant analysis of positron emission tomographic data. *International Journal of Neuroscience, 81*, 1-20.

[12] Azcoitia, I., Sierra, A., & Garcia-Segura, L. M. (1999). Localization of estrogen receptor beta-immunoreactivity in astrocytes of the adult rat brain. *Glia, 26*, 260-267.

[13] Babstock, D., Malsbury, C. W., & Harley, C. W. (1997). The dorsal locus coeruleus is larger in male than in female Sprague-Dawley rats. *Neuroscience Letters, 224*, 157-160.

[14] Barker, J. L., & Owen, D. G. (1986). Electrophysiological pharmacology of GABA and diazepam in cultures SNC neurons. In R.W. Olsen, & J. C.Ventor (Eds.), *Benzodiazepine/GABA Receptors and Chloride Channels: Structural and Functional Properties.* (pp. 135). New York: Alan R. Liss.

[15] Barrick, T. R., Mackay, C. E., Prima, S., Maes, F., Vandermeulen, D., Crow, T. J., & Roberts, N. (2005). Automatic analysis of cerebral asymmetry: an exploratory study of the relationship between brain torque and planum temporale asymmetry. *Neuroimage, 24*, 678-691.

[16] Baxter, L. C., Saykin, A. J., Flashman, L. A., Johnson, S. C., Guerin, S. J., Babcock, D. R, & Wishart, H. A. (2003). Sex differences in semantic language processing: a functional MRI study. *Brain and Language, 84*, 264-272.

[17] Beatty, W. W. (1979) Gonadal hormones and sex differences in nonreproductive behaviors in rodents: organizational and activational differences. *Hormones and Behavior, 12*, 112-163.

[18] Beatty, W. W. (1992). Gonadal hormones and sex differences in nonreproductive behaviors. In A. A. Gerall, H. Moltz, & I. L. Ward (Eds.), *Handbook of behavioral neurobiology, Vol 11*, Sexual differentiation (pp. 85- 128). New York: Plenum Press.

[19] Beaumont, J. G., Mayes, A. R., & Rugg, M. D. (1978). Asymmetry in EEG alpha coherence and power: effects of task and sex. *Electroencephalography and Clinical Neurophysiology, 445*, 393-401.

[20] Benca, R. M., Obermeyer, W. H., Larson, Ch. L., Yun, B., Dolski, I., Kleist, K. D., Weber, S. M., & Davidson, R. J. (1999). EEG alpha power and alpha power asymmetry in sleep and wakefulness. *Psychophysiology, 36*, 430-436.

[21] Bereiter, D. A., Cioffi, J. L., & Bereiter, D. F. (2005). Oestrogen receptor-immunoreactive neurons in the trigeminal sensory system of male and cycling female rats. *Archives of oral biology, 50*, 971-979.

[22] Berger-Sweeney, J. (1998). The effects of neonatal basal forebrain lesions on cognition: towards understanding the developmental role of the cholinergic basal forebrain. *International Journal of Developmental Neuroscience, 16*, 603-612.

[23] Berman, K. F., Schmidt, P. J., Rubinow, D. R., Danaceau, M. A., Van Horn, J. D., Esposito, G., Ostrem, J. L., & Weinberger, D. R. (1997). Modulation of cognition-specific cortical activity by gonadal steroids: a positron-emission tomography study in women. *Proceedings of the National Academy of Sciences of the United States of America, 94*, 8836-88341.

[24] Bernal, J., Guzmán-Florez, C., Alcaraz, M., García-Castells, E., Juárez, J., & Solís, S. (1989). Dimorfismo sexual en el desarrollo de la conducta de juego del mono verde. In A. Estrada, R. López-Wilchis, & R. Coates-Estrada (Eds.), *Primatología en México: Comportamiento, ecología aprovechamiento, y conservación de primates* (pp. 98–107). México, DF: Universidad Autónoma Metropolitana.

[25] Beyer, C., Damm, N., Brito, V., & Kuppers, E. (2002). Developmental expression of progesterone receptor isoforms in the mouse midbrain. *Neuroreport, 13*, 877-880.

[26] Bishop, K. M., & Wahlsten, D. (1997). Sex differences in the human corpus callosum: myth or reality? *Neuroscience and biobehavioral reviews, 21*, 581-601.

[27] Bitran, D., & Dowd, J. A. (1996). Ovarian steroids modify the behavioural and neurochemical responses of the central benzodiazepine receptor. *Psychopharmacology, 125*, 67-73.

[28] Bitran, D., Purdy, R. H., & Kellogg, C. K. (1993). Anxiolytic effects of progesterone is associated with increased in cortical allopregnanolone and $GABA_A$ receptor function. *Pharmacology Biochemistry and Behavior, 45*, 423-428.

[29] Bitran, D., Shiekh, M., & McLeod, M. (1995). Anxiolytic effect of progesterone is mediated by the neuroesteroid allopregnanolone at brain GABA (A) receptors. *Journal of neuroendocrinology, 7*, 171-177.

[30] Blanch, R. J., Brennan, D., Condon, B., Santosh, C., & Hadley, D. (2004). Are there gender-specific neural substrates of route learning from different perspectives? *Cerebral cortex, 14*, 1207-1213.

[31] Bland, H. B., & Wishaw, J. K., (1976). Generators and topography of hippocampal theta (RSA) in the anaesthetized and freely moving rat. *Brain Research, 118*, 259-280.

[32] Blood, A. J., Zatorre, R. J., Bermudez, P., & Evans, A. C. (1999). Emocional responses to pleasant and unpleasant music correlate with activity in paralimbic brain regions. *Nature Neuroscience, 2*, 382-387.

[33] Blurton-Jones, M. M., & Tuszynski, M. H. (2002). Estrogen receptor-beta colocalizes extensively with parvalbumin-labeled inhibitory neurons in the cortex, amygdala, basal forebrain, and hippocampal formation of intact and ovariectomized adult rats. *The Journal of comparative neurology, 452*, 276-287.

[34] Blurton-Jones, M. M., & Tuszynski, M. H. (2006). Estradiol-induced modulation of estrogen receptor-beta and GABA within the adult neocortex: a potential transsynaptic mechanism for estrogen modulation of BDNF. *The Journal of comparative neurology, 499*, 603-612.

[35] Blurton-Jones, M. M., Roberts, J. A., Tuszynski, M. H. (1999). Estrogen receptor immunoreactivity in the adult primate brain: neuronal distribution and association with p75, trkA, and choline acetyltransferase. *The Journal of comparative neurology, 405*, 529-542.

[36] Bogic, L., Gerlach, J. L., & McEwen, B. S. (1988). The ontogeny of sex differences in estrogen-induced progesterone receptors in rat brain. *Endocrinology, 122*, 2735-2741.

[37] Bowers, C. A., & LaBarba, R. C. (1988). Sex differences in the lateralization of spatial abilities: a spatial component analysis of extreme group scores. *Brain and cognition, 8*, 165-177.

[38] Bremner, J. D., Soufer, R., McCarthy, G., Delaney, R., Staib, L. H., Duncan, J. S., & Charney. D. S. (2001). Related gender differences in cognitive and neural correlates of remembrance of emotional words. *Psychopharmacology Bulletin, 35*, 55-78.

[39] Broverman, D. N., Vogel, W., Klaiber, E., Majcher, D., Shea, D., & Paul, V. (1981). Changes in cognitive task performance across the menstrual cycle. *Journal of comparative and physiological psychology, 95*, 646-654.

[40] Brugger, P., Milicevic, A., Regard, M., & Cook, N. D. (1993). Random-number generation and the premenstrual cycle: preliminary evidence for a premenstrual alteration of frontal lobe functioning. *Perceptual and motor skills, 77*, 915-921.

[41] Buffett-Jerrott, S. E., & Stewart, S.H. (2002). Cognitive and sedative effects of benzodiazepine use. *Current pharmaceutical design, 8*, 45-58.

[42] Bullock, Th. H. & Basar, E. (1988). Comparison of ongoing compound field potentials in the brains of invertebrates and vertebrates. *Brain Research. Review, 13*, 57-75.

[43] Buzsaki, G. (2006). Rhythms of the Brain. New York: Oxford University Press.

[44] Cahill, L., Uncapher, M., Kilpatrick, L., Alkire, M. T., & Turner, J. (2004). Sex-related hemispheric lateralization of amygdale function in emotionally influenced memory: an fMRI investigation. *Learning and Memory, 11*, 261-266.

[45] Campanella, S., Rossignol, M., Mejias, S., Joassin, F., Maurage, P., Debatisse, D., Bruyer, R., Crommelinck, M., & Guerit, J. M. (2004). Human gender differences in an emotional visual oddball task: an event-related potentials study. *Neuroscience Letters, 367*, 14-18.

[46] Canli, T., Desmond, J. E., Zhao, Z., & Gabrieli, J. D. (2002). Sex differences in the neural basis of emotional memories. *Proceedings of the National Academy of Sciences of the United States of America, 99*, 10789-10794.

[47] Carrer, H. F., Araque, A., & Buno, W. (2003). Estradiol regulates the slow Ca2+-activated K+ current in hippocampal pyramidal neurons. *Journal of Neuroscience, 23*, 6338-6344.

[48] Chesler, E. J., & Juraska, J. M. (2000). Acute administration of estrogen and progesterone impairs the acquisition of the spatial morris water maze in ovariectomized rats. *Hormones and behavior, 38*, 234-242.

[49] Coenen, A. M. L., & van Luijtelaar, E. L. J. M. (1991). Pharmacological dissociation of EEG and behavior: a basic problem in sleep-wake classification. *Sleep, 14*, 464-465.

[50] Coney, J. (2002). Lateral asymmetry in phonological processing: relating behavioral measures to neuroimaged structures. *Brain and Language, 80*, 355-365.

[51] Coppola, R., & Herrmann, W. M. (1987). Psychotropic drug profiles: comparison by topographic maps of absolute power. *Neuropsychobiology, 18*, 97-104.

[52] Cordoba Montoya, D. A., Carrer, H. F. (1997). Estrogen facilitates induction of long term potentiation in the hippocampus of awake rats. *Brain Research, 778*, 430-438.

[53] Corsi-Cabrera, M., Arce, C., Ramos, J., & Guevara, M. A. (1997). Effect of spatial ability and sex on inter- and intrahemispheric correlation of EEG activity. *Electroencephalography and Clinical Neurophysiology, 102*, 5-11.

[54] Corsi-Cabrera, M., Galindo-Vilchis, L., del-Río-Portilla, Y., Arce, C, & Ramos-Loyo, J. (2007). Within subject reliability and inter-session stability of EEG power and coherent activity in women evaluated monthly over 9 months. *Clinical Neurophysiology, 118*, 9-21.

[55] Corsi-Cabrera, M., Gutiérrez, S., Ramos, J., & Arce, C. (1988). Interhemispheric correlation of EEG activity during successful and unsuccessful cognitive performance. *International Journal of Neuroscience, 39*, 253-259.

[56] Corsi-Cabrera, M., Herrera, P., & Malvido, M. (1989). Correlation between EEG and cognitive abilities: sex differences. *International Journal of Neuroscience, 45*, 133-141.

[57] Corsi-Cabrera, M., Juárez, J., Ponce-de-León, M., Ramos, J., & Velázquez, P. N. (1992). EEG activity during estral cycle in the rat. *Electroencephalography and Clinical Neurophysiology, 83*, 265-269.

[58] Corsi-Cabrera, M., Ramos, J., & Meneses, S. (1989). Effect of normal sleep and sleep deprivation on interhemispheric correlation during subsequent wakefulness in man. *Electroencephalography and clinical neurophysiology, 72*, 305-311.

[59] Corsi-Cabrera, M., Ramos, J., Arce, C., Guevara, M. A., Ponce-de-León, M., & Lorenzo, I. (1992). Changes in the waking EEG as a consequence of sleep and sleep deprivation. *Sleep, 15*, 550-555.

[60] Corsi-Cabrera, M., Ramos, J., Guevara, M. A., Arce, C., & Guriérrez, S. (1993). Gender difference in the EEG during cognitive activity. *International Journal of Neuroscience, 72*, 257-264.

[61] Corsi-Cabrera, M., Sánchez, A. I., del-Río-Portilla, Y., Villanueva, Y., & Pérez-Garci, E. (2003). Effects of 38 h of total sleep deprivation on the waking EEG in women: sex differences. *International Journal of Psychophysiology, 50*, 213-224.

[62] Corsi-Cabrera, M., Solís-Ortiz, S., & Guevara, M. A. (1997). Stability of inter- and intrahemispheric correlation in women. *Electroncephalography and Clinical Neurophysiology, 102*, 248-255.

[63] Corsi-Cabrera, M., Ugalde, E., del-Río-Portilla, I. Y., & Fernández-Guasti, A. (2000). Organizational and activational effects of gonadal steroid hormones on the EEG of male and female rats. *Developmental Psychobiology, 37*, 194-207.

[64] Coull, J. T., Frith, C. D., & Dolan, R. J. (1999). Dissociating neuromodulatory effects of diazepam on episodic memory encoding and executive function. *Psychopharmacology, 145*, 213-222.

[65] Coull, J. T., Middleton, H. C., Robbins, T. W., Sahakian, B. J. (1995). Contrasting effects of clonidine and diazepam on tests of working memory and planning. *Psychopharmacology (Berl). 120*, 311-321.

[66] Creutz, L. M. & Kritzer, M. F. (2002). Estrogen receptor-beta immunoreactivity in the midbrain of adult rats: regional, subregional, and cellular localization in the A10, A9, and A8 dopamine cell groups. *Journal of Neuroscience Research, 446*, 288-300.

[67] Davatzikos, C., & Bryan, R. N. (2002). Morphometric analysis of cortical sulci using parametric ribbons: a study of the central sulcus. *Journal of computer assisted tomography, 26*, 298-307.

[68] Davatzikos, C., & Resnick, S. M. (1998). Sex differences in anatomic measures of interhemispheric connectivity: correlations with cognition in women but not men. *Cerebral Cortex, 8*, 635-640.

[69] Davidson, R. J. (1992). Anterior cerebral asymmetry and the nature of emotion. *Brain and cognition, 20*, 125-151.

[70] De Lacoste-Utamsing, C., & Holloway, R. L., (1982). Sexual dimorphism in the human corpus callosum. *Science, 216*, 1431-1432.

[71] Deakin, J. B., Aitken, M. R., Dowson, J. H., Robbins, T. W., & Sahakian, B. J. (2004). Diazepam produces disinhibitory cognitive effects in male volunteers. *Psychopharmacology, 173*, 88-97.

[72] Deitch, H. R., Mershon, J. L., & Clark, K. E. (2001). Estrogen receptor beta is the predominant isoform expressed in the brain of adult and fetal sheep. *American journal of obstetrics and gynecology, 184*, 1077-1079.

[73] del Río-Portilla, I. Y., Ugalde, E., Juárez, J., Roldán, A., & Corsi-Cabrera, M. (1997). Sex differences in EEG in adult gonadectomized rats before and after hormonal treatment. *Psychoneuroendocrinology, 22*, 627-642.

[74] Diamond, M .C., Ingham, C. A., Johnson, R .E., Bennett, E. L., & Rosenzweig, M. R. J. (1976). Effects of environment on morphology of rat cerebral cortex and hippocampus. *Neurobiology, 7*, 75-85.

[75] Diamond, M. C. (1991). Hormonal effects on the development or cerebral lateralization. *Psychoneuroendocrinology, 16*, 121-129.

[76] Diamond, M. C., Johnson, R. E., Young, D., & Sing, S. S. (1983). Age related morphologic differences in the rat cortex and hippocampus: male-female; right-left. *Experimental Neurology, 81*, 1-13.

[77] Dorion, A. A., Chantome, M., Hasboun, D., Zouaoui, A., Marsault, C., Capron, C., & Duyme, M. (2000). Hemispheric asymmetry and corpus callosum morphometry: a magnetic resonance imaging study. *Neuroscience Research, 36*, 9-13.

[78] Dubb, A., Gur, R., Avants, B., & Gee, J. (2003). Characterization of sexual dimorphism in the human corpus callosum. *Neuroimage, 20*, 512-519.

[79] Eberling, J. L., Wu, C., Haan, M. N., Mungas, D., Buonocore, M., & Jagust, W. J. (2003). Preliminary evidence that estrogen protects against age-related hippocampal atrophy. *Neurobiology of aging. 24*, 725-732.

[80] Edelman, G. M., & Tononi, G. (2000). *A Universe of Consciousness*. New York: Basic Books.

[81] Elliot, M. A., Becker, C., Boucart, M., & Müller, H. J. (2000). Enhanced $GABA_A$ inhibition enhances synchrony coding in human perception. Cognitive Neuroscience *Neurophysiology, 11*, 3403-3407.

[82] Epperson, C. N., Haga, K., Mason, G. F., Sellers, E., Gueorguieva, R., Zhang, W., Weiss, E., Rothman, D. L., & Krystal, J. H. (2002). Cortical gamma-aminobutyric acid levels across the menstrual cycle in healthy women and those with premenstrual dysphoric disorder: a proton magnetic resonance spectroscopy study. *Archives of general psychiatry. 59*, 851-858.

[83] Erdogan, A. R., Dane, S., Aydin, M. D., Ozdikici, M., & Diyarbakirli, S. (2004). Sex and handedness differences in size of cerebral ventricles of normal subjects. *International Journal of Neuroscience, 114*, 67-73.

[84] Fabes, R. A., & Martin, C. J. (1991). Gender and age stereotypes in emotionality. *Personality and Social Psychology Bulletin, 17*, 532-540.

[85] Farabollini, F., Fluck, E., Albonetti, E., & File, S. E. (1996). Sex differences in benzodiazepine binding in the frontal cortex and amygdala of the rat 24 hours after restraint stress. *Neuroscience Letters, 218*, 177-180.

[86] Fernández, T., Harmony, Th., Rodríguez, M., Reyes, A., Marosi, E., & Bernal, J. (1993). Test-retest reliability of EEG spectral parameters during cognitive tasks: I Absolute and relative power. *International Journal of Neuroscience, 68*, 255-261.

[87] Fernández-Guasti, A., & Picazo, O. (1992). Changes in burying behavior during the estrous cycle: effect of estrogen and progesterone. *Psychoneuroendocrinology, 17*, 681-689.

[88] Fernández-Guasti, A., & Picazo, O. (1995). Flumazenil blocks the anxiolytic action of allopregnanolone. *European Journal of Pharmacology, 281*, 113-115.

[89] Fernández-Guasti, A., & Picazo, O. (1997). Anxiolytic actions of diazepam, but not buspirone, are influenced by gender and the endocrine stage. *Behavioural brain research, 88*, 213-218.

[90] Fernández-Guasti, A., & Picazo, O. (1999). Sexual differentiation modifies the allopregnanolone anxiolytic actions in rats. *Psychoneuroendocrinology, 24*, 251-267.

[91] Fernandez-Guasti, A., del Río Portilla, I. Y., Ugalde, E., & Corsi-Cabrera, M. (2003). Diazepam and progesterone produce sexually dimorphic actions on the rat EEG: role of the neonatal sexual differentiation process. *Psychoneuroendocrinology, 28*, 85-100.

[92] Filipek, P. A., Richelme, C., Kennedy, D. N., & Caviness, V. S. Jr. (1994). The young adult human brain: an MRI-based morphometric analysis. *Cerebral Cortex, 4*, 344-360.

[93] Flores-Gutierrez, E. O., Díaz, J. L., Barrios, F. A., Favila-Humara, R., Guevara, M. A., del Río-Portilla, Y., & Corsi-Cabrera, M. (Submitted). The metabolic and electric brain patterns during pleasant and unpleasant emotion induced by music masterpieces.

[94] Flores-Gutierrez, E. O., Díaz, J. L., Barrios, F. A., Guevara, M. A., del Río-Portilla, Y., & Corsi-Cabrera, M. (Submitted). Differential alpha coherence hemispheric patterns in men and women during pleasant and unpleasant musical emotions.

[95] Flores-Gutiérrez, E. O., Díaz, J. L., Guevara, M. A., del Río-Portilla, Y., & Corsi-Cabrera, M. (Submitted). The type and number of emotions induced by music masterpieces differ between men and women.

[96] Flor-Henry, P., Koles, Z. J., & Reddon, J. R. (1987). Age and sex related EEG configurations in normal subjects. In A. Glass (Ed*.), Individual Differences in Hemispheric Specialisation.* (pp. 121-134). New York: Plenum Press.

[97] Foundas, A. L., Faulhaber, J. R., Kulynych, J. J., Browning, C. A., & Weinberger, D. R. (1999). Hemispheric and sex-linked differences in Sylvian fissure morphology: a quantitative approach using volumetric magnetic resonance imaging. *Neuropsychiatry, neuropsychology, and behavioral neurology, 12*, 1-10.

[98] Franklin, M. S., Kraemer, G. W., Shelton, S. E., Baker, E., Kalin, N. H., & Uno, H. (2000). Gender differences in brain volume and size of corpus callosum and amygdala of rhesus monkey measured from MRI images. *Brain Research, 852*, 263-267.

[99] Frederikse, M. E., Lu, A., Aylward, E., Barta, P., & Pearlson, G. (1999). Sex differences in the inferior parietal lobule. *Cerebral Cortex, 9*, 896-901.

[100] Frederikse, M., Lu, A., Aylward, E., Barta, P., Sharma, T., & Pearlson, G. (2000). Sex differences in inferior parietal lobule volume in schizophrenia. *The American journal of psychiatry, 157*, 422-427.

[101] Freeman, E. W., Purdy, R. H., Coutifaris, C., Rickels, K., & Paul, S. M. (1993). Correlation with mood and performance measures following oral progesterone administration to healthy female volunteers. *Neuroendocrinology, 58*, 478-484.

[102] Furst, G. J. (1976). EEG alpha asymmetry and visuospatial performance. *Nature, 260*, 254-255.

[103] Gadea, M., Marti-Bonmati, L., Arana, E., Espert, R., Casanova, V., & Pascual, A. (2002). Dichotic listening and corpus callosum magnetic resonance imaging in relapsing-remitting multiple sclerosis with emphasis on sex differences. *Neuropsychology, 16*, 275-281.

[104] Galea, L. A., Spritzer, M. D, Barker, J. M, Pawluski, J. L. (2006). Gonadal hormone modulation of hippocampal neurogenesis in the adult. *Hippocampus, 16*, 225-232.

[105] Gauthier, P., Jouffray, L., & Gottesmann, R. M. (1980). Electroencephalographic recordings during parachute jump sessions. *Aviation, space, and environmental medicine, 50*, 336-338.

[106] Gee, K. W., Brinton, R. E., Chang, W. C., & McEwen, B. S. (1987). Gamma-aminobutyric aciddependent modulation of the chloride ionophore by steroids in rat brain. *European Journal of Pharmacology, 136*, 419-423.

[107] Georgopoulos, A. P., Whang, K., Georgopoulos, M. A., Tagaris, G. A., Amirikian, B., Richter, W., Kim, S. G., & Ugurbil, K. (2001). Functional magnetic resonance imaging of visual object construction and shape discrimination: relations among task, hemispheric lateralization, and gender. *Journal of Cognitive Neuroscience, 13*, 72-89.

[108] Geschwind, N. (1968). The mechanism of normal pressure hydrocephalus. *Journal of the neurological sciences, 7*, 481-493.

[109] Glick, S. D., Ross, D. A., & Hough, L. B. (1982). Lateral asymmetry of neurotransmitters in human brain. *Brain Research, 234*, 53-63.

[110] Goldstein, J. M., Seidman, L. J., Horton, N. J., Makris, N., Kennedy, D. N., Caviness, V. S. Jr., Faraone, S. V., & Tsuang, M. T. (2001). Normal sexual dimorphism of the adult human brain assessed by in vivo magnetic resonance imaging. *Cerebral Cortex. 11*, 490-497.

[111] Goldstein, J. M., Seidman, L. J., O'Brien, L. M., Horton, N. J., Kennedy, D. N., Makris, N., Caviness, V. S. Jr., Faraone, S. V., & Tsuang, M. T. (2002). Impact of normal sexual dimorphisms on sex differences in structural brain abnormalities in schizophrenia assessed by magnetic resonance imaging. *Archives of general psychiatry, 59*, 154-164.

[112] Gong, G., Jiang, T., Zhu, C., Zang, Y., Wang, F., Xie, S., Xiao, J., & Guo, X. (2005). Asymmetry analysis of cingulum based on scale-invariant parameterization by diffusion tensor imaging. *Human Brain Mapping, 24*, 92-98.

[113] Good, C. D., Johnsrude, I., Ashburner, J., Henson, R. N., Friston, K. J., & Frackowiak, R. S. (2001). Cerebral asymmetry and the effects of sex and handedness on brain structure: a voxel-based morphometric analysis of 465 normal adult human brains. *Neuroimage, 14*, 685-700.

[114] Gorski, R. A., Harlan, R. E., Jacobson, C. D., Shryne, J. E., & Southam, A. M. (1980). Evidence for the existence of sexually dimorphic nucleus in the preoptic area of the rat. *The Journal of Comparative Neurology, 193*, 529-539.

[115] Goy, R. W., Bercovitch, F. B., & McBrair, M. C. (1988). Behavioral masculinization is independent of genital masculinization in prenatally androgenized female rhesus macaques. *Hormones and Behavior, 22*, 552–571.

[116] Grabowski, T. J., Damasio, H., Eichhorn, G. R., & Tranel, D. (2003). Effects of gender on blood flow correlates of naming concrete entities. *Neuroimage, 20*, 940-954.

[117] Greenblatt, D. J., Ehrenberg, B. L., Gunderman, J., Locniskar, A., Scavone, J. M., Harmatz, J. S., & Shader, R.I. (1989). Pharmacokinetic and electroencephalographic study of intravenous diazepam, midazolam and placebo. *Clinical pharmacology and therapeutics, 45*, 356–365.

[118] Greenblatt, D. J., Gan, L., Harmatz, J. S., & Shader, R. I. (2005). Pharmocokinetics and pharmacodynamics of single-dose triazolam: electroencephalography compared with the Digit-Symbol Substitution Test. *British journal of clinical pharmacology, 60*, 244-248.

[119] Greenblatt, D. J., Harmatz, J. S., von Moltke, L. L., Wright, C. E., & Shader, R. I. (2004). Age and gender effects on the pharmacokinetics and pharmacodynamics of triazolam, a cytochrome P450 3A substrate. *Clinical pharmacology and therapeutics, 76*, 467-479.

[120] Greenblatt, D. J., Harmatz, J. S., von Moltke, L. L., Wright, C. E., Durol, A. L., Harrel-Joseph, L. M., & Shader, R. I. (2000). Comparative kinetics and response to the benzodiazepine agonists triazolam and zolpidem: evaluation of sex-dependent differences. *The Journal of pharmacology and experimental therapeutics, 293*, 435-443.

[121] Gron, G., Wunderlich, A. P,, Spitzer, M., Tomczak, R., & Riepe, M. W. (2000). Brain activation during human navigation: gender-different neural networks as substrate of performance. *Nature Neuroscience, 3*, 404-408.

[122] Guerra-Araiza, C., Coyoy-Salgado, A., & Camacho-Arroyo, I. (2002). Sex differences in the regulation of progesterone receptor isoforms expression in the rat brain. *Brain Research Bulletin, 59*, 105-109.

[123] Gundlah, C., Kohama, S. G., Mirkes, S. J., Garyfallou, V. T., Urbanski, H. F., & Bethea, C. L. (2000). Distribution of estrogen receptor beta (ERbeta) mRNA in hypothalamus, midbrain and temporal lobe of spayed macaque: continued expression with hormone replacement. *Brain research. Molecular brain research, 76*, 191-204.

[124] Gunther, W., Steinberg, R., Streck, P., Banquet, J.P., Bscheid, I., Raith, L., Riedel, R., Klages, U., & Stiltz, I. (1991). Brain dysfunction in psychiatric patients during music perception measured by EEG mapping: relation to motor dysfunction and influence of neuroleptic drugs. *European Neuropsychopharmacology, 1*, 143-155.

[125] Guo, X. Z., Su, J. D., Sun, Q. W., & Jiao, B. H. (2001). Expression of estrogen receptor (ER) -alpha and -beta transcripts in the neonatal and adult rat cerebral cortex, cerebellum, and olfactory bulb. *Cell Research, 11*, 321-424.

[126] Gur, R. C., Alsop, D., Glahn, D., Petty, R., Swanson, C. L., Maldjian, J. A., Turetsky, B. I., Detre, J. A., Gee, J., & Gur, R. E. (2000). An fMRI study of sex differences in regional activation to a verbal and a spatial task. *Brain and language, 74*, 157-170.

[127] Gur, R. C., Gunning-Dixon, F., Bilker, W. B., & Gur, R. E. (2002). Sex differences in temporo-limbic and frontal brain volumes of healthy adults. *Cerebral Cortex, 12*, 998-1003.

[128] Gur, R. C., Gur, R. E., Obrist, W. D., Hungerbuhler, J. P., Younkin, D., Rosen, A. D., Skilnick, B. E., & Reivich, M. (1982). Sex and handedness differences in cerebral blood flow during rest and cognitive activity. *Science, 217*, 659-661.

[129] Gur, R. C., Turetsky, B. I., Matsui, M., Yan, M., Bilker, W., Hughett, P., & Gur, R. E. (1999). Sex differences in brain gray and white matter in healthy young adults: correlations with cognitive performance. *Journal of Neuroscience, 19*, 4065-4072.

[130] Gyermek, L., & Stoyka, L. F. (1975). Steroid anesthetics. Anesthesiology, 42, 331-334.

[131] Haefely, E.W., (1989). Pharmacology of the benzodiazepine receptor. *European Archives of Psychiatry & Neurological Sciences, 238*, 294-301.

[132] Hagihara, K., Hirata, S., Osada, T., Hirai, M., & Kato, J. (1992). Expression of progesterone receptor in the neonatal rat brain cortex: detection of its mRNA using reverse transcription-polymerase chain reaction. *The Journal of steroid biochemistry and molecular biology, 41*, 637-640.

[133] Hall, J. (1978). Nonverbal sex differences: Communication accuracy and expressive style. Baltimore: Johns Hopkins University Press.

[134] Hampson, E. (1990). Variation in sex-related cognitive abilities across the menstrual cycle. *Brain and Cognition, 4*, 26-43.

[135] Handa, R. J., Corbier, P., Shryne, J. E., Schoonmaker, J. N., & Gorski, R. A. (1985). Differential effects of the perinatal steroid environment on the sexually dimorphic parameters or the rat brain. *Biology of Reproduction, 32*, 855-864.

[136] Handa, R. J., Hejna, G. M., & Lorens, S. A. (1997). Androgen inhibits neurotransmitter turnover in the medial prefrontal cortex of the rat following exposure to a novel environment. *Brain Research, 751*, 131-138.

[137] Harmony, T., Fernández, T., Rodríguez M., Reyes, A., Marosi. E., & Bernal, J. (1993). Test-retest reliability of EEG spectral parameters during cognitive tasks: II coherence. *International Journal of Neuroscience, 68*, 263-271.

[138] Harrison, L. N., Majewska, D. M., Harrington, W. J. & Barker, L. J. (1987). Structure-activity relationships for steroid interaction with the gama-aminobutyric acidA receptor complex. *Journal of Pharmacology and Experimental Therapeutics, 241*, 346-353.

[139] Harshman, R. A.. Hampson, E., & Berenbaum, S. A. (1983). Individual differences in cognitive abilities and brain organization, part I: sex and handedness differences in ability. *Canadian journal of psychology, 37*, 144-192.

[140] Hausmann, M., & Gunturkun, O. (2000) Steroid fluctuations modify functional cerebral asymmetries: the hypothesis of progesterone-mediated interhemispheric decoupling. *Neuropsychologia, 38*, 1362-1374.

[141] Hawke, R. L., & Welch, R. M. (1985). Major differences in the specificity and regulation of mouse renal cytochrome P-450-dependent monooxygenases. A comparison of xenobiotic and endogenous substrates. *Molecular pharmacology, 28*, 283-289.

[142] Hayashi, S., Hayashi, H., Ueda, H., & Papadopoulos, G. C. (2001). Analysis of transient expression of estrogen receptor-alpha in newborn rat primary auditory cortex. *Hormones and Behavior, 40*, 191-195.

[143] Heiss, W. D., & Herholz, K. (2006). Brain receptor imaging. *Journal of nuclear medicine, 47*, 302-312.

[144] Herbison, A. E., Simonian, S. X., Thanky, N. R., Bicknell, R. J. (2000). Oestrogen modulation of noradrenaline neurotransmission. *Novartis Foundation symposium, 230*, 74-85.

[145] Hernández-González, M., Guevara, M. A., Cervantes, M., Moralí, G., & Corsi-Cabrera, M. (1998). Characteristic frequency bands of cortico-frontal EEG during the sexual interaction of the male rat as a result of factorial analysis. *Journal of Physiology-Paris, 92*, 43-50.

[146] Herrmann, W. M., & Schaerer, E. (1990). Pharmaco-EEG: computer EEG analysis to describe the projection to drug effects on a functional cerebral level in humans. In F. H. Lopes da Silva, W. Storm van Leewen, & A. Rémond, (Eds.), *Clinical applications of computer analysis of EEG and other neurophysiological signals*. (pp. 385-445). Amsterdam: Elsevier.

[147] Heyer, A., Hasselblatt, M., von Ahsen, N., Hafner, H., Siren, A. L., & Ehrenreich, H. (2005). In vitro gender differences in neuronal survival on hypoxia and in 17beta-estradiol-mediated neuroprotection. *Journal of cerebral blood flow and metabolism, 25*, 427-430.

[148] Higgit, A., Fonagy, P., Toone, B., & Shine, P. (1990). The prolonged benzodiazepine withdrawal syndrome: anxiety or hysteria? *Acta psychiatrica Scandinavica, 82*, 165-168.

[149] Hirata, S., Osada, T., Hirai, M., Hagihara, K., & Kato, J. (1992). Expression of estrogen receptor in the rat brain: detection of its mRNA using reverse transcription-polymerase chain reaction. *The Journal of steroid biochemistry and molecular biology, 41*, 583-587.

[150] Hobbs, W. R., Rall, T. W., & Verdoozn, T. A. (1996). Hypnotics and sedatives. In L. S. Goodman, A. Gilman (Eds.), *The pharmacological basis of therapeutics*. (pp. 361-480). New York: McGraw-Hill.

[151] Howard, R. Fenwick, P., Brown, D., & Norton, R. (1992). Relationship between CNV asymmetries and individual differences in cognitive performance, personality and gender. *International journal of psychophysiology: official journal of the International Organization of Psychophysiology, 13*, 191-197.

[152] Hu, Z. Y., Bourreau, E., Jung-Testas, I., Robel, P., & Baulieu, E. E. (1987). Neurosteroids: oligodendrocyte mitochondria convert cholesterol to pregnenolone. *Proceedings of the National Academy of Sciences of the USA, 84*, 8215-8229.

[153] Insausti, R., Juottonen, K., Soininen, H., Insausti, A. M., Partanen, K., Vainio, P., Laakso, M. P., & Pitkanen, A. (1998). MR volumetric analysis of the human entorhinal, perirhinal, and temporopolar cortices. *AJNR American journal of neuroradiology, 19*, 659-671.

[154] Isgor, C., & Sengelaub, D. R. (2003). Effects of neonatal gonadal steroids on adult CA3 pyramidal neuron dendritic morphology and spatial memory in rats. *Journal of neurobiology, 55*, 179-190.

[155] Ishunina, T. A. & Swaab, D. F. (2001), Increased expression of estrogen receptor alpha and beta in the nucleus basalis of Meynert in Alzheimer's disease. *Neurobiology of aging, 22*, 417-426.

[156] Ishunina, T. A., van Heerikhuize, J. J., Ravid, R., & Swaab, D. F. (2003). Estrogen receptors and metabolic activity in the human tuberomamillary nucleus: changes in relation to sex, aging and Alzheimer's disease. *Brain Research, 988*, 84-96.

[157] Ivanova, T., & Beyer, C. (2000). Ontogenetic expression and sex differences of aromatase and estrogen receptor-alpha/beta mRNA in the mouse hippocampus. *Cell and Tissue Research, 300*, 231-237.

[158] Iwata, N., & Mikuni, N. (1980). EEG changes in the conscious rat during immobility induced by psychological stress. *Psychopharmacology, 71*,117-122.

[159] Jacobs, L. F., Gaulin, S. J., Sherry, D. F., Hoffman, G. E. (1990). Evolution of spatial cognition: sex-specific patterns of spatial behavior predict hippocampal size. *Proceedings of the National Academy of Science, USA, 87*, 6349-6352.

[160] Jaeger, J. J., Lockwood, A. H., Van Valin R. D. Jr., Kemmerer, D. L., Murphy, B. W., & Wack, D.S. (1998). Sex differences in brain regions activated by grammatical and reading tasks. *Neuroreport, 9*, 2803-2807.

[161] Jayasundar, R. (2002). Human brain: biochemical lateralization in normal subjects. *Neurology India, 50*, 267-271.

[162] John, E. R. (2001). A field theory of consciousness. *Consciousness and Cognition, 10*, 184- 213.

[163] Johnston, A. L., & File, S. E. (1991). Sex differences in animal tests of anxiety. *Physiology & behavior, 49*, 245–250.

[164] Jordan, K., Wustenberg, T., Heinze, H. J., Peters, M., & Jancke, L. (2002). Women and men exhibit different cortical activation patterns during mental rotation tasks. *Neuropsychologia, 40*, 2397-2408.

[165] Juárez, J., & Corsi-Cabrera, M. (1995). Sex differences in interhemispheric correlation and spectral power of EEG activity. *Brain Research Bulletin, 38*, 149-151.

[166] Juárez, J., Corsi-Cabrera, M., & del Río-Portilla, I. Y. (1995). Effects of prenatal testosterone treatment on sex differences in the EEG activity of the rat. *Brain Research, 694*, 21-28.

[167] Juárez, J., del-Río-Portilla, I., & Corsi-Cabrera, M. (1998). Effects of prenatal testosterone on sex and age differences in behavior elicited by pups. *Developmental Psychobiology, 32*, 121-129.

[168] Jung-Testas, I., Renoir, M., Bugnard, H., Greene, G. L., & Baulieu, E. E. (1992). Demonstration of steroid hormone receptors and steroid action in primary cultures of rat glial cells. *The Journal of steroid biochemistry and molecular biology, 41*,621-631.

[169] Juraska, J. M. (1984). Sex differences in dendritic response to differential experience in the rat visual cortex. *Brain Research, 295*, 27-34.

[170] Juraska, J. M., & Kopcik, J. R. (1988). Sex and environmental influences on the size and ultrastructure of the rat corpus callosum. *Brain Research, 450*, 1-8.

[171] Juraska, J. M., Fitch, J. M., Henderson, C., & Rivers, N. (1985). Sex differences in the dendritic branching of dentate granule cells following differential experience. *Brain Research, 333*, 73-80.

[172] Juraska, J. M., & Markham, J. A. (2004). The cellular basis for volume changes in the rat cortex during puberty: white and gray matter. *Annals of the New York Academy of Sciences. 1021*, 431-435.

[173] Kato, J., & Onouchi, T. (1981). Progesterone receptors in the cerebral cortex of neonatal female rats. *Develompent Neuroscience, 4*, 427-432.

[174] Kato, J., Hirata, S., Nozawa, A., & Yamada-Mouri, N. (1994). Gene expression of progesterone receptor isoforms in the rat brain. *Hormones and behavior. 28*, 454-463.

[175] Kawakami, M., & Sawyer, Ch. H. (1959). Neuroendocrine correlates of changes in brain activity thresholds by sex steroids and pituitary hormones. *Endocrinology, 65*, 652-668.

[176] Kennan, P. A., Stern, R. A., Janowsky, D. S., & Pedersen, C. A. (1992). Psychological aspects of the premenstrual syndrome, I: Cognition and memory. *Psychoneuroendocrinology, 17*, 179-187.

[177] Killgore, W. D., & Yurgelun-Todd, D. A. (2001). Sex differences in amygdala activation during the perception of facial affect. *Neuroreport, 12*, 2543-2547.

[178] Kilpatrick, L. A., Zald, D. H., Pardo, J. V., & Cahill, L. F. (2006). Sex-related differences in amygdala functional connectivity during resting conditions. *Neuroimage, 30*, 452-61.

[179] Kim, J. H., Ellman, A., & Juraska, J. M. (1996). A reexamination of sex differences in axon density and number in the splenium of the rat corpus callosum. *Brain Research, 740*, 47-56.

[180] Kimura, D. (1987). Are men´s and women´s brain really different? *Canadian journal of psychology, 28*, 133-147.

[181] Kimura, D. (1999). *Sex and cognition.* Cambridge: The MIT Press.

[182] Kinoshita, T., Michel, C. H., Yagyu, T., & Lenmann, D. (1994). Diazepam and sulfiride effects on frequency domain EEG source locations. *Neuropsychobiology, 30*, 126-131.

[183] Kleinlogel, H. (1991). Analysis of the vigilance stages in the rat by Fast Fourier Transformation. *Pharmacoelectroencephalography 23*, 197-204.

[184] Knaus, T. A., Bolich, A. M., Corey, D. M., Lemen, L. C., & Foundas, A. L. (2004). Sex-linked differences in the anatomy of the persylvian language cortex: a volumetric MRI study of gray matter volumes. *Neuropsyhology, 18*, 738-747.

[185] Koelsch, S., Fritz, T., V Cramon, D.Y., Muller, K., & Friederici, A. D. (2006). Investigating emotion with music: An fMRI study. *Human Brain Mapping, 27*, 239-250.

[186] Koelsch, S., Maess, B., Grossmann, T., & Friederici, A. D. (2003) Electric brain responses reveal gender differences in music processing. *Neuroreport, 14*, 709-713.

[187] Kondacs, A., & Szabó, M. (1999). Long-term intra-individual variability of the background EEG in normals. *Clinical Neurophysiology, 110*, 1708-1716.

[188] Konishi, M., & Gurney, M. E. (1982). Sexual differentiation of brain and behavior. *Trends in Neuroscience, 5*, 20-23.

[189] Korneyev, A., & Costa, E. (1996). Allopregnanolone (THP) mediates anesthetic effects of progesterone in rat brain. *Hormones and Behavior, 30*, 37-43.

[190] Kovalev, V. A., Kruggel, F., & von Cramon, D. Y. (2003). Gender and age effects in structural brain asymmetry as measured by MRI texture analysis. *Neuroimage, 19*, 895-905.

[191] Kritzer, M. F. (2002). Regional, laminar, and cellular distribution of immunoreactivity for ER alpha and ER beta in the cerebral cortex of hormonally intact, adult male and female rats. *Cerebral Cortex, 12*, 116-128.

[192] Kritzer, M. F. (2004). The distribution of immunoreactivity for intracellular androgen receptors in the cerebral cortex of hormonally intact adult male and female rats: localization in pyramidal neurons making corticocortical connections. *Cerebral Cortex, 14*, 268-280.

[193] Kritzer, M. F. (2006). Regional, laminar and cellular distribution of immunoreactivity for ERbeta in the cerebral cortex of hormonally intact, postnatally developing male and female rats. *Cerebral Cortex, 16*, 1181-1192.

[194] Kritzer, M. F., & Pugach, I. (2001). Administration of tamoxifen but not flutamide to hormonally intact, adult male rats mimics the effects of short-term gonadectomy on the catecholamine innervation of the cerebral cortex. *The Journal of comparative neurology. 431*, 444-459.

[195] Kroboth, P. D., & McAuley, J. W. (1997). Progesterone: does it affect response to drug? *Psychopharmacology bulletin, 33*, 297-301.

[196] Kruijver, F. P., & Swaab, D. F. (2002) Sex hormone receptors are present in the human suprachiasmatic nucleus. *Neuroendocrinology, 75*, 296-305.

[197] Kudwa, A. E., Bodo, C., Gustafsson, J. A, & Rissman, E. F. (2005). A previously uncharacterized role for estrogen receptor beta: defeminization of male brain and behavior. *Proceedings of the National Academy of Sciences of the United States of America, 102*, 4608-4612.

[198] Kuhnemann, S., Brown, T. J., Hochberg, R. B., & MacLusky, N. J. (1994). Sex differences in the development of estrogen receptors in the rat brain. *Hormones and behavior, 28*, 483-491.

[199] Lake, D. A., & Bryden, M. P. (1976). Handedness and sex differences in hemispheric asymmetry. *Brain and Language, 3*, 266-282.

[200] Lam, T. T., & Leranth, C. (2003). Role of the medial septum diagonal band of Broca cholinergic neurons in oestrogen-induced spine synapse formation on hippocampal CA1 pyramidal cells of female rats. *The European Journal of Neuroscience, 17*, 1997-2005.

[201] Lancel, M. (1999). Role of $GABA_A$ receptors in the regulation of sleep: initial sleep responses to peripherally administered modulators and agonists. *Sleep, 22*, 33-44.

[202] Lancel, M., Faulhaber, J., Holsboer, F., & Ruprecht, R. (1996). Progesterone induces changes in sleep comparable to those of agonistic GABAA receptor modulators. *The American journal of physiology, 271*, E763-E772.

[203] Landgren, S. O. (1991). Pregnanolone (3 alpha-hydroxy-5 alpha-pregnane-20-one), a progesterone metabolite, facilitates inhibition of synaptic transmission in the Schaffer collateral pathway of the guinea pig hippocampus in vitro. *Epilepsy research, 10*, 156-165.

[204] Landgren, S., Bäckström, T., & Kalistratov, G. (1978). The effect of progesterone on the spontaneous interictal spike evoked by the application of penicillin to the cat's cerebral cortex. *Journal of Neurological Science, 36*, 119-133.

[205] Lee, T. M., Hummer, D. L., Jechura, T. J., & Mahoney, M. M. (2004). Pubertal development of sex differences in circadian function: an animal model. *Annals of the New York Academy of Sciences, 1021*, 262-275.

[206] Liederman, J. (1995). A reinterpretation of the split-brain syndrome: implications for the function of corticocortical fiber. In R. J. Davidson, & K. Hugdahl, (Eds.), *Brain Asymetry*. (pp. 451-490). A Bradford Book, MIT Press.

[207] Linn, M. C., & Petersen, A. C. (1985). Emergence and characterization of sex differences in spatial ability: a meta-analysis. *Child development, 56*, 1479-1498.

[208] Llinás, R., & Paré, D. (1996). The brain as a closed system modulated by the senses. In R. Llinás, & P. Churchland (Eds.), *The Mind-Brain Continuum: Sensory processes* (pp. 1-28). USA: MIT.

[209] Lord, T., & Taylor, K. Monthly fluctuations in task concentration in female students. (1991). *Perceptual and motor skills, 72*, 435-439.

[210] Luders, E., Narr, K. L., Thompson, P. M., Rex, D. E., Jancke, L., Steinmetz, H., & Toga, A. W. (2004). Gender differences in cortical complexity. *Nature Neuroscience, 7*, 799-800.

[211] Maccoby, E. E., & Jacklin, C. N. (1974). *The Psychology of Sex Differences. Stanford*, CA: Stanford University Press.

[212] Maclusky, N., & Naftolin, F. (1981). Sexual differentiation of the central nervous system. *Science, 211*, 1294–1303.

[213] Maggi, A., & Perez, J. (1985). Role of female gonadal hormones in the CNS: clinical and experimental aspects. *Life sciences, 37*, 893-906.

[214] Maggi, A., & Zucchi, I. (1987). Sexual differentiation of mammalian frontal cortex. *Life Science, 40*, 1155-60.

[215] Majewska, M. D. (1992). Neurosteroids: endogenous bimodal modulators of the GABAA receptor. Mechanisms of action and physiological significance. *Progress in neurobiology, 38*, 379-395.

[216] Majewska, M. D., Harrison, N. L., Schwartz, R. D., Barker, J. L., & Paul, S. M., (1986). Steroid hormone metabolites are barbiturate-like modulators of the GABA receptor. *Science, 232*, 1004-1007.

[217] Mandal, M. K., & Palchoudhury, S. (1985). Perceptual skills in decoding facial affect. *Perceptual and motor skills, 6*0, 96-98.

[218] McCarthy, M. M., Feltzenberg, E., Robbins, A., Pfaff, D. W., & Schwartz-Giblin, S. (1995). Infusions of diazepam and allopregnanolone into the midbrain central brain facilitate open field behavior and sexual receptivity in female rats. *Hormones and behavior, 29*, 279-295.

[219] McCauley, L. D., & Gee, K. W. (1995). Influence of the estrous cycle on the discrimination of apparent neuroactive steroids subtypes on the gamma aminobutyric acidA receptor complex in the rat. *The Journal of pharmacology and experimental therapeutics, 175*, 1412–1417.

[220] McEwen, B. S. (2001). Estrogens effects on the brain: multiple sites and molecular mechanisms. *Journal of applied physiology, 91*, 2785-27801.

[221] McEwen, B. S., Alves, S. E., Bulloch, K., & Weiland, N. G. (1997). Ovarian steroids and the brain: implications for cognition and aging. *Neurology, 48*, 8-15.

[222] McGee, M. G. (1979). Human spatial abilities: psychometric studies and environmental, genetic, hormonal, and neurological influences. *Psychological bulletin, 86*, 889-918.

[223] McGlone, J. (1986). The neuropsychology of sex differences in human brain organization. In G. Goldstein, & R. Tarter (Eds.), *Advances in Clinical Neuropsychology*. (pp. 1-30). New York: Plenum Press.

[224] Meaney, M. J. (1989). The sexual differentiation of social play. *Psychiatric Develoments, 7*, 247-261.

[225] Melchior, C. L., & Ritzmann, R. F. (1994). Pregnenolone and pregnenolone sulfate alone and with ethanol in mice on the plus maze. *Pharmacology, biochemistry, and behaviour, 48*, 893-897.

[226] Meyerson, B. J., Eliasson, M., & Hetta, J. (1980). Sexspecific orientation in female and male rats: development and effects of early endocrine manipulation. In A. M. Kaye & M. Kaye (Eds.), *Development of responsiveness to steroid hormones: advances in the biosciences (Vol. 25, pp. 451-460)*. Oxford, England: Pergamon Press.

[227] Miller, T. P., Taylor, J. L., & Tinklenberg, J. R. (1988). A comparison of assessment techniques measuring the effects of methylphenidate, secobarbital, diazepam and diphenhydramine in abstinent alcoholics. *Neuropsychobiology, 19*, 90-96.

[228] Miranda, R. C., & Toran-Allerand, C. D. (1992). Developmental expression of estrogen receptor mRNA in the rat cerebral cortex: a nonisotopic in situ hybridization histochemistry study. *Cerebral Cortex, 2*, 1-15.

[229] Mitchell, T. N., Free, S. L., Merschhemke, M., Lemieux, L., Sisodiya, S. M., & Shorvon, S. D. (2003). Reliable callosal measurement: population normative data confirm sex-related differences. *AJNR American journal of neuroradiology, 24*, 410-418.

[230] Moguilewsky, M., & Raynaud, J. P. (1979). Estrogen-sensitive progestin-binding sites in the female rat brain and pituitary. *Brain Research, 164*, 165-175.

[231] Mohamed, M. K., & Abdel-Rahman, A. A. (2000). Effect of long-term ovariectomy and estrogen replacement on the expression of estrogen receptor gene in female rats. *Eur J Endocrinol, 142*, 307-314.

[232] Mok, W. M., & Krieger, N. R. (1990). Evidence that 5alpha-pregnan-3alpha-ol-20-one is the metabolite responsible for progesterone anesthesia. *Brain Research, 533*, 4245.

[233] Mok, W. M., Herschkowitz, S., & Krieger, N. R. (1991). In vivo studies identify 5α-pregnan-3α-ol- 20-ona-as an active anesthetic agent. *Journal of Neurochemistry, 57*, 1296-1301.

[234] Mong, J. A. & Blutstein, T. (2006). Estradiol modulation of astrocytic form and function: implications for hormonal control of synaptic communication. *Neuroscience, 138*, 967-975.

[235] Morrow, A. L., Pace, J. R., Purdy, R. H. & Paul, S. M. (1990). Characterization of steroid interaction with gama-amino-butyric acid receptor-gated chloride ion channels: evidence for multiple recognition sites. *Molecular Pharmacology, 37*, 263-270.

[236] Myhrer, T., & de Groot, D.M. (1988). Immobility-related hippocampal theta activity in rats during reaction to novelty. *Scandinavian journal of psychology, 29*, 214-22.

[237] Nanbu, Y., Wada, Y., Kadoshima, R., Kitazawa, S., Futamata, H., Hashimoto, T., Koshino Y. (1995). Sex-related differences in EEG coherence in normal young adults-evaluation during rest and photic stimulation. Rinsho byori. *The Japanese journal of clinical pathology. 43*, 713-717.

[238] Nikulin, V. V., & Brismar, T. (2004). Long-range temporal correlations in alpha and beta oscillations: effect of arousal level and test-retest reliability. *Clinical Neurophysiology, 115*, 1896-1908.

[239] Nopoulos, P., Flaum, M., O'Leary, D., & Andreasen, N. C. (2000). Sexual dimorphism in the human brain: evaluation of tissue volume, tissue composition and surface anatomy using magnetic resonance imaging. *Psychiatry research, 98*, 1-13.

[240] Norberg, L., Backstrom, T., Wahlstrom, G. (1999). Anaesthetic effects of pregnanolone in combination with allopregnanolone, thiopental, hexobarbital and fluorazepam: an EEG study in the rat. British journal of anaesthesia. 82, 731.

[241] Nowak, S. M., & Maczynski, T. (1981). Trait anxiety is reflected in EEG alpha response to stress. *Electroencephalography and Clinical Neurophysiology, 52*, 175-191.

[242] Nunez, J. L., & Juraska, J. M. (1998). The size of the splenium of the rat corpus callosum: influence of sex hormones, sex ratio, and neonatal cryoanesthesia. *Developmental Psychobiology, 33*, 295-303.

[243] Nunez, J. L., Huppenbauer, C. B., McAbee, M. D., Juraska, J. M., DonCarlos, L. L. (2003). Androgen receptor expression in the developing male and female rat visual and prefrontal cortex. *Journal of neurobiology, 56*, 293-302.

[244] Nunez, P. L. (1995). Quantitative states of the cortex. In P. L. Nunez (Ed.), *Neocortical dynamics and human EEG rhythms*. (pp. 3-67). New York: Oxford University Press.

[245] Ohtani-Kaneko, R. (2006). Mechanisms underlying estrogen-induced sexual differentiation in the hypothalamus. *Histology and Histopathology, 21*, 317-324.

[246] O'Keefe, J. A., & Handa, R. J. (1990). Transient elevation of estrogen receptors in the neonatal rat hippocampus. *Brain Research Developmental Brain Research, 57*, 119-127.

[247] O'Keefe, J. A., Pedersen, E. B., Castro, A. J., & Handa, R. J. (1993). The ontogeny of estrogen receptors in heterochronic hippocampal and neocortical transplants demonstrates an intrinsic developmental program. *Brain Research, Developmental Brain Research, 75*, 105-12.

[248] Olesen, K. M., & Auger, A. P. (2005). Sex differences in Fos protein expression in the neonatal rat brain. *J Neuroendocrinol, 17*, 255-261.

[249] Oppenheim, G. (1984). A case of rapid mood cycling with estrogen: implications for therapy. *Journal of Clinical Psychiatry, 45*, 34-35.

[250] Orikasa, C., McEwen, B. S., Hayashi, H., Sakuma, Y., & Hayashi, S. (2000). Estrogen receptor alpha, but not beta, is expressed in the interneurons of the hippocampus in prepubertal rats: an in situ hybridization study. *Brain Research Developmental Brain Research, 120*, 245-254.

[251] Osterlund, M. K., Gustafsson, J. A., Keller, E., & Hurd, Y. L. (2000). Estrogen receptor beta (ERbeta) messenger ribonucleic acid (mRNA) expression within the human forebrain: distinct distribution pattern to ERalpha mRNA. *The Journal Clinical Endocrinology and metabolism, 85*, 3840-3846.

[252] Osterlund, M. K., Keller, E., & Hurd, Y. L. (2000). The human forebrain has discrete estrogen receptor alpha messenger RNA expression: high levels in the amygdaloid complex. *Neuroscience, 95*, 333-342.

[253] Palva, E. S. (1985). Gender-related differences in diazepam effects on performance. *Medical biology, 63*, 92-95.

[254] Panek, D. U., & Dixon, W. R. (1986). Effect of continuous intraventricular estrogen or catechol estrogen treatment on catecholamine turnover in various brain regions. *The Journal of pharmacology and experimental therapeutics, 236*, 646-652.

[255] Patchev, A. V., Gotz, F., & Rohde, W. (2004). Differential role of estrogen receptor isoforms in sex-specific brain organization. *The FASEB journal, 18*, 1568-1570

[256] Paul, S. M., & Purdy, R. H. (1992). Neuroactive steroids. FASEB Journal 6, 2311-2322.

[257] Pawlak, J., & Beyer, C. (2005). Developmental expression of MNAR mRNA in the mouse brain. *Cell and tissue research, 320*, 545-549.

[258] Pedersen, E. B., O'Keefe, J. A., Handa, R. J., & Castro, A. J. (1992). Estrogen receptors are present in neocortical transplants. *Journal of neural transplantation plasticity, 3*, 135-8.

[259] Pellow, S., Chopin, P., File, S. E., & Briley, M. (1985). Validation of open: closed arm entries in an elevated plus-maze as a measure of anxiety in the rat. *Journal of neuroscience methods, 14*, 149-167.

[260] Perez, S. E., Chen, E. Y., & Mufson, E. J. (2003). Distribution of estrogen receptor alpha and beta immunoreactive profiles in the postnatal rat brain. *Brain Research Developmental Brain Research 145*, 117-39.

[261] Pérez-Garci, E., del-Río-Portilla, Y., Guevara, M. A., Arce, C., & Corsi-Cabrera, M. (2001). Paradoxical sleep is characterized by uncoupled gamma activity between frontal and perceptual cortical regions. *Sleep, 24*, 118-126.

[262] Perlman, W. R., Tomaskovic-Crook, E., Montague, D. M., Webster, M. J., Rubinow, D. R., Kleinman, J. E., & Weickert C. S. (2005). Alteration in estrogen receptor alpha mRNA levels in frontal cortex and hippocampus of patients with major mental illness. *Biology Psychiatry, 58*, 812-824. Epub 2005 Aug 22.

[263] Peters, J. A., Kirkness, E. F., & Callachan, H. (1988). Modulation of the $GABA_A$ receptor by depressant barbiturates and pregnane steroids. *British journal of pharmacology, 94*, 1257-1269.

[264] Petsche, H., & Etlinger, S. C. (1998). EEG and thinking: power and coherence analysis of cognitive processes. Vienna: Austrian Academy of Sciences.

[265] Petsche, H., & Stumpf, C. (1960). Topographic and toposcopic study of origin and spread of the regular synchronized arousal pattern in the rabbit. *Electroencephalography Clinical Neurophysiology, 12*, 589-600.

[266] Phillips, M. D., Lowe, M. J., Lurito, J. T., Dzemidzic, M., & Mathews, V. P. (2001). Temporal lobe activation demonstrates sex-based differences during passive listening. *Radiology, 220*, 202-207.

[267] Picazo, O., & Fernández-Guasti, A. (1995). Anti-anxiety effects of progesterone and some of its reduced metabolites: an evaluation using the burying behavior test. *Brain Research, 680*, 135-141.

[268] Pigott, A. T. (1999). Gender differences in the epidemiology and treatment of anxiety disorders. *The Journal of clinical psychiatry, 60*, 4-15.

[269] Podell, K., Lovell, M., & Goldberg, E. (2001). Lateralization of frontal lobe functions. In S. P. Salloway, P .F. Malloy, J. D. Duffy (Eds.), *The frontal lobes and neuropsyhciatric illness*. (pp. 83-99). London: American Psychiatric Publishing.

[270] Posner, M. I., & Driver, J. (1992). The neurobiology of selective attention. *Current opinion in neurobiology, 2*,165-169.

[271] Primus, R. J., & Kellogg, C. K. (1990). Gonadal hormones during puberty organize environmentrelated social interaction in the male rat. *Hormones and Behavior, 24*, 311-323.

[272] Puia, G., Santi, M. R., & Vicini, S. (1990). Neurosteroids act on recombinant human $GABA_A$ receptors. *Neuron, 4*, 759-765.

[273] Pujol, J., Lopez-Sala, A., Deus, J., Cardoner, N., Sebastian-Galles, N., Conesa, G., & Capdevila, A. (2002). The lateral asymmetry of the human brain studied by volumetric magnetic resonance imaging. *Neuroimage, 17*, 670-679.

[274] Rahman, Q., Andersson, D., & Govier, E. (2005). A specific sexual orientation-related difference in navigation strategy. *Behavioral Neuroscience, 119*, 311-316.

[275] Raisman, G., & Field, P. M. (1971). Sexual dimorphism in the preoptic area of the rat. *Science, 173*, 731-733.

[276] Ramos, J., & Corsi-Cabrera, M. (1989). Does brain electrical activity react to music? *International Journal of Neurosciencie, 47*, 351-357.

[277] Ramos, J., Corsi-Cabrera, M., Guevara, M. A., & Arce. C. (1993). EEG activity during cognitive performance in women. *International Journal of Neuroscience. 69*, 185-195.

[278] Ramos-Loyo, J., & Sánchez-Loyo, L. M. (Submitted) Sex differences in brain electrical coherent activity during virtual navigation before and after training.

[279] Rappelsberger, P., & Petsche, H. (1988). Probability mapping: power and coherence analyses of cognitive processes. *Brain Topography, 1*, 46-54.

[280] Ravizza, T., Veliskova, J., & Moshe, S. L. (2003). Testosterone regulates androgen and estrogen receptor immunoreactivity in rat substantia nigra pars reticulata. *Neuroscience Letters, 338*, 57-61.

[281] Ray, W. J., Newcombe, N., Semon, J., & Cole, P. M. (1981). Spatial abilities, sex differences and EEG functioning. *Neuropsychologia, 19*, 712-722.

[282] Reddy, D. S., & Kulkarni, S. K. (1996). Role of $GABA_A$ and mitochondrial diazepam binding inhibitor receptors in the anti-stress activity of neurosteroids in mice. *Psychopharmacology, 128*, 280-292.

[283] Reynolds, B., Richards, J. B., Dassinger, M., de Wit, H. (2004). Therapeutic doses of diazepam do not alter impulsive behavior in humans. *Pharmacology, biochemistry, and behavior. 79*, 17-24.

[284] Rhyu, I. J., Cho, T. H., Lee, N. J., Uhm, C. S., Kim, H., & Suh, Y.S. (1999). Magnetic resonance image-based cerebellar volumetry in healthy Korean adults. *Neuroscience Letters, 270*, 149-152.

[285] Riekkinen, P., Buzsaki, G., Riekkinen, P. Jr., Soininen, H., & Partanen, J. (1991). The cholinergic system and EEG slow waves. *Electroencephalography and clinical neurophysiology, 78*, 89-96.

[286] Roberts, J. E., & Bell, M. A. (2000). Sex differences on a mental rotation task: variations in electroencephalogram hemispheric activation between children and college students. *Developmental neuropsychology, 17*, 199-223.

[287] Rodgers, R. J., & Johnson, J.T. (1998). Behaviourally selective effects of neuroactive steroids on plus maze anxiety in mice. *Pharmacology, biochemistry, and behaviour, 59*, 221-232

[288] Rodríguez-Sierra, J. F., Hagley, M. T. & Hendricks, S. E. (1986). Anxiolytic effects of progesterone are sexually dimorphic. *Life Science, 38*, 1841-1845.

[289] Romano-Torres, M., Borja-Lascurain, E., Chao-Rebolledo, del Río-Portilla, Y., & Corsi-Cabrera, M. (2002). Effects of diazepam on EEG power and coherent activity: sex differences. *Psychoneuroendocrinology, 27*, 821-833.

[290] Romeo, R. D., McCarthy, J. B., Wang, A., Milner, T. A., & McEwen, B. S. (2005). Sex differences in hippocampal estradiol-induced N-methyl-D-aspartic acid binding and ultrastructural localization of estrogen receptor-alpha. *Neuroendocrinology, 81*, 391-399.

[291] Roof, R. L., & Havens, M. D. (1992). Testosterone improves maze performance and induces development of a male hippocampus in females. *Brain Research, 572*, 310-313.

[292] Rossell, S. L., Bullmore, E. T., Williams, S. C., & David, A. S. (2002). Sex differences in functional brain activation during a lexical visual field task. *Brain and Language, 80*, 97-105.

[293] Rubinow, D. R., & Schmidt, P. J. (2006). Gonadal steroid regulation of mood: the lessons of premenstrual syndrome. *Frontiers in neuroendocrinology, 27*, 210-216.

[294] Rupprecht, R. (1997). The neuropsychopharmacological potential of neuroactive steroids. *Journal of psychiatric research, 31*, 297–314.

[295] Sainsbury, R. S., & Montoya, C. P. (1984). The relationship between type2 theta and behavior. *Physiology & behavior, 33*, 621-626.

[296] Salat, D., Ward, A., Kaye, J. A., & Janowsky, J. S. (1997). Sex differences in the corpus callosum with aging. *Neurobiology of aging, 18*, 191-197.

[297] Saleh, T. M., Connell, B. J., & Cribb, A. E. (2005). Sympathoexcitatory effects of estrogen in the insular cortex are mediated by GABA. *Brain Research, 1037*, 114-122.

[298] Sandhu, S., Cook, P., & Diamond, M. C. (1986). Rat cerebral cortical estrogen receptors: male-female, right-left. *Experimental Neurology, 92*, 186-196.

[299] Scharfman, H. E., & MacLusky, N. J. (2006). The influence of gonadal hormones on neuronal excitability, seizures, and epilepsy in the female. *Epilepsia, 47*, 1423-1440.

[300] Schienle, A., Schafer, A., Stark, R., Walter, B., & Vaitl, D. (2005). Gender differences in the processing of disgust- and fear-inducing pictures: an fMRI study. *Neuroreport, 16*, 277-280.

[301] Schirmer, A., & Kotz, S. A. (2003). ERP evidence for a sex-specific Stroop effect in emotional speech. *Journal of cognitive neuroscience, 15*, 1135-1148.

[302] Schlaepfer, T. E., Harris, G. J., Tien, A. Y., Peng, L., Lee, S., & Pearlson, G. D. (1995). Structural differences in the cerebral cortex of healthy female and male subjects: a magnetic resonance imaging study. *Psychiatry research, 61*, 129-135.

[303] Schlenker, E. H. & Hansen, S. N. (2006). Sex-specific densities of estrogen receptors alpha and beta in the subnuclei of the nucleus tractus solitarius, hypoglossal nucleus and dorsal vagal motor nucleus weanling rats. *Brain Research, 1123*, 89-100.

[304] Schlosser, R., Hutchinson, M., Joseffer, S., Rusinek, H., Saarimaki, A., Stevenson, J., Dewey, S. L., & Brodie, J. D. (1998). Functional magnetic resonance imaging of human brain activity in a verbal fluency task. *Journal of neurology, neurosurgery, and psychiatry, 64*, 492-498.

[305] Schmidt, L. A., & Trainor, L. J. (2001). Frontal brain electrical activity (EEG) distinguishes valence and intensity of musical emotions. *Cognition and Emotion, 15*, 487-500.

[306] Schneider, F., Habel, U., Kessler, C., Salloum, J. B., & Posse, S. (2000). Gender differences in regional cerebral activity during sadness. *Human Brain Mapping, 9*, 226-38.

[307] Schumacher, M., (1990). Rapid membrane effects of steroid hormones: an emerging concept in neuroendocrinology. *Trends in Neuroscience, 13*, 359-362.

[308] Selye, H. (1942). Correlations between the chemical structure and the pharmacological actions of the steroids. *Endocrinology, 30*, 437-453.

[309] Seymoure, P., & Juraska, J. M. (1992). Sex differences in cortical thickness and the dendritic tree in the monocular and binocular subfields of the rat visual cortex at weaning age. *Developmental Brain Research, 69*, 185-189.

[310] Seymoure, P., & Juraska, J. M. (1997). Vernier and grating acuity in adult hooded rats: the influence of sex. *Behavioral Neuroscience, 111*, 792-800.

[311] Shapleske, J., Rossell, S. L., Woodruff, P. W., & David, A. S. (1999). The planum temporale: a systematic, quantitative review of its structural, functional and clinical significance. Brain research. *Brain research reviews, 29*, 26-49.

[312] Shaw, J.C. (1984). Correlation and coherence analysis of the EEG: a selective tutorial review. *International Journal of Psycophysiology, 1*, 255-266.

[313] Shaywitz, B. A., Shaywitz, S. E., Pugh, K. R., Constable, R. T., Skudlarski, P., Fulbright, R. K, Bronen, R. A., Fletcher, J. M., Shankweiler, D. P., & Katz, L. (1995). Sex differences in the functional organization of the brain for language. *Nature, 373*, 607-609.

[314] Sheng, Z., Kawano, J., Yanai, A., Fujinaga, R., Tanaka, M., Watanabe, Y., & Shinoda, K. (2004). Expression of estrogen receptors (alpha, beta) and androgen receptor in serotonin neurons of the rat and mouse dorsal raphe nuclei; sex and species differences. *Neuroscience Research, 49*, 185-196.

[315] Shughrue, P. J, Scrimo, P. J., & Merchenthaler, I. (2000). Estrogen binding and estrogen receptor characterization (ERalpha and ERbeta) in the cholinergic neurons of the rat basal forebrain. *Neuroscience, 96*, 41-49.

[316] Shughrue, P. J., & Dorsa, D. M. (1994). Estrogen and androgen differentially modulate the growth-associated protein GAP-43 (neuromodulin) messenger ribonucleic acid in postnatal rat brain. *Endocrinology, 134*, 1321-1328.

[317] Shughrue, P. J., Dellovade, T. L., & Merchenthaler, I. (2002). Estrogen modulates oxytocin gene expression in regions of the rat supraoptic and paraventricular nuclei that contain estrogen receptor-beta. *Progress in brain research, 139*,15-29.

[318] Shughrue, P. J., Lane, M. V., & Merchenthaler, I. (1997). Comparative distribution of estrogen receptor-alpha and -beta mRNA in the rat central nervous system. *The Journal of comparative neurology, 388*, 507-525.

[319] Shughrue, P. J., Sar, M., & Stumpf, W. E. (1992). Progestin target cell distribution in forebrain and midbrain regions of the 8-day postnatal mouse brain. *Endocrinology. 130*, 3650-3659.

[320] Shughrue, P. J., Scrimo, P. J., & Merchenthaler, I. (1998). Evidence for the colocalization of estrogen receptor-beta mRNA and estrogen receptor-alpha immunoreactivity in neurons of the rat forebrain. *Endocrinology, 139*, 5267-570.

[321] Shughrue, P. J., Stumpf, W. E., Elger, W., Schulze, P. E., & Sar, M. (1991). Progestin receptor cells in the 8-day-old male and female mouse cerebral cortex: autoradiographic evidence for a sexual dimorphism in target cell number. *Endocrinology, 128*, 87-95.

[322] Sierra, J. C., Luna-Villegas, G., Buela-Casal, G., & Fernández-Guardiola, A. (1997). The assessment of residual effects of a single dose of diazepam on visually-defined EEG patterns, *Journal of psychopharmacology, 11*, 367–372.

[323] Simerly, R. B., Chang, C., Muramatsu, M., & Swanson, L. W. (1990). Distribution of androgen and estrogen receptor mRNA-containing cells in the rat brain: an in situ hybridization study. *The Journal of comparative neurology, 294*, 76-95.

[324] Singer, W. (1990). Search for coherence: a basic principle of cortical self-organization. *Concepts in Neuroscience, 1*, 1-26.

[325] Slob, A. K., Huizer, T., & Van Der Werff Ten Bosch, J. J. (1986). Ontogeny of sex differences in open-field ambulation in the rat. *Physiological Behavior, 37*, 313-315.

[326] Smith, M. D., Jones, L. S., & Wilson, M. A. (2002). Sex differences in hippocampal slice excitability: role of testosterone. *Neuroscience, 109,* 517-530.

[327] Sodersten, P. (1984). Sexual differentiation: do males differ from females in behavioral sensitivity to gonadal hormones? *Progress in brain research, 61*, 257-270.

[328] Solís-Ortiz, S., & Corsi-Cabrera, M. (2002). EEG pattern of anxiety along the menstrual cycle. *Revista Mexicana de Psicología, 19*, 187-195.

[329] Solís-Ortiz, S., Guevara, M. A., & Corsi-Cabrera, M. (2004). Performance in a test demanding prefrontal functions is favored by early luteal phase progesterone: an electroencephalographic study. *Psychoneuroendocrinology, 29*, 1047-1057.

[330] Solís-Ortiz, S., Ramos, J., Arce, C., Guevara, M. A., & Corsi-Cabrera, M. (1994). EEG oscillations during menstrual cycle. *International Journal of Neuroscience, 76*, 279-292.

[331] Solum, D. T., & Handa, R. J. (2001). Localization of estrogen receptor alpha (ER alpha) in pyramidal neurons of the developing rat hippocampus. Brain Research Developmental *Brain Research, 128*, 165-75.

[332] Squires, R. F., & Braestrup, C. (1977). Benzodiazepine receptors in rat brain. *Nature, 166*, 732-734.

[333] Starkstein, S. E., Ginsberg, S., Shnayder, L., Bowersox, J., Mersey, J. H., Robinson, R. G., & Moran, T. H. (1989). Developmental and hormonal factors in the sexually dimorphic, asymmetrical response to focal cortical lesions. *Brain Research, 478*, 16-23.

[334] Steinmetz, H., Staiger, J. F., Schlaug, G., Huang, Y., & Jancke, L. (1995). Corpus callosum and brain volume in women and men. *Neuroreport, 6*, 1002-1004.

[335] Steriade, M., & McCarley, R.W. (1990). *Brainstem control of wakefulness and sleep.* Plenum Press, New York.

[336] Suganthy, J., Raghuram, L., Antonisamy, B., Vettivel, S., Madhavi, C., & Koshi, R. (2003). Gender- and age-related differences in the morphology of the corpus callosum. *Clinical anatomy 16*, 396-403.

[337] Sumida, H., Nishizuka, M., Kano, Y., & Arai, Y. (1993). Sex differences in the anteroventral periventricular nucleus of the preoptic area and in the related effects of androgen in prenatal rats. *Neuroscience Letters, 151*, 41-44.

[338] Szymusiak, R. (1995). Magnocellular nuclei of the basal forebrain: substrates for sleep and arousal regulation. *Sleep, 18*, 478-500.

[339] Tallman, J. F., & Gallager, D. W. (1985). The GABAergic system: a locus of benzodiazepine action. *Annual Review of Neuroscience, 8*, 21-44.

[340] Terasawa, E., & Timiras, P. S. (1968). Electrical activity during the estrous cycle of the rat: cyclic changes in limbic structures. *Endocrinology, 83*, 207-216.

[341] Teyler, T. J., Vardaris, R. M., Lewis, D. & Rawitch, A. B. (1980). Gonadal steroids on excitability of hippocampal pyramidal cells. *Science, 209*, 1017-1019.

[342] Thomsen, T., Hugdahl, K., Ersland, L., Barndon, R., Lundervold, A., Smievoll, A. I., Roscher, B. & E., Sundberg, H. (2000). Functional magnetic resonance imaging (fMRI) study of sex differences in a mental rotation task. Medical science monitor. *International Medical Journal of Experimental and Clinical Research, 6*, 1186-1196.

[343] Timofeev, I., Grenier, F., Bazhenov, M., Houweling, A. R., Sejnowski, T. J., Steriade, M. (2002). Short- and medium-term plasticity associated with augmenting responses in cortical slabs and spindles in intact cortex of cats in vivo. *Journal of physiology, 542*, 583-598.

[344] Tononi, G., Sporns, O., & Edelman, G. M. (1992). Reentry and the problem of integrating multiple cortical areas: simulation of dynamic integration in the visual system. *Cerebral Cortex, 2*, 310-335.

[345] Toran-Allerand, C. D. (2000). Novel sites and mechanisms of oestrogen action in the brain. *Novartis Foundation symposium, 230*, 56-69.

[346] Toran-Allerand, C. D., Miranda, R. C., Hochberg, R. B., & MacLusky, N. J. (1992). Cellular variations in estrogen receptor mRNA translation in the developing brain: evidence from combined [125I] estrogen autoradiography and non-isotopic in situ hybridization histochemistry. *Brain Research, 576*, 25-41.

[347] Tucker, D. M., Roth, D. L., & Bair, T. B. (1986). Functional connections among cortical regions: topography of EEG coherence. *Electroncephalography and Clinical Neurophysiology, 63*, 242-250.

[348] Ugalde, E., del-Río-Portilla, Y., Juárez, J., Roldán, A., Guevara, M. A., & Corsi-Cabrera, M. (1998). Effect of diazepam and sex hormones on EEG of gonadectomized male and female rats. *Psychoneuroendocrinology, 23*, 701-712.

[349] Vanderwolf, C. H. (1992). The electrocorticogram in relation to physiology and behavior: a new analysis. *Electroencephalogr. Clinical Neurophysiology, 82*, 165-175.

[350] Vega-Matuszczyk, J., Fernandez-Guasti, A., & Larsson, K. (1988). Sexual orientation, proceptivity, receptivity in the male rat as function of neonatal hormone manipulation. *Hormones and Behavior, 22*, 362-378.

[351] Vikingstad, E. M., George, K. P., Johnson, A. F., & Cao, Y. (2000). Cortical language lateralization in right handed normal subjects using functional magnetic resonance imaging. *Journal of the neurological sciences, 175*, 17-27.

[352] Wager, T. D., Phan, K. L., Liberzon, I., & Taylor, S. F. (2003). Valence, gender, and lateralization of functional brain anatomy in emotion: a meta-analysis of findings from neuroimaging. *Neuroimage, 19*, 513-531.

[353] Wagner, C. K., Pfau, J. L., De Vries, G. J, & Merchenthaler, I. J. (2001). Sex differences in progesterone receptor immunoreactivity in neonatal Mouse brain depend on estrogen receptor alpha expression. *Journal of neurobiology, 47*, 176-182.

[354] Wang, J., Cheng, C. M., Zhou, J., Smith, A., Weickert, C. S., Perlman, W. R., Becker, K. G., Powell, D., & Bondy, C. A. (2004). Estradiol alters transcription factor gene expression in primate prefrontal cortex. *Journal of Neuroscience Research, 76*, 306-314.

[355] Warren, S. G., & Juraska, J. M. (1997). Spatial and nonspatial learning across the rat estrous cycle. *Behav Neurosci. 111*, 259-266.

[356] Warren, S. G., Humphreys, A. G., Juraska, J. M., Greenough, W. T. (1995). LTP varies across the estrous cycle: enhanced synaptic plasticity in proestrus rats. *Brain Research, 703*, 26-30.

[357] Weiland, N. G., Orikasa, C., Hayashi, S., & McEwen, B.S. (1997). Distribution and hormone regulation of estrogen receptor immunoreactive cells in the hippocampus of male and female rats. *J Comp Neurol, 388*, 603-612.

[358] Weiss, E., Siedentopfc, C. M., Hofera, A., Deisenhammerd, E. A., Hoptmanb, M. J., Kremserc, C., Golaszewskic, S., Felberc, S., Fleischhackera, W. W., & Delazer, M. (2003). Sex differences in brain activation pattern during a visuospatial cognitive task: a functional magnetic resonance imaging study in healthy volunteers. *Neuroscience Letters, 344*, 169–172

[359] Wieland, S., Belluzzi, J. D., Stein, L., & Lan, N. C. (1995). Comparative behavioural characterization of the neuroactive steroids 3 alpha-OH, 5 alpha-pregnan-20-one and 3 alpha-OH, 5 beta-pregnan-20-one in rodents. *Psychopharmacology, 118*, 65-71.

[360] Wieland, S., Lan, N. C., Mirasedeghi, S., & Gee, K. W., (1991). Anxiolytic activity of progesterone metabolite 5-alpha pregnan-3 alpha ol-20 one. *Brain Research, 565*, 263-268.

[361] Witelson, S. F. (1989). Hand and sex differences in the isthmus and girus of human corpus callosum: a postmorten morphological study. *Brain, 112*, 199-235.

[362] Wolley, D. E., & Timiras, P. S. (1962). The gonad brain relationship: effects of female sex hormones on electroshock convulsion in the rat. *Endocrinology, 70*, 196-209.

[363] Wong, M., & Moss, R. L. (1991). Electrophysiological evidence for a rapid membrane action of the gonadal steroid, 17beta-estradiol, on CA1 pyramidal neurons of the rat hippocampus. *Brain Research, 543*, 148-152.

[364] Wu, F. S., Gibbs, T. T., & Farb, D. H. (1990). Inverse modulation of 7-aminobutyric acid and glycine induced currents by progesterone. *Molecular Pharmacology, 37*, 597-602.

[365] Xiao, L., & Jordan, C.L. (2002). Sex differences, laterality, and hormonal regulation of androgen receptor immunoreactivity in rat hippocampus. *Hormones and behavior, 42*, 327-336.

[366] Yonkers, K. A., Kando, J. C., Cole, J. O., & Blumenthal, S. (1992). Gender differences in pharmacokinetics and pharmacodynamics of psychotropic medication. *The American journal of psychiatry, 149*, 587-595.

[367] Zhang, J. Q., Cai, W. Q., Zhou, D. S., & Su, B. Y. (2002). Distribution and differences of estrogen receptor beta immunoreactivity in the brain of adult male and female rats. *Brain Research, 935*, 73-80.

In: Psychoneuroendocrinology Research Trends
Editor: Martina T. Czerbska, pp. 73-115

ISBN: 978-1-60021-665-7

Chapter II

THE NEUROENDOCRINOLOGY OF TESTOSTERONE – SOCIOSEXUAL BEHAVIOR RELATIONS

George T. Taylor*, Joshua Dearborn and Susan Fortenbury

Behavioral Neuroscience, Univ. of Missouri, St. Louis, MO 63121 USA.

ABSTRACT

Steroids produced in the periphery have long been of interest to reproductive biologists. The emergence of neuroendocrinology as a dominant subfield of behavioral studies appeared to demote steroids to a lesser status than the exciting neuropeptides that clearly interacted with many regions of the brain. Seminal findings in recent years, however, have propelled the study of steroids back onto center stage. In this chapter we identify the seminal findings and describe why they are revolutionary. These new findings have demonstrated steroids have the capacity to modify the structure and function of brain regions both related to and unrelated to reproduction. The consequences on behavior are likely to be profound. This chapter focuses on one steroid, testosterone, because it occupies a strategic position in the metabolic cascade of the sex steroids. It is a position that can influence the actions of androgens and estrogens, as well as the adrenal precursors of testosterone and the recently identified neurosteroids, on the brain and behavior. We believe future research will reveal testosterone to have behavioral effects extending well beyond those related to reproduction. Still, the massive literature on testosterone – sociosexual behavior provides a solid background on which to begin that work.

* Correspondence concerning this article should be addressed to George T. Taylor, Ph.D. Professor of Psychology & Director of Behavioral Neuroscience, University of Missouri - St. Louis, One University Blvd. St. Louis, MO 63121 U.S.A. Tel: (314) 516-5475; Fax: (314) 516-5392; E-mail: geot@UMSL.edu.

SECTION I. INTRODUCTION

Reproductive function, most notably sexual behavior, has been the meat and potatoes of behavioral endocrinology for over a century. Androgens, and especially testosterone (TS), have been front and center for much of the century.

The first experimental manipulation of the endocrine system often cited in textbooks is of a demonstration of castrated roosters being implanted with testes. The treatment restored the color and size of their combs, as well as fully restored crowing, fighting and sexual behaviors. Later, an aging researcher reported rejuvenation of sexual prowess with an implant of an aqueous solution from crushed testes of experimental animals injected into his own abdominal cavity (Wilson 2000). Few can watch today's television commercials on topical TS without smiling and thinking how little times have changed.

These early studies point to four important features that defined much of the research in endocrinology for the next decades. First, there was a clear focus on practical therapeutic application to humans. Understanding the basic biology of sex hormones was only of secondary interest. Second, the studies demonstrated a fundamental research paradigm of endocrinology – ablation and hormone replacement as the experimental manipulation with functional restoration as the outcome measure. Third, the studies focused on the relation of a hormone to function and dysfunction of systems related to reproduction. A special focus was on sociosexual behaviors, defined as behaviors related to successful reproduction. Finally, the hormones on which most attention was placed were the sex steroids. It was TS and males in the early days, but today there has been at least equal focus on estrogens and progesterone in females.

Later still came the recognition that a non-sex steroid, corticosteroid, was intimately involved in the response to stressors. The bind that tied corticosteroids to sex steroids were the findings that stress could suppress reproductive physiology and function. Without doubt, steroids were firmly entrenched into the field of behavioral endocrinology (Beach 1981).

Behavioral neuroendocrinology emerged as the dominant subfield with a handful of seminal findings in recent decades These results have made it clear that steroids do more than simply maintain peripheral structures and reflexively activate subcortical brain regions to initiate regulatory behaviors. The seminal findings suggest sex steroids do more, much more.

One of the more satisfying outcomes is to give contemporary legitimacy to a behavioral biologist with interest in the steroids. The new findings also allowed researchers to move beyond the simple correlational studies of circulating TS and behavior. We now can understand better puzzling findings from the earliest days in behavioral endocrinology. More important, the seminal findings have thrust the brain into a central position in the study of steroids.

Our goal in this chapter will be to address the current status of TS and behavior in relation to the recent findings. We will use reproductive behaviors as the exemplary literature because that is surely the most complete. There are already excellent reviews of endocrine - sociosexual behavior relations (Archer 1994, Baum 2002, Meisel and Sachs 1994, Meston, 2000). Rather than presenting another similar review, we have organized topics to emphasize the changes in thinking induced by the seminal findings as they have impacted the study of TS. We characterize these findings as "revolutionary" for several reasons. 1) The findings were more or less unexpected. 2) They introduced a level of complexity for the study of steroid - behavior interactions that few would have envisioned in the early days of endocrinology. 3) They provided a searchlight to guide our study of the fundamental mechanisms in the brain responsible for the steroid – behavior correlations.

It is our hope that the reader who persists to the end of this chapter will better appreciate the neuroendocrine complexities researchers have encountered in the study of a seemingly simple hormone – behavior interaction. Emphasis will be primarily on male animal models and men, although the emerging role of TS on female behaviors will be discussed where appropriate. Finally, the chapter may also prove helpful to students and researchers in behavioral neuroendocrinology because similar complexities likely apply to other hormone – behavior relations.

I.1. Sexual Behaviors can be Separated into Motivation and Performance Behaviors

A first step in digesting the large literature on sexual behavior is to distinguish its two separate components, motivation and performance. Sexual motivation refers to measures of the male's interest to engage in sexual activity, sometimes referred to as sexual arousal or libido. Sexual performance is the effectiveness of the male to initiate and complete a copulatory bout. The two components of sexual behavior involve distinct, quantifiable behaviors.

Successful reproduction requires the male to be sufficiently motivated to find a receptive female. Clever methods have been adopted to quantify intensity of sexual motivation in laboratory animals. One example is a small treadmill on which a male mouse must continue to run to stay nearby the stationary female (Craigen and Bronson 1982). More commonly, the measure of motivation is latency to initiate a mount of the female (Meisel and Sachs 1994) or time spent near an inaccessible estrous female (Taylor, *et al* 1991). Sexual motivation in females can be assessed also by time spent near a reproductive male or by recording solicitations, a species-specific set of behaviors used in the pre-copulatory phase (Taylor, *et al* 1989).

Copulation can be more easily observed than motivation in animal models. There are a number of distinct behaviors that can be independently measured, most notably latencies and frequencies of mounting, intromissions and ejaculations (Taylor and Weiss 1987). The single behavioral marker of receptivity in the female is lordosis, the bent-back positioning by the female, exposing her genitals and allowing the male to penetrate her.

In humans, the most common measures of both sexual interest and activity are self-reports, either from recall or a diary. There are a number of sophisticated recording instruments available to researchers, but they are intrusive and have limited use. Self-reports remain most convenient and most popular.

I.2. Distinct Brain Pathways have Evolved for Sexual Motivation and Performance in Males

The two components of sexual behavior also involve different underlying neural systems. An exemplary experiment (Everitt 1990) demonstrated the different neural systems underlying sexual motivation and performance. Male rats pressed a bar to deliver a receptive female rat into his chamber. After training, the medial preoptic area (MPOA) of the hypothalamus was lesioned in one group of males. The animals continued to bar press for the female but no longer would attempt to mount and copulate with her. Both amygdalae were lesioned in a second group. These males no longer bar pressed to produce the receptive female. If she was placed inside the chamber, however, the male readily copulated. The suggestion is clear that an intact MPOA is necessary for normal copulatory performance whereas the amygdala is intimately involved in sexual motivation.

That experiment, and many others before and afterwards (Pfaus 1996), have demonstrated that the CNS ultimately holds the key to understanding sexual function in both males and females. The structures in the brain most often cited as being directly related to the control of male sexual behavior are the olfactory tubercles, corticomedial nuclei of the amygdaloid complex, bed nucleus of the stria terminalis (BNST), and the hypothalamus, especially the medial preoptic area (Baum 2002, Cooke, *et al* 1999b, Dominguez and Hull 2005).

Circulating levels of TS and the other sex hormones readily cross the blood brain barrier and, joined by a host of neurotransmitters and neuropeptides involved in reproduction, exert profound influence on the activity of these structures (Hull, *et al* 1999, McEwen, *et al* 1995). The evidence is overwhelming that TS is critical for the organization and activation of sexual motivation and performance of males. We present in the next section five findings (highlighted in bold and listed as II.1, II.2 and so forth) that changed, indeed revolutionized, the study of steroid physiology and behavior.

SECTION II. FINDINGS THAT REVOLUTIONIZED THE STUDY OF NEUROENDOCRINOLOGY OF TESTOSTERONE – BEHAVIOR RELATIONS

II.1. Fetal Testosterone Organizes Male Brain Circuitry

The influence of TS on the brain begins early. Beginning in the prenatal phase and continuing as the brain and periphery develop in the neonate, TS or its absence determines sexual differentiation. Early exposure to TS permanently masculinizes the male for an array of behavioral and physiological functions. The fetal and postnatal periods of testicular activity have profound effects on male sexual differentiation, whereas in the female early sexual development occurs without influence of ovarian function. In the absence of circulating TS, the mammalian brain and urogenital systems are feminized and de-masculinized.

Other data confirm male perinatal physiology being more reactive than that of females to sex hormones. Testosterone is being released in the male fetus long before the hypothalamic – pituitary - gonadal (HPG) axis has matured. Still, the HPG axis is fully functional in the male rat during the last days of gestation (Huhtaniemi 2000). In the first hours after birth, only the male experiences a burst of TS that increases circulating titers to near adult levels for a short period of time (Timiras and Cons 1982). This dramatic increase may or may not have consequences for later adult sociosexual behaviors (McGivern, *et al* 1998). By contrast, the fetal ovary is in a quiescent state, and its endocrine activity begins in earnest only with the first signs of puberty.

Sex steroidal influences lay relatively dormant in the postnatal animal until puberty. Then the same sex hormones, along with surges of gonadotropins and neuropeptides, activate the circuits organized during fetal development. This two phase influence of TS on behavior is the now classic organizational – activational model (Arnold and Gorski 1984, Phoenix, *et al* 1959). Fetal organization of sexually dimorphic brain morphology and circuitry can be observed in behaviors prior to puberty. Examples are the sexually dimorphic play and cognitive behaviors of prepubertal animals (Hodes and Shors 2005, Taylor 1980a). Clearer differences emerge with the activation of the circuitry by pubertal hormones. The results are gender-typical adult sociosexual behaviors, as well as influences on an array of non-reproductive behaviors.

The organizational – activational concept remains a powerful guide and is treated today as dogma by most neuroendocrinologists (Hall, *et al* 2004). However, there are several refinements of the original organizational – activational model that are based largely on the rodent brain (Ball and Balthazart 2006). For example, a key step suggested by rodent studies is that metabolism of TS to estradiol is necessary to induce a masculine organization of the brain. There is little evidence, however, to indicate that step is necessary in primates. Testosterone is likely directly responsible for the masculinization of the primate brain (Breedlove and Hampson 2002, Cooke, *et al* 1999a). Nonetheless, these findings point out the importance of the metabolic cascade for the synthesis of steroids for understanding the influence of TS on sociosexual behaviors.

II.2. Estradiol is a Metabolite of Testosterone

The metabolic cascade of steroidal hormones is remarkably complex and details remain unclear (Ball and Balthazart 2006, Occhiato, *et al* 2004). One basic feature is that the two primary metabolites of TS are estradiol (E2) and dihydrotestosterone (DHT). Thus, TS may directly activate steroid-sensitive tissues or be metabolized to either E2 or DHT which in turn activates the tissue. Whether TS or its metabolites activate the target cell is dependent upon the presence or absence of specific enzymes. Aromatase is responsible for converting TS to E2 and 5alpha-reductase converts TS to DHT (Payne and Hales 2004).

In tissue containing aromatase, and that includes many brain regions, the result is that TS is a prohormone in adult males. The direct effect on function is after TS is metabolized to E2 and E2 binds the estrogen receptor (Adkins-Regan 1981, Cooke, *et al* 1998). Evidence comes from the many reports with experimental animals showing E2 alone could restore sexual behaviors of castrated adult males (Baum 2002). For example, castrated male rats were injected either with TS, E2 or DHT. Only DHT failed completely to restore the animals' sexual behaviors (Johnston and Davidson 1972).

More recent work has confirmed that it is the conversion of TS to estrogen that is the critical step (Ball and Balthazart 2006). Blocking aromatase activity or injecting an anti-estrogen into the MPOA blocked the activation of male-typical sexual behavior by TS. Administering an estrogen directly to the MPOA in castrated animals activated male-typical sexual behavior.

Not to be outdone, DHT had its own surprise. It turns out to be a more potent androgen than TS. Upon conversion of TS to DHT in many peripheral sites, DHT binds the androgen receptor with higher affinity (Martini, *et al* 1990).

DHT also can magnify the effects of TS, as exemplified with its role in maturation of the penis at puberty. During adulthood, DHT also has primary responsibility for maintaining normal function of many of the secondary sex characteristics, as well as pathological conditions. Benign prostatic hyperplasia, prostate cancer, acne, androgenic alopecia in men, and hirsutism in women are all related to DHT production (Occhiato, *et al* 2004).

Originally, it seemed that DHT and E2 had divided the androgenic labor (Bonsall, *et al* 1992, Martini, *et al* 1990). The idea was of DHT being the active metabolite in peripheral tissues and E2 being the metabolite in the brain. Thus, E2 was the steroid most responsible for behavior in males. Supporting data came from administering DHT to a castrated rat. When given peripherally, DHT was less effective than TS in activating sexual behavior in castrated male rats (Butera and Czaja 1989). Integrity of the penis was maintained, but DHT did little to restore sexual behavior. Administering E2 did the opposite for a castrated male. Penile morphology was disturbed, but sexual motivation was intact (Sachs, *et al* 1984).

Such parsimony is rare in endocrinology, and it was not to last long. In rodents, it appears TS influences sexual behaviors of the males after it is converted to E2 while, in the brains of primates, it is TS interacting directly with the AR that underlies male sexual behaviors (Michael, *et al* 1987). We now know also that DHT is found in abundance in the brain (Bonsall, *et al* 1985), along with concentrations of the 5alpha-reductase enzyme converting TS to DHT (Cooke, *et al* 2003). Clearly, DHT influences brain structure and function. The functional dichotomy for DHT and E2 was officially put to rest with findings that E2 plays a critical role in male peripheral structures and functions (Akingbemi 2005, Sharpe 1998).

II.2.1. There are two Metabolic Pathways for Synthesizing Testosterone

TS is the major androgenic steroid in circulation, with the Leydig cells of the testes as the principal source of the hormone in males. The adrenal glands supplement TS synthesis by secreting the hormone precursors androstenedione (ANDRO), dehydroepiandrosterone (DHEA) and its sulfate (DHEAS). Adrenal steroids are converted to TS in peripheral tissues, often in the skin or fat (Payne and Hales 2004).

The role of the adrenal precursors of TS, especially DHEA, in normal physiology is of considerable recent interest (Beck and Handa 2004). Development of a mature adrenal gland, known as adrenarche, typically begins in children between 6 and 8 years of age, i.e., well before the onset of puberty. One functional outcome is the appearance of pubic hair.

Interestingly, the dramatic increase in levels of the adrenal precursors of TS occurs without corresponding increases in the other adrenal steroids, corticosteroids and mineralocorticoids. Peak levels of DHEA are found in adult humans during the 20s and 30s and decline progressively thereafter.

Steroidogenesis of androgens, ovarian steroids and adrenal corticosteroids involves many of the same precursors. Two pathways are used for the synthesis of TS. The typical, preferred pathway is cholesterol –> pregnenolone –> progesterone –> 17alpha-hydroxyprogesterone –> ANDRO –> TS (Lieberman and Prasad 1990). A secondary route however is from pregnenolone –> DHEA –> ANDRO –> TS.

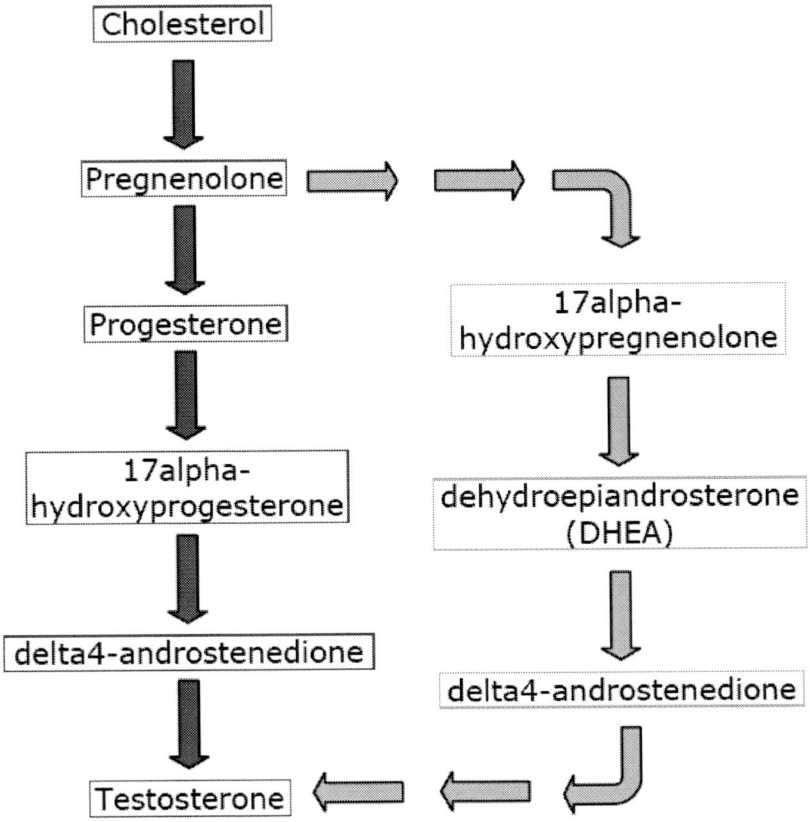

Figure 1. Metabolic cascade for the synthesis of testosterone and its potent metabolites, estradiol and dihydrotestosterone. The primary pathway is represented by the blue arrows.

II.2.2. Supplements of the Adrenal Precursor Androstenedione do not Increase Testosterone

A fundamental question about steroid metabolism is under what conditions is the secondary pathway used in place of the preferred pathway? We do not know. We do know, however, that there are surprising outcomes from manipulating precursor levels that may involve switching between the two pathways. For example, ANDRO supplements would seem to predict increased levels of TS by providing more precursor for increased production of TS. The enzyme responsible for converting ANDRO to TS, 17beta-hydroxysteroid dehydrogenase, is found in most body tissues. Androstenedione is also produced by some plants and has been marketed as a product for increasing blood TS concentrations, that is, as a natural alternative to anabolic steroid use. At least that was surely the thinking of baseball player Mark McGwire and other professional athletes who were self-administering ANDRO.

However, the interconversion of ANDRO and TS and metabolism to other bioactive steroids is complex. The results are often surprising. For example, supplements of a TS precursor may not have the expected effects on TS or its functions (Arlt, *et al* 1999, King, *et al* 1999). Oral ANDRO supplements revealed none of the expected increases in serum TS nor did it promote muscle strength (Birchard 1999). We also were surprised by our findings when castrated male rats were administered ANDRO. In TS restored castrates, concomitant ANDRO treatments suppressed androgen-sensitive behaviors (Taylor, *et al* 1994).

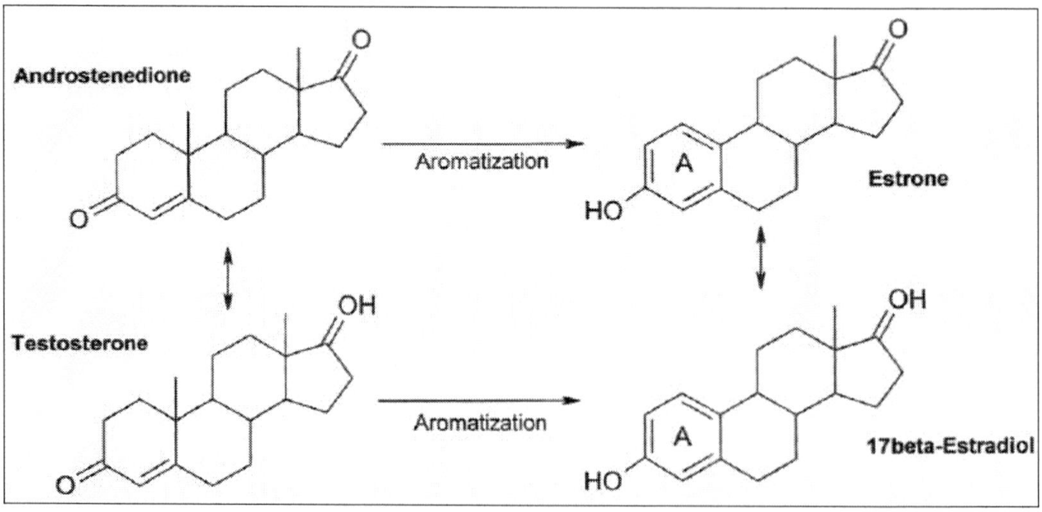

Figure 2. Metabolic cascade for the two possible pathways for the metabolism of androstenedione (ANDRO).

The key may be that in addition to serving as a precursor to TS, ANDRO is closely connected to estrogens (King, *et al* 1999). Androstenedione can be converted directly to an estrogen, specifically estrone, thereby completely bypassing testosterone (Labrie, *et al* 1998, Welt, *et al* 2006). Alternatively, because TS is also aromatized to E2, it also is possible that increased production of TS following ANDRO administration may result in increased aromatization, which would further attenuate any increase in the blood TS concentration (King, *et al* 1999). Sure enough, a study of young men with normal TS values confirmed the increased estrogen levels. The only steroids to increase with oral ANDRO supplements were serum E2 and estrone (King, *et al* 1999).

II.2.3. Female Sexuality May be More Sensitive to the Adrenal Precursor DHEA

Androstenedione, however, has not received the research attention lavished on the other major TS precursor synthesized by the adrenal glands. DHEA appears to be a more potent precursor and perhaps serves both reproductive and non-reproductive functions.

DHEA and its sulphate DHEAS are easily detected in circulation of humans, although only with some difficulty in rodent models. DHEAS serves as reservoir for DHEA and the conversion is via cellular sulfatase. DHEA can then serve as a precursor for a number of other steroids, including ANDRO, TS, and E2 (Chen, *et al* 2005).

One set of findings that has focused attention on DHEA and DHEAS is that both show a progressive decline in adults with age that corresponds to physical aging. Circulating levels of DHEA and DHEAS are dramatically reduced in aged primates, including humans.

There has been considerable discussion recently over whether to consider DHEA a hormone. At this time, DHEA is more appropriately referred to as a steroid, a prohormone or an androgen precursor (Dhatariya 2005, Labrie, *et al* 2005). It is not an androgen nor is it a hormone. DHEA does not bind to either the estrogen or androgen receptor and, therefore, it has no direct androgenic activity. Unlike other major steroids, a receptor for DHEA or one of its metabolites has not been definitively isolated in the reproductive system or in the brains of animal models (Johannsson, *et al* 2002). Consequently, the problem of proponents of DHEA therapies is to identify a cellular mechanism of action for DHEA (Widstrom and Dillon 2004). One possibility is that DHEA exerts functional effects by neuromodulation and improvement in endothelial cell function (Piroli, *et al* 2002). Another is DHEA may act as a neurosteroid by binding a cell membrane-bound receptor in the brain. Unfortunately however, this potential receptor is yet to be isolated (Widstrom and Dillon 2004).

Still, the best evidence is that DHEA plays an important role in the periphery by its role as a precursor for the peripheral synthesis of TS, DHT, and E2. Levels of DHEA predict androgen and estrogen concentrations in peripheral tissues, although they have little effect on circulating levels of the sex hormones (Chen, *et al* 2005). Other possible influences on the periphery include enhancement of the immune system, antagonism of corticosteroids, and involvement in cellular pathology and even obesity. Williams (Williams 2000) has reviewed the literature on each of these possible functions as well as critiquing the mechanisms suggested for DHEA influences on the periphery.

More relevant here is to consider the role of DHEA in sociosexual behaviors. Interestingly, there has been more enthusiasm for a DHEA – sexuality relation in females than in males. This may reflect the notable paucity of findings supporting a clear role for DHEA in males. For example, a study of sexual behaviour in young healthy eugonadal men also assayed adrenal and testicular sex steroids. Serum DHT concentration was the only independent hormonal predictor of their sexuality (Mantzoros, *et al* 1995).

In an experiment with animals, we treated castrated rats daily for 4 wks with adrenal steroids in the presence or absence of restorative doses of TS (Taylor, *et al* 1994). DHEA, DHEAS, and ANDRO were administered as cyclodextrin complexes to mimic the pharmacodynamics of the endogenous products. DHEA supplements alone were unable to maintain integrity of sociosexual responses and androgen target tissues after castration (Beyer, *et al* 1973). More remarkable was that DHEA suppressed the normal restorative functions of TS therapy to the castrated animals.

The story is quite different in females, especially in postmenopausal women. That low levels of circulating DHEA have been demonstrated in women with diminished libido and other symptoms of sexual dysfunction has suggested the use of DHEA as a supplement to increase androgenic activity (Saltzman and Guay 2006). Acute DHEA exposure, however, induced no changes on subjective (self-report) or physiological (vaginal photoplethysmograph) sexual responses to erotic stimuli in premenopausal women (Meston and Heiman 2002). Chronic DHEA treatment was more successful in women with androgen deficiency due to hypopituitarism. Oral DHEA raised the serum levels of DHEAS to normal age-related reference ranges, but only slightly increased TS titers. Despite subnormal TS levels, the DHEA treated women increased their initiation of sexual activity (Johannsson, *et al* 2002).

II.2.4. Testosterone is Important for Sexual Motivation in Women

Testosterone appears to have been adopted by the female endocrine system for some aspects of successful reproduction. Aggression is possibly one behavioral example (Dabbs and Hargrave 1997), but sexual motivation is more likely. Although not without controversy, TS seems to play an important role in sexual interest exhibited by the female (Federman 2004).

Most of the supporting data comes from studies of women because it proves more difficult to assess initiation of a sexual encounter in animal models. One reason is that non-human females are solicitous of sex only when in estrus, which makes separating the effects of TS from other sex hormones difficult. Moreover, there are surprisingly few studies of animal models in which previously untreated adult females are treated with TS and behavior is measured. An exception is an experiment by Van de Poll and colleagues (VandePoll, *et al* 1988). Female rats treated for 14 days with exogenous TS spent more time near a male than untreated control females. Interestingly, urinary marking by the females increased immediately upon injection of TS.

The steroid metabolic cascade revealing that E2 is a metabolite of TS suggested two conclusions. Testosterone, the prototypical masculine hormone, is a requisite precursor for the estrogens found in females (Goldstein, *et al* 2004). And, that one may find detectable amounts of unmetabolized TS in circulation of females.

Subsequent assays of serum confirmed the existence of TS in females. Indeed, levels of TS were surprisingly high, higher even than circulating estrogens. Although at levels that are approximately 10% of the titers found in males, women typically have more absolute amounts of TS in circulation than E2 (Miller, *et al* 2001, Miller 2002). This is partly because E2 is synthesized at low rates. Androgens are quantitatively the predominant sex steroid in women, circulating in the micromolar and nanomolar concentration range, compared with picomolar levels of estrogens (Sarkola, *et al* 1999, Young, *et al* 2000).

In the male, the majority of TS production occurs in the testes. In the female, 25% of TS is produced in the adrenals, 25% in the ovaries, and 50% in the peripheral tissues from precursors released from the adrenals and ovaries (Bachmann 2002).

Serum levels of TS follow a pattern of increases and decreases across the menstrual cycle that somewhat mimic the rise and fall of estrogens (Dougherty, *et al* 1997, Miller, *et al* 2001). Peak levels are found at mid-cycle and remain detectable throughout the cycle.

The next question is whether TS activates brain processes underlying reproduction and non-reproduction function in females. The answer suggested is yes for both functions. For example, data from a recent study of ovariectomized rats were that hippocampal neurons responded to TS in a similar manner to the influence of E2 on dendritic morphology in the hippocampus, a brain region critical for learning and memory (Leranth, *et al* 2004a).

Other findings suggested an impact of circulating TS on female sexual behavior. A correlational study of premenopausal women with or without complaints of sexual dysfunction revealed lower adrenal precursors and TS of the latter group (Guay, *et al* 2004). Indeed, the predominant symptom of women with androgen deficiency is loss of sexual desire (Davis 1999). Low TS levels may accompany a variety of conditions in women of reproductive age, including hypothalamic amenorrhea, hypopituitarism, premature ovarian failure, and premenstrual syndrome, as well as glucocorticosteroid or oral estrogen therapies (Kalantaridou and Calis 2006). Consequences of androgen insufficiency include various peripheral changes such as increased vasomotor flushing, loss of bone and muscle mass and replacement of muscle with adipose tissue, as well as a pattern of declining sexual functioning. Along with decreased sexual motivation, there is a reduction of sexual fantasy and enjoyment; , diminished sexual arousability, and decreased vaginal vasocongestion in response to erotic stimuli (Bachmann 2002).

The more common cause of androgen decreases is simply aging. Serum androgen levels exhibit a progressive stepwise decline from 20 to 49 years of age (Guay, *et al* 2004). Women in their forties have approximately half the level of circulating total TS as that of women in their twenties (Dennerstein, *et al* 2002).

The best evidence for a TS – sexual motivation relation is the value of TS therapy on the sexual vitality of postmenopausal women (Shifren 2004, Somboonporn, *et al* 2005). Although the evidence for TS as fundamental to low sexual interest in younger women has proved more inconclusive (Guay, *et al* 2004, Nyunt, *et al* 2005), there is greater support from androgen treatments of postmenopausal women (Davis 1999).

For example, Sherwin et al. (Sherwin, *et al* 1985) assessed sexual functioning in menopausal women administered TS over a 3 months period. It was clear that exogenous androgen enhanced the intensity of sexual desire and arousal and the frequency of sexual fantasies. However, the results were limited to sexual motivation rather than sexual activity. This may be a reflection only that sexual activity involves non-endocrine factors such as marital status, partner interest and past experiences.

Postmenopausal women volunteers were administered either oral estrogens with or without concomitant androgen for eight weeks in a double-blind study. Sexual function was assessed with a questionnaire. Results were that sexual desire, satisfaction and frequency increased significantly more in the group receiving androgen than in women receiving by estrogen only or estrogen-progestin therapy (Sarrel, *et al* 1998). Finally, in a single-blind study TS was administered by implants for 2 years to postmenopausal women. Results indicated significantly greater scores for sexual activity, satisfaction, pleasure, and orgasm compared with controls (Davis, *et al* 1995).

An unanswered question is whether TS directly increases arousability in women or if it must be first converted to E2 (Bancroft 2005). In the latter case, it would suggest that estrogen therapy would have the same effect on sexual motivation. The studies cited above suggest that not to be the case. Sexuality of postmenopausal women was greater with a combination of TS and estrogen compared to women receiving estrogen alone (Davis, *et al* 1995, Dobs, *et al* 2002). More recently (Davis, *et al* 2006) there was a report that an aromatase inhibitor in combination with TS therapy had no significant effect on sexual restoration in women. The suggestion is that it is TS acting on the androgen receptor and that is the key to sexual motivation in women.

II.3. Androgen Receptors are Found in Structures Both Related and Unrelated to Reproduction

Molecular biology has allowed identification of receptors for steroidal hormones and for localizing concentrations of the receptors in tissues throughout the body. Estrogen receptors (ER) and androgen receptors (AR) were found in the periphery and brains of both genders of common laboratory animals. Surprisingly the receptors appear in brain regions responsible for reproduction as well as in brain regions with little relation to reproduction (Nyborg, 1994).

It came as a further surprise to many of us to learn that there is no specific TS receptor. The receptor is the more generic androgen receptor (AR). As already noted, TS serves both as an active hormone and as a prohormone. Genetic evidence indicates that TS and DHT work via a common intracellular receptor (Wilson, *et al* 2002). Testosterone has good affinity for the AR, but, interestingly, DHT has a notably higher affinity for the androgen receptor. Either directly or after conversion to E2 or DHT, TS interacts with sex steroid receptors to produce both peripheral and central activation that underlies sociosexual behaviors in the male.

In the periphery, steroid receptors are found in a remarkable variety of tissues, from muscle and bone to cranial motor nuclei to the cornea and retina of the eye (Wickham, *et al* 2000). Some of the influences of TS on these structures appear to require aromatization, bone for example, and others not, muscle for example (Bronwyn, *et al* 2003). The greatest concentration of AR, however, is where one would expect to find it, in the reproductive tract. Estrogen receptors are often found co-localized with the AR, for example in the testes (Carreau, *et al* 2003). The result is that TS and its metabolites are integral in maintaining the structure and function of peripheral structures involved in normal male sexual function (Goldstein, *et al* 2004). More specifically, TS directly impacts sexual behaviors by its effects on peripheral neurons that trigger and complete copulatory reflexes (Keast 1999). The influence on genital organ sensitivity plays an important role in sexual arousal and orgasm.

II.3.1. Distribution of Androgen Receptors in the Brain

In the rodent brain, conversion of androgens into estrogens by aromatase is a key mechanism by which TS regulates many physiological and behavioral processes, including the activation of male sexual behavior, brain sexual differentiation and negative feedback effects of steroid hormones on gonadotropin secretion (Balthazart and Ball 1998). Therefore, cells that express steroid hormone receptors are integral to the action of gonadal hormones and represent essential components of the mechanism underlying masculinization of the brain.

Neurons with ARs in their nuclei are distributed widely in many regions of the brain including the preoptic area of the hypothalamus, septal area-bed nucleus, amygdala, and brain stem (Bonsall, *et al* 1985). ARs are also concentrated in areas of the brain that are not primarily involved in reproduction, such as the basal forebrain, hippocampus, caudate putamen, midbrain raphe and brainstem locus coeruleus (McEwen 2001).

The brain regions most closely related to male-specific sexual behaviors are located in various limbic structures and in the olfactory system. Most critical are the hypothalamus and amygdala. Testosterone and its metabolites interact with steroid receptors in hypothalamic limbic structures, where they elicit conscious perception and pleasurable reactions by influencing the release of specific neurotransmitters.

Although traditionally believed to be only in subcortical regions, ARs are also found in cortical and subcortical brain regions unrelated to reproduction, notably the hippocampus. AR immunoreactivity and binding are very similar in rat hippocampus as in the sexual behavior-dependent hypothalamus. ARs are concentrated in the CA1 region, an area that plays a major role in learning and memory (Xiao and Jordan 2002). That ARs are present also on axons and dendrites in cerebral cortex highlights the likelihood that androgens play an important and novel extra-nuclear role in neuronal function (DonCarlos, 2003}. This lays the groundwork for other surprises that have changed the way neuroendocrinologists think about sex steroids and behavior.

II.3.2. Gender Differences in Steroid Receptors Predict Sexual Behavior Differences but not Dimorphic Non-reproductive Behaviors

When neuroendocrinologists talk about "brain sex" or a brain being organized as masculine or feminine, it could mean many things. It could mean males and females having more neurons in certain brain areas, leaving those areas larger. It could mean having the same number of neurons but the neurons are connected differently leaving neurotransmitter pathways wired differently. It could also refer to the different numbers of steroid receptors in masculine and feminine brains with those receptors being distributed differently. Or it could refer to a brain region that metabolizes the same hormones differently. The net result of all these possibilities is that the same hormone, whether it be TS or estrogen or another hormone, delivered to the brain of an adult can lead to sexually dimorphic behaviors (Resko and Roselli 1997, Simerly 2002).

Still, much of the interest has centered on size differences and on receptor differences, likely because they are the easiest to quantify. Only a couple of size differences exist, most famously is the clearly larger size of the higher vocal center (or HVC) in male songbirds (Nottebohm and Arnold 1976). There are reports of subtle size differences in brain regions between men and women. The amygdala and hypothalamus appear to be larger in men. Both regions are also the same regions that contain high concentrations of steroid receptors (Hamann 2005).

It is notable that the both androgen and estrogen receptors are detected in the same brain regions in males and female animal models. ARs are found in neuroanatomical areas of both genders in the medial hypothalamic area, ventromedial and dorsomedial nucleus, preoptic area, arcuate nucleus, amygdala and regions of the hippocampus, arcuate-median eminence and BNST, with the greatest AR density reported in the preoptic and hypothalamic areas, septum and amygdala of these animals.

Receptor numbers differ in the expected direction in some brain regions. The two subtypes of estrogen receptor (ER), designated as ER-alpha and ER-beta, are distributed differently in male and female brains (Shughrue, *et al* 2000). Although ER-beta is found in the hippocampus of male and female rats, females have higher concentrations (Zhang, *et al* 2002).

Using 4 strains of male and female mice Lu (Lu, *et al* 1998) found sexually dimorphic receptor concentrations in various brain regions. Males had a higher absolute number of AR than females. Concentrations of AR were higher in males in brain regions associated with male-typical sexual behaviors, e.g., BNST and MPOA. AR decreased significantly with castration. Indeed, most differences in AR numbers between the sexes are detected in hypothalamic nuclei and other structures related to reproduction (Chen and Tu 1992, Lu, *et al* 1998).

In many non-reproductive structures, however, there are surprisingly few or no sex differences in distribution and concentrations of ERs and ARs (Michael, *et al* 1995, Scott, *et al* 2000). Moreover, where there may be sex differences, there is no clear functional logic for the existence of different receptor numbers. An example is the AR findings from autopsies of a small number of men and women between 20 years and 39 years (Fernandez-Guasti, *et al* 2000). Again, ARs were found in high numbers in both male and female hypothalamus. However, men had higher AR concentrations in the suprachiasmatic nucleus of the hypothalamus, a region generally agreed to be related to diurnal activity rather than sexual behavior.

The presence of large numbers of ER and AR in both sexes would seem to pose interpretation problems for relating dimorphic functions to the steroid receptor. High concentrations of ERs in males can be explained, however, by the metabolism of TS into E2 in brain tissues to interact with ERs. Notably, amounts of aromatase are higher in many brain regions of male animal models than of females (Lu, *et al* 1998, Roselli and Resko 1997). The presence of high concentrations of ARs in the brains of females is more puzzling (Xiao and Jordan 2002).

II.4. Steroids act through Genomic AND Non-genomic Mechanisms

Steroid hormones are unique endocrine products in the way they interact with nuclear DNA. This capacity allows for many of the distinctive features of steroids to induce the long-lasting changes that are observed, for example, the surge of TS with the oncoming breeding season in feral animals or the onset of puberty in humans.

In the past decade another mechanism for influences on brain and behavior has been identified. In addition to the well-established nuclear action of steroid receptors, membrane-associated receptors for gonadal steroid hormones have been discovered. This finding has revolutionized the way we think of steroids, recognizing rapid actions of steroids that could not be attributed to transcriptional regulation. Considerable evidence has now accumulated to support the existence of these nonnuclear receptors. Our ideas have now been broadened to recognize gonadal steroid receptor action via extranuclear sites.

II.4.1. The Classic Mechanism for Steroidal Influence is Genomic and Slow

It has been known for some time that TS and the other steroids have direct influence on DNA transcription in the nucleus of target cells. The result is that the genomic effects are slow because it requires time for the multiple steps to initiate DNA activation and subsequent protein synthesis to initiate functional changes. Genomic changes, consequently, are long-lasting.

Steroidal hormones belong to a ligand-activated nuclear receptor family. Members of this family have been defined by their ability to bind to specific DNA sequences in promoter regions of hormone responsive genes.

Unbound TS in circulation enters a cell and binds the androgen receptor (AR) to form a steroid ligand. Prior to confronting TS, the AR is loosely bound to a protein in the cellular cytoplasm or nucleus. Surprisingly, the exact location of the steroid-hormone receptor is not known. What is known is that TS diffuses into the cell and binds the receptor and the TS – AR complex then diffuses into the cell nucleus. The TS - AR complexes bind to specific sequences of DNA called steroid-response elements, as well as to nuclear-receptor coactivators or repressors (Gruber, *et al* 2002). Upon binding to these hormone response elements in the DNA, the receptors are capable of exerting powerful modulatory effects on transcription of specific genes.

The outcomes are prominent changes in neural morphology and function. Testosterone surges are linked to an increase in neuron somal size, neuritic growth, plasticity and synaptogenesis in both motoneurons of the spinal nucleus of the bulbocavernosus muscle in the reproductive system of males and several populations of pelvic autonomic neurons (Bialek, *et al* 2004). The genomic influences of TS are now known to extend beyond the reproductive system. Changes in lateralized function of the hemispheres and hippocampal activation are but two examples.

Castration of adult rats is reported to induce a decrease of dendritic spines in the CA1 region of the hippocampus (Leranth, *et al* 2004b). Castration also reduced numbers of acetylcholine neurons in some cortical areas, and 4 weeks of exogenous TS restored the cholinergic pathways in the adult rats. These data suggest that the presence or absence of TS profoundly influenced the cholinergic population in cortical and subcortical areas of the adult rat brain (Nakamura, *et al* 2002). These and other data have led to the speculation that TS is a neuroprotective agent (Frye and McCormick 2000).

One final word on androgen receptors and it has a familiar ring. There is more complexity than researchers had hoped. The concept of a single hormone activating a single receptor is being replaced by a more complex model (Vanderschueren and Bouillon 2000). The picture emerging is of the original hormonal signal undergoing transformation in the target tissue from processes such intracellular enzyme activities, receptor concentration and function, and post-receptor cross-talk. The result is modulation of the original signal and fine-tuning of the biological response.

II.4.2. A second Mechanism for Steroidal Influence is Non-genomic and Fast

Steroids are usually identified as genomic regulators. Recently a body of evidence has accumulated demonstrating specific membrane effects more akin to neurotransmitter activities. That is, steroids can help coordinate neural firings by interacting with both membrane and intracellular receptors. The resulting rapid modulation of cellular activity is measured in seconds or minutes and suggests a non-genomic role for certain steroids. Rapid steroid synthesis and metabolism have been found to exist in the CNS, and downstream effects can be seen in both the central and peripheral nervous systems.

The mechanism for rapid, non-genomic action is suggested by the findings that ARs are present on axons and dendrites within the mammalian central nervous system. AR expression in axons was identified in the rat brain at the light microscopic level and details confirmed at the ultrastructural level. AR-immunoreactive axons were observed primarily in the cerebral cortex and were rare in regions where nuclear AR expression is abundant. This suggests that there are separate functional systems for the genomic and non-genomic influence of the sex steroids. The observation that ARs are present in axons and dendrites highlights the possibility that TS plays an important and novel extra-nuclear role in neuronal function (DonCarlos, *et al* 2003).

Current levels of circulating steroids can induce changes in neural morphology that would appear to be too fast to be genomic. The best example comes from estrogen and female rats. The rapid rise and fall of estrogen during the 4-day estrous cycle of f rats is accompanied by equally rapid increases and decreases in numbers of dendrites in certain hippocampal regions (Woolley 1998). Significantly, those are the same regions responsible for certain types of memory.

The implications that steroids have fast, membrane-mediated influences on the brain function are nothing short of revolutionary. That steroids can have acute, momentary actions opens a new avenue of thinking about the brain. The pharmaceutical industry has been among the first to recognize the potential of non-genomic steroidal actions. A prominent example is that sex steroids may contribution to brain plasticity.

II.4.3. Steroids Play a Role in Brain Plasticity

Neuroscientists use the word "plasticity" to refer to brain tissues that maintain the capacity for structural changes in mature animals. That TS and the other sex steroids could contribute to brain plasticity is a remarkable development for several reasons.

To set the scene, recall the original organizational - activational model. The suggestion is that organizational effects are structural and irreversible, whereas activational effects were functional and reversible. The thought was that once fetal development and its sensitive periods to TS had passed, the organized morphology of the brain would not change with changes in circulating adult hormones.

The activational component, on the other hand, was considered labile and modifiable, say by castration or hormone supplements. The hormonal influences that activated a function after puberty, sexual motivation for instance, were completed without changing integrity of the long-ago established structures. As long as those masculinized brain tissues have sufficient current levels of TS from the adult testes, we would see normal male sexual functioning. Remove the testes and TS and we would observe sexual behaviors beginning a downhill slide. Brain anatomy does not change with castration, but the behavior activated by TS does.

In general, the structural – functional distinction between organization – activation roles of the model have been largely upheld today. But, the line between function and structure has been blurred in the brain. Current research on the effects of gonadal steroids on the brain and spinal cord indicates that these agents can have profound trophic effects on many aspects of neuronal functioning. As example, TS induced structural remodeling of the neuromuscular junctions of the peripheral nervous system in the male rat reproductive tract (Gordon, *et al* 1990). Cranial nerves also show morphological sensitivity to androgens. Some of these effects may be genomic and some non-genomic. Regardless of the mechanism of action, the consequences are notable on cell survival, growth and metabolism, elaboration of processes, synaptogenesis, and neurotransmission (Jones, *et al* 1999).

Still, the best evidence that some structures in the brain maintain plasticity to hormones is the capacity of hormones to change the physical structure of the adult hippocampus. Specifically, features of hippocampal anatomy can be molded by current levels of ovarian hormone in cycling female rats. The initial findings were of neural morphology rapidly changing with fluctuations of endogenous sex hormones in adult female rats (Woolley, *et al* 1993). More recently, ovariectomized rats administered TS induced similar increases in dendritic spines (Leranth, *et al* 2004a). Male brains may or may not undergo similar changes. These data highlight both the plasticity of the adult brain and the importance of hormone fluctuations on the brain during the activational phase (Breedlove and Jordan 2001).

There are good reasons, however, to believe that the male brain will show less plasticity than female brains, and that TS will be less likely to be as important of a player as ovarian hormones. A key difference between male and female brains is that only the females undergo dramatic endocrine fluctuations over days. It is probably accurate to suggest that even neuroendocrinologists have not fully appreciated the dramatic differences between a fluctuating endogenous hormonal stimulus experienced by females and a more or less constant one experienced by males (Taylor and Pitha 1988).

II.5. The Brain Synthesizes its own Steroids

We have known for decades that steroid hormones influence the brain. That is to say, a sex hormone released by the far-away gonads is pumped in the blood to the brain where it binds a steroid receptor. But, it has been recognized only in the past decade or so that the brain appears to be producing its own steroids. This has led, among other things, to a confusion of terms.

Terminology in the steroid hormone literature has not shown ideal consistency. The result is confusion about the definitions of steroids in the periphery and the CNS (Stoffel-Wagner 2001). We have already introduced one example, DHEA, which is referred to erroneously as an androgen or as a hormone.

"Biologically active steroids" are steroids that have the ability to bind to a steroid receptor in the periphery or the CNS. Examples are TS, E2, progesterone and corticosteroid. "Neuroactive steroids" are those substances that bind to a steroid receptor in the CNS. All the aforementioned steroids also meet the criterion for neuroactive steroids.

But the brain synthesizes its own steroids for local use in the brain. These became known as "neuroactive steroids" (Baulieu 1998), i.e., steroids used by the brain for both reproductive and non-reproductive functions. The most important feature is that neurosteroids are synthesized directly in the brain. The neuroactive steroids are not the familiar steroids, and include DHEA and progesterone metabolites such as allopregnanolone. Likely neurosteroids use non-genomic mechanisms to quickly modify brain activity and behavior (Rupprecht and Holsboer 1999). The terminology details may seem like splitting of hairs, but it is an important distinction because there are revolutionary implications in the concept of a neurosteroid.

One implication is that, rather than relying entirely on the peripheral endocrine organs for its hormones, the brain has evolved the capacity to synthesize its own steroids (Corpechot, *et al* 1981, Mellon 1994). This can be demonstrated by removing the peripheral organs, adrenals and gonads, and the amounts of neurosteroid in the brain do not change (Baulieu 1997). Economically, neurosteroids are synthesized mostly from the same precursors and enzymes as used for peripheral synthesis.

Another implication is perhaps more profound and has the potential to open a new chapter for steroidal influence on brain and behavior (Engel and Grant 2001). The mammalian brain has adapted both peptides and steroids used in reproduction for other neural functions. The ancient hormones synthesized in the periphery now have a new role. Rather than evolving completely new substances for complex behaviors such as memory or mood, steroids have been hijacked by the brain for other uses (Belelli and Lamber 2005, Birzniece, *et al* 2006).

SECTION III. CIRCULATING TESTOSTERONE AND SOCIOSEXUAL BEHAVIORS

III.1. Testosterone in Circulation shows wide between and within Individual Differences

A basic tenet of endocrinology is that endocrine products are released from glands to be carried by the bloodstream to target tissues some distance away. Sometimes this bloodstream is nearby and served by a specially evolved vascular source, a so-called portal system. The direct portal connection between hypothalamus and anterior pituitary is an important example. Much more commonly, the blood delivery system is general circulation, and that is almost always the source obtained for analyses of the steroids. There are myriads of methods to collect blood from experimental animals (Weiss, *et al* 2000), including the one familiar to human patients.

Once blood is drawn from men and assayed, the serum reveals only 2% found in the bloodstream is so-called unbound or free TS. That is, the portion of TS that is not bound to a protein, either albumin or sex hormone binding globulin (SHBG). The free fraction has traditionally been viewed as the source of biologically active TS that could enter cells and interact with a receptor. For example, aging men experience both reduced free TS and increased levels of SHBG (Swerdloff and Wang 2004). Much has been made of this observation. This includes a comparison of bound and free TS in patients with probable Alzheimer's disease with the conclusion that only free TS is significantly lower in male patients (Moffat, *et al* 2004).

Now it seems that albumin bound TS can be uptaken by cells and dissociated from the protein to interact with the receptor. Moreover, the bound form is a type of reservoir of TS that may actually facilitate the entry of hormone into the cell (Goldstein and Sites 2002). The result is that a considerable portion of the TS in circulation actually is in a readily bioactive form in males. The anxiety of researchers and clinicians over bound and free TS is probably unwarranted (Davidson, *et al* 1982). Serum TS levels and free TS are often highly correlated in healthy males (Stalenheim, *et al* 1998).

III.1.1. Within Individual Assays revealed that Testosterone is Released in Pulses throughout the Day and Night

In the mid-1970s an important discovery was made about the way that gonads release their hormones into circulation. Release is not by slow leakage but in a series of episodic bursts throughout the day (Bartke and Dalterio 1975, Sodersten, *et al* 1983). Soon, the findings were extended to other endocrine glands, including those outside the HPG axis (Linkowski, *et al* 1985, Penny 1985, Samuels, *et al* 1990). This pattern became known as pulsatile release. The term continues to be used to distinguish circadian pulses from the gradual increases in hormone known as surges that occur, for example, in males of seasonal breeding species or with the advent of puberty in boys.

Recognition of the ubiquity of pulsatile release of endocrine products forced researchers to better control sampling protocols. Most prominent was to measure hormone levels at the same time of the day. With that control in place, however, it was possible to make comparisons among and within males.

Little TS is stored in the testes and newly synthesized TS is quickly released into the bloodstream in 6-10 pulses of different amplitudes per day in men. The pattern is variable both between males and over days in the same male. Still, in men there is a fairly reliable TS pulse early in the morning, around daybreak. The pulse-generator responsible for the pattern of pulsatile release of TS is unknown, but we do know it is more than simply the HPG axis. Specifically, neurons projecting from various subcortical brain regions influence the amplitude and frequency of hypothalamic gonadotropin releasing hormone (GnRH) released from the hypothalamus into the pituitary (Abbud, *et al* 1999, Scott, *et al* 1997).

One result of this pulsatile feature is that hormone sensitive tissue responds best to aperiodic stimulation. Indeed, a continuous stream of hormone can produce a paradoxical suppression of target tissue activity (Southworth, *et al* 1991). That feature has made its way into hormone therapies as GnRH agonists are used to suppress the downstream release of luteinizing hormone and, thus, TS (Matthews, *et al* 1998, Rosler and Witztum 1998).

III.1.2. Differences in Endogenous Testosterone between Males do not Explain Differences in Sexual Behavior

Reports of within individual variability were quickly followed by findings of considerable variability between individual males. Testosterone values in healthy young males of many species, including humans, range over values that differ significantly. In men the lower limit for normative values is 300 ng/ dl but the values may range to 900 ng/ dl or more (Howell and Shalet 2001). Upon reflection, individual differences in TS were not surprising. Males within a social group differ, of course, along many physical and behavioral continua. Some of these differences are androgen dependent, including sexual behaviors. Yet, the surprise was that in normal adult males the wide individual variability in circulating TS levels did not seem to be linked in any meaningful way with individual differences in any aspects of sexual behavior (Meston and Frohlich 2000, Schiavi and White 1976).

Of course, a certain amount of TS is necessary for normal sexuality. Castrated animal models and men (Greenstein, *et al* 1995) typically experience dramatic decreases in all measures of interest and performance. In gonadally intact males, hypogonadism is also associated with low level of male sexual behavior (Gray, *et al* 1981, Sternbach 1998).

Yet repeatedly the findings from studies of intact laboratory animals and men are that individual differences in endogenous TS levels and measures of sexual behaviors are not significantly related (Damassa, *et al* 1977, Meston and Frohlich 2000). Phoenix and Chambers for instance have made an exhaustive study of the sexual behaviors of rhesus monkeys and laboratory rats. One dominant theme is their conclusion that circulating TS is not the key factor in individual differences in the sexual behaviors of intact young or old males (Chambers and Phoenix 1992). Even administering exogenous hormone to elevate TS titers of so-called sexually sluggish young male monkeys failed to improve their performance (Phoenix and Chambers 1986).

The same absence of a TS – sexual behavior relation appears to hold for non-traditional animal models. A study of male and female meerkats (Carlson, *et al* 2004) reported no rank-related differences in circulating levels of TS among males. Dominant, breeding males and females, however, did have significantly higher cortisol levels than the subordinates, non-breeding meerkats.

Similar conclusions have been reported for men (Brown, *et al* 1978, Schiavi and White 1976, Zitzmann and Nieschlag 2001). The former study employed collection of two samples taken a week apart from 101 young adult men. Serum TS concentrations in the two samples were highly correlated, indicating good reliability and consistency. However, there was no correlation with their self-reported frequencies of sexual behaviors. The conclusion was that differences among men in circulating TS concentration within the normal range do not account for differences in sexual interest or activity (Brown, *et al* 1978).

One finds data in the literature suggesting a positive correlation of TS and sexuality (Knussmann, *et al* 1986). However, the bulk of the data supports the original conclusions. There is little correlation between circulating levels of the hormone and performance measures.

There are many possible explanations for the absence of correlation between resting TS and sexual behavior. One possibility is differences in nocturnal hormone pulses that are not typically measured. After finding the usual absence of a correlation between TS levels and copulation in rhesus monkeys, the researchers measured nocturnal hormone titers (Michael, *et al* 1984). They found that the most sexually active males had less of a nadir of the circadian rhythm at night, thus maintaining a higher level of TS during sleep. The implication is that, considered over the entire day, the TS levels were higher in the more sexually active male.

Another possibility is that the relation is in the opposite direction, that is, sexual activity increases TS in males rather than TS increasing the behavior. Sure enough, sexually experienced male rats have larger testes and higher TS levels (Frankel 1984, Taylor, *et al* 1985) compared to virgin males. However, the increases of TS appear to be only observed with chronic sexual contact. Testosterone is not released with copulation. Measurement of various hormones in male rhesus monkeys following copulation revealed no acute increases in circulating TS (Phoenix, *et al* 1977). Only cortisol increased in the monkeys. Although it was with masturbation, the same result for TS was reported after orgasm in men. Only prolactin increased (Kruger, *et al* 2003).

Interestingly this same relation, or absence of a relation, may not hold for females. As noted in a previous section, it is likely that TS is responsible for libido in women. Although there are far fewer studies of women than men, resting levels of TS appear to be related to self-reports of sexual interest in women (Sherwin, *et al* 1985).

III.1.3. Rather than a Dose-response Model, a threshold Model may be more accurate for the Testosterone - Behavior Relations

These findings led to the conclusion that the relation of TS and sexual behavior, and likely all behaviors, were better described by a threshold hormone model than a dose-response model. Simply the idea is that, once a threshold is passed, increases in circulating TS will have no further effects on behavior (Buena, *et al* 1993, Ewing, *et al* 1979, Isidori, *et al* 2005).

One prediction from the threshold model is that TS restored castrates will show similar sexuality as before the surgery. That is, exogenous TS to a formerly sexually active castrated male will restore his sexual behavior, but only to his previous level. The data support that prediction (Phoenix and Chambers 1986).

Additional evidence for the threshold model is that when TS levels are manipulated to ensure high or low TS values within the normal male range, sexuality of the two groups are the same (Buena, *et al* 1993, Taylor, *et al* 1985). In the former study, two groups of healthy men were treated for 9 weeks with a GnRH agonist to slightly suppress or exogenous TS to slightly increase TS serum levels. Sexual function evaluated by daily dairies revealed no significant differences in overall scores or in subcategories of intensity of sexual feelings or sexual activity between the two groups. The authors' conclusion was that a minimal value of TS above threshold was sufficient to fully restore sexual behavior in men (Buena, *et al* 1993).

The threshold model remains a dominant concept today despite mounting evidence that some behaviors, including aspects of sexual behaviors, may show a dose-response relation to TS (Gray, *et al* 2005). The threshold vs. dose-response issue continues to define much of the research in this area, both basic research and applications to clinical populations. More recently the debate has included the effects of TS supplements to eugonadal men, defined as healthy males with normal endocrine functions (O'Connor, *et al* 2004). That controversy surrounds the consequences from use and abuse of anabolic steroids, as well as the value of supplement of TS to healthy men. Those issues will be addressed in the last sections of this chapter.

An important feature of the literature reviewed above is to emphasize that the typical paradigm is a study of baseline or steady-state hormone. Researchers refer to this as "current levels" or "prevailing levels" of hormone in serum. Better perhaps is "resting levels" because it focuses on the key concept that the values reported are the serum steroidal titers of a male in a normal, unstimulated condition. And the threshold model seems to most accurately describe current TS levels – sociosexual behavior relations in males.

III.2. Acute and Chronic Changes in TS Induce Behavioral Changes

The conclusion that resting TS levels of healthy males are seldom informative in explaining individual differences in behavior would seem to place a cloud on further research on TS – behavior relations. There is, however, another way to ask questions about the relation. Rather than resting levels, for instance, do changes in TS induce reliable changes in behavior? Given the proclivity of experimentalists and clinicians to manipulate, it is not surprising that the changing hormone literature is much larger, with more interesting outcomes, than the literature on resting hormone titers.

III.2.1. Momentary Changes in Testosterone can Change Sexual Function

As discussed at length above, a basic principle of steroid endocrinology is that the primary pathway for influencing behavior is genomic. The result is that, in most situations, TS and the other steroids are unlikely to show a momentary relation to behaviors. Rather hormones help set the conditions that a behavior will occur in the presence of a specific stimulus (Wade 2006). This is because most steroid effects require some period of time to influence the genome.

There have been clever attempts to identify the time period. One example in humans was with women exposed to a dose of TS while watching pornographic films. Sublingual intake of exogenous TS was followed by a sharp increase in plasma TS levels within 15 minutes. The circulating levels declined gradually over the next 90 minutes to reach baseline values. Only 3 to 4 hours after reaching peak levels was there a significantly increased genital responsiveness (Tuiten, *et al* 2000). These data suggest behavioral changes can occur within a few hours, although the time period is likely longer for other steroid – behavior relations.

The time delays of the classic genomic activity of steroids raised the puzzle of how it was possible for sudden changes in TS to have an immediate effect on physiology and behavior (Sachs 1997). Such momentary changes have been demonstrated in a variety of experimental paradigms.

Males of many species experience a rapid increase in TS levels to the appearance of a receptive female, her odors or even an environment in which the male had previously copulated (Fukada, *et al* 1988, Huang, *et al* 1989, Taylor 1980b). Those males may also exhibit an immediate increase in sociosexual behaviors (Taylor, *et al* 1984). These results also point out the important concept of hormone – behavior relations being a two-way street. Environments and behavior can modify the endocrine milieu which, then, can influence the behavior sensitive to the hormone (Aluja and Torrubia 2004).

More recent studies have confirmed the influence of the environment, as well as providing additional neurophysiological evidence of a rapid activation of neural activity to receptive females. For example, using nuclear fos immunoreactivity as a marker of neuronal activation (Kelliher, *et al* 1999), male rats were exposed to the soiled bedding of the female. Those males showed significantly more neuronal fos immunoreactivity than clean-bedding control males in the nucleus accumbens, medial amygdala, BNST and MPOA. Even greater neuronal fos responses were detected in these brain regions in males allowed to mate.

Molecular changes in the brain with acute exposure to exogenous TS include an experiment in which androgen receptor concentrations were examined in castrated adult male rats and gonadally intact females (Xiao and Jordan 2002). Animals received an injection of either TS propionate or vehicle 2 hours prior to sacrifice. AR receptor numbers within the CA1 region of the hippocampus increased in both genders to the acute TS injection. Interestingly, the increases were greater in the left hippocampus than the right, but only in the males, suggesting a sex difference in laterality of AR in hippocampal structure and function.

Still, the most important conclusion for our discussion is that the rapid rise in AR numbers point to a non-genomic effect of steroids in the hippocampus. Additional evidence comes from findings that a single acutely stressful event quickly enhanced dendritic spine density in the male hippocampus but reduced density in the female hippocampus (Shors, *et al* 2001).

Rapid behavioral changes to acute changes in circulating steroids have been reported in other paradigms. These include increased aggression in male rats to acute increases in corticosteroids (Haller, *et al* 2000), and an immediate response by women to a recently formed metabolite of TS. This is in the form of attraction to the odors of androstenol from men that is found in the apocrine glands located in the axillae of the skin in the armpits and pubic region (Kohl, *et al* 2001).

A more titillating behavioral example is the so-called Coolidge effect (Dewsbury 1981). This refers to the reinstitution of copulation in sexually satiated males with the introduction of a new, unmated female. In this paradigm, a male is allowed to copulate to exhaustion, identified by the male refusing to attempt to initiate another copulation. However, removal of that female and replacement with a novel, receptive female is quickly followed by renewed, vigorous copulation. This procedure can be repeated several times with a new female and the same male, with the same result.

The implication is of a momentary rise in TS followed by an immediate behavioral change. The most likely explanation is for non-genomic activities of steroid hormones. We know now that steroids, including androgens, have the capacity for membrane-induced activation of neurotransmitter pathways underlying behavior (Frye 2001). Steroid receptors on the presynaptic and postsynaptic neurons can quickly influence neural activity and, thus, behavior.

Another mechanism for fast changes in sexual responsiveness of the male changes to momentary changes in TS is an indirect effects. We have demonstrated that female rats prefer a male with high TS levels. Intact females given a choice between inaccessible castrated males restored with different amounts of TS preferred to spend time near the higher TS male (Taylor, *et al* 1982). It is possible that the adaptive value of a sudden pulse of TS in the male is that the female will be more responsive to the male, thus increasing his sexual prowess.

III.2.2. Chronic Changes in Testosterone are Likely to Change Sexual Function

Most of the research on changing levels of hormone, however, has looked at long term TS exposure. This is in recognition that typically steroid hormonal changes slowly modify sexual behavior.

Chronic hormonal changes can be either from natural changes or experimentally induced changes. Most often the changes are to increase circulating hormone. Puberty or increases during a breeding season represent natural elevation of TS. Exogenous administration to castrated animals or hypogonadal men represent the most studied groups for behavioral changes accompanying chronic increases in TS titers. Natural declines with aging are a notable instance of decreases in circulating TS.

An interesting study group are men convicted of felonies for sexually deviant behaviors. Often they are pedophiles. Studies of surgical castration in a large study of 104 sex offenders found highly significant decreases in sexual arousal or performance in 90% of the men. Further studies of recidivism indicated a significant decrease in sex offenders following castration (Stone, *et al* 2000).

Still, it is notable that castrated males do not immediately lose sexual capacity (Baum, *et al* 1986). In male rats, the decline in copulatory performance takes a time course that is the reverse of the normal sequence of mounting without intromission, multiple intromissions followed by ejaculation. With castration of the rat, the first copulatory response to be lost is ejaculation and the last to be lost is mounting (Davidson 1966). Actually the very first sexual response, sexual motivation, is the last to be lost, and may never disappear entirely in castrated males, most notably men. The same is observed with a pharmacological agent that blocked the conversion of TS to E2 in the brains of male rats. All aspects of sexual behavior were eventually disrupted, but again the ordering of behavioral disruption was the sequence of ejaculation to sexual motivation (Bonsall, *et al* 1992).

The point is that a sudden and dramatic decrease in TS requires time to impact most androgen – sensitive structures and, especially, behaviors. In the following sections, we will consider paradigms that underline the importance of significant changes in circulating TS and the impact on sexual behaviors.

III.2.3. Aging Represents a Natural Chronic Decrease in Testosterone and Sexual Function

In assessing the behavioral consequences of changes in circulating hormone levels and sexual behavior, a good place to start is with the aging male literature. Studies with non-human animals and humans have repeatedly confirmed the informal observations of many of us; aging is accompanied by a decrease in sexual behavior. Sexual performance declines earlier and more dramatically with aging than does sexual interest, but both performance and motivation show declines (Bancroft 2005, Taylor, *et al* 1996). There is also a progressive and systematic decrease in systemic TS in males (Lupo-DiPrisco and Dessi-Fulgheri 1980, Seidman 2003). Endocrine aging is less dramatic than the menopause of women, but aging exacts a toll on males and androgen-sensitive functions.

In a study of gonadally intact 22-24 months old male rats in good general health, we used multiple measures of sociosexual behavior and reproductive physiology to demonstrate that even healthy aging is accompanied by endocrine declines (Taylor, *et al* 1996). The untreated old animals showed clear decrements on all 13 measures of hypothalamic – pituitary – testicular function. This included reduced TS titers and decreased sexual activity.

The same is reported for a retrospective and prospective evaluation of sexual function and behavior in 77 healthy married men aged 45 to 74 years. There were significant age-related decreases in sexual desire, sexual arousal and activity, and increases in erectile problems. Aging also was negatively correlated with TS (Schiavi, *et al* 1991).

III.2.4. Hormone Therapy to Aged Males Increases Sexual Behaviors

There may or may not (Haren, *et al* 2002, Mazur, *et al* 2002) be a clearly identifiable causal relation between the two events, but a logical step for treatment of low sex hormone levels and reduced sexuality accompanying aging is hormone supplementation (Hayes 2000). There is understandable reluctance for clinicians to recommend TS therapy for aging men, given the well-known androgen sensitivity of the vulnerable prostate gland (Hijazi and Cunningham 2005). Nonetheless, given the success in experimental animals, TS supplementation for aging men remains an attractive therapy (Swerdloff and Wang 2004).

Clearly, androgen sensitive reproductive structures in older male animal models maintain sensitivity to additional TS (Chambers, *et al* 1991, Chubb and Desjardins 1984). The bulk of the data confirm benefits of TS to older males (Davidson, *et al* 1982, Schiavi, *et al* 1997). Treated males do not experience complete recovery to the behaviors of young males, but treated old males show more sexual interest and, to a lesser extent, increased copulatory capabilities.

Benefits of androgen therapy are accrued best in aged animal models and humans that are free of the diseases of old age (Wespes and Schulman 2002) and in aging males suffering the lowest TS levels (Corona, *et al* 2004). Not surprisingly, supplementation during middle age may be most effective. Testosterone supplements to 13 months old male rats, an ontogenetic period that models middle age in men, successfully increased most aspects of sexual behavior (Gray, *et al* 1981).

On rare occasion, experiments have been devised to increase endogenous circulating TS. In one experiment, we took advantage of the findings cited earlier that young males experience acute increases in TS to the presence of receptive females. We first asked if the same would be observed in old rats, and the results confirmed a significant rise in TS in old males (Taylor, *et al* 1996). Using the same receptive female exposure paradigm, we compared chronic endogenous and exogenous TS increases on two groups of healthy 22-24 months rats. Old male rats were administered either exogenous (ExT) supplementations daily for 6 weeks or were exposed to brief daily exposures to an inaccessible estrous female for additional episodes of endogenous (EnT) hormone over the same time period. Neither treatment restored our healthy old male rats to levels of sexual behavior approximating those of untreated young adults. Nonetheless, EnT males showed greater increases in sexual motivation and performance than ExT males. Our conclusions included that aged sexual behaviors of healthy old rats retain notable capacity, particularly, for endogenous TS activation of sociosexual behaviors (Taylor, *et al* 1996).

III.2.5. Testosterone to Hypogonadal Young Males Increases Sexual Behavior

Success is often reported with chronic TS supplementation in young hypogonadal men. Insufficient production of TS results in various sexual deficits affecting both sexual interest and sexual performance. Testosterone therapy commonly improves both significantly.

For example, men 25 – 40 yrs of age were diagnosed with severe hypogonadism by having a mean TS level of 35 ng/ dl, compared to the minimal level of 300 ng/ dl in eugonadal males (Burris, *et al* 1992). Patients were injected with 200 mg TS enanthate intramuscularly every 2 weeks for up to a year. Results were striking. Nocturnal erections in the hypogonadal men increased steadily during hormone replacement and were in the normal range with 6 to 12 months of treatment. The hypogonadal men also reported increases in several aspects of sexual activity, including sexual interest and the number of spontaneous erections.

It is interesting to highlight that this was clearly a genomic effect of the steroids. Serum TS concentration increased immediately and finally reached the upper range of normal levels. Yet, the behaviors only gradually changed with the immediate changes in circulating TS (Burris, *et al* 1992).

Other studies have reported similarly significant improvements in sexual function in hypogonadal young men treated with testosterone (Arver, *et al* 1996, Ashmed, *et al* 1988). A recent meta-analysis (Isidori, *et al* 2005) of men with TS levels below 300 ng/ dl receiving chronic TS therapy concluded that hormone treatment was moderately effective, likely because some were moderately hypogonadal. Even with this analysis, the men reported improvements in the number of nocturnal erections, sexual thoughts and motivation, number of successful intercourses, erectile function and overall sexual satisfaction.

Adolescent boys with delayed puberty also appear to benefit from TS supplementation, however, the benefits are not as clear as in young men. Depo-testosterone was administered to delayed pubertal boys in 3-month blocks alternating with placebo. Although there were modest increases in measures of sexuality, there were significant hormone effects on physical aggressive behaviors and aggressive impulses. A causal correlation is difficult to dispute because numerous studies have shown that withdrawal of exogenous TS to hypogonadal or castrated males is followed by a rapid and marked decrease in sexual interest and activity that is reinstated in a few weeks with TS replacement therapy (Meston and Frohlich 2000).

Nonetheless, there are data to jolt us back to the recognition that sexuality is dependent on factors other than normal androgens in circulation (OCarroll and Bancroft 1984). Hormonal supplementation that results in normal TS values does not always restore libido and erectile function in hypogonadal males. Moreover, many hypogonadal men are capable of sexual erections. Almost a third of men receiving effective anti-androgen therapy for prostate disease can develop erections when tested with erotic stimulation (Morales and Heaton 2003).

Previous findings with normal males may provide a clue to understanding at least some of the exceptions. In one experiment, we treated gonadally intact male rats to increase TS levels in one group by exogenous TS injections and decreased TS effectiveness by administering an antiandrogen to a second group. We found that decreased TS potency had minimal effects on sexual and aggressive behaviors despite profound suppressive effects on reproductive organs and glands (Taylor, *et al* 1985).

In an elegant experiment with healthy young men (Buena, *et al* 1993), two groups were treated with a GnRH agonist that suppresses endogenous TS production. Both groups also received TS replacement by a sustained release, long-acting microcapsule formulation to restore serum TS levels at low and high ends of the normal male range, respectively. There were no differences between the two groups on any measure of libido or copulatory performance. The most plausible explanation is that the threshold for sexual behavior is considerably below the circulating TS value of normal males (Bhasin, *et al* 2001). This may explain why some partially hypogonadal men continue to have normal sexual function and the absence of correlation between sexuality in males with serum T levels either in the high or low normal ranges.

III.2.6. Supplements that Produce Supraphysiological Increases in Testosterone do not Change Behaviors of Healthy Young Males

Inspired by the upsurge in abuse of anabolic steroids there is now a substantial literature on the behavioral outcomes of supraphysiological increases in TS in gonadally intact young males. Here, we are defining supraphysiological as being in the highest values (circa 1000 ng/ dl), or slightly above these levels, of TS found in untreated, healthy men.

There are various reasons to predict that additional TS to men with normal endocrine function, the so-called eugonadal condition, would not change sociosexual behaviors. One is that androgen receptors in most tissues are either saturated or downregulated at physiological TS concentrations (Antonio, *et al* 1999). Additional TS would have no more binding sites.

Also, as cited in the previous section, the evidence points to the level of TS required for sexual interest and activity in adult males is lower than normal males' circulating levels of

TS. The suggestion is that most males already are at high functional levels of circulating TS. Finally, the threshold model discussed earlier certainly predicts that exogenously induced TS changes above the threshold level would not be expected to influence sexual interest or behavior (Schiavi and Segraves 1995).

The data from men serving as subjects in biomedical experiments provide considerable support for the prediction. A meta-analysis concluded that supplements of TS had no effect on erectile function in eugonadal men compared to placebo (Isidori, *et al* 2005). In a similar paradigm, healthy young men with normal levels of TS were administered supplemental hormone to boost levels to supraphysiological TS values. The men showed an increased self-reported interest in sex during TS treatment. These changes, however, were not reflected in any change in overt sexual behavior, which in eugonadal men may be more determined by sexual relationship factors (Anderson, *et al* 1992).

Our work with gonadally intact male rats provides additional support. Additional hormone does not uniformly increase copulation in sexually experienced males, although TS therapy increased sexual performance of virgin males (Taylor, *et al* 1985).

The conclusion is that supplementation that increases TS to physiological, albeit high physiological, levels is unlikely to increase sexual performance. However, the treatment can increase sexual interest. Supplements that increase TS levels to pharmacologic levels may be quite another story.

III.2.7. Steroid Abuse Raises Testosterone to Pharmacological Levels with Increases in Aggressiveness and Perhaps Sexual Motivation

An important feature of the supraphysiological research cited above is that the TS supplements to eugonadal men were of minimal to modest dosages and of short durations. This is dictated by the consideration of human researchers bound by ethical considerations to keep the TS dosages to the participants in ranges that are unlikely to endanger their health. Even many researchers using animal models attempt to maintain as near physiological titers of steroid as possible to provide data that can be generalized to natural populations.

There are people outside the usual biomedical research protocols, however, who experience chronic, and greatly exaggerated, pharmacologic levels of circulating TS. These are, of course, professional or amateur athletes and others who abuse anabolic steroids. These compounds are more accurately identified as androgenic anabolic steroids (AAS) to indicate they have the significant androgenic effects of TS and muscle building (anabolic) features. AAS are synthetic derivatives of TS. Indeed, many AAS were developed by pharmaceutical firms as possible TS therapies with different characteristics (Clark and Henderson 2003).

The most potent AAS are injected to avoid the dangers of first-pass metabolism, with the result that androgen titers are chronically high and pharmacological. It is not uncommon for these values to be 20x or more of the highest, physiological values. Because these individuals are engaged in criminal activity, they are off-limits to responsible researchers. Thus, there are only informal, non-systematic data available on people abusing steroids.

Adding to the uncertainty of information on anabolic steroid abuse is that results of the supraphysiological biomedical studies cited in the previous section may not accurately reflect the outcomes of the high pharmacological dosages common among abusers of steroids (Yates, *et al* 1999). Fortunately, there are systematic data available from animal models administered high doses to mimic the pharmacological levels of steroid abusers (Blanco, *et al* 1997).

The behavioral results from these animal experiments suggest that AAS are more likely to elevate aggression than sexuality. For example, injection of TS propionate over 10 weeks to intact adult male rats increased inter-male aggression significantly with less effect on copulation (Lumia, *et al* 1994). Aggression was also increased toward the source of a tail pinch in intact male rats treated with high doses of TS for 12 weeks (McGinnis, *et al* 2002). Finally, a study of a highly strain of mice administered TS observed no further increases in their intact adults. Females of the strain, however, showed more aggression to the same dosage of TS (Bronson 1996). A recent review (Clark and Henderson 2003) of the animal literature concluded that AAS has minimal effects on sexual behavior of intact rats, although there were a few exceptions for animals treated chronically with very high dosages.

The available human data suggests that very high doses of TS may, in fact, increase both some aspects of sociosexual behaviors (Lukas 1993). Sexual performance in the form of impotency may be a significant risk factor of pharmacological dosages (Eklof, *et al* 2003), but heightened aggressiveness and libido may be increased. A 2-week, double-blind placebo-controlled crossover study administered high doses of methyltestosterone to 20 young, healthy volunteers. Results were statistically significant increases in sexual arousal, energy and irritability (Su, *et al* 1993). Self-reports by AAS abusing athletes have suggestion a similar conclusion (Clark and Henderson 2003). These results point to AAS agents that induce long-term pharmacologic elevation of systemic TS almost surely increase aggressive behaviors and also may increase sexual motivation. Sexual performance, however, is likely to suffer declines.

CONCLUSION

To summarize, we offer the following general principles of testosterone – sociosexual behavior relations. 1) The evidence is overwhelming that testosterone is critical for the organization and activation of sexual responses of the male. 2) In males, and likely also in females, testosterone is responsible for sexual motivation. 3) Probably the testosterone metabolites, DHT and especially E2, are ultimately responsible for much of the activation of brain system underlying the sociosexual behaviors of males. 4) Resting testosterone levels are unlikely to be related to individual differences in behavior, but momentary and, especially, chronic changes in testosterone in circulation are directly associated with changes in sociosexual behaviors. 5) Steroid supplements that increase circulating testosterone to high physiological levels are likely to improve the sexual well-being of hypogonadal and aged men, but anabolic steroids that produce pharmacologic testosterone levels are unlikely to produce beneficial behavioral outcomes. 6) The distribution of steroid receptors in brain areas unrelated to reproduction, along with the non-genomic effects of neuroactive steroids and neurosteroids, point to the likelihood that we have only scratched the surface of the impact steroids have on brain morphology and function, and behavior. The neuroendocrinology of testosterone and the other steroids are likely to have a most interesting future.

REFERENCES

Abbud, R.A., Ameduri, R.K., Rao, J.S., Nett, T.M. & Nilson, J.H. (1999) Chronic hypersecretion of luteinizing hormone in transgenic mice selectively alters responsiveness of the alpha-subunit gene to gonadotropin-releasing hormone and estrogens. *Molecular Endocrinology, 13,* 1449-1459.

Adkins-Regan, E. (1981) Hormone specificity, androgen metabolism, and social behavior. *American Zoologist, 21,* 257-271.

Akingbemi, B.T. (2005) Estrogen regulation of testicular function. *Reproductive Biology and Endocrinology, 3,* doi: 10.1186/1477-7827-1183-1151.

Aluja, A. & Torrubia, R. (2004) Hostility-aggressiveness, sensation seeking, and sex hormones in men: Re-exploring their relationship. *Neuropsychobiology, 50,* 102-108.

Anderson, R.A., Bancroft, J. & Wu, F.C. (1992) The effects of exogenous testosterone on sexuality and mood of normal men. *Journal of Clinical Endocrinology & Metabolism, 75,* 1503-1507.

Antonio, J., Wilson, J.D. & George, F.W. (1999) Effects of castration and androgen treatment on androgen-receptor levels in rat skeletal muscles. *Journal of Applied Physiology, 87,* 2016–2019.

Archer, J. (1994) Testosterone and aggression. In: *The Psychobiology of Aggression* (ed. by M. Hillbrand & N.J. Pallone), pp. 3-25. The Haworth Press, New York.

Arlt, W., Haas, J., Callies, F., Reincke, M., Hubler, D., Oettel, M., Ernst, M., Schulte, H. & Allolio, B. (1999) Biotransformation of oral dehydroepiandrosterone in elderly men: significant increase in circulating estrogens. *Journal of Clinical Endocrinology & Metabolism, 84,* 2170-2176.

Arnold, A.P. & Gorski, R.A. (1984) Gonadal steroid induction of structural sex differences in the central nervous system. *Annual Review of Neuroscience, 7,* 413-442.

Arver, S., Dobs, A.S., Meikle, A.W., Allen, R.P., Sanders, S.W. & Mazer, N.A. (1996) Improvement of sexual function in testosterone deficient men treated for 1 year with a permeation enhanced testosterone transdermal system. *Journal of Urology, 155,* 1604-1608.

Ashmed, S.F., Boucher, A., Manni, A., Santen, R., Bartholomew, M. & Demers, L. (1988) Transdermal testosterone therapy in the treatment of male hypogonadism. *Journal of Clinical Endocrinology & Metabolism, 66,* 546-551.

Bachmann, G.A. (2002) The hypoandrogenic woman; pathophysiologic overview. *Fertility and Sterility, 77,* S72-S76

Ball, G.F. & Balthazart, J. (2006) Androgen metabolism and the activation of male sexual behavior: It's more complicated than you think! *Hormones and Behavior, 49,* 1-3.

Balthazart, J. & Ball, G.F. (1998) New insights into the regulation and function of brain estrogen synthase (aromatase). *Trends in Neuroscience, 21,* 243-249.

Bancroft, J. (2005) The endocrinology of sexual arousal. *Journal of Endocrinology, 186,* 411-427.

Bartke, A. & Dalterio, S. (1975) Episodic secretion of testosterone in laboratory mice. *Science., 100,* 1-5.

Baulieu, E.E. (1997) Neurosteroids: of the nervous system, by the nervous system, for the nervous system. *Recent Progress in Hormone Research, 52,* 1-32.

Baulieu, E.E. (1998) Neurosteroids: a novel function of the brain. *Psychoneuroendocrinology, 23,* 963-987.

Baum, M., Melamed, E. & Globus, M. (1986) Dissociation of the effects of castration and testosterone replacement on sexual behavior and neural metabolism of dopamine in the male rat. *Brain Research Bulletin, 16,* 145-148.

Baum, M.J. (2002) Neuroendocrinology of sexual behavior in the male. In: *Behavioral Endocrinology, 2nd Edition* (ed. by J. Becker, M. Breedlove & D. Crews), pp. 153-203. The M.I.T. Press, Cambridge, MA.

Beach, F.A. (1981) Historical origins of modern research on hormones and behavior. *Hormones & Behavior, 15,* 325-376.

Beck, S.G. & Handa, R.J. (2004) Dehydroepiandrosterone (DHEA): A misunderstood adrenal hormone and spine-tingling neurosteroid? *Endocrinology, 145,* 1039-1041.

Belelli, D. & Lamber, J.J. (2005) Neurosteroids: endogenous regulators of the GABA-A receptor. *Nature Reviews of Neuroscience 6,* 565-566.

Beyer, C., Larsson, K., Perez-Placios, G. & Morali, G. (1973) Androgen structure and male sexual behavior in the castrated rat. *Hormones & Behavior, 4,* 99-107.

Bhasin, S., Woodhouse, L., Dzekov, C. & Others, X. (2001) Testosterone dose-response relationships in healthy young men. *American Journal of Physiology: Endocrinology & Metabolism, 281,* E1172-E1181.

Bialek, M., Zaremba, P., Borowicz, K.K. & Czuczwar, S.J. (2004) Neuroprotective role of testosterone in the nervous system. *Polish Journal of Pharmacology, 56,* 509-518.

Birchard, K. (1999) Body-building supplement fails to strengthen muscle and may harm health. *Lancet, 353,* 1943-1944.

Birzniece, V., Backstrom, T., Johansson, M.A. & Others, X. (2006) Neuroactive steroid effects on cognitive functions with a focus on the serotonin and GABA systems. *Brain Research Reviews, 51,* 212-239.

Blanco, C.E., Popper, P. & Micevych, P. (1997) Anabolic-androgenic steroid induced alterations in choline acetyltransferase messenger RNA levels of spinal cord motoneurons in the male rat. *Neuroscience, 78,* 873-882.

Bonsall, R., Clancy, A. & Michael, R. (1992) Effects of the nonsteroidal aromatase inhibitor, fadrozole, on sexual behavior in male rats. *Hormones & Behavior, 26,* 240-254.

Bonsall, R.W., Rees, H. & Michael, R. (1985) The distribution, nuclear uptake and metabolism of (3H) dihydrotestosterone in the brain, pituitary gland and genital tract of the male rhesus monkey. *Journal of Steroid Biochemistry, 23,* 389-398.

Breedlove, S.M. & Hampson, E. (2002) Sexual differentiation of the brain and behavior. In: *Behavioral Endocrinology, Second Edition* (ed. by J.B. Becker, S.M. Breedlove, D. Crews & M.M. McCarthy), pp. 75-116. The MIT Press, Cambridge, MA.

Breedlove, S.M. & Jordan, C.L. (2001) The increasingly plastic, hormone-responsive adult brain. *Proceedings of the National Academy of Sciences, 98,* 2956-2957.

Bronson, F.H. (1996) Effects of prolonged exposure to anabolic steroids on the behavior of male and female mice. *Pharmacology Biochemistry & Behavior, 53,* 329–334.

Bronwyn, A.L., Crawford, C., Liu, P.Y., Kean, M.T., Bleasel, J.F. & Handelsman, D.J. (2003) Randomized placebo-controlled trial of androgen effects on muscle and bone in men requiring long- term systemic glucocorticoid treatment. *Journal of Clinical Endocrinology and Metabolism, 88,* 3167-3176.

Brown, W.A., Monti, P. & Corriveau, D. (1978) Serum testosterone and sexual activity and interest in men. *Archives of Sexual Behavior, 7,* 97-103.

Buena, F., Swerdloff, R.S., Steiner, B.S., Lutchmansingh, P., Peterson, M.A., Pandian, M.R., Galmarini, M. & Bhasin, S. (1993) Sexual function does not change when serum testosterone levels are pharmacologically varied within the normal male range. *Fertility & Sterility, 59,* 1118-1123.

Burris, A.S., Banks, S.M., Carter, S., Davidson, J.M. & Sherins, R.J. (1992) A long-term, prospective study of the physiologic and behavioral effects of hormone replacement in untreated hypogonadal men. *Journal of Andrology, 13,* 297-304.

Butera, P.C. & Czaja, J.A. (1989) Activities of sexual behavior in male rats by combined subcutaneous and intracranial treatments of 5 alpha-dihydrotestosterone. *Hormones & Behavior, 23,* 92-105.

Carlson, A.A., Young, A.J., Russell, A.F., Bennett, N.C., McNeilly, A.S. & Clutton-Brock, T. (2004) Hormonal correlates of dominance in meerkats (Suricata suricatta). *Hormones & Behavior, 46,* 141-150.

Carreau, S., Lambard, S., Delalande, C. & Others, X. (2003) Aromatase expression and role of estrogens in male gonad: a review. *Reproductive Biology and Endocrinology, 1,* 211-223.

Chambers, K.C. & Phoenix, C.H. (1992) Sexual behavior and serum levels of prolactin, testosterone, and estradiol in young and old rhesus males. *Physiology & Behavior, 52,* 13-16.

Chambers, K.C., Thornton, J. & Roselli, C. (1991) Age-related deficits in brain androgen binding and metabolism, testosterone, and sexual behavior of male rats. *Neurobiology of Aging, 12,* 123-130.

Chen, F., K., K., E., B., J., F. & Others, X. (2005) Direct agonist/antagonist functions of dehydroepiandrosterone. *Endocrinology, 146,* 4568-4576.

Chen, T. & Tu, W. (1992) Sex differences in estrogen and androgen receptors in hamster brain. *Life Sciences, 50,* 1639-1647.

Chubb, C. & Desjardins, C. (1984) Testicular function and sexual activity in senescent mice. *American Journal of Physiology, 247,* 569-573.

Clark, A.S. & Henderson, L.P. (2003) Behavioral and physiological responses to anabolic-androgenic steroids. *Neuroscience and Biobehavioral Reviews, 27,* 413-436.

Cooke, B., Hegstrom, C.D., Villeneuve, L.S. & Breedlove, S.M. (1999a) Sexual differentiation of the vertebrate brain: principles and mechanisms. *Frontiers in Neuroendocrinology, 20,* 323-362.

Cooke, B.M., Breedlove, S.M. & Jordan, C.L. (2003) Both estrogen receptors and androgen receptors contribute to testosterone-induced changes in the morphology of the medial amygdala and sexual arousal in male rats. *Hormones & Behavior, 43,* 336-346.

Cooke, B.M., Tabibnia, G. & Breedlove, S.M. (1999b) A brain sexual dimorphism controlled by adult circulating androgens. *Proceedings of the National Academy of Sciences (U.S.A.), 96,* 7538-7540.

Cooke, P.S., Buchanan, D.L., Lubahn, D.B. & Cunha, G.R. (1998) Mechanism of estrogen action: Lessons from the estrogen receptor-alpha knockout mouse. *Biology of Reproduction, 59,* 470-475.

Corona, G., Mannucci, E., Mansani, R. & Others, X. (2004) Aging and pathogenesis of erectile dysfunction. *International Journal of Impotence Research, 16,* 395-402.

Corpechot, C., Robel, P., Axelson, M., Sjovall, J. & Baulieu, E. (1981) Characterization and measurement of dehydroepiandrosterone sulfate in the rat brain. *Proceedings of the National Academy of Science, 78,* 4704-4707.

Craigen, W. & Bronson, F. (1982) Deterioration of the capacity for sexual arousal in aged male mice. *Biology of Reproduction, 26,* 869-874.

Dabbs, J. & Hargrave, K.R. (1997) Age, testosterone, and behavior among female prison inmates. *Psychosomatic Medicine, 59,* 477-480.

Damassa, D.A., Smith, E.R., Tennent, B. & Davidson, J.M. (1977) The relationship between circulating testosterone levels and male sexual behavior in rats. *Hormones & Behavior, 8,* 275-286.

Davidson, J.M. (1966) Characteristics of sex behaviour in male rats following castration. *Animal Behaviour, 14,* 266-272.

Davidson, J.M., Kwan, M. & Greenleaf, W.J. (1982) Hormonal replacement and sexuality in men. *Journal of Clinical Endocrinology and Metabolism, 11,* 599-623.

Davis, S.R. (1999) The therapeutic use of androgens in women. *Journal of Steroid Biochemistry & Molecular Biology, 69,* 177-184.

Davis, S.R., Goldstat, R., Papalia, M.A., Shah, S., Kulkarni, J., Donath, S. & Bell, R.J. (2006) Effects of aromatase inhibition on sexual function and well-being in postmenopausal women treated with testosterone: a randomized, placebo-controlled trial. *Menopause, 13,* 37-45.

Davis, S.R., McCloud, P., Strauss, B.J. & Burger, H. (1995) Testosterone enhances estradiol's effects on postmenopausal bone density and sexuality. *Maturitas, 21,* 227-236.

Dennerstein, L.M.B., Randolph, J., Taffe, J., Dudley, E. & Burger, H. (2002) Hormones, mood, sexuality, and the menopausal transition. *Fertility and Sterility, 77,* S42-S48.

Dewsbury, D.A. (1981) Effects of novelty on copulatory behavior: The Coolidge effect and related phenomena. *Psychological Bulletin, 89,* 464-482.

Dhatariya, K. (2005) Self-reported sexual function in women and androgen levels. *JAMA, 294,* 2167-2168.

Dobs, A.S., Nguyen, T., Pace, C. & Roberts, C.P. (2002) Differential effects of oral estrogen versus oral estrogen-androgen replacement therapy on body composition in postmenopausal women. *Journal of Clinical Endocrinology & Metabolism, 87,* 1509-1516.

Dominguez, J.M. & Hull, E.M. (2005) Dopamine, the medial preoptic area, and male sexual behavior. *Physiology & Behavior, 86,* 356-368.

DonCarlos, L.L., Garcia-Ovejero, D., Sarkey, S., Garcia-Segura, L.M. & Azcoitia, I. (2003) Androgen receptor immunoreactivity in forebrain axons and dendrites in the rat. *Endocrinology, 144,* 3632-3638.

Dougherty, D.M., Bjork, J.M., Moeller, F.G. & Swann, A.C. (1997) The influence of menstrual-cycle phase on the relationship between testosterone and aggression. *Physiology & Behavior, 62,* 431-435.

Eklof, A.C., Thurelius, A.M., Garle, M., Rane, A. & Sjoqvist, F. (2003) The anti-doping hot-line, a means to capture the abuse of doping agents in the Swedish society and a new service function in clinical pharmacology. *European Journal of Clinical Pharmacology, 59,* 571-577.

Engel, S.R. & Grant, K.A. (2001) Neurosteroids and behavior. *International Review of Neurobiology, 46,* 321-348.

Everitt, B.J. (1990) Sexual motivation: A neural and behavioral analysis of the mechanisms underlying appetitive and copulatory responses in male rats. *Neuroscience and Biobehavioral Reviews, 14,* 217-232.

Ewing, L., Gorski, R., Sbordane, R., Tyler, J., Desjardins, C. & Robaire, B. (1979) Testosterone-estradiol filled polydimethylsiloxane subdermal implants: Effect on fertility and masculine sexual and aggressive behavior of male rats. *Biology of Reproduction, 21,* 765-772.

Federman, D.D. (2004) Three facets of sexual differentiation. *New England Journal of Medicine, 350,* 323-324.

Fernandez-Guasti, A., Kruijver, F.P., Fodor, M. & Swaab, D.F. (2000) Sex differences in the distribution of androgen receptors in the human hypothalamus. *Journal of Comparative Neurology, 425,* 422-435.

Frankel, E. (1984) Plasma testosterone levels are higher in old mating rats than in old non-mating rats. *Experimental Gerontology, 16,* 345-348.

Frye, C.A. (2001) The role of neurosteroids and non-genomic effects of progestins and androgens in mediating sexual receptivity of rodents. *Brain Research Reviews, 37,* 201-222.

Frye, C.A. & McCormick, C.M. (2000) Androgens are neuroprotective in the dentate gyrus of adrenalectomized female rats. *Stress, 3,* 185-194.

Fukada, S., Nagashima, H., Marioka, K. & Aoki, J. (1988) Fluctuation in peripheral serum testosterone levels within a day with age and by sexual stimulation in male beagle dogs bred indoors. *Experimental Animals, 37,* 381-386.

Goldstein, I., Traish, A., Kim, N. & Munarriz, R. (2004) The role of sex steroid hormones in female sexual function and dysfunction. *Clinical Obstetrics And Gynecology, 47,* 471-484.

Goldstein, J.S. & Sites, C.K. (2002) Selective modulation of sex steroids. *Ageing Research Reviews, 1,* 17-28.

Gordon, R.J., Breedlove, M.S., Bernstein, S. & Lichtman, J.W. (1990) Neuromuscular junctions shrink and expand as muscle fiber size is manipulated: in vivo observations in the androgen-sensitive bulbocavernosus muscle of mice. *Journal of Neuroscience, 10,* 2660-2671.

Gray, G.D., Smith, E., Dorsa, D. & Davidson, J.M. (1981) Sexual behavior and testosterone in middle-aged rats. *Endocrinology, 109,* 1597-1604.

Gray, P.B., Singh, A.B., Woodhouse, L.J. & Others, X. (2005) Dose-dependent effects of testosterone on sexual function, mood, and visuospatial cognition in older men. *Journal of Clinical Endocrinology & Metabolism, 90,* 3838-3846.

Greenstein, A., Plymate, S.R. & Katz, P.G. (1995) Visually stimulated erection in castrated men. *Journal of Urology, 153,* 650-652.

Gruber, C.J., Tschugguel, W., Schneeberger, C. & Huber, J.C. (2002) Production and actions of estrogens. *New England Journal of Medicine, 346,* 340-352.

Guay, A., Munarriz, R., Jacobson, J., Talakoub, L., Traish, A., Quirk, F., Goldstein, I. & Spark, R. (2004) Serum androgen levels in healthy premenopausal women with and without sexual dysfunction: Part A. Serum androgen levels in women aged 20-49 years with no complaints of sexual dysfunction. *International Journal of Impotence Research, 16,* 112-120.

Hall, C.M., Jones, J.A., Meyer-Bahlburg, H.F., Dolezal, C., Coleman, M., Foster, P., Price, D.A. & Clayton, P.E. (2004) Behavioral and physical masculinization are related to genotype in girls with congenital adrenal hyperplasia. *Journal of Clinical Endocrinology & Metabolism, 89,* 419-424.

Haller, J., Millar, S., van de Schraaf, J., de Kloet, R.E. & Kruk, M.R. (2000) The active phase-related increase in corticosterone and aggression are linked. *Journal of Neuroendocrinology, 12,* 431-436.

Hamann, S. (2005) Sex differences in the responses of the human amygdala. *Neuroscientist, 11,* 288-293.

Haren, M.T., Morley, J.E., Chapman, I.M., O'Loughlin, P.D. & Wittert, G.A. (2002) Defining 'relative' androgen deficiency in aging men: how should testosterone be measured and what are the relationships between androgen levels and physical, sexual and emotional health?. *Climacteric, 5,* 15-25.

Hayes, F.J. (2000) Testosterone - fountain of youth or drug of abuse? *Journal of Clinical Endocrinology & Metabolism, 85,* 3020-3023.

Hijazi, R.A. & Cunningham, G.R. (2005) Andropause: Is androgen replacement therapy indicated for the aging male? . *Annual Review of Medicine, 56,* 117-137.

Hodes, G.E. & Shors, T.J. (2005) Distinctive stress effects on learning during puberty. *Hormones and Behavior, 48,* 163-171.

Howell, S. & Shalet, S. (2001) Testosterone deficiency and replacement. *Hormone Research, 56 (Suppl. 1),* 86-92.

Huang, J., Chang, L., Ho, D. & Ho, L. (1989) The effect of sexual activity on androgen metabolism, androgen-receptor translocation and glandular weight of ventral prostates of rats. *Endocrinology, 120,* 505-510.

Huhtaniemi, I. (2000) Mutations of gonadotrophin and gonadotrophin receptor genes: what do they teach us about reproductive physiology? *Journal of Reproduction and Fertility, 119,* 173–186.

Hull, E.M., Lorrain, D.S., Du, J., Matuszewich, L., Lumley, L., Putnam, S.K. & Moses, J. (1999) Hormone-neurotransmitter interactions in the control of sexual behavior. *Behavioural Brain Research, 105,* 105-116.

Isidori, A.M., Giannetta, E., Gianfrilli, D. & Others, X. (2005) Effects of testosterone on sexual function in men: results of a meta-analysis. *Clinical Endocrinology, 63,* 381-394.

Johannsson, G., Burman, P., Wiren, L. & Others, X. (2002) Low dose dehydroepiandrosterone affects behavior in hypopituitary androgen-deficient women: a placebo- controlled trial. *Journal of Clinical Endocrinology & Metabolism, 87,* 2046–2052.

Johnston, P. & Davidson, J.M. (1972) Intracerebral androgens and sexual behavior in the male rat. *Hormones & Behavior, 3,* 345-357.

Jones, K.J., Coers, S., Storer, P.D., Tanzer, L. & Kinderman, N.B. (1999) Androgenic regulation of the central glia response following nerve damage. *Journal of Neurobiology, 40,* 560-573.

Kalantaridou, S.N. & Calis, K.A. (2006) Testosterone therapy in premenopausal women. *Seminars in Reproductive Medicine, 24,* 106-114.

Keast, J.R. (1999) The autonomic nerve supply of male sex organs--an important target of circulating androgens. *Behavioural Brain Research, 105,* 81-92.

Kelliher, K.R., Liu, Y.C., Baum, M.J. & Sachs, B.D. (1999) Neuronal Fos activation in olfactory bulb and forebrain of male rats having erections in the presence of inaccessible estrous females. *Neuroscience, 92,* 1025-1033.

King, D.S., Sharp, R.L., Vukovich, M.D., Brown, G.A., Reifenrath, T.A., Uhl, N.L. & Parsons, K.A. (1999) Effect of oral androstenedione on serum testosterone and adaptations to resistance training in young men: a randomized controlled trial. *The Journal of the American Medical Association (JAMA), 281,* 2020-2028.

Knussmann, R., Christiansen, K. & Couwenbergs, C. (1986) Relations between sex hormone levels and sexual behavior in men. *Archives of Sexual Behavior, 15,* 429-445.

Kohl, J.V., Atzmueller, M., Fink, B. & Grammer, K. (2001) Human pheromones: Integrating neuroendocrinology and ethology. *Neuroendocrinology Letters, 22,* 309-321.

Kruger, T.H., Haake, P., Chereath, D., Knapp, W., Janssen, O.E., Exton, M.S., Schedlowski, M. & Hartmann, U. (2003) Specificity of the neuroendocrine response to orgasm during sexual arousal in men. *Journal of Endocrinology, 177,* 57-64.

Labrie, F., Belanger, A., Luu-The, V., Labrie, C., Simard, J., Cusan, L., Gomez, J. & Candas, B. (1998) DHEA and the intracrine formation of androgens and estrogens in peripheral target tissues: Its role during aging. *Steroids, 63,* 322-328.

Labrie, F., Luu-The, V., Belanger, A., Lin, S.X., Simard, J., Pelletier, G. & Labrie, C. (2005) Is dehydroepiandrosterone a hormone? *Journal of Endocrinology, 187,* 169-196.

Leranth, C., Hajszan, T. & MacLusky, N.J. (2004a) Androgens increase spine synapse density in the CA1 hippocampal subfield of ovariectomized female rats. *Journal of Neuroscience, 24,* 495- 499.

Leranth, C., Prange-Kiel, J., Frick, K.M. & Horvath, T.L. (2004b) Low CA1 spine synapse density is further reduced by castration in male non-human primates. *Cerebral Cortex, 14,* 503-510.

Lieberman, S. & Prasad, V. (1990) Heterodox notions on pathways of steroidogenesis. *Endocrine Rev., 11,* 469-493.

Linkowski, P., Mendlewicz, J., Leclercq, R. & Others, X. (1985) The 24-hour profile of adrenocorticotropin and cortisol in major depressive illness. *Journal of Clinical Endocrinology & Metabolism, 61,* 429-438.

Lu, S.F., McKenna, S.E., Cologer-Clifford, A., Nau, E.A. & Simon, N.G. (1998) Androgen receptor in mouse brain: sex differences and similarities in autoregulation. *Endocrinology, 139,* 1594-1601.

Lukas, S.E. (1993) Current perspectives on anabolic-androgenic steroid abuse. *Trends in Pharmacological Science, 14,* 61-68.

Lumia, A.R., Thorner, K.M. & McGinnis, M.Y. (1994) Effects of chronically high doses of the anabolic androgenic steroid, testosterone, on intermale aggression and sexual behavior in male rats. *Physiology & Behavior, 55,* 331-335.

Lupo-DiPrisco, C. & Dessi-Fulgheri, F. (1980) Endocrine and behavioral modifications in aging male rats. *Hormone Research, 12,* 149-160.

Mantzoros, C., Georgiadis, E. & Trichopoulos, D. (1995) Contribution of dihydrotestosterone to male sexual behaviour. *British Medical Journal, 310,* 1289-1291.

Martini, L., Celotti, F., Lechuga, M., Melcangi, R., Motta, M., Negri-Cesi, P., Poletti, A. & Zoppi, S. (1990) Androgen metabolism in different target tissues. *Annals New York Academy of Sciences, 595,* 184-198.

Matthews, K.A., Berga, S.L., Owens, J.F. & Flory, J.D. (1998) Effects of short-term suppression of ovarian hormones on cardiovascular and neuroendocrine reactivity to stress in women. *Psychoneuroendocrinology, 23,* 307-322.

Mazur, A., Mueller, U., Krause, W. & Booth, A. (2002) Causes of sexual decline in aging married men: Germany and America. *International Journal of Impotence Research, 14,* 101-106.

McEwen, B.S. (2001) Estrogens effects on the brain: multiple sites and molecular mechanisms. *Journal of Applied Physiology, 91,* 2785-2801.

McEwen, B.S., Gould, E., Orchinik, M., Weiland, N. & Woolley, C.S. (1995) Oestrogens and the structural and functional plasticity of neurons: implications for memory, ageing and neurodegenerative processes. In: *Non-reproductive actions of sex steroids* (ed. by C.R. Brock & J.A. Goode), pp. 52-73. John Wiley, New York.

McGinnis, M.Y., Lumia, A.R., Breuer, M.E. & Possidente, B. (2002) Physical provocation potentiates aggression in male rats receiving anabolic androgenic steroids. *Hormones & Behavior, 41,* 101-110.

McGivern, R.F., Handa, R.J. & Raum, W.J. (1998) Ethanol exposure during the last week of gestation in the rat: inhibition of the prenatal testosterone surge in males without long-term alterations in sex behavior. *Neurotoxicology & Teratology, 20,* 483-490.

Meisel, R.L. & Sachs, B.D. (1994) The physiology of male reproductive behavior. In: *The Physiology of Reproduction, Volume 2* (ed. by E. Knobil & J. Neill), pp. 3-106. Raven Press, Inc., N.Y.

Mellon, S.H. (1994) Neurosteroids: Biochemistry, modes of action, and clinical relevance. *Journal of Clinical Endocrinology & Metabolism, 78,* 1003-1008.

Meston, C.M. & Frohlich, P.F. (2000) The neurobiology of sexual function. *Archives of General Psychiatry, 57,* 1012-1030.

Meston, C.M. & Heiman, J.R. (2002) Acute dehydroepiandrosterone effects on sexual arousal in premenopausal women. *Journal of Sex & Marital Therapy, 28,* 53-60.

Michael, R.P., Bonsall, R.W. & Zumpe, D. (1987) Testosterone and its metabolites in male cynomologus monkeys (Macaca fascicularis behavior and biochemistry. *Physiology & Behavior, 40,* 527-537.

Michael, R.P., Clancy, A.N. & Zumpe, D. (1995) Distribution of androgen receptor-like immunoreactivity in the brains of cynomolgus monkeys. *Journal of Neuroendocrinology, 7,* 713-719.

Michael, R.P., Zumpe, D. & Bonsall, R.W. (1984) Sexual behavior correlates with the diurnal plasma testosterone range in intact male rhesus monkeys. *Biology of Reproduction, 30,* 652-657.

Miller, K., Sesmilo, G., Schiller, A., Schoenfeld, D., Burton, S. & Klibanski, A. (2001) Androgen deficiency in women with hypopituitarism. *Journal of Clinical Endocrinology & Metabolism, 86,* 561-567.

Miller, K.K. (2002) The androgen deficiency syndrome in women. *Endocrinology Rounds, 1,* 1-8.

Moffat, S.D., Zonderman, A.B., Metter, E.J., Kawas, C., Blackman, M.R., Harman, S.M. & Resnick, S.M. (2004) Free testosterone and risk for Alzheimer disease in older men. *Neurology, 62,* 188-193.

Morales, A. & Heaton, J.P. (2003) Hypogonadism and erectile dysfunction: pathophysiological observations and therapeutic outcomes. *BJU International, 92,* 896-899.

Nakamura, N., Fujita, H. & Kawata, M. (2002) Effects of gonadectomy on immunoreactivity for choline acetyltransferase in the cortex, hippocampus, and basal forebrain of adult male rats. *Neuroscience, 109,* 473-485.

Nottebohm, F. & Arnold, A.P. (1976) Sexual dimorphism in vocal control areas of the songbird brain. *Science, 194,* 211–213.

Nyunt, A., Stephen, G., Gibbin, J., Durgan, L., Fielding, A.M., Wheeler, M. & Price, D.E. (2005) Androgen status in healthy premenopausal women with loss of libido. *Journal of Sex & Marital Therapy, 31,* 73-80.

O'Carroll, R.E. & Bancroft, J. (1984) Testosterone therapy for low sexual interest and erectile dysfunction in men: a controlled study. *British Journal of Psychiatry, 145,* 146-151.

O'Connor, D.B., Archer, J. & Wu, F.C.W. (2004) Effects of testosterone on mood, aggression, and sexual behavior in young men: a double-blind, placebo- controlled, cross-over study. *Journal of Clinical Endocrinology & Metabolism, 89*, 2837-2845.

Occhiato, E.G., Guarna, A., Danza, G. & Serio, M. (2004) Selective non-steroidal inhibitors of 5alpha-reductase type 1. *Journal of Steroid Biochemistry & Molecular Biology, 88,* 1-16.

Payne, A.H. & Hales, D.B. (2004) Overview of steroidogenic enzymes in the pathway from cholesterol to active steroid hormones. *Endocrine Reviews, 25,* 947-970.

Penny, R. (1985) Episodic secretion of melatonin in pre-and post-pubertal girls and boys. *Journal of Clinical Endocrinology & Metabolism, 60,* 751-756.

Pfaus, J.G. (1996) Homologies of animal and human sexual behaviors. *Hormones & Behavior, 30,* 187-200.

Phoenix, C.H. & Chambers, K.C. (1986) Threshold for behavioral response to testosterone in old castrated male rhesus macaques. *Biology of Reproduction. 35,* 918-926.

Phoenix, C.H., Dixson, A.F. & Resko, J.A. (1977) Effects of ejaculation on levels of testosterone, cortisol, and luteinizing hormone in peripheral plasma of rhesus monkeys. *Journal of Comparative & Physiological Psychology, 91,* 120-127.

Phoenix, C.H., Goy, R.W., Gerall, A.A. & Young, W.C. (1959) Organizing action of prenatally administered testosterone propionate on the tissues mediating mating behavior in the female guinea pig. *Endocrinology, 65,* 369-382.

Piroli, G.G., Pietranera, L., Grillo, C.A. & DeNicola, A.F. (2002) Dehydroepiandrosterone regulation of prolactin gene expression in the anterior pituitary does not depend on galanin induction. *Neuroendocrinology Letters, 23,* 321-324.

Resko, J.A. & Roselli, C.E. (1997) Prenatal hormones organize sex differences of the neuroendocrine reproductive system: observations on guinea pigs and nonhuman primates. *Cellular & Molecular Neurobiology, 17,* 627-648.

Roselli, C.E. & Resko, J.A. (1997) Sex differences in androgen-regulated expression of cytochrome P450 aromatase in the rat brain. *Journal of Steroid Biochemistry & Molecular Biology, 61,* 365-374.

Rosler, A. & Witztum, E. (1998) Treatment of men with paraphilia with a long-acting analogue of gonadotropin-releasing hormone. *New England Journal of Medicine, 338,* 416-422.

Rupprecht, R. & Holsboer, F. (1999) Neuropsychopharmacological properties of neuroactive steroids. *Steroids, 64,* 83-91.

Sachs, B.D. (1997) Erection evoked in male rats by airborne scent from estrous females. *Physiology & Behavior, 62,* 921-924.

Sachs, B.D., Glater, G.B. & O'Hanlon, J.K. (1984) Morphology of the erect penis in rats under various gonadal hormone conditions. *Anatomical Record, 210,* 45-52.

Saltzman, E. & Guay, A. (2006) Dehydroepiandrosterone therapy as female androgen replacement. *Seminars in Reproductive Medicine, 24,* 97-105.

Samuels, M., Veldhuis, J., Henry, P. & Ridgway, E. (1990) Pathophysiology of pulsatile and copulsatile release of thyroid-stimulating hormone, luteinizing hormone, follicle-stimulating hormone, and alpha-subunit. *Journal of Clinical Endocrinology & Metabolism, 71,* 425-432.

Sarkola, T., Makisalo, H., Fukunaga, T. & Eriksson, C.J. (1999) Acute effect of alcohol on estradiol, estrone, progesterone, prolactin, cortisol, and luteinizing hormone in premenopausal women. *Alcoholism: Clinical & Experimental Research, 23,* 976-982.

Sarrel, P., Dobay, B. & Wiita, B. (1998) Estrogen and estrogen-androgen replacement in postmenopausal women dissatisfied with estrogen-only therapy. Sexual behavior and neuroendocrine responses. *Journal of Reproductive Medicine, 43,* 847-856.

Schiavi, R.C., Schreiner, P., White, D. & Mandeli, J. (1991) The relationship between pituitary-gonadal function and sexual behavior in healthy aging men. *Psychosomatic Medicine, 53,* 363-374.

Schiavi, R.C. & Segraves, R.T. (1995) The biology of sexual function. *Psychiatric Clinics of North America, 18,* 7-23.

Schiavi , R.C. & White, D. (1976) Androgens and male sexual function: a review of human studies. *Journal of Sex & Marital Therapy, 2,* 214-228.

Schiavi, R.C., White, D., Mandeli, J. & Levine, A.C. (1997) Effect of testosterone administration on sexual behavior and mood in men with erectile dysfunction. *Archives of Sexual Behavior, 26,* 231-241.

Scott, C., Kuehl, D., Ferreira, S. & Jackson, G. (1997) Hypothalamic sites of action for testosterone, dihydrotestosterone, and estrogen in the regulation of luteinizing hormone secretion in male sheep. *Endocrinology, 138,* 3686-3694.

Scott, C.J., Tilbrook, A.J., Rawson, J.A. & Clarke, I.J. (2000) Gonadal steroid receptors in the regulation of GnRH secretion in farm animals. *Animal Reproduction Science, 60-61,* 313-326.

Seidman, S.N. (2003) Testosterone deficiency and mood in aging men: pathogenic and therapeutic interactions. *World Journal of Biological Psychiatry, 4,* 14-20.

Sharpe, R.M. (1998) The roles of oestrogen in the male. *Trends in Endocrinology & Metabolism, 9,* 371-377.

Sherwin, B.B., Gelfand, M.M. & Brender, W. (1985) Androgen enhances sexual motivation in females: a prospective, crossover study of sex steroid administration in the surgical menopause. *Psychosomatic Medicine, 47,* 339-351.

Shifren, J.L. (2004) The role of androgens in female sexual dysfunction. *Mayo Clinic Proceedings, 79 (Supplement),* S19-S24.

Shors, T.J., Chua, C. & Falduto, J. (2001) Sex differences and opposite effects of stress on dendritic spine density in the male versus female hippocampus. *Journal of Neuroscience, 21,* 6292-6297.

Shughrue, P.J., Scrimo, P.J. & Merchenthaler, I. (2000) Estrogen binding and estrogen receptor characterization (ERalpha and ERbeta) in the cholinergic neurons of the rat basal forebrain. *Neuroscience, 96,* 41-49.

Simerly, R.B. (2002) Wired for reproduction: organization and development of sexually dimorphic circuits in the mammalian forebrain. *Annual Review of Neuroscience, 25,* 507-536.

Sodersten, P., Enertoh, P. & Petersson, A. (1983) Episodic secretion of luteinizing hormone and androgen in male rats. *Journal of Endocrinology, 97,* 145-153.

Somboonporn, W., Davis, S., Seif, M.W. & Bell, R. (2005) Testosterone for peri- and postmenopausal women. *Cochrane Database of Systematic Reviews, 2005,* DOI: 10.1002/14651858.CD14004509.pub14651852.

Southworth, M.B., Matsumoto, A., Gross, K., Soules, M. & Bremner, W. (1991) The importance of signal pattern in the transmission of endocrine information: Pituitary gonadotropin responses to continuous and pulsatile gonadotropin-releasing hormone. *Journal of Clinical Endocrinology and Metabolism, 72,* 1286-1289.

Stalenheim, E.G., Eriksson, E., von Knorring, L. & Wide, L. (1998) Testosterone as a biological marker in psychopathy and alcoholism. *Psychiatry Research, 77,* 79-88.

Sternbach, H. (1998) Age-associated testosterone decline in men: Clinical issues for psychiatry. *American Journal of Psychiatry, 155,* 1310-1318.

Stoffel-Wagner, B. (2001) Neurosteroid metabolism in the human brain. *European Journal of Endocrinology, 145,* 669-679.

Stone, T.H., Winslade, W.J. & Klugman, C.M. (2000) Sex offenders, sentencing laws and pharmaceutical treatment: a prescription for failure. *Behavioral Sciences & the Law, 18,* 83-110.

Su, T.P., Pagliaro, M., Schmidt, P.J., Pickar, D., Wolkowitz, O. & Rubinow, D.R. (1993) Neuropsychiatric effects of anabolic steroids in male normal volunteers. *JAMA The Journal of the American Medical Association, 269,* 2760-2764.

Swerdloff, R.S. & Wang, C. (2004) Androgens and the ageing male. *Best Practice & Research Clinical Endocrinology & Metabolism, 18,* 349-362.

Taylor, G., Bardgett, M., Farr, S., Humphrey, W., Womack, S. & Weiss, J. (1996) Aging of the brain-testicular axis: Reproductive systems of healthy old males with or without endocrine stimulation. *Proceedings of the Society of Experimental Biology & Medicine, 211,* 69-75.

Taylor, G.T. (1980a) Fighting in juvenile rats and the ontogeny of agonistic behavior. *Journal of Comparative & Physiological Psychology, 94,* 953-961.

Taylor, G.T. (1980b) Sex pheromones and aggressive behavior in male rats. *Animal Learning & Behavior, 8,* 485-490.

Taylor, G.T., Bardgett, M.E., Griffin, M.G. & Weiss, J. (1989) Reproductive behavior and physiology of nulliparous female rats after sexual experience. *Behavioural Processes, 19,* 31-46.

Taylor, G.T., Griffin, M.G. & Bardgett, M.E. (1991) Search for a male contraceptive: the effect of gossypol on sexual motivation and epididymal sperm. *Journal of Medicine, 22,* 29-44.

Taylor, G.T., Haller, J. & Regan, D. (1982) Female rats prefer an area vacated by a high testosterone male. *Physiology & Behavior, 28,* 953-958.

Taylor, G.T., Haller, J., Rupich, R. & Weiss, J. (1984) Testicular hormones and intermale aggressive behaviour in the presence of a female rat. *Journal of Endocrinology, 100,* 315-321.

Taylor, G.T. & Pitha, J. (1988) Hormone-cyclodextrin complexes in behavioral endocrinology: Simulating natural episodic release during therapy. *Biomaterials, Artificial Cells & Artificial Organs, 15,* 759-760.

Taylor, G.T., Scherrer, J., Weiss, J. & Pitha, J. (1994) Endocrine interactions: Adrenal steroids and precursors. *American Journal of Physiology, 266,* E676-E681.

Taylor, G.T. & Weiss, J. (1987) Behaviour and fecundity of female rats mated with preferred and non-preferred males. *Animal Behaviour, 35,* 115-121.

Taylor, G.T., Weiss, J. & Rupich, R. (1985) Suprathreshold manipulations of testosterone and reproductive functioning in gonadally intact sexually experienced and inexperienced male rats. *Physiology & Behavior, 35,* 735-739.

Timiras, P. & Cons, J.M. (1982) Hormones during prenatal and neonatal development. In: *Hormones in Development and Aging* (ed. by A. Vernadakis & P. Timiras), pp. 25-79. Spectrum Publications: Medical & Scientific Books, New York.

Tuiten, A., Van Honk, J., Koppeschaar, H., Bernaards, C., Thijssen, J. & Verbaten, R. (2000) Time course of effects of testosterone administration on sexual arousal in women. *Archives of General Psychiatry, 57,* 149-153.

VandePoll, N.E., Taminiau, M.S., Endert, E. & Louwerse, A.L. (1988) Gonadal steroid influence upon sexual and aggressive behavior of female rats. *International Journal of Neuroscience, 47,* 271-286.

Vanderschueren, D. & Bouillon, R. (2000) Estrogen deficiency in men is a challenge for both the hypothalamus and pituitary. *Journal of Clinical Endocrinology & Metabolism, 85,* 3604-3606.

Wade, J. (2006) Relationships among hormones, brain and behavior: Exceptions in search of a rule? *Hormones and Behavior, 49,* 577-579.

Weiss, J., Taylor, G.T., Zimmermann, F. & Nebenahl, K. (2000) Collection and analyses of biological fluids. In: *Handbook of experimental laboratory animals* (ed. by F. Zimmermann), pp. 485-510. Academic Press, New York.

Welt, C.K., Jimenez, Y., Sluss, P.M., Smith, P.C. & Hall, J.E. (2006) Control of estradiol secretion in reproductive ageing. *Human Reproduction, 21,* 2189-2193.

Wespes, E. & Schulman, C.C. (2002) Male andropause: myth, reality, and treatment. *International Journal of Impotence Research, 14,* S93-98.

Wickham, L.A., Gao, J., Toda, I., Rocha, E.M., Ono, M. & Sullivan, D.A. (2000) Identification of androgen, estrogen and progesterone receptor mRNAs in the eye. *Acta Ophthalmologica Scandinavica, 78,* 146-153.

Widstrom, R.L. & Dillon, J.S. (2004) Is there a receptor for dehydroepiandrosterone or dehydroepiandrosterone sulfate? *Seminars in Reproductive Medicine, 22,* 289-298.

Williams, J.R. (2000) The effects of dehydroepiandrosterone on carcinogenesis, obesity, the immune system, and aging. *Lipids, 35,* 325-331.

Wilson, J.D. (2000) Endocrinology: Survival as a discipline in the 21st century? *Annual Review of Physiology,* 947–950.

Wilson, J.D., Leihy, M.W., Shaw, G. & Renfree, M.B. (2002) Androgen physiology: unsolved problems at the millennium. *Molecular & Cellular Endocrinology, 198,* 1-5.

Woolley, C.S. (1998) Estrogen-mediated structural and functional synaptic plasticity in the female rat hippocampus. *Hormones & Behavior, 34,* 140-148.

Woolley, C.S., McEwen, B.S. & Beatty, W.W. (1993) Roles of estradiol and progesterone in regulation of hippocampal dendritic spine density during the estrous cycle in the rat. *Journal of Comparative Neurology, 336,* 293-306.

Xiao, L. & Jordan, C.L. (2002) Sex differences, laterality, and hormonal regulation of androgen receptor immunoreactivity in rat hippocampus. *Hormones & Behavior, 42,* 327-336.

Yates, W.R., Perry, P.J., MacIndoe, J., Holman, T. & Ellingrod, V. (1999) Psychosexual effects of three doses of testosterone cycling in normal men. *Biological Psychiatry, 45,* 254-260.

Young, E.A., Midgley, A.R., Carlson, N.E. & Brown, M.B. (2000) Alteration in the hypothalamic-pituitary-ovarian axis in depressed women. *Archives of General Psychiatry, 57,* 1157-1162.

Zhang, J.Q., Cai, W.Q., Zhou de, S. & Su, B.Y. (2002) Distribution and differences of estrogen receptor beta immunoreactivity in the brain of adult male and female rats. *Brain Research, 935,* 73-80.

Zitzmann, M. & Nieschlag, E. (2001) Testosterone levels in healthy men and the relation to behavioural and physical characteristics: facts and constructs. *European Journal of Endocrinology, 144,* 183-197.

In: Psychoneuroendocrinology Research Trends
Editor: Martina T. Czerbska, pp. 117-148
ISBN: 978-1-60021-665-7

Chapter III

SALIVARY ALPHA-AMYLASE AS A MARKER FOR STRESS

Urs M. Nater*
Department of Psychiatry and Behavioral Sciences, Emory University School of Medicine, Atlanta, Georgia 30322, USA.

ABSTRACT

Description and evaluation of new biomarkers in psychoneuroendocrinology is a constantly evolving field. Salivary biomarkers have received special attention since they are readily accessible and easily obtained. Salivary alpha-amylase has been proposed to be a sensitive biomarker for stress-related changes in the body, and a growing body of research is accumulating showing the validity and reliability of this parameter. This chapter attempts to describe salivary alpha-amylase as an emerging biomarker for stress and provides an overview of the current literature on stress-related alterations in salivary alpha-amylase. It is critically discussed how salivary alpha-amylase might reflect changes of the autonomic nervous system. New approaches in the measurement of salivary alpha-amylase and potential sources of confounding measurement results are identified. Finally, current and future fields of the application of salivary alpha-amylase measurement are outlined.

INTRODUCTION

In recent years, measures obtained from body fluids have become increasingly important in psychoneuroendocrinological research. The three body fluids which are most easily obtainable for analysis are plasma, urine and saliva. However, some problems with regard to measurement might invalidate the results one wants to gain. For example, to obtain a sufficient number of plasma samples either several venepunctures or insertion of an indwelling catheter is demanded. Through their traumatic nature, these procedures might cause by themselves a certain degree of stress and therefore confound results. Furthermore, the presence of trained medical personnel is required. In contrast, urine samples can be obtained non-invasively and are painless. But problems may arise due to incomplete voiding of urine from the bladder, leading to the mixing of urine formed during different time spans. The temporal resolution of changes in parameters gained by urine cannot be satisfactorily displayed, since the filling of the bladder requires some time. Also, a certain amount of privacy has to be assured for the subjects to collect samples. This intimacy coupled with urinary sampling often acts as a deterrent to subjects.

On the other hand, saliva samples can be collected in a completely non-traumatic manner at discrete time points from subjects in places and situations which do not require privacy. In the early seventies, Brown suggested changes of saliva parameters to be regarded as an "index of specific states of psychopathology, and as a sensitive measure in the evaluation of the effectiveness of psychotherapy and chemotherapy in psychiatric patients" (Brown, 1970, p. 66). Some effort has been undertaken to establish salivary parameters as useful measures in psychophysiology. However, apart from the analysis of hormones in saliva, such as cortisol and DHEA (see e.g. Kirschbaum & Hellhammer, 1994; Vining & McGinley, 1987), not many other salivary components have been taken into consideration as meaningful physiological markers in psychoneuroendocrinological research.

A specific area of interest within psychoneuroendocrinological research is focused on stress research. Stress is not only a ubiquitous phenomenon determining the course of modern life, but may occasionally become detrimental if encountered and/or experienced in too high a dose or too frequently. It is therefore the goal of stress research to better understand this burden and to develop measures to assess it. However, stress has proven to be an elusive concept. There are many different concepts of stress that have been formulated in the last decades. Concomitantly, operationalization of stress-related changes in both mind and body is not easy to develop. Since there is a need for the examination of stress, academic interest and research attention are focused on this phenomenon. In stress research, subjects might be examined either in the field or in the laboratory. Wherever the setting is chosen, valid and reliable measures for changes that are associated with stress must be applied. But not only validity and reliability are an issue; also simple handling and easy obtainment of a stress measure is of utmost importance. Therefore, a wide array of possible parameters indicating stress-related changes in subjects has been proposed. Some of them have disappeared into oblivion, others have endured decades and are still used. Although there are a number of parameters used as indicators for changes of the sympathetic nervous system, for example, these parameters do not correlate well with each other (Grassi & Esler, 1999). Thus, there seem to be differential processes that are responsible for this observation. Since stress is a multi-faceted phenomenon, it requires a multidimensional measurement approach. As a consequence, the canon of psychobiological parameters should be enlarged. One parameter

* Correspondence concerning this article should be addressed to Urs M. Nater, PhD, Department of Psychiatry and Behavioral Sciences, Emory University School of Medicine, 101 Woodruff Circle, Atlanta, Georgia 30322, USA. Tel: (404) 712-8515, Fax: (404) 727-3233, Email: unater@emory.edu

that has been suggested to reflect stress-related changes in our body is the salivary enzyme alpha-amylase. Salivary alpha-amylase release is known to be elicited by activation of the autonomic nervous system which controls the salivary glands. These function both as the site of production and storage of salivary alpha-amylase. Physiological stressors as well as psychological stressors have been used to show stress-related changes in salivary alpha-amylase. However, despite the existence of numerous studies examining this relationship, conflicting results have been reported. Some of these inconsistencies might be explained by the different stressors used, others by methodological factors such as sample size, measurement issues, etc.

The aim of this chapter is to undertake a careful assessment of whether salivary alpha-amylase is suited as a parameter worth being incorporated in a canon of psychoneuroendocrinological parameters measuring stress.

What is Salivary Alpha-Amylase

Alpha-amylase is one of the major organic constituents of saliva. The enzyme was first described by Leuchs in saliva in 1831. It was later found in serum by Magendie in 1846 and in urine by Cohnheim in 1863. In 1929, Elman's description of the association of elevated levels of amylase in blood with acute pancreatitis led to greater interest in the measurement and clinical utility of alpha-amylase (Zakowski & Bruns, 1985). Salivary alpha-amylase (α-1,4-α-D-glucan 4-glucanohydrolase; EC 3.2.1.1) is one of the most important enzymes in saliva. It consists of two families of isoenzymes of which one set is glycosylated and the other contains no carbohydrate. The isoenzymes contain a varying number of amino groups and they can, therefore, be separated using proper electrophoretic techniques. The molecular weight of the glycosylated form is about 57'000, whereas the molecular weight of the nonglycosylated form is about 54'000. Alpha-amylase accounts for 40 to 50% of the total salivary gland-produced protein, most of the enzyme being synthesized in the parotid gland (80% of the total) (Makinen, 1989; Zakowski & Bruns, 1985).

Alpha-amylase is a calcium-containing metalloenzyme that hydrolyzes the alpha-1,4 linkages of starch to glucose and maltose. Traditionally, this enzyme was thought to be mainly involved in the initiation of the digestion of starch in the oral cavity. However, evidence obtained over the last several years has shown that alpha-amylase has an important bacterial interactive function. A review of its protective activities can be obtained from Scannapieco, Torres and Levine (1993). Other than pancreatic and salivary alpha-amylase, alpha-amylase activity has also been reported in a variety of other normal tissues and fluids. These include lung, sweat, leukocytes and thrombocytes, colostrum and milk, tears, tonsils, thyroid, liver, bile, endometrium, semen, fallopian tubes, amniotic fluid, cervical mucosa, intestinal mucosa, prostate gland, seminal vesicles, and the female genital tract (Kasperczyk et al., 2001; Zakowski & Bruns, 1985).

Pharmacologically Induced Alpha-Amylase Secretion

Classic neurotransmitters and specific bioactive peptides serve as the main stimuli of salivary alpha-amylase secretion. In acinar cells, release of salivary components is under control of neuronal stimuli. Acinar cells are innervated by both the sympathetic and the parasympathetic branches of the autonomic nervous system (ANS). Autonomic nerves are adjacent to both acinar and ductal cells, suggesting that they may have a role in regulating functional responses in all salivary cell types. In the following section, studies in rats and humans are presented that used blocking or stimulating pharmacological agents or electrical stimuli to better understand alpha-amylase secretion.

In a study on slices of the parotid gland in rats, Batzri and his co-workers found that conditions which leave the alpha-receptor dormant but fully activate beta-adrenergic receptors caused almost total secretion of amylase from the slices (Batzri & Selinger, 1973; Batzri, Selinger, Schramm, & Robinovitch, 1973; Selinger, Batzri, Eimerl, & Schramm, 1973). These findings are among the first indications for the specific involvement of the ANS in the secretion process of salivary alpha-amylase, with a specific focus on beta-adrenergic mechanisms. Later studies were adding electrical stimulation of the salivary glands to the research of amylase secretion processes. Anderson et al. (1984) examined contributions of the two branches of the ANS to salivary alpha-amylase secretion. They found that sympathetic stimulation in unconscious rats led to the secretion of parotid saliva characterized by low salivary flow rate and high total protein and amylase concentrations, whereas parasympathetic stimulation induced a rich flow of saliva with low protein content, with mean concentrations of amylase being approximately 1% of those in sympathetically stimulated saliva. After performing adrenalectomy, the authors observed significant reductions of amylase levels after parasympathetic stimulation of the parotid gland. The authors interpreted these findings by suggesting that circulating catecholamines (originating from the adrenals) might play a specific role in alpha-amylase secretion. Asking (1985) set out to compare sympathetic and parasympathetic stimulation of the parotid gland in the rat both separately and concomitantly. As in the previously mentioned study, sympathetic stimulation caused a slow flow of saliva containing amylase in very high concentration; the parasympathetic saliva had a low concentration of amylase. However, after combined sympathetic and parasympathetic stimulation, amylase secretion was highly increased, much higher than the separate stimulation concentrations taken together. By using the beta-1-antagonist pafenolol it was shown that the higher alpha-amylase secretion due to sympathetic stimulation superimposed on parasympathetic background stimulation was elicited via beta-1-adrenoceptors, thus further characterizing the important role of beta-adrenergic involvement (Asking, 1985). In a subsequent study (Asking & Gjorstrup, 1987), these findings were replicated in that sympathetic and parasympathetic stimulation led to equal alpha-amylase concentrations and that the combination of both was drastically enhancing alpha-amylase concentrations. The authors were specifically interested in salivary alpha-amylase synthesis in the rat parotid gland. Synthesis was stimulated by both parasympathetic and sympathetic impulses. The authors found that amylase secreted on sympathetic activation consisted of pre-formed amylase, stored in granules, that was not replaced during the secretory activity under these conditions. Ongoing sympathetic stimulation led to a diminution of the alpha-amylase concentration in saliva. Asking and Proctor (1989) studied the effects of prolonged parasympathetic nerve stimulation in rat parotid glands on alpha-amylase content in saliva and glands. The salivary alpha-amylase concentration and total output of alpha-amylase throughout a stimulatory period of 120 min were measured, as well as glandular concentrations of amylase. The results suggest that amylase stores are much more rapidly replenished by synthesis during parasympathetic than during sympathetic activity, whereas sympathetic nerve excitation causes a pronounced loss of amylase-containing acinar granules. No such loss could be detected on parasympathetic stimulation, although the output of amylase was about as large as that produced during sympathetic stimulation. Further examining which adrenergic mechanisms are involved in alpha-amylase secretion, another study (Skov Olsen et al., 1988) found that stimulation of the beta-adrenergic receptors in rats increased the concentration of amylase in saliva by a factor of 30, while stimulation of the alpha-adrenergic receptors increased the concentration of amylase in saliva by a factor of 10. Experiments with several substances showed that alpha-amylase secretion in saliva was mainly contributed by beta-1-adrenergic mechanisms. Following-up on these previous studies, Busch et al. set out to investigate the differences in release of salivary alpha-amylase by the parotid and the submandibular gland in rats. They found that submandibular amylase did not respond with an increase to the administration of isoproterenol, a beta-adrenergic agonist, whereas parotid amylase did. This effect was inhibited by the selective beta-1-antagonist atenolol but not by the beta-2-antagonist butoxamine (Busch, Sterin-Borda, & Borda, 2002).

These findings show the most important involvement of the autonomic nervous system, and the sympathetic nervous system in particular, in the release of salivary alpha-amylase with beta-adrenergic mechanisms as the main contributing factor in salivary alpha-amylase secretion. However, findings from animal studies are not readily transferable to humans. Furthermore, stimulation of the sympathetic nerves in isolation, as it was used in the previously described studies as a paradigm to elicit autonomic activation, may not be considered a physiological reflection of secretory processes in vivo, although superimposition of sympathetic stimulation during continuous parasympathetic excitation may constitute a more physiological approach. In humans, both approaches are not feasible. Thus, the use of pharmacological agents may be a more ideal methodological approach to determine the role of the autonomic nervous system in the secretion of alpha-amylase in humans, as the following studies show.

Speirs et al. provoked in their study a sympathetic response by either immersing the subjects up to the waist in cold water (4-5° C) or by administering isoprenaline and propanolol (both are beta-adrenergic blockers). Salivary flow rate was held constant by constantly dipping citric acid on the tongue. Exposure to cold water and isoprenaline raised salivary alpha-amylase concentrations in the parotid gland, whereas propranolol led to a reduction of amylase concentrations. The authors could show that these changes were not attributed to changes in salivary flow rates. These results offered first evidence for the sympathetic control of amylase secretion in humans (Speirs, Herring, Cooper, Hardy, & Hind, 1974). As it has been shown in a number of the aforementioned studies in animals, stimulation of beta-adrenergic receptors modulates the synthesis and the release of salivary alpha-amylase. Laurikainen et al. studied the effects of timolol maleate, a widely used beta-blocking agent, on the quantity and quality of saliva secretion controlled by beta-adrenoceptors in 8 healthy male humans (Laurikainen, Laurikainen, Tenovuo, Kaila, & Vilja, 1988). No significant changes were seen in the salivary flow rate of either parotid or whole resting salivas after drug administration. Alpha-amylase concentrations in parotid saliva decreased significantly after drug intake whereas no such changes were observed in the control experiment, thus corroborating similar findings from rat studies. In a study by Nederfors and co-workers, the effects of therapeutic doses of the selective beta-1-antagonist atenolol and the non-selective beta-antagonist propranolol on stimulated glandular saliva were investigated in 19 human subjects. The two substances did not affect whole salivary flow rate. However, atenolol, but not propranolol, resulted in a decrease of parotid amylase in the morning, and in a decrease in both morning and lunch time submandibular/sublingual alpha-amylase (Nederfors, Ericsson, Twetman, & Dahlof, 1994). The authors were able to replicate their findings in a well-controlled study on human hypertensive subjects with the selective beta-1-antagonist metoprolol (Nederfors & Dahlof, 1996). Thus, these studies demonstrate the importance of beta-adrenergic mechanisms of alpha-amylase secretion in humans, as has already been shown in rats. Following-up on these results, van Stegeren et al. conducted a placebo-controlled double-blind study using propanolol in a stress-rest protocol (van Stegeren, Rohleder, Everaerd, & Wolf, 2006). Thirty participants were allocated to a rest condition and a stressful brain scanning procedure. While the placebo group showed a substantial increase in salivary alpha-amylase due to the stress test, the amylase response in the propanolol group was attenuated. The authors therefore suggest that salivary alpha-amylase might serve as an indicator for beta-adrenergic activity. Finally, Ehlert et al. proposed that salivary alpha-amylase increases might reflect the interaction of stress-dependent sympathetic and parasympathetic stimulation via central nervous noradrenergic input (Ehlert, Erni, Hebisch, & Nater, 2006). To examine this hypothesis, they assessed the indirect effect of yohimbine hydrochloride, an alpha-2-adrenergic receptor antagonist, on salivary alpha-amylase release in a randomized placebo-controlled study in 13 healthy men. The results showed significant increases under the condition of yohimbine compared to placebo for salivary flow rate and salivary alpha-amylase concentrations.

The results from studies on animals and humans taken together indicate that the autonomic nervous system plays a powerful role in the secretion of salivary alpha-amylase, with contributions of both alpha-adrenergic and beta-adrenergic mechanisms. These findings suggest that alpha-amylase might be regarded as an indirect indicator of autonomic activation.

Psychological Stress Induced Alpha-Amylase Secretion

As outlined in the previous section of this chapter, the release of salivary alpha-amylase is governed by activation of the autonomic nervous system. Thus, during psychological stress when autonomic activation is high, an increase in salivary alpha-amylase may be expected.

The following section aims to describe findings on salivary alpha-amylase responses due to psychological stress.

Early Findings

The notion that salivary alpha-amylase may serve as indicator of stress reaches as far back as the late 70s of the 20th century. In a seminal study, Gilman and colleagues were exposing their subjects to eight days of hyperbaric pressure (Gilman, Fischer, Biersner, Thornton, & Miller, 1979). As a result, they found increased concentrations of salivary alpha-amylase. The authors concluded that these increases were probably not only due to the hyperbaric exposure with its subsequent effect on the autonomic nervous system, but also due to the psychologically stressful nature of the procedure itself. This was probably the first time salivary alpha-amylase was related to psychological stress. In the 80s, Donald Morse and his group undertook a series of experiments with the goal to examine the impact of stress on various salivary parameters. In two of their studies, they induced stress by the anticipation of endodontic (root canal) therapy, and relaxation was applied by the use of hypnosis, meditation, and local anaesthesia. The authors found stress-related changes in several salivary parameters, such as decreased volume, increase in visually-determined opacity, decreases in pH, and increased protein. Opposing findings were observed during induction of relaxation (Morse, Schacterle, Esposito, Furst, & Bose, 1981; Morse, Schacterle, Furst, & Bose, 1981). In both studies, salivary alpha-amylase was measured. In a subsequent study, Morse and co-workers used mean protein values in saliva as an indicator of stress levels. Dental students were studied during an examination period of ten weeks. Half of the students taking part in the experiment were taught meditation techniques developed by Morse himself. Samples of whole, unstimulated saliva were collected twice a week. Results showed that non-meditators displayed increased salivary protein before examinations with decreased levels after the test, whereas in meditators no marked changes in protein levels occurred (Morse et al., 1981). These results were found again in another study of this group, as reported in a review by Morse et al. (Morse et al., 1982). Increased salivary alpha-amylase levels were found in both studies after the examinations. Morse and his team also conducted a study in ten subjects to evaluate the effect of stress (anticipation of nonsurgical endodontic treatment) on the secretions of the sublingual and the submandibular glands. Using again a simple meditation technique, the authors were able to show significant increases in alpha-amylase under relaxation (Morse, Schacterle, Esposito et al., 1983). In a follow-up study, the parotid gland was also examined. Subjects had to relax and subsequently to chew relaxedly. Alpha-amylase activity increased over six and a half times from the relaxation alone to the relaxed chewing condition (Morse, Schacterle, Zaydenberg et al., 1983). In one subject, the authors examined in a single-case study all three major glands together, and were able to replicate their findings from the earlier two studies (Morse et al., 1984).

Taken together, the findings from Morse and his colleagues indicate that alpha-amylase activity declines due to stress whereas it rises during relaxation. Morse reasoned that the sublingual and minor glands are predominantly composed of mucous cells that are activated by the sympathetic division of the ANS which is responsible for the production of large-molecular weight, glycoprotein-rich, low-volume saliva. The parotid glands are predominantly composed of serous cells that are activated by the parasympathetic division of the ANS leading to an alpha-amylase-rich, high-volume, free-flowing, high-bicarbonate saliva. Morse regarded this physiological fact as responsible for the augmentation of amylase due to relaxation (Morse, Schacterle, Furst, Zaydenberg, & Pollack, 1989). However, the studies of Morse had to deal with the methodological flaw that no control condition or random allocation to treatment groups was used. Furthermore, the "relaxation condition" always followed immediately after the "stress condition" suggesting the increases in amylase might be attributed to a stress response that occurred after the stressor. Since the conclusion by Morse was that salivary parameters were valid and reliable indicators for psychological stress and relaxation, it was attempted to re-evaluate this statement and to compare the usefulness of salivary indicators with other known psychophysiological parameters (Borgeat, Chagon, & Legault, 1984). In their study, Borgeat et al. exposed 16 subjects to a relaxation (progressive muscle relaxation sensu Jacobson), and to a stress (written exercises, and mental arithmetic task) session. In addition to salivary parameters, such as alpha-amylase, frontal muscle EMG, skin conductance, heart rate, and skin temperature were measured. The interventions resulted in clear and distinct reaction patterns in the "classical" psychophysiological parameter, whereas no salivary changes (alpha-amylase among others) could be observed.

Is Salivary Alpha-Amylase an Indicator for Plasma Catecholamines?

The relationship between psychological stress and salivary alpha-amylase was not examined for the next ten years. Other stress related factors in saliva, such as salivary cortisol, gained more scientific attraction (Kirschbaum & Hellhammer, 1989, 1994). Interest in salivary alpha-amylase as a stress marker sparked again after the publication of Chatterton and colleagues' results of increases of salivary alpha-amylase in a variety of stressful conditions (Chatterton, Vogelsong, Lu, Ellman, & Hudgens, 1996). They undertook a series of studies in which subjects were exposed to physical (running, exercise, exposure to heat and cold) and psychological (examination) stressors. The observation of a rise in salivary alpha-amylase due to physical and (in some cases) psychological stressors, and increases observed in plasma catecholamines (norepinephrine and epinephrine) due to these stressors, led Chatterton to the assumption that the two parameters might react concomitantly to stress and that they can thus be measured as a substitute for each other. Indeed, the authors found significant correlations between salivary alpha-amylase and plasma norepinephrine, as well as for epinephrine ($r = .64$, $r = .49$, respectively) in the exercise conditions of their study. The authors thus suggested that salivary alpha-amylase might be an indicator for plasma catecholamines (specifically norepinephrine).

As a consequence, several studies were performed which measured salivary alpha-amylase as an indicator for either epinephrine or norepinephrine. Studying subjects preparing for skydiving, Chatterton et al. (1997), for example, found increased alpha-amylase, which was measured in place of norepinephrine, prior to the jump from the airplane in contrast to control subjects who did not jump. The highest levels were observed right after landing. In another study, chronic stress levels were compared in two groups. One consisted of female subjects experiencing adverse pregnancy outcomes, the other comprised females with a favourable outcome. Neither psychological nor hormonal stress markers differentiated between the two groups. Again, the authors measured alpha-amylase as an indicator for adrenergic activity (Milad, Klock, Moses, & Chatterton, 1998). To examine the influence of physiological stress indicators on inadequate milk production in mothers of preterm infants, several stress hormones as well as alpha-amylase as an indicator for norepinephrine was measured. However, alpha-amylase was not related to stress and milk production (Chatterton et al., 2000). Using a stressful video game as a possibility to induce laboratory stress, Skosnik et al. measured alpha-amylase as an indicator for norepinephrine. They found a significant increase in alpha-amylase after the 15 minutes-stressor (Skosnik, Chatterton, Swisher, & Park, 2000). In a small study with 10 subjects, the influence of a high-fidelity trauma management simulation was examined. Significant levels of salivary alpha-amylase (measured instead of norepinephrine) during the simulation in contrast to a baseline period were found, suggesting the stressful consequence of this training (Xiao, Via, Kyle, Mackenzie, & Burton, 2000). Finally, the impact of noise in eleven nurses was examined. Alpha-amylase was determined as a measure of hormonal stress reaction. The results indicated large variations in alpha-amylase which was not significantly affected by noise levels (Morrison, Haas, Shaffner, Garrett, & Fackler, 2003). These studies have all in common that salivary alpha-amylase was measured as substitute for plasma norepinephrine. However, this notion was merely based on the initial findings of Chatterton et al. (1996). A close inspection of this study shows that in the psychological stress condition (examination) only a small and non-significant correlation ($r = 0.17$) was observed. The results from that study indicate that during physiological stress experience, increases in both catecholamines and salivary alpha-amylase share similar mechanisms. In contrast, during psychological stress the two parameters seem to partly dissociate. In a recent study using a standardized psychological stress protocol, the plasma catecholamines norepinephrine and epinephrine were measured concomitantly with salivary alpha-amylase levels (Nater et al., 2006). However, no associations between these parameters were observed. Thus, a direct association between plasma catecholamines and salivary alpha-amylase in the body's response to psychological stress seems to be doubtful.

Salivary Alpha-Amylase is Sensitive to Psychological Stress

In recent years, a number of studies have been performed showing that salivary alpha-amylase is a highly sensitive parameter in the context of a variety of different psychological stress protocols. The following section attempts to give a brief overview of these studies.

In a study of 28 subjects exposed to an academic examination considered difficult, Bosch et al. measured several salivary parameter including alpha-amylase (Bosch et al., 1996). Whole unstimulated saliva was taken 30 minutes before the examination, two weeks later, and six weeks later. Results indicate an increase in concentration and output of alpha-amylase during the stress condition (30 minutes before the examination, anticipation stress), while the salivary flow rate did not change. Interestingly, Bosch et al. were able to show anticipatory stress-dependent alterations in alpha-amylase. In a separate analysis, the authors found that the amount and severity of critical life events was related to alpha-amylase activity, thus suggesting that everyday stress also contributes to the stress-dependent changes in salivary parameters observed in the 1996 study (Bosch et al., 1998). In a further study, the same group used a laboratory task to induce acute stress (Bosch, De Geus, Veerman, Hoogstraten, & Nieuw Amerongen, 2003). In 32 subjects, whole unstimulated saliva was measured before, during and after an active memory task, a passive watching of a gruesome video, and a control condition. Alpha-amylase output differed significantly between the three conditions with the highest levels found during the passive video task. A decrease in alpha-amylase output was found in the active memory task, suggesting that it depends on the nature of the stressor and the active or passive coping possibilities of the subjects what the physiological reaction of the body is. Another group also used a stressful video and a relaxing video to induce both stress and rest conditions in 83 healthy participants in order to directly compare the effects of both conditions on salivary cortisol, alpha-amylase and salivary flow rate responses in whole unstimulated saliva (Takai et al., 2004). The stressful video induced marked increases in both cortisol and alpha-amylase, but not salivary flow rate, whereas watching the relaxing video resulted in no changes of salivary cortisol and flow rate, but in significantly decreased alpha-amylase concentrations. Interestingly, amylase peak levels and state anxiety measures correlated highly ($r = 0.535$). A similarly high correlation between state anxiety and salivary alpha-amylase was found in another study. Noto and colleagues examined salivary alpha-amylase levels in 10 healthy female participants during a mental arithmetic task (Noto, Sato, Kudo, Kurata, & Hirota, 2005). The task resulted in significant increases in both state anxiety and salivary alpha-amylase, with a correlation of $r = 0.589$ between the two variables.

A number of studies have been conducted to assess the usefulness of salivary alpha-amylase as a biomarker for psychologically induced stress using a standardized psychosocial stress test, i.e. the Trier Social Stress Test or TSST (Kirschbaum, Pirke, & Hellhammer, 1993). This test comprises a mock job interview and a mental arithmetic task, both performed in front of an audience, leading to marked increases in both psychological and physiological stress indicators. Rohleder et al. found in a pilot study of 12 healthy subjects (Rohleder, Nater, Wolf, Ehlert, & Kirschbaum, 2004) marked increases in salivary alpha-amylase due to the TSST. These findings were corroborated in a subsequent study of 24 healthy subjects which were exposed to the TSST and a rest condition (Nater et al., 2005). The TSST led to significant increases in salivary alpha-amylase, while the rest condition did not impact amylase levels. In a third study, this study design was used in 30 healthy subjects, again showing marked increases due to the TSST (Nater et al., 2006). Subsequently, other studies also used the TSST and measured salivary alpha-amylase as a stress marker. Rohleder and colleagues have repeatedly shown salivary alpha-amylase increases as a response to the TSST (Rohleder et al., 2006; Rohleder, Wolf, Maldonado, & Kirschbaum, 2006). Tu et al. found stress-induced salivary alpha-amylase increases in the TSST in postpartum mothers (Tu, Lupien, & Walker, 2006), Nierop et al. found that pregnant women showed lower alpha-amylase responses compared to non-pregnant women in the TSST (Nierop et al., 2006), and Gordis et al. found increased salivary alpha-amylase responses to the TSST in adolescents (Gordis, Granger, Susman, & Trickett, 2006).

Salivary alpha-amylase increases were also shown as a response to other stressful conditions, such as experience of medical procedures (Yamaguchi, Takeda, Onishi, Deguchi, & Higashi, 2006), adverse musical stimuli in men (Nater, Abbruzzese, Krebs, & Ehlert, 2006), the cold pressor test (West, Granger, Kivlighan, & Hurston, 2006), achievement and interpersonal stress (Stroud, Handwerger, Granger, & Kivlighan, 2006), and use of a noise burst and infant arm restraint in depressive mothers (Shea et al., 2006).

In conclusion, the discrepant results between earlier and newer findings regarding the nature of the effect on salivary alpha-amylase may be attributed to the nature of the different stressors and stress paradigms used in those studies. The results of these studies show that changes in alpha-amylase due to psychological stress were found repeatedly. Thus, alpha-amylase can be regarded as a good indicator of stress-related body changes. However, it is not yet quite clear which mechanisms underlie the psychological stress-dependent changes of salivary alpha-amylase.

Methodological Considerations

Several methodological factors have to be taken into account when measuring alpha-amylase in saliva in a psychoneuroendocrinological context. Before designing an experimental protocol, decisions have to be made as to how saliva is collected and stored. Furthermore, biochemical determination requires the choice for the adequate method. Finally, there are a multitude of factors that might influence concentrations of salivary alpha-amylase, and specific issues will be discussed in the following sections.

Collection and Preparation of Samples

There is a wide array of methods for collection of saliva. Extensive overviews of techniques measuring saliva and salivation are available in the literature (Navazesh, 1993; White, 1977). Techniques can basically be divided into two groups: one group comprises methods for collecting whole saliva and the other one for collecting saliva from individual salivary glands (Birkhed & Heintze, 1989; Navazesh, 1993).

Whole saliva may be collected either in a stimulated or unstimulated manner. Unstimulated whole saliva may be collected by the so-called draining method. With this method, the subject is in the sitting position with the head tilted forward. Saliva is accumulated in the mouth. After an initial swallow, the subject allows saliva to drain continuously from the lower lip into a funnel in a plastic container. At the end of the collection period, the subject expectorates residual saliva from the mouth. In contrast, stimulated whole saliva may be stimulated by mechanical stimuli (an alternative might be gustatory stimuli; however, these dissolve in whole saliva which might confound the results). Stimuli such as parafilm, paraffin wax, neutral gum base, and rubber bands have been employed in a multitude of studies. According to Navazesh (1993), these are preferable because they will not absorb any saliva. However, if the medium absorbing saliva is centrifuged together with saliva, no saliva would be lost (see below). The chewing frequency may be standardized. The amount of saliva secreted is more accurately determined by weight than by volume. Therefore, collected saliva can be weighed after completion of the study, subtracting the weight of the collection vessel. This measurement results in the salivary flow rate, which is indicated by milligram per time unit (mg/min).

Whole stimulated saliva may also be collected by the Salivette device (Sarstedt). The device comprises a plastic vessel, a centrifugation tube, and cotton swab. The use of salivettes may currently be viewed as the gold standard in research examining cortisol levels in saliva (Kirschbaum & Hellhammer, 1999). However, the Salivette itself might influence changes of amylase. In a recent study, the reliability of Salivette measurements was shown for salivary alpha-amylase (Granger et al., 2006). In another study, the Salivette collection method was compared to the passive drooling method in a stress experiment (Rohleder, Wolf, Maldonado, & Kirschbaum, 2006). The results show that stress exposure resulted in similar salivary alpha-amylase responses in both the Salivette and the passive drooling condition. The same result was obtained in a small study of n = 8, finding similar salivary alpha-amylase levels in both collection methods (Granger et al., 2006). Salivettes may also be used to measure whole unstimulated saliva. Cotton swabs can be placed under the tongue (sublingual swab method) instead of chewing. White was able to show high reliability for this measurement technique (White, 1977). However, saliva volume contributed by the parotid glands may be underestimated. In a recent study, the method suggested by White was compared with chewing on cotton rolls (Nater, 2004). Results show that the unstimulated sublingual collection of saliva resulted in a mean activity of 32.87 U/ml, whereas the stimulated whole saliva collection resulted in a mean activity of 212.77 U/ml. This difference was highly significant. Thus, the two collection strategies are not comparable with regard to the amylase activity measured. To measure amylase activity, it is recommended not to use the unstimulated sublingual cotton roll method, because the amylase contribution of the parotid glands may not reach the collection device, whereas the amylase contribution of the sublingual glands seem to be too small.

In contrast to the collection of whole saliva, glandular-derived saliva may be obtained, depending on the research questions (Veerman, van den Keybus, Vissink, & Nieuw Amerongen, 1996). Direct parotid saliva can be collected by using Lashley cup devices (Lashley, 1916). The Lashly cup is an advanced development of the Carlson-Crittenden cup (Carlson & Crittenden, 1910), which consists of a metal cup utilizing vacuum to hold the cup to the cheek without cannulation or direct suction. An advanced design of the Carlson-Crittenden cup for collection of parotid fluid was presented by Sproles and Schaeffer (Sproles & Schaeffer, 1974). Their cup is fabricated from nylon and makes use of similar principles applied in the Carlson-Crittenden cup. Ericson and Nordlund devised a new collection device, called SLURP, for parotid saliva. The device has good correlation in secretion rates with the classical Lashley-cup (Ericson & Nordlund, 1993). However, these devices are not feasible in studies which examine subjects in an ambulatory environment, e.g. field studies. Furthermore, they cannot be used in laboratory studies where short-time measurement (1 min) is necessary, since collection of saliva by these devices cannot be completed in short time periods.

Biochemical Determination of Alpha-Amylase Activity

There are more than 200 published methods of analyzing amylase. There is currently no standard or reference method for measurement of total alpha-amylase activity, or even of a standard unit activity definition (Zakowski & Bruns, 1985). Reliable normal values are available only for a few salivary enzymes, but not for amylase (Makinen, 1989). Alpha-amylase is probably the enzyme with the most published methods for assay of its catalytic activity. In a study of six different methods for alpha-amylase determination, Gubern et al. found significant differences between methods (Gubern, Canalias, & Gella, 1995). Apart from this study, no published findings on comparability of alpha-amylase assays exist. Thus, the choice of the kits to be used is up to the investigator and may depend on economic factors.

Stability of Samples

Alpha-amylase is generally stable at short term storage (a few days) at 4° C and 20° C and at long term storage at 20° C (a few weeks or months) (Söderling, 1989). A recent study has shown that storing samples at room temperature, 4°C, or -80°C for 96 hours did not change salivary alpha-amylase concentrations significantly (Granger et al., 2006). It is however recommended to store samples at -70° C if prolonged storage is anticipated (6-12 months), as has been shown for serum alpha-amylase (Zakowski & Bruns, 1985). Furthermore, a recent study has shown that repeated freezing and thawing of saliva samples did not affect alpha-amylase concentrations (Granger et al., 2006). Temporal stability of baseline resting values was determined in a recent study (Nater, 2004). Values of basal measurements of two consecutive days correlated highly ($r = 0.95$), thus indicating a high test-retest validity of salivary alpha-amylase measurements. Further studies should determine whether longer time periods might have an impact on amylase variability.

Circadian Fluctuations of Alpha-Amylase

Daily oscillations characterize the secretion of almost all endogenous substances. These rhythms are driven by a circadian pacemaker, the suprachiasmatic nucleus in the hypothalamus. It is well established that both salivary flow rate and saliva composition vary rhythmically over a 24-hour period (for a review see Dawes, 1974). Studies in the parotid gland of the rat, for example, have found minimum alpha-amylase concentrations at 05.00 hours and maximum concentrations at 17.00 hours. Similar findings were obtained in humans. Whole saliva was collected in a total of 14 subjects in a study by Jenzano and colleagues. Samples were obtained at 08.00, 11.00, 14.00, and 17.00 hours on two different days. For salivary alpha-amylase, significant changes were observed during the day, with a peak in the early evening and a trough in the morning at 08.00 (Jenzano, Brown, & Mauriello, 1987). In another study, daily fluctuations in saliva viscosity, flow rate, and alpha-amylase were examined in 30 subjects. Unstimulated and stimulated whole saliva were collected 5 times during a 12-hour period. No significant changes during the day could be found for stimulated saliva whereas for unstimulated saliva flow rates changed in a significant manner. Concomitant measurement of alpha-amylase activity showed significant variations within the subjects during the collection period. Lowest values were found in the morning (Rantonen & Meurman, 1998). In another study, patients with diabetes mellitus and healthy controls were examined with respect to the diurnal rhythm of alpha-amylase activity. Unstimulated whole saliva was collected in the morning (07.30-08.00 hours) and in the early evening (17.30-18.00 hours). Peaks of alpha-amylase activity were found at the latter time point. However, the two groups did not differ from each other (Artino et al., 1998). Rohleder et al. found in a small study of 17 subjects a diurnal pattern of salivary alpha-amylase with a trough in the morning and a steady increase throughout the day (Rohleder, Nater, Wolf, Ehlert, & Kirschbaum, 2004). In contrast, Yamaguchi et al. (2006) found that there were no distinct changes in salivary alpha-amylase activity over the course of the day at all. In a recent study, Nater et al. (2007) describe the diurnal pattern of salivary alpha-amylase and its determinants. Saliva samples were collected immediately after waking-up, 30 and 60 minutes later, and each full hour between 09.00-21.00 hours by 76 healthy volunteers (44 women, 32 men). Compliance was controlled by electronic monitors. In order to control factors which might influence the diurnal profile of salivary alpha-amylase (such as acute stress, mood, food, or body activity), at each sampling time point the subjects filled out a diary examining the activities they had carried out during the previous hour. Salivary alpha-amylase activity showed a distinct diurnal profile pattern with a pronounced decrease within 60 minutes after awakening and a steady increase of activity during the course of the day. Multilevel-modelling showed a relative independence of diurnal salivary alpha-amylase from acute stress and other factors, but significant associations with chronic stress and mood.

Despite the congruence of most of these findings, no attempts have been made to elucidate the underlying mechanisms responsible for circadian release patterns in humans. However, mechanisms of circadian variation of amylase were investigated in rats. In one study, animals were subjected to a variety of experimental conditions (fast, reversed photoperiod, constant light or darkness, treatment with reserpine and alpha-methyl-p-tyrosine). The rhythm of alpha-amylase secretion was changed by the photoperiod shifted in time (lights on from 18.00-08.00 hours instead of 08.00-18.00 hours). The rhythm disappeared completely when rats were exposed to constant light or darkness for 15 days or to the two substances given. Thus, the circadian rhythm of alpha-amylase in the parotid gland of rats seems to be endogenous, apparently controlled by the photoperiod under direct control of the sympathetic nervous system (Bellavia, Sanz, Chiarenza, Sereno, & Vermouth, 1990). Interestingly, alpha-amylase has recently been proposed to be a biomarker for sleep drive (Seugnet, Boero, Gottschalk, Duntley, & Shaw, 2006). Taken together, the fact that a diurnal rhythm has been described allows for planning studies on specific times of day. It is recommended to either controlling for or restricting studies measuring salivary alpha-amylase to one time of day (i.e. morning vs. afternoon).

Sex Differences in Alpha-Amylase Activity

Due to the differential make-up of endocrine systems in men and women, a myriad of studies have been conducted in order to examine sex differences in hormones. However, not many studies exist that looked into sex-specific alterations of alpha-amylase activity. Many studies have shown a higher flow rate for men than for women, although this sex difference is not always statistically significant. This holds true for both resting and stimulated saliva as well as for various age groups. This sex difference may be due to the smaller size of the salivary glands in females. It was also found that the flow rate of stimulated saliva was highest during the secretory phase of the menstrual cycle (for a review, see Birkhed & Heintze, 1989). However, none of these studies measured alpha-amylase.

In one study, the impact of low-dose hormonal contraceptives was examined. Results show that there is no influence on alpha-amylase activity (Laine, Pienihakkinen, Ojanotko-Harri, & Tenovuo, 1991). Not many studies exist which have examined basal or reactive differences in alpha-amylase activity between men and women. A recent study examined potential sex-specific differences in basal salivary alpha-amylase concentrations (Nater, 2004). The experiment was carried out in 53 volunteers (26 males, 27 females). All female participants took part in their luteal phase (i.e. approx. 18th day of the menstrual cycle) to minimize endocrine variations. Intake of hormonal contraceptives was an exclusion criterion. Results indicate that men and women did not differ in basal alpha-amylase activity. These results are reflected in a recent study, in which no sex differences were shown in the diurnal variation of salivary alpha-amylase (Nater et al., 2007). While the influence of sex on salivary alpha-amylase seems to be minor in basal conditions, no study so far has looked into stress-related alpha-amylase differences between men and women.

In some studies, the impact of pregnancy on alpha-amylase concentrations has been the subject of investigation. In a study in female rats, amylase activity in different tissues was measured during the menstrual cycle. For serum and salivary amylase activity no changes were found during the menstrual cycle in most tissues. The only changes were observed in the ovaries (Kasperczyk et al., 2001). In a study in humans, saliva was collected with a Carlson-Crittenden device (see above), under citric acid stimulation, in 107 pregnant women, 9 puerperal and 7 non-pregnant controls. No significant changes were found in salivary flow rate, pH and alpha-amylase levels between these groups (D'Alessandro, Curbelo, Tumilasci, Tessler, & Houssay, 1989). In another human study, unstimulated whole saliva was assessed, and total protein concentration, alpha-amylase activity, sialic acid content, and calcium and phosphate concentrations were evaluated. Forty-five healthy primigravid women, and 15 non-pregnant women as controls were included in the study. A higher alpha-amylase activity at 10 and 21 weeks of gestation compared with the controls and with pregnant women at 40 weeks was observed (Salvolini, Di Giorgio, Curatola, Mazzanti, & Fratto, 1998). Taken together, contrasting findings seem to exist in the literature on salivary alpha-amylase changes during pregnancy or lactation (Laine et al., 1988).

Age

Age differences in salivary alpha-amylase were examined in only a few studies. In one study (Aguirre, Levine, Cohen, & Tabak, 1987), alpha-amylase concentrations in stimulated parotid saliva were studied in 128 healthy subjects. The subjects were divided into three age groups: a) 23-39, b) 40-59, and c) 60-84 years old. No differences between the three groups were found. The aim of another study was to evaluate changes in the concentration of certain components of unstimulated whole saliva during aging. Total protein concentration, alpha-amylase activity, sialic acid content, and calcium and phosphorus concentrations were assessed in 100 healthy subjects of both genders, aged between 10 and 80 years, who were subdivided into four groups according to their age: 10-25 years, 26-40 years, 41-65 years, and 66-80 years. Other than sialic acid, the concentrations of the components studied were not affected by age and, accordingly, no differences in alpha-amylase were observed (Salvolini et al., 1999). These findings suggest that age does not influence salivary alpha-amylase. However, this notion is based on only very few studies.

Influence of Smoking

Smoking has a detrimental effect on a variety of physiological factors. Salivary components may be affected as well (Macgregor, 1989). In a study in 72 subjects, Callegari and Lami (1984) found highly significant differences in total alpha-amylase (serum and salivary) and in salivary amylase between smokers and non-smokers. In contrast, a study in a total of 346 subjects did not find significant differences in smokers and non-smokers regarding amylase activity (Nagaya & Okuno, 1993). Comparing baseline amylase activity levels in smokers and non-smokers, Zuabi et al. were able to show that there were no significant differences between the two groups (Zuabi et al., 1999). However, smoking of a single cigarette can induce significant reduction of gluthatione concentration (Zappacosta, Persichilli, De Sole, Mordente, & Giardina, 1999). The authors were also able to show a significant inhibition of alpha-amylase activity after one cigarette. Interestingly enough, one hour after smoking cessation, enyzme activity returned to baseline levels (Zappacosta et al., 2002). An in vitro study showed that three hours of cigarette smoke exposition led to a significant reduction of amylase activity by 33.8% in saliva (Nagler et al., 2000).

In the aforementioned study on diurnal changes of salivary alpha-amylase (Nater et al., 2007), it was shown that having smoked prior to saliva sample collection did not influence salivary alpha-amylase activity. In contrast to the diurnal profile, the wake-up response of salivary alpha-amylase seemed to be influenced in habitual smokers, with a more pronounced response in smokers compared to non-smokers. It may be speculated that cigarettes smoked during the first 30 minutes of the day have a stronger influence on salivary alpha-amylase activity than cigarettes smoked on a later time of day. Taken together, there is some evidence that smoking might influence salivary alpha-amylase, and should therefore be controlled for.

Food and Drinks

Some food products may increase or decrease alpha-amylase activity. A high carbohydrate diet has been correlated with increased amylase activity (Squires, 1953). The presentation of oral stimuli such as citric acid leads to elevation of parotid salivary flow rate, amylase activity, and amylase secretion rate within seconds (Froehlich, Pangborn, & Whitaker, 1987). Udupa et al. were able to show that sweet preparations and fried foods were able to exert an inhibitory effect on enzyme activity in a group of ten children (Udupa, Prabhakar, & Tandon, 1989). A recent study (Nater et al., 2007) found no effects of eating and drinking in a field study of factors determining diurnal salivary alpha-amylase changes, although salivary alpha-amylase tended to be associated with drinking sugar-containing soft-drinks. Thus, ingestion of food should be avoided prior to studies measuring alpha-amylase.

Moreover, two recent studies have shown that intake of caffeine has a significant impact on salivay alpha-amylase levels, and should thus be avoided in studies measuring amylase (Bishop, Walker, Scanlon, Richards, & Rogers, 2006; Klein, Whetzel, Ritter, & Granger, 2006).

Special care should be taken to avoid alcohol intake before saliva collection. Acute alcohol consumption in 24 healthy volunteers resulted in a significant decrease of both salivary alpha-amylase secretion and concentration (Enberg, Alho, Loimaranta, & Lenander-Lumikari, 2001). However, influence on amylase seems to be dependent on the alcohol dose. Light to moderate use of alcohol did not affect basal levels of alpha-amylase activity in a study of 346 subjects (Nagaya & Okuno, 1993). In contrast, parotid saliva samples from 24 alcoholic subjects were analyzed for changes in flow rate, and composition. Mean stimulated parotid saliva flow rate was significantly lower in alcoholic subjects than in matched control subjects. Reduction in parotid saliva flow rate was associated with significant decreases in total protein and amylase secretion in this group of patients. These data suggest that chronic alcohol ingestion is associated with significant changes in parotid saliva secretion and its composition (Dutta, Orestes, Vengulekur, & Kwo, 1992).

Therefore, experimental subjects should refrain from intake of alcohol prior to alpha-amylase assessment.

Brushing Teeth

Brushing teeth may have an effect on saliva components. One study specifically examined the impact of tooth brushing on total protein and alpha-amylase (Hoek, Brand, Veerman, & Amerongen, 2002). No changes were found in those parameters. However, contamination of saliva with serum should be considered as a possible confounding factor in the analysis of salivary components. A recent study has shown that gingival bleeding can affect reliable measurements of a variety of salivary hormones (Kivlighan, Granger, & Schwartz, 2005; Kivlighan et al., 2004). Thus, it is advisable not to brush teeth prior to an experiment to avoid blood contamination of saliva.

Effects of Exercise on Salivary Alpha-Amylase

A variety of studies have examined the effects of exercise on salivary alpha-amylase. Gilman and co-workers observed a significantly higher concentration of salivary alpha-amylase during exercise in comparison to a control period. The authors discussed salivary alpha-amylase as a valid measure of sympathetic activity via adrenergic receptors (Gilman, Thornton, Miller, & Biersner, 1979). In another study, ten 1 minute samples of parotid saliva at a constant flow rate 1-2 hours before exercise, immediately after running 3-8 miles and 3 hours after exercise were collected in ten subjects. Exercise caused a marked elevation in total protein concentration. However, no specifications were made about alpha-amylase concentrations in this paper, although it might be deducted that alpha-amylase was elevated, as well (Dawes, 1981). The impact of a 2-hour long cross-country race was examined in a study by Nexo and co-workers. In 25 subjects, stimulated whole saliva was collected before and after running. A marked increase in amylase was found after the race, with a median increase that was almost sevenfold compared to the beginning of the race (Nexo, Hansen, & Konradsen, 1988). In two studies reported by Chatterton et al. (1996), salivary alpha-amylase was determined in subjects participating in a running and a bicycle exercise task. Both protocols resulted in increases of salivary alpha-amylase. The aim of another study was to investigate salivary components during and after a 2-hour marathon. Twenty subjects followed the protocol, a total of three samples (before, after, and 1 hour after the marathon) were taken. Alpha-amylase activity increased significantly due to the strenuous activity, with high levels still 1 hour after the end of the marathon (Ljungberg, Ericson, Ekblom, & Birkhed, 1997). In another study, alpha-amylase in saliva was used as an indicator for anaerobic threshold in subjects performing a treadmill exercise test. Since the correlation between amylase and anaerobic threshold was almost 1 ($r = .93$), the authors concluded that amylase was a good and valid measure for the anaerobic threshold (Calvo et al., 1997). In 42 triathletes, several salivary parameters in unstimulated whole saliva were measured before and after a race consisting of swimming, cycling and running. During the triathlon, the mean amylase activity increased significantly (Steerenberg et al., 1997). Another study examined eight well-trained athletes during a high-performance 60-minute cycle exercise task. Unstimulated whole saliva was collected before exercise, immediately after the task, and 1, 2.5, 5 and 24 hours after exercise. After exercise, a significant increase in alpha-amylase activity was found in contrast to pre-exercise (Walsh et al., 1999). Similar results were obtained in a more recent study (Bishop, Walker, Scanlon, Richards, & Rogers, 2006), which examined the influence of caffeine intake on salivary alpha-amylase measures due to exercise. The authors found that intake of caffeine (in contrast to a placebo condition) resulted in an additional increase. Finally, a study examining the impact of rowing ergometer on salivary alpha-amylase levels showed increased levels due to the exercise (Kivlighan & Granger, 2006). Interestingly, performance was positively associated with alpha-amylase measures in this study.

Taken together, during exercise, salivary levels of alpha-amylase may be increased since saliva secretion is then mainly evoked by the action of adrenergic mediators. Exercise is known to increase sympathetic activity, and the high protein level in saliva following exercise may be due to increased beta-adrenergic activity in the salivary glands. As a consequence, exercise should be avoided prior to experimental procedures.

Personality

The influence of personality on alpha-amylase (among other salivary components) was examined in a study by Costa et al. The authors studied stimulated parotid saliva and personality factors in 390 men by means of the Cattell Sixteen Personality Factor questionnaire and the Eysenck Personality Inventory. The authors found a significant correlation between amylase and Eysenck's introversion scale in the magnitude of $r = .14$ ($p < .01$) (Costa, Chauncey, Rose, & Kapur, 1980). However, it is not surprising that, due to the big sample size, this small correlation and therefore small effect size ($R^2 = .02$) became highly significant. The role of anxiety as a personality trait and its influence on salivary parameters was investigated in a study by Moret and colleagues. Using Cattell's Anxiety Scale, a total of 295 subjects was examined. After completion of the questionnaire, 88 subjects were randomized in three classes of basal anxiety (apathetic, normally anxious, hyperanxious). The three groups did not differ in any of the salivary variables measured, including amylase activity (Moret, Coudert, Bejat, Robin, & Lissac, 1993). In a study by Gordon et al., the role of salivary parameters in seasickness susceptibility and personality factors was investigated. Twenty-nine highly susceptible subjects were compared to 25 non-susceptible subjects. Results indicate that alpha-amylase activity was significantly higher in the susceptible group (resting saliva). However, no significant differences were found in the measurements of the personality traits introversion and neuroticism. Elevated amylase activity values were explained by the higher sympathetic tone expressed by people susceptible for motion sickness and seasickness (Gordon et al., 1994). In the study by Bosch and co-workers mentioned above, the authors correlated the scores of the trait-anxiety scale with alpha-amylase output, resulting in a significant relationship ($r = .38$; $p < .05$). The correlation between the Worry Scale of the Dutch version of Spielberger's Test Anxiety Inventory and amylase was also significant for alpha-amylase concentration ($r = 32$, $p < .05$), and a trend for alpha-amylase output was found ($r = .30$; $p = .06$) (Bosch et al., 1996).

Applications of Salivary Alpha-Amylase Measurement

While the previous sections attempted to describe the physiological basis of salivary alpha-amylase secretion and dealt with methodological issues of salivary alpha-amylase measurement, this section is devoted to previous and future possibilities of research applications for measurement of salivary alpha-amylase.

Salivary Alpha-Amylase as a Diagnostic Parameter

Salivary alpha-amylase has been used as a diagnostic variable in previous research. In one study (Davidson, Koheil, & Forstner, 1978), e.g., salivary alpha-amylase was determined in patients with cystic fibrosis (CF), control patients, and patients with Shwachman's syndrome. The authors found that patients with CF had increased basal levels of circulating salivary alpha-amylase in contrast to patients with Shwachman's syndrome and controls. Interestingly, the increase was interpreted by the authors as a reflection of an autonomic hyperstimulation of salivary secretion in these patients. This was among the first studies to measure salivary alpha-amylase in a circumscribed patient population.

A number of studies have been performed in patients with eating disorders. In an initial study, serum alpha-amylase levels were measured in 56 underweight anorectics, 24 weight-recovered anorectics, 23 normal-weight bulimics, and 31 volunteer women. Normal-weight bulimic patients had significantly higher admission serum amylase values than controls. The authors observed that modest increases of serum amylase values appear to be a consequence of binge-vomit behavior and suggested that serial serum amylase determination may be useful in monitoring the degree of patient abstinence in therapeutic programs (Gwirtsman et al., 1989). In a further study (Scheutzel & Gerlach, 1991), alpha-amylase activity was measured in the serum and saliva of 45 patients with eating disorders and in 30 normal controls. Of the 45 patients evaluated, 12 had restrictive anorexia nervosa, 13 were bulimic anorectics and 20 had bulimia nervosa. In all these groups, the mean alpha-amylase values in serum and saliva were higher than those of the control group. Increased salivary alpha-amylase concentrations might be attributed to hypertrophy of the parotid glands, which is a condition often found in patients with eating disorders. Consequently, the parotid salivary secretory patterns in 28 bulimics were determined in order to investigate the functional abnormality in the glands. The salivary amylase activity was increased in both the resting and stimulated states in bulimics (Riad, Barton, Wilson, Freeman, & Maran, 1991).

Another study investigated the clinical relevance of alpha-amylase level monitoring as an objective measure in diagnosis and assessment of treatment response in bulimia nervosa (Kronvall, Fahy, Isaksson, & Theander, 1992). Thirty-three subjects with bulimia nervosa had serum levels of total and salivary amylase monitored during an 8-week treatment trial. At the beginning of treatment, the average total amylase level was within the upper limits of normal, whereas average salivary amylase levels were abnormally high. During the course of treatment, there was a significant reduction in the average salivary isoenzyme to within the normal range. Significant reductions in amylase levels were recorded in patients with good treatment outcome, but not in those with poor outcome. Amylase levels were not significantly correlated with severity of bulimic symptoms. These results do not seem to justify the use of alpha-amylase assays as a routine diagnostic or monitoring test in eating disorders, but the authors conclude that salivary alpha-amylase monitoring may provide useful clinical information at least in selected cases.

Eating disorders are not the only conditions related to changes in alpha-amylase levels. Both pancreatic and salivary isoenzymes were examined in a study with 12 Alzheimer patients looking for the possibility of amylase as a marker of M3 activity. Overall alpha-amylase results were not significant; however, a trend for salivary amylase was found

(Sramek, Cutler, Hurley, & Seifert, 1995). In another study, diurnal changes in alpha-amylase activity were examined in a group of patients with diabetes mellitus and a healthy control group. Alpha-amylase might be involved in the high incidence of dental complications in patients with diabetes mellitus, such as gingival and periodontal diseases, and an increased number of caries. However, comparison of the two groups shows that there are no differences in daily fluctuations of the enzyme (Artino et al., 1998). In 15 patients with chronic atopic dermatitis, aged 4 to 11, parotid secretory response to intra-oral 0.1 ml citric acid was measured. Protein concentration as well as amylase activity was significantly decreased in children with atopic dermatitis. These findings were interpreted by the authors as a reflection of impaired beta-adrenergic mediated responses in atopic dermatitis (Crespi et al., 1982). Unstimulated whole saliva was examined in a group of patients with systemic lupus erythematosus (SLE) and control subjects. Alpha-amylase activity was determined. No differences between the two groups were observed (Ben-Aryeh, Gordon, Szargel, Toubi, & Laufer, 1993).

Taken together, salivary alpha-amylase has been used as a diagnostic parameter in a variety of conditions, however with limited success. In recent years, to our best knowledge, no attempts have been made to clarify the usefulness of alpha-amylase in the conditions discussed above. Based on the knowledge that has been accumulated about the role of stress and its underlying physiological mechanisms in the secretion of alpha-amylase, measuring alpha-amylase concentrations might serve, however, as an index for pathological dysregulations of the autonomic nervous system in specific clinical and subclinical conditions. In some psychiatric disorders, autonomic dysregulations are present. In post-traumatic stress disorder (PTSD), for example, a heightened autonomic activity has been found, going together with the cardinal symptom of hyperarousal (Shalev & Rogel-Fuchs, 1993). Elevated catecholamine levels have been shown (Southwick et al., 1993). Measurement of alpha-amylase might add additional information on autonomic changes occurring in these patients. Another psychiatric disorder in which autonomic changes occur is phobia and anxiety (Lang, Davis, & Ohman, 2000). Several studies have shown that autonomic variables are dysregulated in anxiety patients in comparison to healthy subjects (Coupland, Wilson, Potokar, Bell, & Nutt, 2003; Friedman et al., 1993; Laederach-Hofmann, Mussgay, Buchel, Widler, & Ruddel, 2002). Measurement of alpha-amylase might provide non-invasive measurement of autonomic functioning without inducing discomfort by the assessment technique. But alpha-amylase may be also measured in a non-psychiatric context. In a variety of somatic disorders, autonomic dysregulations are present. Exaggerated autonomic responses to different stimuli can be observed in hypertensive patients (al'Absi & Wittmers, 2003; Fredrickson, Tuomisto, & Bergman-Losman, 1991), and in patients with HIV (Cole et al., 2001). It might be of some interest to measure alpha-amylase concentrations in these groups of patients. Measurement might be applied to ambulatory settings, such as field studies, when autonomic activity has to be measured in a simple and non-invasive manner. As a consequence of findings of increased or attenuated alpha-amylase in clinical populations, the effects of psychotherapy, e.g. stress management training (Gaab et al., 2003) may be measured by biological parameters such as salivary alpha-amylase (La Marca, Nater, Gaab, & Ehlert, 2006).

Conclusion

The aim of this chapter was to evaluate the role of alpha-amylase activity in saliva as a potential indicator for stress responses of the human body. Although this role was already considered two decades ago, no attempts were made to scientifically elaborate it until the mid-nineties. A number of studies have shown that changes in alpha-amylase are indeed dependent on stressful stimuli of either physiological or psychological nature. However, methodological issues have prevented researchers from reaching a definitive conclusion. It was found that alpha-amylase in saliva responds very sensitively to psychosocial stress. But the question remains what the biological meaning behind this phenomenon might be? It might be worthwhile reconsidering the function of alpha-amylase, with the main property of this enzyme being its digestive action. Alpha-amylase hydrolyzes starch to glucose and maltose, initiating the digestion of starch in the oral cavity. Alpha-amylase has also been shown to have a bacterial interactive function (Scannapieco, Torres, & Levine, 1993). However, more studies are needed that examine cellular processes that occur after initiation of stress-related responses in the oral environment. It is difficult to infer an important function of short-term increases in alpha-amylase, because the biological meaning of a transient rise in anti-bacterial action of the enzyme remains elusive. However, it might be useful to the body when energy is set free by the increased digestive action due to stress. Physiological stress reactions comprise orchestrated actions within the whole body, setting the organism in a state of overall preparedness to engage in fight or flight. Thus, increases in amylase activity may be one of many actions to activate the resources of the body to cope with a stressful event or a threat to homeostasis. However, this explanation only applies to reactions to short-term acute stressors. Further studies are needed that examine changes of long-term alpha-amylase concentrations. If clinical or subclinical conditions that go together with an increase of activity in the autonomic nervous system result in a chronically elevated concentration of alpha-amylase, a sustained higher amylase activity and therefore digestive action might prove detrimental to oral health.

The studies summarized in this chapter clearly show that alpha-amylase is heightened in a state of stress, i.e. autonomic activation. This finding stands in line with results obtained in other studies that have found increases of alpha-amylase when the organism is exposed to either physiological or psychological stressors. Therefore, as a general conclusion, increases in salivary alpha-amylase might reflect changes in the autonomic nervous system. The release of alpha-amylase by acinar cells in the salivary glands is regulated by neuronal pathways. Acinar cells are innervated by both sympathetic and parasympathetic nerve fibers. A collaboration of parasympathetic and sympathetic inputs leads to an increased release of alpha-amylase via classic neurotransmitters (Turner & Sugiya, 2002). As shown in studies conducted on the rat parotid gland in vivo, parasympathetic stimulation evokes an output of saliva that has a large volume and a low protein concentration, while sympathetic stimulation has the opposite effect, causing the release of saliva that has a relatively small volume and high protein concentration (Garrett, 1999). Thus, during psychological stress, when autonomic activation is high, an increase in alpha-amylase can be observed. However, there is no evidence as to which branch of the autonomic nervous system is predominant in such a reaction. It would therefore be interesting to compare alpha-amylase with other sympathetic markers. Contradictory to the study of Chatterton et al. (1996), later studies did not find any relationship between the two catecholamines norepinephrine and epinephrine, and alpha-amylase (Ehlert, Erni, Hebisch, & Nater, 2006; Nater et al., 2006). This indicates that alpha-amylase is probably not an indicator for peripheral activation of catecholamines, although there are concomitant changes that are very similar to each other. Additional research is needed to elucidate the underlying physiological mechanisms of a stress-induced increase in alpha-amylase concentrations in more detail.

Future studies have to conclusively determine the fate of alpha-amylase as a parameter that may become part of a canon of biological stress markers. It seems to be of great interest to elucidate the mechanisms that are responsible for elevations in alpha-amylase as a response to stress. Although there is a variety of studies that have examined physiological mechanisms of amylase production and secretion in animals, studies in humans are scarce. Especially the use of pharmacological agents that inhibit or activate the autonomic nervous system might prove useful in this matter (e.g. beta-agonists such as isoproterenol, alpha-2-antagonists such as yohimbine). With these substances, more detailed insight into the autonomic branches responsible for increases in alpha-amylase might be gained. More invasive techniques are available in animal models. Experimental damaging of nerve fibres or direct stimulation of nerve cells is not feasible in a basic research context. It might be interesting, however, to experiment on electrical stimulation techniques in awake or anaesthesized humans, e.g. in the clinical context of a dental hospital. Measurement of direct sympathetic nerve activity via microneurography is considered the most accurate technique to reflect sympathetic activation (Grassi & Esler, 1999). Apart from peripheral measurements, the relationship between central parameters and alpha-amylase changes might prove very interesting. Since cerebrospinal fluid concentrations of norepinephrine reflect other mechanisms than peripheral catecholamines, it would be worthwhile to compare both amylase and central norepinephrine (Goldstein, 1998).

Summarizing, the future yields a multitude of possibilities to employ measurement of salivary alpha-amylase in different research areas. If further evaluated, salivary alpha-amylase might become a parameter playing an important role in the canon of biological stress markers.

REFERENCES

Aguirre, A., Levine, M. J., Cohen, R. E., & Tabak, L. A. (1987). Immunochemical quantitation of alpha-amylase and secretory IgA in parotid saliva from people of various ages. *Arch Oral Biol, 32*(4), 297-301.

al'Absi, M., & Wittmers, L. E. (2003). Enhanced adrenocortical responses to stress in hypertension-prone men and women. *Annals of Behavioral Medicine, 25*, 25-33.

Anderson, L. C., Garrett, J. R., Johnson, D. A., Kauffman, D. L., Keller, P. J., & Thulin, A. (1984). Influence of circulating catecholamines on protein secretion into rat parotid saliva during parasympathetic stimulation. *J Physiol, 352*, 163-171.

Artino, M., Dragomir, M., Ionescu, S., Badita, D., Nita, V., & Chitoi, E. (1998). Diurnal behaviour of some salivary parameters in patients with diabetes mellitus (protein concentration, amylase activity, density)--note I. *Rom J Physiol, 35*(1-2), 79-84.

Asking, B. (1985). Sympathetic stimulation of amylase secretion during a parasympathetic background activity in the rat parotid gland. *Acta Physiol Scand, 124*(4), 535-542.

Asking, B., & Gjorstrup, P. (1987). Synthesis and secretion of amylase in the rat parotid gland following autonomic nerve stimulation in vivo. *Acta Physiol Scand, 130*(3), 439-445.

Asking, B., & Proctor, G. B. (1989). Parasympathetic activation of amylase secretion in the intact and sympathetically denervated rat parotid gland. *Q J Exp Physiol, 74*(1), 45-52.

Batzri, S., & Selinger, Z. (1973). Enzyme secretion mediated by the epinephrine -receptor in rat parotid slices. Factors governing efficiency of the process. *J Biol Chem, 248*(1), 356-360.

Batzri, S., Selinger, Z., Schramm, M., & Robinovitch, M. R. (1973). Potassium release mediated by the epinephrine -receptor in rat parotid slices. Properties and relation to enzyme secretion. *J Biol Chem, 248*(1), 361-368.

Bellavia, S. L., Sanz, E. G., Chiarenza, A. P., Sereno, R., & Vermouth, N. T. (1990). Circadian rhythm of alpha-amylase in rat parotid gland. *Acta Odontol Latinoam, 5*(1), 13-23.

Ben-Aryeh, H., Gordon, N., Szargel, R., Toubi, E., & Laufer, D. (1993). Whole saliva in systemic lupus erythematosus patients. *Oral Surg Oral Med Oral Pathol, 75*(6), 696-699.

Birkhed, D., & Heintze, U. (1989). Salivary secretion rate, buffer capacity, and pH. In J. O. Tenovuo (Ed.), *Human saliva: Clinical chemistry and microbiology* (Vol. 1, pp. 25-73). Boca Raton: CRC Press.

Bishop, N. C., Walker, G. J., Scanlon, G. A., Richards, S., & Rogers, E. (2006). Salivary IgA responses to prolonged intensive exercise following caffeine ingestion. *Med Sci Sports Exerc, 38*(3), 513-519.

Borgeat, F., Chagon, G., & Legault, Y. (1984). Comparison of the salivary changes associated with a relaxing and with a stressful procedure. *Psychophysiology, 21*(6), 690-698.

Bosch, J. A., Brand, H. S., Ligtenberg, A. J., Bermond, B., Hoogstraten, J., & Nieuw Amerongen, A. V. (1998). The response of salivary protein levels and S-IgA to an academic examination are associated with daily stress. *Journal of Psychophysiology, 12*, 384-391.

Bosch, J. A., Brand, H. S., Ligtenberg, T. J., Bermond, B., Hoogstraten, J., & Nieuw Amerongen, A. V. (1996). Psychological stress as a determinant of protein levels and salivary-induced aggregation of Streptococcus gordonii in human whole saliva. *Psychosom Med, 58*(4), 374-382.

Bosch, J. A., De Geus, E. J., Veerman, E. C., Hoogstraten, J., & Nieuw Amerongen, A. V. (2003). Innate secretory immunity in response to laboratory stressors that evoke distinct patterns of cardiac autonomic activity. *Psychosom Med, 65*(2), 245-258.

Brown, C. C. (1970). The parotid puzzle: a review of the literature on human salivation and its applications to psychophysiology. *Psychophysiology, 7*(1), 65-85.

Busch, L., Sterin-Borda, L., & Borda, E. (2002). Differences in the regulatory mechanism of amylase release by rat parotid and submandibular glands. *Arch Oral Biol, 47*(10), 717-722.

Callegari, C., & Lami, F. (1984). Cigarette smoking and salivary amylase activity. *Gut, 25*(8), 909.

Calvo, F., Chicharro, J. L., Bandres, F., Lucia, A., Perez, M., Alvarez, J., et al. (1997). Anaerobic threshold determination with analysis of salivary amylase. *Can J Appl Physiol, 22*(6), 553-561.

Carlson, A. J., & Crittenden, A. L. (1910). The relation of ptyalin concentration to the diet and to the rate of secretion of the saliva. *American Journal of Physiology, 26*, 169.

Chatterton, R. T., Jr., Hill, P. D., Aldag, J. C., Hodges, K. R., Belknap, S. M., & Zinaman, M. J. (2000). Relation of plasma oxytocin and prolactin concentrations to milk production in mothers of preterm infants: influence of stress. *J Clin Endocrinol Metab, 85*(10), 3661-3668.

Chatterton, R. T., Jr., Vogelsong, K. M., Lu, Y. C., Ellman, A. B., & Hudgens, G. A. (1996). Salivary alpha-amylase as a measure of endogenous adrenergic activity. *Clin Physiol, 16*(4), 433-448.

Chatterton, R. T., Jr., Vogelsong, K. M., Lu, Y. C., & Hudgens, G. A. (1997). Hormonal responses to psychological stress in men preparing for skydiving. *J Clin Endocrinol Metab, 82*(8), 2503-2509.

Cole, S. W., Naliboff, B. D., Kemeny, M. E., Griswold, M. P., Fahey, J. L., & Zack, J. A. (2001). Impaired response to HAART in HIV-infected individuals with high autonomic nervous system activity. *Proc Natl Acad Sci U S A, 98*(22), 12695-12700.

Costa, P. T., Jr., Chauncey, H. H., Rose, C. L., & Kapur, K. K. (1980). Relationship of parotid saliva flow rate and composition with personality traits in healthy men. *Oral Surg Oral Med Oral Pathol, 50*(5), 416-422.

Coupland, N. J., Wilson, S. J., Potokar, J. P., Bell, C., & Nutt, D. J. (2003). Increased sympathetic response to standing in panic disorder. *Psychiatry Res, 118*(1), 69-79.

Crespi, H., Armando, I., Tumilasci, O., Levin, G., Massimo, J., Barontini, M., et al. (1982). Catecholamines levels and parotid secretion in children with chronic atopic dermatitis. *J Invest Dermatol, 78*(6), 493-497.

D'Alessandro, S., Curbelo, H. M., Tumilasci, O. R., Tessler, J. A., & Houssay, A. B. (1989). Changes in human parotid salivary protein and sialic acid levels during pregnancy. *Arch Oral Biol, 34*(10), 829-831.

Davidson, G. P., Koheil, A., & Forstner, G. G. (1978). Salivary amylase in cystic fibrosis: a marker of disordered autonomic function. *Pediatr Res, 12*(10), 967-970.

Dawes, C. (1974). Rhythms in salivary flow rate and composition. *Int J Chronobiol, 2*(3), 253-279.

Dawes, C. (1981). The effects of exercise on protein and electrolyte secretion in parotid saliva. *J Physiol, 320*, 139-148.

Dutta, S. K., Orestes, M., Vengulekur, S., & Kwo, P. (1992). Ethanol and human saliva: effect of chronic alcoholism on flow rate, composition, and epidermal growth factor. *Am J Gastroenterol, 87*(3), 350-354.

Ehlert, U., Erni, K., Hebisch, G., & Nater, U. (2006). Salivary alpha-amylase levels after yohimbine challenge in healthy men. *J Clin Endocrinol Metab, 91*(12), 5130-5133.

Enberg, N., Alho, H., Loimaranta, V., & Lenander-Lumikari, M. (2001). Saliva flow rate, amylase activity, and protein and electrolyte concentrations in saliva after acute alcohol consumption. *Oral Surg Oral Med Oral Pathol Oral Radiol Endod, 92*(3), 292-298.

Ericson, T., & Nordlund, A. (1993). A new device for collection of parotid saliva. *Ann N Y Acad Sci, 694*, 274-275.

Fredrickson, M., Tuomisto, M., & Bergman-Losman, B. (1991). Neuroendocrine and cardiovascular stress reactivity in middle-aged normotensive adults with parental history of cardiovascular disease. *Psychophysiology, 28*, 656-664.

Friedman, B. H., Thayer, J. F., Borkovec, T. D., Tyrrell, R. A., Johnson, B. H., & Columbo, R. (1993). Autonomic characteristics of nonclinical panic and blood phobia. *Biol Psychiatry, 34*(5), 298-310.

Froehlich, D. A., Pangborn, R. M., & Whitaker, J. R. (1987). The effect of oral stimulation on human parotid salivary flow rate and alpha-amylase secretion. *Physiol Behav, 41*(3), 209-217.

Gaab, J., Blattler, N., Menzi, T., Pabst, B., Stoyer, S., & Ehlert, U. (2003). Randomized controlled evaluation of the effects of cognitive-behavioral stress management on cortisol responses to acute stress in healthy subjects. *Psychoneuroendocrinology, 28*(6), 767-779.

Garrett, J. R. (1999). Effects of autonomic nerve stimulations on salivary parenchyma and protein secretion. In J. R. Garrett, J. Ekström & L. C. Anderson (Eds.), *Neural mechanisms of salivary gland secretion. Front. Oral Biol.* (Vol. 11, pp. 59-79). Basel: Karger.

Gilman, S. C., Fischer, G. J., Biersner, R. J., Thornton, R. D., & Miller, D. A. (1979). Human parotid alpha-amylase secretion as a function of chronic hyperbaric exposure. *Undersea Biomed Res, 6*(3), 303-307.

Gilman, S. C., Thornton, R., Miller, D., & Biersner, R. (1979). Effects of exercise stress on parotid gland secretion. *Horm Metab Res, 11*(7), 454.

Goldstein, D. S. (1998). Catecholamines in the periphery. Overview. *Adv Pharmacol, 42*, 529-539.

Gordis, E. B., Granger, D. A., Susman, E. J., & Trickett, P. K. (2006). Asymmetry between salivary cortisol and alpha-amylase reactivity to stress: relation to aggressive behavior in adolescents. *Psychoneuroendocrinology, 31*(8), 976-987.

Gordon, C. R., Ben-Aryeh, H., Spitzer, O., Doweck, I., Gonen, A., Melamed, Y., et al. (1994). Seasickness susceptibility, personality factors, and salivation. *Aviat Space Environ Med, 65*(7), 610-614.

Granger, D. A., Kivlighan, K. T., Blair, C., El-Sheikh, M., Mize, J., Lisonbee, J. A., et al. (2006). Integrating the measurement of salivary alpha-amylase into studies of child health, development, and social relationships. *Journal of Personal and Social Relationships, 23*, 267-290.

Grassi, G., & Esler, M. (1999). How to assess sympathetic activity in humans. *J Hypertens, 17*(6), 719-734.

Gubern, G., Canalias, F., & Gella, F. J. (1995). Determination of alpha-amylase activity: methods comparison and commutability study of several control materials. *Clin Chem, 41*(3), 435-438.

Gwirtsman, H. E., Kaye, W. H., George, D. T., Carosella, N. W., Greene, R. C., & Jimerson, D. C. (1989). Hyperamylasemia and its relationship to binge-purge episodes: development of a clinically relevant laboratory test. *J Clin Psychiatry, 50*(6), 196-204.

Hoek, G. H., Brand, H. S., Veerman, E. C., & Amerongen, A. V. (2002). Toothbrushing affects the protein composition of whole saliva. *Eur J Oral Sci, 110*(6), 480-481.

Jenzano, J. W., Brown, C. K., & Mauriello, S. M. (1987). Temporal variations of glandular kallikrein, protein and amylase in mixed human saliva. *Arch Oral Biol, 32*(10), 757-759.

Kasperczyk, S., Brzoza, Z., Kasperczyk, A., Beck, B., Duiban, H., & Mertas, A. (2001). The changes of alpha-amylase activity in serum and different tissues of female rat during sex cycle--isoelectrofocusing studies of alpha-amylase. *Med Sci Monit, 7*(1), 49-53.

Kirschbaum, C., & Hellhammer, D. H. (1989). Salivary cortisol in psychobiological research: an overview. *Neuropsychobiology, 22*(3), 150-169.

Kirschbaum, C., & Hellhammer, D. H. (1994). Salivary cortisol in psychoneuroendocrine research: recent developments and applications. *Psychoneuroendocrinology, 19*(4), 313-333.

Kirschbaum, C., & Hellhammer, D. H. (1999). Hypothalamus-Hypophysen Nebennierenrindenachse. In C. Kirschbaum & D. H. Hellhammer (Eds.), *Psychoendokrinologie und Psychoimmunologie. Enzyklopädie der Psychologie, Serie 1, Biologische Psychologie, Band 3* (pp. 79-140). Göttingen: Hogrefe.

Kirschbaum, C., Pirke, K. M., & Hellhammer, D. H. (1993). The 'Trier Social Stress Test'--a tool for investigating psychobiological stress responses in a laboratory setting. *Neuropsychobiology, 28*(1-2), 76-81.

Kivlighan, K. T., & Granger, D. A. (2006). Salivary alpha-amylase response to competition: relation to gender, previous experience, and attitudes. *Psychoneuroendocrinology, 31*(6), 703-714.

Kivlighan, K. T., Granger, D. A., & Schwartz, E. B. (2005). Blood contamination and the measurement of salivary progesterone and estradiol. *Horm Behav, 47*(3), 367-370.

Kivlighan, K. T., Granger, D. A., Schwartz, E. B., Nelson, V., Curran, M., & Shirtcliff, E. A. (2004). Quantifying blood leakage into the oral mucosa and its effects on the measurement of cortisol, dehydroepiandrosterone, and testosterone in saliva. *Horm Behav, 46*(1), 39-46.

Klein, L. C., Whetzel, C. A., Ritter, F. E., & Granger, D. A. (2006). Effects of caffeine and stress on salivary alpha-amylase in young men: a salivary biomarker of sympathetic activity - Abstract. *Psychosomatic Medicine, 68*(1), A4.

Kronvall, P., Fahy, T. A., Isaksson, A., & Theander, S. (1992). The clinical relevance of salivary amylase monitoring in bulimia nervosa. *Biological Psychiatry, 32*(2), 156-163.

La Marca, R., Nater, U. M., Gaab, J., & Ehlert, U. (2006). Does cognitive-behavioral stress management training have a long-term effect on salivary alpha-amylase reactivity to acute stress? - Abstract. *International Journal of Behavioral Medicine, 13* (Suppl.), 300-301.

Laederach-Hofmann, K., Mussgay, L., Buchel, B., Widler, P., & Ruddel, H. (2002). Patients with erythrophobia (fear of blushing) show abnormal autonomic regulation in mental stress conditions. *Psychosom Med, 64*(2), 358-365.

Laine, M., Pienihakkinen, K., Ojanotko-Harri, A., & Tenovuo, J. (1991). Effects of low-dose oral contraceptives on female whole saliva. *Arch Oral Biol, 36*(7), 549-552.

Laine, M., Tenovuo, J., Lehtonen, O. P., Ojanotko-Harri, A., Vilja, P., & Tuohimaa, P. (1988). Pregnancy-related changes in human whole saliva. *Arch Oral Biol, 33*(12), 913-917.

Lang, P. J., Davis, M., & Ohman, A. (2000). Fear and anxiety: animal models and human cognitive psychophysiology. *J Affect Disord, 61*(3), 137-159.

Lashley, K. S. (1916). The human salivary reflex and its use in psychology. *Psychological Reviews, 23*, 446.

Laurikainen, K., Laurikainen, E., Tenovuo, J., Kaila, T., & Vilja, P. (1988). Effects of a beta-blocking agent, timolol maleate, on saliva in healthy volunteers. *Scand J Dent Res, 96*(2), 121-127.

Ljungberg, G., Ericson, T., Ekblom, B., & Birkhed, D. (1997). Saliva and marathon running. *Scand J Med Sci Sports, 7*(4), 214-219.

Macgregor, I. D. (1989). Effects of smoking on oral ecology. A review of the literature. *Clin Prev Dent, 11*(1), 3-7.

Makinen, K. K. (1989). Salivary enzymes. In J. O. Tenovuo (Ed.), *Human saliva: Clinical chemistry and microbiology* (Vol. 2, pp. 93-119). Boca Raton: CRC Press.

Milad, M. P., Klock, S. C., Moses, S., & Chatterton, R. (1998). Stress and anxiety do not result in pregnancy wastage. *Hum Reprod, 13*(8), 2296-2300.

Moret, S., Coudert, J. L., Bejat, C., Robin, O., & Lissac, M. (1993). The influence of basal anxiety on unstimulated parotid and submandibular saliva. *Arch Oral Biol, 38*(9), 751-754.

Morrison, W. E., Haas, E. C., Shaffner, D. H., Garrett, E. S., & Fackler, J. C. (2003). Noise, stress, and annoyance in a pediatric intensive care unit. *Crit Care Med, 31*(1), 113-119.

Morse, D. R., Schacterle, G. R., Esposito, J. V., Chod, S. D., Furst, M. L., DiPonziano, J., et al. (1983). Stress, meditation and saliva: a study of separate salivary gland secretions in endodontic patients. *J Oral Med, 38*(4), 150-160.

Morse, D. R., Schacterle, G. R., Esposito, J. V., Furst, M. L., & Bose, K. (1981). Stress, relaxation and saliva: a follow-up study involving clinical endodontic patients. *J Human Stress, 7*(3), 19-26.

Morse, D. R., Schacterle, G. R., Furst, L., Zaydenberg, M., & Pollack, R. L. (1989). Oral digestion of a complex-carbohydrate cereal: effects of stress and relaxation on physiological and salivary measures. *Am J Clin Nutr, 49*(1), 97-105.

Morse, D. R., Schacterle, G. R., Furst, M. L., & Bose, K. (1981). Stress, relaxation, and saliva: a pilot study involving endodontic patients. *Oral Surg Oral Med Oral Pathol, 52*(3), 308-313.

Morse, D. R., Schacterle, G. R., Furst, M. L., Brokenshire, J., Butterworth, M., & Cacchio, J. (1981). Examination-induced stress in meditators and non-meditators as measured by salivary protein changes. *Stress, 2*, 20-24.

Morse, D. R., Schacterle, G. R., Furst, M. L., Goldberg, J., Greenspan, B., Swiecinski, D., et al. (1982). The effect of stress and meditation on salivary protein and bacteria: a review and pilot study. *J Human Stress, 8*(4), 31-39.

Morse, D. R., Schacterle, G. R., Martin, J. S., DiPonziano, J., Zaydenberg, M., Esposito, J. V., et al. (1984). Stress and relaxation: evaluation by questionnaire, salivary changes and physiological responses in a trained meditator. *J Oral Med, 39*(3), 142-147.

Morse, D. R., Schacterle, G. R., Zaydenberg, M., Furst, M. L., Pollack, R. L., Moskow, A., et al. (1983). Salivary volume and amylase activity. I: Relaxation versus relaxed chewing. *J Am Soc Psychosom Dent Med, 30*(3), 85-96.

Nagaya, T., & Okuno, M. (1993). No effects of smoking or drinking habits on salivary amylase. *Toxicol Lett, 66*(3), 257-261.

Nagler, R., Lischinsky, S., Diamond, E., Drigues, N., Klein, I., & Reznick, A. Z. (2000). Effect of cigarette smoke on salivary proteins and enzyme activities. *Arch Biochem Biophys, 379*(2), 229-236.

Nater, U. M. (2004). *The role of salivary alpha-amylase in stress research.* Göttingen: Cuvillier.

Nater, U. M., Abbruzzese, E., Krebs, M., & Ehlert, U. (2006). Sex differences in emotional and psychophysiological responses to musical stimuli. *Int J Psychophysiol, 62*(2), 300-308.

Nater, U. M., La Marca, R., Florin, L., Moses, A., Langhans, W., Koller, M. M., et al. (2006). Stress-induced changes in human salivary alpha-amylase activity -- associations with adrenergic activity. *Psychoneuroendocrinology, 31*(1), 49-58.

Nater, U. M., Rohleder, N., Gaab, J., Berger, S., Jud, A., Kirschbaum, C., et al. (2005). Human salivary alpha-amylase reactivity in a psychosocial stress paradigm. *Int J Psychophysiol, 55*(3), 333-342.

Nater, U. M., Rohleder, N., Schlotz, W., Ehlert, U., Kirschbaum, C. (2007). Determinants of the diurnal course of salivary alpha-amylase. *Psychoneuroendocrinology, 32,* 392-401.

Navazesh, M. (1993). Methods for collecting saliva. *Ann N Y Acad Sci, 694*, 72-77.

Nederfors, T., & Dahlof, C. (1996). Effects on salivary flow rate and composition of withdrawal of and re-exposure to the beta 1-selective antagonist metoprolol in a hypertensive patient population. *Eur J Oral Sci, 104*(3), 262-268.

Nederfors, T., Ericsson, T., Twetman, S., & Dahlof, C. (1994). Effects of the beta-adrenoceptor antagonists atenolol and propranolol on human parotid and submandibular-sublingual salivary secretion. *J Dent Res, 73*(1), 5-10.

Nexo, E., Hansen, M. R., & Konradsen, L. (1988). Human salivary epidermal growth factor, haptocorrin and amylase before and after prolonged exercise. *Scand J Clin Lab Invest, 48*(3), 269-273.

Nierop, A., Bratsikas, A., Klinkenberg, A., Nater, U. M., Zimmermann, R., & Ehlert, U. (2006). Prolonged Salivary Cortisol Recovery in Second-Trimester Pregnant Women and Attenuated Salivary {alpha}-Amylase Responses to Psychosocial Stress in Human Pregnancy. *J Clin Endocrinol Metab, 91*(4), 1329-1335.

Noto, Y., Sato, T., Kudo, M., Kurata, K., & Hirota, K. (2005). The relationship between salivary biomarkers and state-trait anxiety inventory score under mental arithmetic stress: a pilot study. *Anesth Analg, 101*(6), 1873-1876.

Rantonen, P. J., & Meurman, J. H. (1998). Viscosity of whole saliva. *Acta Odontol Scand, 56*(4), 210-214.

Riad, M., Barton, J. R., Wilson, J. A., Freeman, C. P., & Maran, A. G. (1991). Parotid salivary secretory pattern in bulimia nervosa. *Acta Otolaryngol, 111*(2), 392-395.

Rohleder, N., Nater, U. M., Wolf, J. M., Ehlert, U., & Kirschbaum, C. (2004). Psychosocial stress-induced activation of salivary alpha-amylase: an indicator of sympathetic activity? *Ann N Y Acad Sci, 1032*, 258-263.

Rohleder, N., Wolf, J. M., Herpfer, I., Fiebich, B. L., Kirschbaum, C., & Lieb, K. (2006). No response of plasma substance P, but delayed increase of interleukin-1 receptor antagonist to acute psychosocial stress. *Life Sci, 78*(26), 3082-3089.

Rohleder, N., Wolf, J. M., Maldonado, E. F., & Kirschbaum, C. (2006). The psychosocial stress-induced increase in salivary alpha-amylase is independent of saliva flow rate. *Psychophysiology, 43*(6), 645-652.

Salvolini, E., Di Giorgio, R., Curatola, A., Mazzanti, L., & Fratto, G. (1998). Biochemical modifications of human whole saliva induced by pregnancy. *Br J Obstet Gynaecol, 105*(6), 656-660.

Salvolini, E., Mazzanti, L., Martarelli, D., Di Giorgio, R., Fratto, G., & Curatola, G. (1999). Changes in the composition of human unstimulated whole saliva with age. *Aging (Milano), 11*(2), 119-122.

Scannapieco, F. A., Torres, G., & Levine, M. J. (1993). Salivary alpha-amylase: role in dental plaque and caries formation. *Crit Rev Oral Biol Med, 4*(3-4), 301-307.

Scheutzel, P., & Gerlach, U. (1991). [Alpha-amylase isoenzymes in serum and saliva of patients with anorexia and bulimia nervosa]. *Z Gastroenterol, 29*(7), 339-345.

Selinger, Z., Batzri, S., Eimerl, S., & Schramm, M. (1973). Calcium and energy requirements for K + release mediated by the epinephrine -receptor in rat parotid slices. *J Biol Chem, 248*(1), 369-372.

Seugnet, L., Boero, J., Gottschalk, L., Duntley, S. P., & Shaw, P. J. (2006). Identification of a biomarker for sleep drive in flies and humans. *Proc Natl Acad Sci U S A, 103*(52), 19913-19918.

Shalev, A. Y., & Rogel-Fuchs, Y. (1993). Psychophysiology of the posttraumatic stress disorder: from sulfur fumes to behavioral genetics. *Psychosom Med, 55*(5), 413-423.

Shea, A. K., Steiner, M., Brennan, P., Walker, E., Newport, D. J., & Stowe, Z. N. (2006). Maternal depression and salivary alpha-amylase response to stress in their infants - Abstract. *Psychosomatic Medicine, 68*(1), A5.

Skosnik, P. D., Chatterton, R. T., Jr., Swisher, T., & Park, S. (2000). Modulation of attentional inhibition by norepinephrine and cortisol after psychological stress. *Int J Psychophysiol, 36*(1), 59-68.

Skov Olsen, P., Kirkegaard, P., Rasmussen, T., Magid, E., Poulsen, S. S., & Nexo, E. (1988). Adrenergic effects on secretion of amylase from the rat salivary glands. *Digestion, 41*(1), 34-38.

Söderling, E. (1989). Practical aspects of salivary analyses. In J. O. Tenovuo (Ed.), *Human saliva: Clinical chemistry and microbiology* (Vol. 1, pp. 1-24). Boca Raton: CRC Press.

Southwick, S. M., Krystal, J. H., Morgan, C. A., Johnson, D., Nagy, L. M., Nicolaou, A., et al. (1993). Abnormal noradrenergic function in posttraumatic stress disorder. *Arch Gen Psychiatry, 50*(4), 266-274.

Speirs, R. L., Herring, J., Cooper, W. D., Hardy, C. C., & Hind, C. R. (1974). The influence of sympathetic activity and isoprenaline on the secretion of amylase from the human parotid gland. *Arch Oral Biol, 19*(9), 747-752.

Sproles, A., & Schaeffer, L. D. (1974). An advanced design of the Carlson-Crittenden cup for collection of parotid fluid. *Biomater Med Devices Artif Organs, 2*(1), 95-101.

Squires, B. (1953). Human salivary amylase secretion in relation to diet. *Journal of Physiology, 49*, 153-156.

Sramek, J. J., Cutler, N. R., Hurley, D. J., & Seifert, R. D. (1995). The utility of salivary amylase as an evaluation of M3 muscarinic agonist activity in Alzheimer's disease. *Prog Neuropsychopharmacol Biol Psychiatry, 19*(1), 85-91.

Steerenberg, P. A., van Asperen, I. A., Van Nieuw Amerongen, A., Biewenga, J., Mol, D., & Medema, G. (1997). Salivary levels of immunoglobulin A in triathletes. *Eur J Oral Sci, 105*, 305-309.

Stroud, L. R., Handwerger, K., Granger, D. A., & Kivlighan, K. T. (2006). Salivary alpha-amylase stress reactivity in children and adolescents: validity, associations with cortisol, and links to behavior - Abstract. *Psychosomatic Medicine, 68*(1), A5.

Takai, N., Yamaguchi, M., Aragaki, T., Eto, K., Uchihashi, K., & Nishikawa, Y. (2004). Effect of psychological stress on the salivary cortisol and amylase levels in healthy young adults. *Arch Oral Biol, 49*(12), 963-968.

Tu, M. T., Lupien, S. J., & Walker, C. D. (2006). Multiparity reveals the blunting effect of breastfeeding on physiological reactivity to psychological stress. *J Neuroendocrinol, 18*(7), 494-503.

Turner, R. J., & Sugiya, H. (2002). Understanding salivary fluid and protein secretion. *Oral Dis, 8*(1), 3-11.

Udupa, S. L., Prabhakar, A. R., & Tandon, S. (1989). Alpha-amylase inhibitors in foodstuffs. *Food Chemistry, 34*, 95-101.

van Stegeren, A., Rohleder, N., Everaerd, W., & Wolf, O. T. (2006). Salivary alpha amylase as marker for adrenergic activity during stress: effect of betablockade. *Psychoneuroendocrinology, 31*(1), 137-141.

Veerman, E. C., van den Keybus, P. A., Vissink, A., & Nieuw Amerongen, A. V. (1996). Human glandular salivas: their separate collection and analysis. *Eur J Oral Sci, 104*(4 (Pt 1)), 346-352.

Vining, R. F., & McGinley, R. A. (1987). The measurement of hormones in saliva: possibilities and pitfalls. *J Steroid Biochem, 27*(1-3), 81-94.

Walsh, N. P., Blannin, A. K., Clark, A. M., Cook, L., Robson, P. J., & Gleeson, M. (1999). The effects of high-intensity intermittent exercise on saliva IgA, total protein and alpha-amylase. *J Sports Sci, 17*(2), 129-134.

West, S. G., Granger, D. A., Kivlighan, K. T., & Hurston, K. L. (2006). Salivary alpha-amylase response to the cold pressor is corraled with cardiac markers of sympathetic activation - Abstract. *Psychosomatic Medicine, 68*(1), A4-A5.

White, K. D. (1977). Salivation: a review and experimental investigation of major techniques. *Psychophysiology, 14*(2), 203-212.

Xiao, Y., Via, D., Kyle, R., Mackenzie, C. F., & Burton, P. (2000). Stress with simulated trauma management measured by salivary amylase. *Anesthesiology, 93*(34), 1226.

Yamaguchi, M., Deguchi, M., & Miyazaki, Y. (2006). The effects of exercise in forest and urban environments on sympathetic nervous activity of normal young adults. *J Int Med Res, 34*(2), 152-159.

Yamaguchi, M., Takeda, K., Onishi, M., Deguchi, M., & Higashi, T. (2006). Non-verbal communication method based on a biochemical marker for people with severe motor and intellectual disabilities. *J Int Med Res, 34*(1), 30-41.

Zakowski, J. J., & Bruns, D. E. (1985). Biochemistry of human alpha amylase isoenzymes. *Crit Rev Clin Lab Sci, 21*(4), 283-322.

Zappacosta, B., Persichilli, S., De Sole, P., Mordente, A., & Giardina, B. (1999). Effect of smoking one cigarette on antioxidant metabolites in the saliva of healthy smokers. *Arch Oral Biol, 44*(6), 485-488.

Zappacosta, B., Persichilli, S., Mordente, A., Minucci, A., Lazzaro, D., Meucci, E., et al. (2002). Inhibition of salivary enzymes by cigarette smoke and the protective role of glutathione. *Hum Exp Toxicol, 21*(1), 7-11.

Zuabi, O., Machtei, E. E., Ben-Aryeh, H., Ardekian, L., Peled, M., & Laufer, D. (1999). The effect of smoking and periodontal treatment on salivary composition in patients with established periodontitis. *J Periodontol, 70*(10), 1240-1246.

In: Psychoneuroendocrinology Research Trends
Editor: Martina T. Czerbska, pp. 149-160

ISBN: 978-1-60021-665-7

Chapter IV

BORDERLINE PERSONALITY DISORDER, GENDER AND SEROTONIN: DOES ESTROGEN PLAY A ROLE?

M. Catherine DeSoto
University of Northern Iowa

ABSTRACT

The effect of estrogen on brain function and behavior has been well established in both humans and other animals. Although the changes in hormone levels that occur as part of the normal menstrual cycle in women influence neurochemistry, the changes themselves have not been systematically considered as variables of interest. As a redress, current literature is reviewed and it is proposed that the degree of natural estrogen flux is itself an individual difference variable worthy of study. An illustration of how such a model is possible is presented based on what is known about brain function focusing on speculation that serotonin receptors of subtype 1A could be a plausible vehicle for the effects of estrogen flux to occur. The literature regarding serotonin function and borderline personality disorder among women is reviewed and how the proposed model might account for various discrepancies in the research on borderline personality disorder and serotonin system reactivity is considered. As a whole, the theoretical model presented in which estrogen changes themselves can aggravate borderline personality disorder symptomology is shown to have preliminary support from several lines of research.

The effect of estrogen on brain function and behavior can now be characterized as well established in both humans and other animals. Although the changes in hormone levels that occur as part of the normal menstrual cycle in women influence neurochemistry, the changes themselves have not been systematically considered as variables of interest. As a redress, current literature is reviewed and it is proposed that the degree of natural estrogen flux is itself an individual difference variable worthy of study. An illustration of how such a model is possible is presented based on what is known about brain function focusing with speculation that the serotonin system and estrogenic effects on serotonin receptors in particular could be a plausible vehicle for the effects of estrogen flux to occur.

Despite the fact that research on the disorder accounts for approximately 40% of all publications on the topic of personality disorders (Widiger & Frances, 1989), the etiology of BPD and the factors that modify its expression are still poorly understood. Various research findings from a variety of fields have suggested the possibility that human beta-17 estradiol (hereafter referred to as estradiol, or simply estrogen) plays a significant role in the course and etiology of borderline personality disorder (BPD). Recent research has suggested that a previously uninvestigated aspect of a women's hormonal profile—individual differences in the degree of estrogen fluctuation -- may be specifically associated with symptoms of BPD (DeSoto, Geary, Hoard, Sheldon & Cooper, 2003). The plausibility of a possible underlying mechanism of this link -- estrogenic effects on a serotonin receptor-- in the subject of this paper. Thus, the following discussion presents a theoretical overview of how this might occur neurochemically and details how such a hypothesis fits nicely with existing literature.

First, a review of some research findings that set the stage for the hypothesis will be presented followed by a discussion of a proposed theoretical model of how estrogen might affect BPD. Finally, an overview of the results of initial testing of the hypothesis will be presented.

Setting the Stage

This will be divided into three sections. First, some pertinent research regarding the life course of BPD is reviewed; there is a gender difference favoring women which emerges around adolescence, followed by a worsening of symptoms in the late forties and early fifties for women sufferers. Second, a brief review of estrogen's multiple effects on the serotonin system. And, finally, a review of the research that borderline personality has been linked to serotonin functioning in a variety of ways.

Life Course and Gender Differences in Borderline Personality

As stated, BPD is more common among women and there is evidence that the sex difference in prevalence is more pronounced during times of rapid hormonal fluctuations, that is, during adolescence (Myers, Burkett & Otto, 1993) and the late forties (Bardenstein & McGlashen, 1988). Whenever a sex difference exists for a particular trait and emerges or becomes larger around puberty, the role of sex hormones in the expression of the sex difference, though not proven, is often seen as a distinct possibility (Seeman, 1997). Of some interest to this question, Myers et al. (1993) reported an unusually high rate of BPD (47%) among their sample of inpatient female adolescents. Conversely, none of the inpatient males satisfied the diagnostic criteria. Pertinent to the question of whether this difference might be explainable by some other difference among the male and female adolescent inpatient populations, Grilo et al. (1996) found that the rate of BPD differed among males and females in this age group of inpatients, but no other personality, demographic or diagnostic variable distinguished the two groups. Women and men commonly exhibit a lessening of symptoms during the later twenties, thirties and early forties. But, starting in the middle to late forties, the course of those with BPD begins to differ for women and men. Bardenstein and McGlashen (1988) reported that among persons diagnosed with BPD, women who were in their late forties and early fifties exhibited a greater amount of symptoms than did the who were in their forties and fifties. Although these findings might relate to life events (such as the "empty nest syndrome" or marital break up that are sometimes seen as more salient in a woman's life), hormonal fluctuations could play a role as well and should be ruled out without cause. In sum, we have a sex difference that did not previously exist until hormonal fluctuations begin and which may worsen for women during the time women go through a second time of rapidly shifting hormone levels. In general, this type of pattern is certainly at least suggestive of a role for hormonal factors in the etiology of borderline personality, but the importance of female sex hormones has only begun to be been thoroughly investigated. Results of the initial studies will be reviewed after an overview of estrogen and serotonin and a discussion of the model that may account for the findings is presented.

Estrogen and Serotonin

Research conducted over the past decade has demonstrated that estrogen has a clear-cut effect on the functioning of the serotonin system (Bethea, Pecins-Thompson, Schutzer & Gundlah, 1998; Fink et al., 1998). For example, Bethea and colleagues (see Bethea et al., 1998 for a review) have demonstrated that sex steroids, including estrogen, directly affect the function of the serotonin system on a variety of levels, including the functioning of both afferent and efferent serotonin neurons as well as receptor cites. It was first reported several decades ago that serotonin as well as the principal metabolite of serotonin, 5-HIAA, are higher during high estrogen phases of the menstrual cycle (Fludder & Tonge, 1975). Of course, since so many hormones change across a natural menstrual cycle, one could only speculate from this type of finding that estrogen was the hormone causing the effect on the serotonin system. However, more recent research has clarified the relationship considerably by documenting an increase in serotonin metabolite following exogenous estrogen supplementation (Lippert et al., 1996; Mueck et al., 1997). Although some details have yet to be worked out, it can at least be said that rising estrogen levels down regulate certain serotonin receptors (specifically 5-HT_{1A} receptors, Osterlund & Hurd, 1999) while up-regulating 5-HT_{2A} serotonin receptors (Sumner & Fink, 1995).

GENDER, BPD AND CHALLENGE STUDIES OF SEROTONIN FUNCTION

The relationship between BPD and the neurotransmitter serotonin has been investigated using challenge studies. Challenge studies are the gold standard to determine if an individual's central nervous system responds normally to the particular neurotransmitter of interest. To assess serotonin responsivity, an agent that triggers the release of serotonin (for example, fenfluramine) is administered. Specific hormone levels whose release is mediated by serotonin (such as prolactin) are then directly measured. A reduced response is indicative of a problem with the serotonin system, suggesting it is not as responsive as is typically found. Such challenge studies have directly compared personality disordered males, including a sub-set diagnosed with BPD, to males with affective disorders and to a non-disordered control group (Coccaro et al., 1989; Trestman, Coccoro & Temple, 1992). The prolactin responses of participants with BPD were significantly different from other personality disordered participants, suggesting that reduced responsiveness in the serotonin system may be a hallmark of borderline personality.

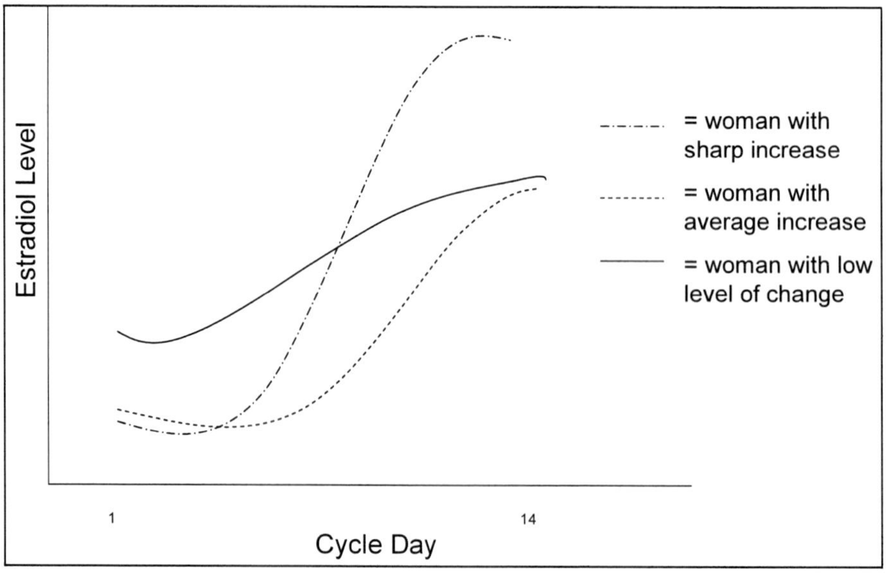

Figure 1. Women with high, low and average amount of estrogen fluctuation. Note that the idea of level of FLUCTUATION is different that the absolute level by comparing the average flux line with the low level of change profile. The absolute level would be the same at mid cycle, but the amount of change that has occurred is different.

In general, women appear to have more active serotonin systems than men, as measured by the level of serotonin metabolite (Bucht et al., 1981), and greater responsiveness to serotonin as measured by challenge studies (Goodwin et al., 1994). However, serotonin responsiveness does seem to vary as a function of menstrual cycle (Su et al., 1997). Commonly, men have been the preferred subjects in challenge studies of serotonin reactivity in relation to various psychopathologies to avoid the confounding effects of varying levels of hormones that women normally experience across a month (since as above these hormonal changes affect serotonin function). Ironically, as is argued in this chapter, it could be that these natural fluctuations in hormones are not a confound at all – but may themselves be an important factor in understanding the myriad causes that culminate in a diagnosis of BPD. Figure 1 illustrates what is meant by individual differences in the degree of hormonal fluctuation (or hormone flux).

Proposed Link between Estrogen, Serotonin and Borderline Personality

Research suggests that a rise in estrogen leads both to an up-regulation of 5-HT_2 receptors as well as a decrease in serotonin 5-HT_{1A} receptor levels. This suggests estrogen may lower 5 -HT_2 bringing forth up regulation, and raise 5-HT_{1A} levels, at least initially. First, focus on 5-HT_{1A} levels, although similar logic would apply to 5-HT_2. If estrogen raises 5-HT_{1A} and then the receptors are then down regulated as an adaptation all is well. But if the estrogen that brought forth the increase in 5-HT_{1A} is abruptly withdrawn the adaptation would be a problem. The combination of decreased receptor sites and lower 5-HT_{1A} serotonin levels would translate to an even more pronounced paucity of serotonin. In other words, if a sharp reduction in estrogen follows (as normally occurs in cycling women) the brain's deprivation of 5-HT_{1A} would conceivably be doubled: less estrogen coupled with less responsive receptors. Although this may be something that most women's serotonin system handles with relative ease, if one's serotonin system is already somewhat dysfunctional, then the series of up regulations and down regulations could be problematic. See Figure 2 for an illustration. It seems plausible that among those with BPD, serotonin responsiveness is somehow defective. Since estrogen changes affect serotonin responsiveness, it seems likely that repeated estrogen fluctuations themselves could exacerbate the problem in such a context.

It is not far fetched at this point to at least speculate that the effect of estrogen variation on the serotonin system among women with BPD might be expected to be especially pronounced. This idea is evaluated next. For example, during those times in the menstrual cycle where rises in estrogen occur (late follicular and mid-luteal) and which are normally associated with an increase in serotonin reactivity, women with BPD might show an unusually large increase in serotonin function. Similarly, times of low estrogen, which are normally associated with a decline in serotonin functioning, might be associated with a larger decline among persons with BPD.

As noted above, serotonin response is usually found to be reduced in patients with BPD. However, as mentioned many of the studies that investigate borderline personality from a neuroendcrine perspective have used only men, even though the majority of those suffering from the disorder are women (Martial et al., 1997). It is, to me, somewhat surprising that results about neuroendocrine function taken from men would be assumed to generalize to a disorder that is primarily women. Clearly, the hormonal functioning would not be identical for men and women. Even if at some point in the past one might argue that the differences sex hormones (such as estrogen) would not be relevant for conclusions about brain function, this time has passed. As reviewed above, the influence of estrogen on brain function is certainly beyond dispute. Before results from males are generalized to females, or vice versa, the burden of proof is to show there are not differences for the variables of interest in males and females brain responses. In challenge studies where women are included, women who have BPD do NOT always show the blunted response that is typical with men (Hollander, 1994). This point should not be underestimated. It suggests that there is something different in the neurochemistry of males and females who exhibit the behaviors of borderline personality. One possibility is that basic differences in serotonin system of males and females are pertinent. Another possibility is that perhaps serotonin function plays an etiological role for borderline personality only among men. A third possibility, and one which this author favors, is that the inconsistent findings with women are because serotonin responsiveness varies across the menstrual cycle, perhaps particularly so among women with borderline personality. It seems plausible that among those with borderline personality, serotonin responsiveness is somehow defective. Since estrogen changes affect serotonin responsiveness, it seems likely that repeated estrogen fluctuations themselves could exacerbate the problem in such a context. This is the fundamental idea that this chapter advances.

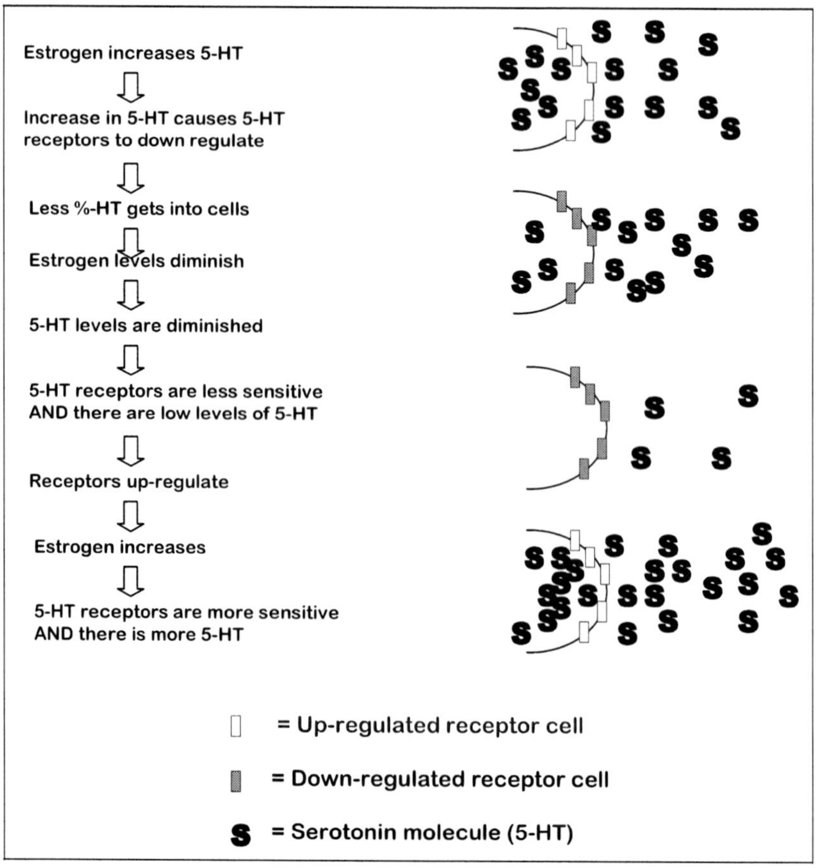

Figure 2. A series of estrogen surges and withdrawals could alternatively flood and deprive the neurons with estrogen if the effects of up and down regulation are taken into account.

If women with the diagnosis of BPD have any sort of preexisting problem with their serotonin systems, any substance that taxes this system might be a further exacerbation. If the amount of such a substance was constantly changing, it might be especially hard for the serotonin system to maintain stability. The changing levels of estrogen, which as reviewed above do affect serotonin function, might put such an additional strain on the serotonin system of person with any weakness in the serotonin system. If the estrogen fluctuations are having the effect of overtaxing the system at times and depriving the system at other times, it would be expected that there would be times when serotonin responsivity is higher than normal, as well as times when it is blunted compared to persons without BPD. If so, then failing to control for time in cycle in a challenge study would be expected to cancel out any actual differences in serotonin responsiveness and result in a null finding, as has been found (Hollander, et al. 1994). Note that if this is the explanation (times of high and low reactivity cancelling each other out) for the unexpected null effects, a greater amount of variance (as in the standard deviations of the female sample) might be expected. McBride at al., 1994 in studying depressed persons with a history of suicidal ideation and/or behavior (of which a number were co-morbid for BPD), did not find a difference in mean values for indices of serotonin function -- but did find that the patient group had more variance on the measures than did controls. On the other hand, if one controls for time in cycle, differences in serotonin system reactivity should be detectable such that an exaggeration of the typical pattern will appear for women with a diagnosis of BPD. Indeed an accelerated response within the serotonin system has been demonstrated among women with BPD in at least one fenfluramine challenge study which attempted to control for estrogen effects by measuring all women during the luteal phase of the menstrual cycle (Martial et al, 1997). Although this study referred to the luteal phase as a time of low estrogen, the luteal phase is more compared to the cycle as a whole a time of high estrogen That is, even though the levels are not as high as estrogen levels common the day prior to ovulation, the 200-300 pg/ml levels of estrogen which occur during the luteal phase are not indicative of a time of low estrogen (see for examples Thompson, Sergejew & Kulkarni, 2000; Kimura, 1999; Hallonquist, et al., 1994; Mead & Hampson, 1997).

Although suggestive, the above review provides only speculation. Although further work is called for on the subject, initial investigations into the role of estrogen fluctuations on the symptoms of borderline personality have supported the idea that estrogen changes are themselves a variable of interest in understanding the etiology of borderline personality. Next, a brief review of previous research that has investigated the potential role of estrogen flux itself as a variable of interest in understanding borderline personality symptoms.

TESTS OF THE MODEL

The first of three studies on the topic examined the relation between BPD symptoms and menstrual cycle in a between-subjects design (DeSoto, et al., 2003). The Personality Assessment Inventory- Borderline Features scale was used, a scale that has been validated with both clinical and community samples (Morey, 1991). The proportion of women with clinically significant borderline features (i.e., overall PAI-BOR scores > 37; Morey, 1991) were compared. This difference between the high and low estrogen group women was statistically significant, ($\chi^2 = 4.1$, $p < .05$). It was particularly interesting to me to note that among the high estrogen group (which was defined as being in day seven to day fourteen in her cycle) there was a difference between the women who would be expected to have just begun the increase in estrogen, compared to the women who were at a stage where the estrogen was still increasing, but the slope of the change would be about the same as the preceding days. In other words, this comparison is capturing the aspectthis author is most interested in: the change in estrogen status. See figure 2 for clarity. The results revealed that 35% of women who were at a point in her cycle marked by "rising estrogen" (see figure 3) endorsed clinically significant levels of BPD symptoms. Women at low estrogen phases were less likely to report significant borderline features (15%). The proportion of women with clinically significant borderline features (i.e., overall PAI-BOR scores > 37; Morey, 1991) was highest for women who were likely to be experiencing the largest change in estrogen levels ($\chi^2 = 4.1$, $p < .05$). The effect size for this analysis using three divisions in the cycle was larger than when just the high versus low split was used, which, as above, was also significant.

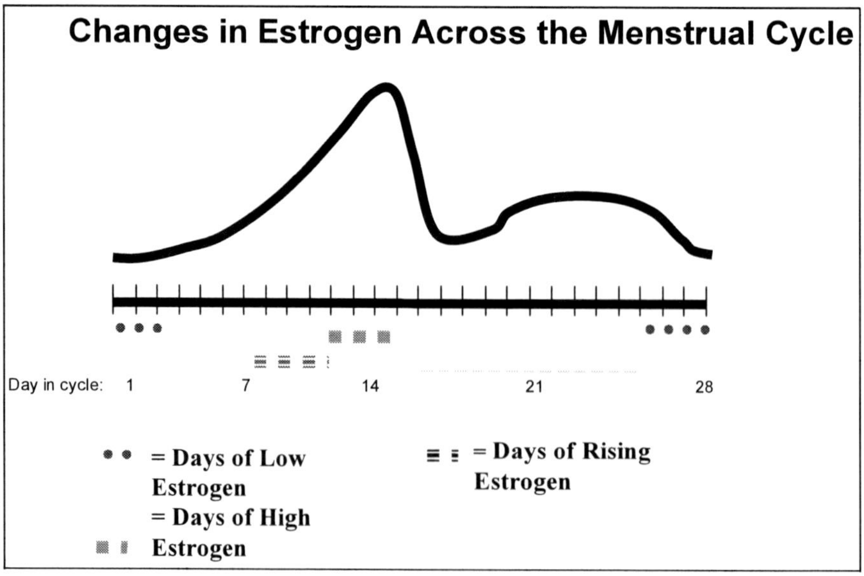

Figure 3. Typical changes in estrogen across a menstrual cycle. Results first compared low estrogen days to days when estrogen is high and/or rising, but then also compared specifically three groups : women at days when estrogen is rising, women at days when estrogen is high, and women when estrogen is low.

A second study employed a within-subjects design to explore further the potential relation between fluctuations in estrogen levels and BPD symptoms and to do so by directly measuring estrogen levels via salivary assay four times across one month. Women whose hormonal profiles showed more variation across the month did tend to report more symptoms. Specifically, for each women a calculation of how much her estrogen level changes over the month (essentially just the standard deviation of her estrogen levels across the month squared) was made. This number was correlated to her score on the Personality Assessment

Features – Borderline scale. The variability of estradiol was significantly correlated with overall PAI-BOR scores ($r = .45$, $p < .01$; DeSoto, et al., 2003).

Previous research also suggests that women who are taking estrogen in the form of birth control pills have more symptoms of borderline personality than would otherwise be expected, and studies comparing women who do and do not begin using estrogen-containing oral contraceptives strongly suggest that it is not pre-existing differences that result in the differences, but rather that commencing the use of estrogen (in the form of oral contraceptives) increases the level of BPD symptoms. It should be noted that this effect appear only to be relevant to women who were above the mean beforehand on borderline symptomology (DeSoto et al., 2003), there is no reason to view this a general contra-indicator for pill use among typical women. However, this finding may suggest that women who are high on symptoms to begin, perhaps relating to some sort of already present vulnerability in the serotonin system, could find there serotonin sytems further taxed in the manner outlined in Figure 2. If so, then OC use would not be a problem for the vast majority of women, but could be potentially contra-indicated for those with a positive history of borderline personality symptoms.

It should be noted that, to this author's knowledge, these are the only studies that have been conducted on this topic. These three studies all suggest a link between estrogen fluctuations and symptoms of borderline personality. Although the speculations about the possible neural mechanisms that might underlie the connection are and must be tentative at this early stage of investigation, they do provide a plausible vehicle for which the effects of estrogen might occur. The speculation may or may not be correct, but the three studies do indicate that something about estrogen appears to be affecting something about borderline symptomology.

Lastly, virtually all drugs used to treat BPD, including monoamine-oxidase (MAO) inhibitors as well as selective serotonin reuptake inhibitors, result in a down regulation of 5-HT_2 receptors, (recall down regulation is likely associated with an increase in the neurotransmitter of interest, see Feldman, Meyer, & Quenzer, 1997 for a tutorial). This is an effect opposite of that of estrogen. This may highlight a mechanism for the effects of estrogen to occur. The above reviewed studies suggest that rising levels of estrogen are associated with an exacerbation of borderline personality symptoms, and at least one known influence of estrogen -- up-regulation of 5-HT type 2 receptors -- is negatively related to symptom improvement.

In summary, the preliminary studies suggest the existence of a previously unknown relationship between estrogen and symptoms of borderline personality, and highlights the importance of add ional research on this connection. As a whole, there is ample reason to lend some credence to the idea that, at least in some women, fluctuations in hormone levels may exacerbate any pre-existing borderline personality tendencies via some neural mechanism still to be elucidated, but which seem likely to involve the serotonin system in one form or another.

POSSIBILITY OF CLINICAL APPLICATIONS

Treatment of women with the disorder could be expected to be improved by a more precise understanding of the mechanisms that underlie the symptom expression in women. For example, it may be that the dosages of drugs that treat BPD, which often focus on inhibiting serotonin reuptake, could regularly be altered across the month based on an individual women's hormonal profile, and these dosages might become routinely re-evaluated in the late 40's when symptoms of BPD worsen among women. In addition, if it is true that there is something about the physiology of women that makes women more influenced by borderline personality symptoms during the years leading up to menopause, counseling efforts among women with a history of borderline personality symptoms could be proactively stepped up during these years. Presumably, if treatment improved, the impact of BPD on the lives of those who have been diagnosed would be lessened.

SUMMARY OF BPD-ESTROGEN CONCEPTUALIZATION

The link between estrogen and the symptoms of borderline personality is assumed to of a nature that would worsen any predisposition for the sets of behaviors that are recognized as BPD, and not necessarily one of direct cause. That is, if links exist between estrogen levels and BPD symptoms, it would bolster the claim that hormonal fluctuations are important in understanding the course and sex difference patterns, but should not be taken to mean that hormonal fluctuations actually causing BPD. Essentially, the idea is that even normal levels of hormonal fluctuations would probably aggravate symptoms of BPD among women who have developed the disorder due to a combination of inherited biological predisposing factors and environmental influences. On the other hand, women prone to especially sharp fluctuations of estrogen might be more likely to develop enough symptoms to be diagnosed given the right predisposing factors of environmental influence, if compared to women with less sharp fluctuations. And, even in the case of those lacking sufficient symptoms to be diagnosed, especially pronounced fluctuations could increase BPD symptoms to a point approaching the diagnostic criteria. In terms of future research, it will be of considerable interest to determine if a clinical sample of women actually diagnosed with the disorder show more variation is estrogen levels than do those without the disorder.

ACKNOWLEDGMENTS

I wish to thank Melinda Collingwood for her comments on an earlier version of this manuscript.

REFERENCES

American Psychiatric Association. (1994). *Diagnostic and statistical manual of mental disorders (4th ed.)*. Washington, DC: Author.

Bardenstein, K. A., & McGlashen, T. H. (1988). The natural course of a residentially treated borderline sample: Gender differences. *Journal of Personality Disorders, 2(1)*, 69-83.

Bethea, C. L., Pecins-Thompson, M., Schutzer, W. E., Gundlah, C., & Lu, Z. N. (1998). Ovarian steroids and serotonin neural function. *Molecular Neurobiology, 18 (2)*, 87-123.

Bucht, G., Adolfsson, R., Gottfries, C. G., Roosa, B. E., & Winblad, B. (1981). Distribution of 5-hydroxytryptamine and 5-hydroxyindoleacetic acid in human brain in relation to age, drug influence, agonal status and circadian variation. *Journal of Neural transmission General Section, 51*, 185-203.

Derogatis, L. R., & Melisaratos, N. (1983). The Brief Symptom Inventory: An introductory report. *Psychological Medicine, 13*, 595-605.

DeSoto, M. C., Geary, D.C., Hoard, M.K., Sheldon, M., & Cooper, M. L. (2003). Estrogen variation, oral contraceptives and borderline personality. *Psychoneuroendocrinology, 28*. 751-766.

Feldman, R. S., Meyer, J. S., & Quenzar, L. F. (1997). *Principles of neuropsychopharmocology*. Sunderland, MA: Sinauer Associates, Inc.

Fink, G., Sumner, B. E., Rosie, R., Grace, O. & Quinn, J.P. (1996). Estrogen control of central neurotransmission: Effect on mood mental state & memory. *Cellular and Molecular Neurobiology 16 (3)*, 325-344.

First, M. B., Gibbon, M., Spitzer, R. L., Williams, J. B., & Benjamin, L. (1997). *Structured Clinical interview for DSM-IV personality disorders*. American Pschiatric Publishing, Washington, D.C.

Fludder, J. M., & Tonge, S. R. (1975). Variations in the concentrations of monoamines and their metabolites in eight regions of rat brain during the estrous cycle: A basis for interactions between hormones and psychotrophic drugs. *Journal of Pharmacuitical Pharmacology, 27*.

Geary, D. C., DeSoto, M. C., Hoard, M. K., Sheldon, M. S., Cooper, L. C. (In press). Estrogens and the menstrual cycle: Relation to jealousy patterns and sexual behavior. *Human Nature*.

Goodwin, G. M., Murray, C. L., & Bancroft, J. (1994). Oral d-fenfluramine and neuroendocrine challenge: Problems with the 30 mg dose in men. *Journal of affective disorders, 30*, 117-122.

Grilo, C. M., Becker, D. F., Fehon, D. C., Walker, M. L., Edell, W. S., & McGlashen, T. H. (1996). Gender differences in personality disorders in psychiatrically hospitalized adolescents. *American Journal of Psychiatry, 153 (8)*, 1089-1091.

Hallonquist, J. D., Seeman, M. V., Lang, M., & Rector, N. A. (1993). Variation in symptom severity over the menstrual cycle of schizophrenics. *Biological Psychiatry, 33(3)*, 207-209.

Hollander, E., Stein, D. J., DeCaria, C. M., Cohen, L., Saoud, J. B., Skodol, A. E., Kellmen, D., Rosnick, L., & Oldham, J. M. (1994). Serotonergic sensitivity in borderline personality disorder: Preliminary findings. *American Journal of Psychiatry, 151 (2)*, 277-280.

Kimura, D. (1999). *Sex and cognition*. Cambridge, MA: Bradford/MIT Press.

Knight, J. W., & Callahan, J. C. (1989). *Preventing birth: Contemporary methods and related moral controversies*. Salt Lake City, UT: University of Utah Press.

Lippert, T. H., Filshe, M., Mueck, A.O.,Seeger, H. & Zwirner, m. (1996). Serotonin metabolite excretion after postmenopausal estradiol therapy. *Maturitas, 24*. 37-41.

Martial, J., Paris, J., Leyton, M., Zweig-Frank, H., Schwartz, G., Teboul, E., Thavundayil, J., Larue, S. Ng Ying Kin, N., & Vasavan Nair, N. (1997). Neuroendocrine study of serotonin function in female borderline personality patients: A pilot study. *Biological Psychiatry, 42*. 737-739.

McBride, P.A., Brown, R.P., DeMeo, M., Keilp, J., Mieczkowski, T., & Mann, J. J. (1994). The relationship of platelet 5-HT2 receptor indices. Major depressive disorder, personality traits and suicidal behavior. *Biological Psychiatry, 35 (5)*. 295-308.

Mead, L. A. & Hampson, E. (1997). Turning bias is influenced by phase of the menstrual cycle. *Hormones and Behavior, 31 (1)*. 65-74.

Morey, L. C. (1991. *The personality assessment inventory: Professional manual*. Odessa, FL: Psychological Assessment Resources.

Mueck, A. O., Seefer, H., Kabpohl-Butz, S., Tiechmann, A. T., & Lippert T. H. (1997). Influence of norethesitone acetate and estradiol on the serotonin system of post-menopausal women. *Hormone Metabolism Research, 29*. 80-83.

Myers, W. C., Burkett, R. C., & Otto T. A. (1993). Conduct disorder and personality disorders in hospitalized adolescents. *Journal of Clinical Psychiatry, 54*, 21-26.

Mueck, A. O., Seefer, H., Kabpohl-Butz, S., Tiechmann, A. T., & Lippert T. H. (1997). Influence of norethesitone acetate and estradiol on the serotonin system of post-menopausal women. *Hormone Metabolism Research, 29*. 80-83.

Shirtcliff, E. A., Granger, D. A., Schwartz, E. B., and Curran, M. J., Booth, A., Overman, W. H. (2000). Assessing estradiol in biobehavioral studies using saliva and blood spots. Simple radioimmunoassay protocols, reliability, and comparative validity. *Hormones and Behavior, 38*, 137-147.

Seeman, M. V. (1997). Psychopathology in women and men: Focus on female hormones. *American Journal of Psychiatry, 154 (12)*, 1641-1647.

Su T. P., Schmidt P. J., Danaceau M., Murphy D. L., & Rubinow D. R.(1997). Effect of menstrual cycle phase on neuroendocrine and behavioral responses to the serotonin agonist m-chlorophenylpiperazine in women with premenstrual syndrome and controls. *Journal of Clinical Endocrinology & Metabolism, 82 (4),* 1220-1228.

Thompson, K., Sergejew, A. & Kulkarni, J. (2000). Estrogen affects cognition in women with psychosis. *Psychiatry-Research. 94 (3)*. 201-209.

Trestman, R. L., Coccaro, E. F., & Temple, J. (1992). Impulsivity and serotonin in borderline personality disorder. Paper presented at the 145th annual meeting of the American Psychiatric Association, Washington, D.C.

In: Psychoneuroendocrinology Research Trends
Editor: Martina T. Czerbska, pp. 161-172

ISBN: 978-1-60021-665-7

Chapter V

DEPRESSION, COGNITIVE DEFICITS AND POOR QUALITY OF LIFE IN METHAMPHETAMINE USE: NEURAL SUBSTRATES AND THE IMPACT OF SEX AND HIV STATUS

***Sarah Cooper*[1]*, Kayla Hatt*[1] *and Amy Wisniewski*[2,*]**
[1]Drake University College of Pharmacy and Health Science and
[2]College of Arts and Sciences, Des Moines, Iowa 50311

ABSTRACT

Despite recent legislative action to limit production and use, the abuse of methamphetamine (MA) has continued to increase. MA is a highly addictive stimulant that is both easy and cost effective to manufacture. As a result, the illicit use of MA is a growing problem in the United States. Immediate physical effects of MA use such as hypertension, hyperthermia, tachycardia, tachypnea, anxiety and paranoia are well documented, but the longer term physical and mental adverse effects of MA use are not as well defined. The importance of identifying the mental and physical effects of long-term MA use is essential for improved treatment. This chapter reviews the results of recent MA research in the areas of quality of life (QOL), depression and cognitive impairment. It also considers recent evidence that MA users exhibit atypical structural and functional brain characteristics that may underlie the suite of adverse cognitive and mental health characteristics associated with this drug of abuse. Finally, the potential additive and synergistic impact of users' sex and HIV status with MA use on physical and mental health disorders are considered.

* Correspondence concerning this article should be addressed to Amy B. Wisniewski, Ph.D. Drake University, Department of Biology, 328 Olin Hall, 1344 27th Street, Des Moines, IA 50311-4505. P: (515) 271-2957; F: (515) 271-3702; amy.wisniewski@drake.edu.

Keywords: addiction, amphetamine, stimulant, AIDS, illicit drugs, brain, gender, estrogen

I. INTRODUCTION

What is Methamphetamine and Why is it a Problem?

Methamphetamine (MA) is a stimulant that is structurally related to amphetamine but exerts stronger effects on the central nervous system (CNS) than amphetamine [1,2]. MA is legally marketed as a pharmaceutical agent to treat attention deficit/hyperactivity disorder (ADHD), narcolepsy and obesity; however, MA abuse is dramatically increasing due to the simplicity and cost effectiveness of its illegal production [3]. In the United States (see Figure 1) illicit MA production originated in the Southwest and has since extended into the Northwest and Midwest regions of the country [2,4].

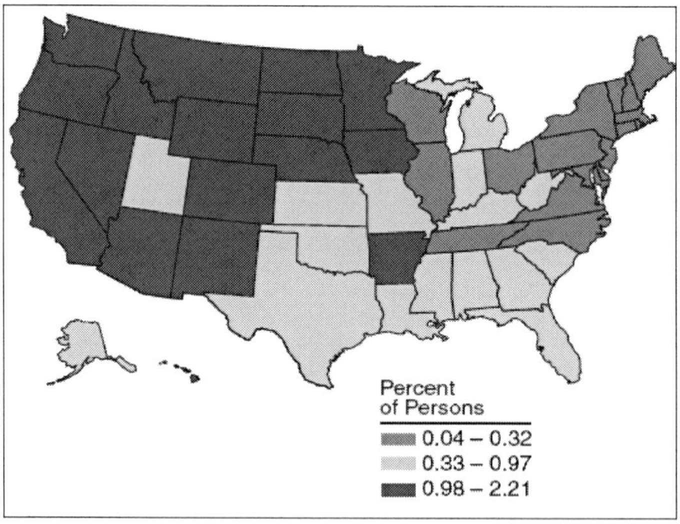

Figure 1. From the National Survey on Drug Use and Health (NSDUH), 2005.

Who Uses MA?

Results from a 2000 National Household Survey on Drug Abuse indicate that approximately 8.8 million Americans used MA at some point [5,6]. Worldwide, it is estimated that 35 million adults are currently using some type of amphetamine-related stimulant including MA [3]. Only marijuana exceeds amphetamine-related stimulant use [7]. The prevalence of methamphetamine abuse is of particular concern in non-metropolitan (~10,000 residents) areas where admission rates for adverse effects associated with MA use exceed those of more-densely populated metropolitan areas.

According to the 2002 Iowa Youth Survey, the most recent epidemiologic data available for methamphetamine use in the state of Iowa, 13% of students in grade 11 report having ever used amphetamine/methampahetamine and 7% reported current use [8]. In contrast, 2% of students in grade 6 reported having ever used MA, and none reported current use [8]. Among college-aged individuals, the reported annual prevalence of methamphetamine use is 2.9% for full-time students and 5.1% for others in the same age group [9]. The Monitoring the Future Study suggests that current trends in MA use by adolescents and young adults did not originate on college campuses. Rather, initiation to MA use started in secondary schools (see Figure 2). Furthermore, the popularity of MA is well-established among young women and gay communities [7].

MA can be inhaled, snorted, injected or swallowed. For some people the preferred route of MA administration shifts from inhalation to intravenous as a result of the faster onset of euphoria following injection [3]. Additionally, injection is favored by younger users [7].

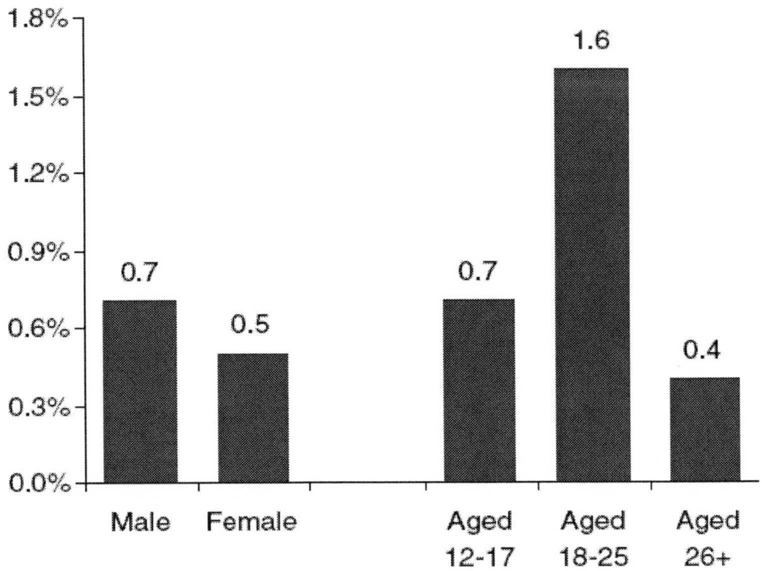

Figure 2. From the NSDUH, 2005.

Physical and Mental Health Adverse Effects of MA Use

MA abuse is of great concern in part due to the severity of physical and mental health adverse effects associated with its use. MA is initially attractive due to improved alertness, wakefulness, and decreased appetite following administration [7]. However, physical health adverse effects of MA use include hypertension, hyperthermia, tachycardia and tachypnea [5]. Commonly reported mental health and neuropsychological adverse effects associated with MA use include depression, anxiety, violent behavior, difficulties with memory and executive function, attention deficits and decreased psychomotor speed [5, 10-15]. MA use is also associated with dental decay known as "Meth Mouth." due to neglect of oral hygiene, xerostomia, decreased fluid intake, shrinking of the gingival tissue and bruxism [16].

Impact of Sex and HIV Status on Adverse Effects of MA Use

Physical and mental health morbidities often differ among illicit drug users according to their sex. For example, women who are both infected with HIV and who inject heroin and/or cocaine reported more depressive symptoms than their male counterparts or individuals affected by either HIV or injection drug use alone [17]. In another study of opiate users (HIV status not reported), women reported more psychiatric symptoms than men [18]. Among recently-detoxified alcoholics women reported a significantly decreased quality of life (QOL) and greater depression than men [19]. MA use is particularly popular among young women [6,12] due to its ability to decrease appetite resulting in weight loss. Also noted previously, intravenous injection is an increasingly popular route of administration of MA, particularly among younger users. Therefore, both sex and HIV status should be considered as potentially important co-variables in understanding the natural history of MA use.

II. Quality of Life

Quality of life (Qol) is a measure of a person's overall sense of their mental and physical well-being. QOL is evaluated on the basis of pain management, mood, energy level, family and social interactions, sexual function, and the ability to work and complete required daily activities. Addiction in general is associated with decreased QOL, perhaps due to the prevalence of mental and physical adverse effects that accompany substance abuse [19-22]. The abuse of stimulants in particular is commonly associated with paranoia, irritability, sleep disturbances/lethargy and anorexia [23]. MA use results in multiple physical health, mental health and social adverse outcomes and is therefore expected to result in decreased QOL. Such results were evident in a group of 106 men and women who use MA and subsequntly reported problems with sleep, depression, paranoia, irritability, eating and interactions with family members [15]. Furthermore, these negative outcomes were associated with frequency of MA use.

A second study comparing a Matrix outpatient model to standard care for treating addiction proposes to study QOL among MA users, but those results are pending [24]. A third study reported decreased QOL among amphetamine users (neither sample size nor duration of amphetamine use was reported); however, the ability to generalize these results specifically to MA users remains untested [21]. Short-term or sporadic MA use may not exert as profound of an impact on QOL as chronic use. Evaluation of QOL among both chronic and sporadic MA users, with attention paid to the amount of MA administered and the duration of use, is needed to better understand the global health needs of affected individuals.

As previously noted, intravenous use is a preferred route of administration for many individuals who use MA. This suggests that a large proportion of MA users are at an increased risk for acquiring both HIV and hepatitis infections. To date, the majority of QOL studies of injection drug users (IDU) who inject heroin and/or cocaine employ the MOS-HIV questionnaire to assess QOL while considering the impact of HIV/AIDS on this outcome. Future studies are needed to assess QOL in MA users who are seropositive for HIV and hepatitis to determine the additive or synergistic impact of MA use and co-morbidities on

QOL. Finally, QOL is often inferior in women when compared to men, regardless of their medical or drug use status [17,19,20,25-28]. Unfortunately, we were unable to find studies that looked for sex differences in QOL among MA users. The potential impact of sex, in addition to HIV status, on QOL among MA users is another scientific question that remains to be studied.

III. Depression

Depression is identified as a leading mental health adverse reaction to MA use and withdrawal [12-15,29,30]. In a study of recently abstinent (4-7 days) MA users, increased depressive symptoms were reported as indicated by Beck Depression Inventory (BDI) scores, compared to non-users [13]. The sample size in the London et al. study was limited (11 men and 6 women participated) and the authors did not state if sex effects were considered in the data analysis of BDI scores. Furthermore, HIV seronegative status was required for participation; therefore, the potentially additive impact of both MA use and HIV on depressive symptoms could not be studied. People who use MA intravenously may have even more pronounced depressive symptoms and increased suicidal tendencies than users who rely on other routes of administration [14], thus the impact of route of administration on depression should also be taken into consideration in future investigations.

A larger study of MA users that included gay and bisexual men found that recent MA use predicted depressive symptoms [14]. Importantly, depressive symptoms did not predict subsequent MA use in this sample. Therefore, while MA use likely results in depressive symptoms, a person with depression will not necessarily abuse MA for self-medication purposes. HIV status was considered in this study, but was not found to be associated with depressive symptoms as indicated by BDI scores. Two additional studies that focus on treatment of MA users reported more depressive symptoms and suicide attempts in affected women than in their male counterparts [12,29]. Similar to the need to consider the impact of HIV status and route of drug administration on MA-related depression is the need to consider how men and women are differentially affected by this drug of abuse in terms of depression.

IV. Cognitive Function

Attention, a key component to cognition, changes with MA use. Recently abstinent (60 days or more) men exhibited impaired selective attention compared to MA naïve men [31]. A second investigation of selective attention in abstinent MA users (both men and women were included) confirmed impaired selective attention following MA use [32]. Recently abstinent MA users (sex not reported) also performed poorly on tests of memory and psychomotor speed [11], as well as attention.

In a study of cognitive performance among 65 men and women who currently use MA (approximately 1/3 of whom injected intravenously), significant deficits in cognition were observed compared to non-users [10]. Specific deficits in verbal and visual memory, psychomotor speed and attention were noted. Furthermore, these deficits were most pronounced in users who administered MA intravenously and who used the drug daily. Although roughly equal numbers of men and women participated in this study, the potential impact of users' sex was not investigated in terms of cognitive function [10].

The additive or synergistic impact of MA use and HIV infection on cognition was studied in a large group of men and women stratified by their MA and HIV status [33]. Men and women who both use MA and are infected with HIV performed worse on a neuropsychological battery of tests compared to individuals who either use MA but are not infected with HIV, or who are infected with HIV but do not use MA. Men and women who neither use MA nor are infected with HIV exhibited the best overall cognitive performance [33]. Specific realms of cognitive performance that were negatively affected by both current MA use and HIV infection include attention, working memory, delayed recall and fine motor movements. Sex differences were not investigated although both men and women participated, perhaps because the majority of participants were male (91%) in the group comprising both MA users and HIV-infected individuals.

V. Brain Structure and Function

Consequences of Current MA Use on Neurotransmitter Systems and Brain Structure

Acute MA administration results in elevated synaptic levels of dopamine (DA) due to decreased DA transporter density in the brain [1,34]. Repeated administration of MA injures dopaminergic projections to the striatum in both humans and nonhuman animal species [26,35]. Multiple doses of MA also damage serotonergic neurons in the striatum, parietal cortex, frontal cortex and hippocampus [1,35]. Regarding brain structure, gray matter deficits in the cingulate gyrus and hippocampus, along with white matter hypertrophy, is evident among individuals who use MA compared to those who do not [37].

Long-Lasting Effects of MA on Brain Structure and Function among Abstinent Users

Long-lasting effects of MA use have been documented in the brain physiology of individuals who have been abstinent from the drug for days to months. For example, abstinent MA users exhibited decreased cerebral blood flow (CBF) to the basal ganglia and right lateral parietal areas, as well as increased CBF to left temporoparietal, left occipital and right posterior parietal regions [38]. Abstinent users also exhibited atypical patterns of glucose metabolism in their limbic and paralimbic regions [13]. Reductions in N-Acetyl [NAA] compounds, a marker of neuronal density and viability, were observed in the frontal lobe and basal ganglia of abstinent MA users along with increased choline-containing [Cho] compounds and myo-inositol [Mi] in frontal gray matter [39]. The [Cho] observations indicate possible glial proliferation in response to neuronal injury due to past MA exposure.

Positron emission tomography (PET) investigations reveal decreased DA transporter (DAT) availability in the striatum and prefrontal cortex of individuals who achieve abstinence from MA use [34,36,40]. This long-term decrease in DAT is associated with persistent deficits in both gross and fine motor function, verbal memory performance [36] and psychiatric symptoms related to MA use [34]. In addition to decreased DAT availability, previous MA use is also associated with decreased DA D_2 receptor availability in the striatum among abstinent individuals [41]. While most PET data collected from abstinent MA users indicate that DA pathways (either via transporters or receptors) are affected by this stimulant, it is worth noting that some PET data also reveal that brain circuits not reliant on DA are also affected by MA exposure in seemingly long-lasting ways. For example, brain glucose metabolism in detoxified MA users is altered in non-dopaminergic areas such as the parietal cortex, as well as in DA pathways such as the striatum and thalamus [42].

Interactions between MA and HIV on Brain Structure and Function

People who use MA are at risk for acquiring HIV as a result of both needle-sharing when administering the stimulant intravenously, as well as from unprotected sex that is frequently reported by individuals who are high on MA [43]. Thus, it is important to consider the potential additive or synergistic effects of MA and HIV on neurotoxicity. MA exposure and HIV infection are each associated with changes in gross brain morphology, as assessed by magnetic resonance imaging (MRI). In humans, increased volumes of the basal ganglia and parietal cortex were observed in both men and women who use MA [44]. Furthermore, these increases were associated with cognitive impairment. In contrast, HIV-infected adults exhibited decreased volumes of their basal ganglia, thalamus, hippocampus, frontal and temporal lobes. Additionally, HIV-infected men and women with the greatest brain volume loss exhibited the most pronounced cognitive impairment [44]. No obvious additive or synergistic effects of MA use and HIV infection were reported for brain structure or cognitive function in this study [33,45,46].

No sex differences in the effects of HIV infection or MA use on brain morphology or cognitive performance were reported [44]; however, this lack of an effect should be cautiously interpreted as only 3 seropositive women and only 5 women currently using MA participated among a total sample of 103 individuals (see Table 1). In contrast to the lack of an interaction between MA and HIV on brain structure/function relationships as measured by MRI, an autopsy study of 20 individuals stratified by their HIV status and MA use history revealed synergistic loss of interneuron number and density in the frontal cortex of individuals affected by both HIV and MA use [47]. Functionally, global cognitive impairment and memory deficits correlated with interneuron loss in this population. Similar to the previously mentioned MRI investigation that included too few women to consider sex differences in outcome measures, the autopsy study included only 2 female participants (1 HIV^+ and no history of MA use, the other both HIV^+ and a positive history of MA use) so that sex differences in interneuron loss and cognitive function could not be considered.

A proton magnetic resonance spectroscopy (^{1}H-MRS) study of 143 adults (45 of whom were women) stratified by their HIV status and history of MA use also revealed an additive effect of MA and HIV on neurotoxicity [38]. Specifically, a history of MA use was independently associated with neuronal injury in the basal ganglia and glial activation in frontal gray matter. HIV infection was independently related to neuronal injury in frontal gray matter and the basal ganglia, as well as to glial activation in white matter. Participants infected with HIV who also have a history of MA use exhibited neuronal injury in frontal white and gray matter as well as in their basal ganglia. These individuals also exhibited glial abnormalities in frontal white matter. Although women were well represented in this study, no discussion of sex differences in neuronal or glial cell markers were mentioned [38].

Table 1. Participants Grouped According to Sex, Methamphetamine and HIV Status. Adapted from Jernigan et al., 2005

	Female	Male
HIV-/MA-	n = 13 (12%)	n = 17 (16.5%)
HIV-/MA+	n = 4 (4%)	n = 17 (16.5%)
HIV+/MA-	n = 2 (2%)	n = 28 (27%)
HIV+/MA+	n = 1 (1%)	n = 21 (20%)

Rodents treated with both MA and HIV transactivator (Tat) protein exhibited greater reduction in DA levels in the striatum than animals treated with either MA or HIV Tat alone (Cass et al., 2003; Nath et al., 2002). Similar to the case in humans, the mechanism underlying combined MA + HIV neurotoxicity in rodents is not understood. Among humans, MA addiction is impacted by medications used to treat HIV/AIDS [3,4]. MA is metabolized through the cytochrome P450 (2D6 isoenzyme) pathway. This creates a problem if patients are currently taking an HIV protease inhibitor, especially ritonavir, an inhibitor of the cytochrome P450 2D6 enzyme [50]. Methamphetamine serum levels have the potential to increase as a result of HIV/AIDS medications [16]. Non-nucleoside analogue reverse transcriptase inhibitors (NNRTI), another key component to the medical management of HIV/AIDS, may exert similar effects in MA users as ritonavir [16].

Does Estrogen Offer Neuroprotection Against MA and HIV?

Interestingly, treatment with estrogen protects against death and mitochondrial toxicity in human brain cell cultures exposed to both MA and HIV proteins, and this protection is blocked when cells are treated with the estrogen receptor antagonist ICI 182,780 [51]. The previously mentioned rodent study employed male animals only; therefore, potential sex differences in the interaction between MA and HIV on DA levels could not be considered [48]. Future investigations of sex differences (and potentially the impact of sex steroids) in the interaction between MA exposure and HIV on brain structure and function in both human and nonhuman animal models are needed to develop maximally effective therapeutic interventions for women and men affected by these increasingly common co-morbid conditions.

Potential for Recovery of Structural and Functional Brain Abnormalities Following Abstinence of MA Use

MA use alters brain structure and function in long-lasting ways even after detoxification from the drug has occurred. Furthermore, the neurotoxic effects of MA appear to worsen in individuals who are also infected with HIV. While these observations are sobering, it is important to note that limited data do indicate that partial neural recovery is possible following prolonged abstinence from MA use. For example, protracted abstinence (> 12 months) results in recovery of MA-induced DAT loss in the striatum [52]. Unfortunately, no improvement in motor function, coordination or memory accompanied the DAT recovery in this preliminary investigation.

VI. Summary and Conclusions

In this chapter we review current studies on the mental health and cognitive adverse effects of MA use in humans. A brief consideration of the neural circuits impacted by MA use is also included. An emphasis on the additive effects of sex and HIV status on the natural history of MA use are considered, with suggestions for including these variables in future studies.

References

[1] Nordahl TE, Salo R, Leamon M (2003). Neuropsychological effects of chronic methamphetamine use on neurotransmitters and cognition: a review. *J Neuropsychiatry Clin Neurosci*; *15*: 317-325.

[2] Meredith CW, Jaffe C, Ang-Lee K, Saxon AJ (2005). Implications of chronic methamphetamine use: a literature review. Harv Rev Psychiatry; *13(3)*: 141-154.

[3] Romanellia F, Smith KM (2006). Clinical effects and management of methamphetamine abuse. *Pharmacotherapy*; *26(8)*: 1148-1156.

[4] Roll JM, Petry NM, Stitzer ML, Brecht ML, Pierce JM, McCann MJ, Blaine J, MacDonald M, DiMaria J, Lucero L, Kellogg S (2006). Contingency management for the treatment of methamphetamine use disorders. *Am J Psychiatry*; *163(11)*: 1993-1999.

[5] Rawson RA, Gonzales R, Brethen P (2002). Treatment of methamphetamine use disorders: an update. Journal of Substance Abuse Treatment; *23*: 145-150.

[6] Cretzmeyer M, Vaughan Sarrazin M, Huber DL, Block RI, Hall JA (2003). Treatment of methamphetamine abuse: research findings and clinical directions. *Journal of Substance Abuse Treatment*; *24*: 267-277.

[7] Ling W, Rawson R, Shoptaw S, Ling W (2006). Management of methamphetamine abuse and dependence. *Curr Psychiatry Rep*; *8(5)*: 345-354.

[8] The Iowa Youth Survey, http://www.iowayouthsurvey.org/

[9] The Monitoring the Future Study, http://www.monitoringthefuture.org/

[10] Simon SL, Domier C, Carnell J, Brethen P, Rawson R, Ling W (2000). Cognitive impairment in individuals currently using methamphetamine. *American Journal on Addictions*; *9*: 222-231.

[11] Kalechstein AD, Newton TF, Green M (2003). Methamphetamine dependence is associated with neurocognitive impairment in the initial phases of abstinence. *Journal Neuropsychiatry Clin Neurosci*; *15*: 215-220.

[12] Hser YI, Evans E, Huang YC (2004). Treatment outcomes among women and men methamphetamine abusers in California. *Journal of Substance Abuse Treatment*; *28*: 77-85.

[13] London ED, Simon SL, Berman SM, Mandelkern MA, Lichtman AM, Bramen J, et al., (2004). Mood disturbances and regional cerebral metabolic abnormalities in recently abstinent methamphetamine abusers. *Arch Gen Psychiatry*; *61(1)*: 73-84.

[14] Peck JA, Reback CJ, Yang X, Rotheram-Fuller E, Shoptaw S (2005). Sustained reductions in drug use and depression symptoms from treatment for drug abuse in methamphetamine-dependent gay and bisexual men. *Journal of Urban Health*; *82*: i100-i108.

[15] Sommers I, Baskin D, Baskin-Sommers A (2006). Methamphetamine use among young adults: health and social consequences. *Addictive Behaviors*; *31*: 1469-1476.

[16] Shoptaw S, Klausner JD, Reback CJ, Tierney S, Stansell J, Hare CB, Gibson S, Siever M, King WD, Kao U, Dang J (2006). A public health response to the methamphetamine epidemic: the implementation of contingency management to treat methamphetamine dependence. *BMC Public Health*; *6*: 214.

[17] Wisniewski AB, Apel S, Selnes OA, Nath A, McArthur JC, Dobs AS (2005). Depressive symptoms, quality of life and neuropsychological performance in HIV/AIDS: the impact of gender and injection drug use. *Journal of NeuroVirology*; *11*: 138-143.

[18] Marsden J, Gossop M, Stewart D, Rolfe A, Farrell M (2000). Psychiatric symptoms among clients seeking treatment for drug dependence. *British Journal of Psychiatry*; *176*: 285-289.

[19] Peters TJ, Millward LM, Foster J (2003). Quality of life in alcohol misuse: comparison of men and women. *Arch Women Ment Health*; *6*: 239-243.

[20] Stein MD, Mulvey KP, Plough A, Samet JH (1998). The functioning and well being of persons who seek treatment for drug and alcohol use. *Journal of Substance Abuse; 10*: 75-84.

[21] Ventegodt S, Merrick J (2003). Psychoactive drugs and quality of life. *The Scientific World Journal*; *3*: 694-706.

[22] Richter D, Eikelmann B, Berger K (2004). Use of the SF-36 in the evaluation of a drug detoxification program. *Qual Life Res*; *13*: 907-914.

[23] Williamson S, Gossop M, Powis B, Griffiths P, Fountain J, Strang J (1997). Adverse effects of stimulant drugs in a community sample of drug users. *Drug and Alcohol Dependence*; *44*: 87-94.

[24] Huber A, Lord RH, Gulati V, Marinelli-Casey P, Rawson R, Ling W (2000). The CSAT methamphetamine treatment program: research design accommodation for "real world" application. *Psychoactive Drugs*; *32*: 149-156.

[25] Linzer M, Spitzer R, Kroenke K, Williams JB, Hahn S, Brody D, deGruy F (1996). Gender, quality of life and mental disorders in primary care: results from the PRIME-MD 1000 Study. *Am J Med*; *101(5)*: 526-533.

[26] Chin MH, Goldman L (1998). Gender differences in 1-year survival and quality of life among patients admitted with congestive heart failure. *Med Care*; *36(7)*: 1033-1046.

[27] Robb JC, Young LT, Cooke RG, Joffe RT (1998). Gender differences in patients with bipolar disorder influence outcome in the medical outcomes survey (SF-20) subscale score. *J Affect Disorder*; *49(3)*: 189-193.

[28] Mrus JM, Williams PL, Tsevat J, Cohn SE, Wu Aw (2005). Gender differences in health-related quality of life in patients with HIV/AIDS. *Qual Life Res*; *14*: 479-491.

[29] Zweben JE, Cohen JB, Christian D, Galloway GP, Salinardi M, Parent D, Iguchi M (2004). Psychiatric symptoms in methamphetamine users. *The American Journal on Addictions*; *13*: 181-190.

[30] McGregor C, Srisurapanont M, Jittiwutikarn J, Laobhripatr S, Wongtan T, White JM (2005). The nature, time course and severity of methamphetamine withdrawal. *Addiction*; *100*: 1320-1329.

[31] Salo R, Nordahl TE, Possin K, Leamon M, Gibson DR, Galloway GP, Flynn NM, Henik A, Pfefferbaum A, Sullivan EV (2002). Preliminary evidence of reduced cognitive inhibition in methamphetamine-dependent individuals. *Psychiatry Research*; *111*: 65-74.

[32] Salo R, Nordahl TE, Moore C, Waters C, Natsuaki Y, Galloway GP, Kile S, Sullivan EV (2005). A dissociation in attentional control: evidence from methamphetamine dependence. *Biological Psychiatry*; *57*: 310-313.

[33] Rippeth JD, Heaton RK, Carey CL, Marcotte TD, Moore DJ, Gonzalez R, Wolfson T, Grant I, The HNRC Group (2004). Methamphetamine dependence increases risk of neuropsychological impairment in HIV infected persons. *Journal Intl Neuropsychological Soc*; *10*: 1-14.

[34] Sekine Y, Iyo M, Ouchi Y, Matsunaga T, Tsukada H, Okada H, Yoshikawa E, Futatsubashi M, Takei N, Mori N (2001). Methamphetamine-related psychiatric symptoms and reduced brain dopamine transporters studied with PET. Am J Psychiatry: *158*: 1206-1214.

[35] Hanson GR, Rau KS, Fleckenstein AE (2004). The methamphetamine experience: a NIDA partnership. *Neuropharmacology*; *47*: 92-100.

[36] Volkow ND, Chang L, Wang GJ, Fowler JS, Leonido-Yee M, Franceschi D, Sedler MJ, Gatley SJ, Hitzemann R, Ding YS (2001a). Association of dopamine transporter reduction with psychomotor impairment in methamphetamine abusers. *Am J Psychiatry*; *158*: 377-382.

[37] Thompson PM, Hayashi KM, Simon SL, Geaga JA, Hing MS, Sui Y, Lee JY, Toga AW, Ling W, London ED (2004). Structural abnormalities in the brains of human subjects who use methamphetamine. *J Neuroscience*; *24*: 6028-6036.

[38] Chang L, Ernst T, Speck O, Grob CS (2005). Additive effects of HIV and chronic methamphetamine use on brain metabolite abnormalities. *Am J Psychiatry*; *162*: 361-369.

[39] Ernst T, Chang L, Leonido-Yee M, Speck O (2000). Evidence for long-term neurotoxicity associated with methamphetamine abuse: A 1H MRS Study. *Neurology*; *54(6)*: 1344-1349.

[40] McCann UD, Wong DF, Yokoi F, Villemagne V, Dannals RF Ricaurte GA (1998). Reduced striatal dopamine transporter density in abstinent methamphetamine and methcathinone users: evidence from positron emission tomography with [^{11}C]WIN-35, 428. *J Neuroscience*; *18*: 8417-8422.
[41] Volkow ND, Chang L, Wang GJ, Fowler JS, Ding YS, Sedler M, Logan J, Franceschi D, Gatley J, Hitzemann R, Gifford A, Wong C, Pappas N (2001b). Low level of brain dopamine D_2 receptors in methamphetamine abusers: association with metabolism in orbitofrontal cortex. *Am J Psychiatry*; *158*: 2015-2021.
[42] Volkow ND, Chang L, Wang GJ, Fowler JS, Franceschi D, Sedler MJ, Gatley SJ, Hitzemann R, Ding YS, Wong C, Logan J (2001c). Higher cortical and lower subcortical metabolism in detoxified methamphetamine abusers. *Am J Psychiatry*; *158*: 383-389.
[43] Urbina A, Jones K (2004). Crystal methamphetamine, its analogues and HIV infection: medical and psychiatric aspects of a new epidemic. *Clinical Infectious Diseases; 38*; 890-894.
[44] Jernigan TL, Gamst AC, Archibald SL, Fennema-Notestine C, Mindt MR, Marcotte TL, Heaton RK, Ellis RJ, Grant I (2005). Effects of methamphetamine dependence and HIV infection on cerebral morphology. *Am J Psychiatry*; *162*: 1461-1472.
[45] Carey CL, Woods SP, Rippeth JD, Gonzalez R, Heaton RK, Grant I (2006). Additive deleterious effects of methamphetamine dependence and immunosuppression on neuropsychological functioning in HIV infection. *AIDS Behavior*; *10*: 185-190.
[46] Cherner M, Letendre S, Heaton RK, Durelle J, Marquie-Beck J, Gragg B, Grant I and the HIV Neurobehavioral Research Center Group (2005). Hepatitis C augments cognitive deficits associated with HIV infection and methamphetamine. *Neurology*; *64*: 1343-1347.
[47] Chana G, Everall IP, Crews L, Langford D, Adame A, Grant I, Cherner M, Lazzaretto D, Heaton R, Ellis R, Masliah E and the HNRC Group (2006). Cognitive deficits and degeneration of interneurons in HIV^+ methamphetamine users. *Neurology*; *67*: 1486-1489.
[48] Cass WA, Harned ME, Peters LE, Nath A, Maragos WF (2003). HIV-1 protein Tat potentiation of methamphetamine-induced decreases in evoked overflow of dopamine in the striatum of the rat. *Brain Research*; *984*: 133-142.
[49] Nath A, Hauser KF, Wojna V, Booze RM, Maragos W, Prendergrast M, Cass W, Turchin JT (2002). Molecular basis for interactions of HIV and drugs of abuse. *JAIDS*: *31*; S62-S69.
[50] Lexicomp online, http://www.lexi.com/web/index.jsp
[51] Turchan J, Anderson C, Hauser KF, Sun Q, Zhang J, Liu Y, Wise PM, Kruman I, Maragos W, Mattson MP, Booze M, Nath A (2001). Estrogen protects against the synergistic toxicity by HIV proteins, methamphetamine and cocaine. *BMC Neuroscience*; *2*:3.
[52] Volkow ND, Chang L, Wang GJ, Fowler JS, Franceschi D, Sedler M, Gatley SJ, Miller E, Hitzemann R, Ding YS, Logan J (2001d). Loss of dopamine transporters in methamphetamine abusers recovers with protracted abstinence. *J Neuroscience*; *21*: 9414-9418.

In: Psychoneuroendocrinology Research Trends
Editor: Martina T. Czerbska, pp. 173-212
ISBN: 978-1-60021-665-7

Chapter VI

CENTRAL AND PERIPHERAL PEPTIDE REGULATION OF HUNGER, SATIETY AND FOOD INTAKE IN EATING DISORDERS

Francesca Brambilla[1,2,*], Palmiero Monteleone[2] and Mario Maj[2]
[1]Department of Mental Health, Sacco Hospital, Milan, Italy;
[2]Department of Psychiatry, Naples University, SUN, Naples, Italy.

ABSTRACT

In major primates and humans hunger, satiety, food intake, food preference and aversion are regulated by a vast array of neurotransmitters, neuropeptides and peripheral peptides. The secretory patterns of most of them have been investigated in Eating Disorders (ED), including Anorexia Nervosa (AN), Bulimia Nervosa (BN) and Binge Eating Disorder (BED), and found to be mostly impaired. Alterations of the orexigenic neuropeptides, including neuropeptide Y, aguti-related peptide, opioid peptides, galanin, vasopressin, and of anorexigenic neuropeptides, including α-melanocyte stimulating hormone, brain derived neurotrophic factor, corticotropin-releasing hormone, thyreotropin-releasing hormone, neurotensin, somatostatin, and oxytocin, have been observed in AN and BN. Peripheral peptides stimulating or inhibiting hunger and satiety include ghrelin, leptin, insulin, cholecystokinin, peptide YY, bombesin, pancreatic polypeptide, gastrin-releasing peptide, neuromedin B, vasoactive intestinal peptide, gastrin, resistin and adiponectin. Their secretory patterns have also been found mostly impaired in ED. These central and peripheral peptides interfere also with the development of psychological aspects whose alterations occur in ED. In fact, they modulate mood, anxiety, aggressiveness, hedonic and rewarding patterns, and cognitive processes, particularly learning and memory. This suggests that neuropeptides and peripheral peptides specifically regulating hunger satiety and food intake may interfere

* Correspondence concerning this article should be addressed to Prof. Francesca Brambilla, Centro di Psiconeuroendocrinologia, Piazza Grandi 3, Milan 20129 Italy. Phone: 0039-02-717350 or 0039-368-3017420; Fax: 0039-02-70122889; e-mail: francesca.brambilla4@tin.it.

not only in the eating pathology of AN and BN, but also in some of their psychological disorders.

INTRODUCTION

Regulation of hunger, satiety, food intake, food preference and aversion, a complex mechanism underlying normal eating, is essential for survival. Although Eating Disorders (ED), including Anorexia Nervosa (AN), Bulimia Nervosa (BN) and Binge Eating Disorder (BED), are pathologies in which the altered feeding component is perhaps only the tip of an iceberg of psychopathological disturbances, impaired feeding in ED and its biochemical background may modulate the development, the course, the prognosis, and the responses to treatments of these disorders. Neurotransmitters, including noradrenalin, dopamine, serotonin, acetylcholine, GABA, glutamate and histamine, are all essential regulators of feeding in its various components. Intriguingly, perturbations in neurotransmitter secretory patterns in ED have been repeatedly reported (Eyny and Horwitz 1998, Wellman 2000, Kelley et al. 2005, Kaye et al. 2003).

A vast array of neuropeptides and peripheral peptides are known to regulate food intake, and alterations in peptide physiology have been observed in ED. Currently, no conclusive data exist as to whether derangement in peptide secretion precede the onset of ED or are the consequence of the nutritional changes that characterize ED. Moreover, it is not known whether secretory derangements are independent phenomena or are linked to one another in a specific circuitry of altered biochemical functions. What is interesting is that some of these peptides regulate psychological behaviors, including cognition and emotions, which are also altered in ED. Therefore, this review includes also several aspects of neuropeptide-peripheral peptide activity not specifically related to eating behaviour, but centered on the altered psychological aspects that consistently emerge in ED and therefore suggestive of a link between altered nutritional regulation and psychopathology of ED.

For simplicity, the neuropeptides and peripheral peptides of interest in ED have been ordered according to whether they are part of central or peripheral secretion, and then further classified in stimulators or inhibitors of hunger and satiety, even mindful that their activities are far more complex than just the regulation of feeding behaviour, and that they probably intervene in a variety of coping and survival functions.

NEUROPEPTIDES STIMULATING HUNGAR AND FOOD INTAKE

Neuropeptide Y (NPY)

Hypothalamic neuropeptide Y is a 36-aminoacid peptide produced in the arcuate nucleus and in the midbrain, whose neurons project to the paraventricular nucleus, the dorsomedial and the ventromedial hypothalamic nuclei, the median eminence and the neurointermediate lobe of the pituitary. NPY is a highly potent stimulator of hunger, preferential carbohydrate intake in particular, and of gastrointestinal motility (Stanley et al. 1985). In turns, NPY secretion is stimulated by carbohydrate-rich meals (Wang et al. 1999). When chronically infused in experimental animals NPY provokes unabated hyperphagia and an obese phenotype. Hypothalamic NPY is co-secreted with another potent orexigenic peptide, the Aguti-related peptide (AGRP), both being antagonists to the secretion of the anorexigenic proopiomelanocortin (POMC) peptides. In turn, NPY secretion is inhibited by POMC peptides which through the release of tonic gamma-amino–butyrric acid (GABA) secrete the anorexigenic melanocortins (Schwartz 2006). NPY secretion is inhibited by peripherally secreted leptin and insulin, an effect that is also implicated in the ability of the two adiposity-related hormones to promote negative energy balance (Schwartz 2006). NPY and its receptors Y-1, Y-2, Y-4, Y-5 have been implicated in a variety of physiological and pathological processes such as modulation of learning and memory, anxiety, sleep, sexual activity and blood pressure. In addition, this peptide intervenes in the regulation of hormonal secretions by inhibiting luteotropic hormone and stimulating corticotrophin releasing hormone. Impairments of the above mentioned parameters have all been reported in AN and BN (Kaye et al. 1990, Cleary et al. 1994, Abizaid et al. 1997, Jimenez Vasquez et al. 2000, Li and Ritter 2004, Saydyk et al. 2004, Primeaux et al. 2005).

In emaciated anorexic patients, NPY concentrations in cerebrospinal fluid (CSF) are significantly elevated, do not normalize after short-term weight restoration, but do so in long-term weight-restored patients. However, weight-restored patients whith persistent amenorrhoea still have significantly raised concentrations of NPY. Persistently elevated NPY levels in the central nervous system (CNS) may contribute to the persistence of amenorrhoea during and after recovery from AN. Increased NPY activity could contribute to the obsessive and paradoxical interest of anorexics in caloric intake and food preparation. Alternatively, chronically elevated NPY could downregulate NPY receptors leading to food refusal (Kaye et al. 1990, Kaye 1996, Baranowska et al. 1997).

In BN, CSF and plasma concentrations of NPY are normal, during both the symptomatic phases and after recovery (Kaye et al. 1990, Gendal et al. 1999).

No data are available on NPY secretion in BED.

Aguti-Related Peptide (AGRP)

Aguti-related peptide is a 49-aminoacid protein that stimulates hunger and food intake. In experimental animals, this peptide is normally expressed in the brain and particularly in the hypothalamic arcuate nucleus, where is co-secreted with NPY. AGRP is an endogenous antagonist of the melanocyte stimulating hormone (α-MSH) and the melanocortin receptors MC3R and MC4R (Cone 1999, Wirth and Giraudo 2000). AGRP and AGRPmRNA have been found in the brain, adrenal glands, lungs, testis and skeletal muscles (Moriya et al 2006). In experimental animals levels of AGRPmRNA in the arcuate nucleus increase in response to food restriction or deprivation (Mizuno et al. 1999, Harrold et al. 1999). AGRP is also involved in neuroendocrine regulation, stimulating the secretion of ACTH, cortisol and prolactin and modulating neuroendocrine responses to inflammation (Shen et al. 2002, Xiao et al. 2003). In humans, AGRP secretion correlates with the degree of obesity (Moriya et al. 2006)

In AN patients, plasma AGRP levels have been found to be significantly raised. Although it has been suggested that AGRP secretion may be inhibited by leptin secretion, no correlations were found between low leptin and high AGRP levels in anorexic patients, or between AGRP levels and psychopathological aspects of the patients (Moriya et al. 2006). Defects in the AGRP gene have been associated with AN and weight loss (Vink et al 2001).

No data are available on AGRP secretion in BN and in BED.

Opioid Peptides (OP)

An involvement of opioid peptides in the short-term stimulation of eating, the regulation of appetite, food intake and energy metabolism has been repeatedly suggested and demonstrated for opioid peptides, and specifically for the proopiomelanocortin-derived (POMC) β-endorphin (β-EP), a 31-aminoacid peptide, and for dynorphin, a 32-aminoacid peptide. Opioid peptides and their k and δ receptor active sites are located in the hypothalamic paraventricular and ventromedial nuclei, the perifornical area, the amygdala, the globus pallidus and the nucleus accumbens, a forebrain area critical for regulating reward-related behaviour in the short-term control of eating behaviour. Regulatory inputs to the nucleus accumbens come from diencephalic and brain stem structure, such as the dorsomedial hypothalamus, the ventral tegmental area, the intermediate region of the solitary tract. In turn, from the nucleus accumbens opioids modulate taste hedonics by activating the ventral striatum (Kelley et al. 2002, Will et al. 2003). All involved in the regulation of appetite, food and water intake and energy metabolism, these areas are believed to play a major role in controlling the hedonic and rewarding aspects of food choice and consumption, and in governing the positive emotional responses to highly palatable food , such as fat, sugar, salt and ethanol (Morley and Blundell 1988). Endogenous opioid function is also associated with alterations in learning and memory, pain perception, anxiety, compulsivity, depressive mood, as well as in several physiological and neuroendocrine functions related to stress reaction, these alterations being all characteristic of patients affected with ED (Olson et al. 1997).

In emaciated AN patients opioid-like activity (OLA), which includes all the molecules expressing activity on the μ receptor, is increased in the CSF (Kaye et al. 1982). However, when the opioid peptides were examined separately, baseline CSF levels of β-endorphin (β-EP) were found to be either normal (Gerner and Sharp 1982) or significantly lower than normal (Baranowska 1990, Kaye 1996), The reduced β-EP concentrations persist in short-term restored patients and normalize in long-term recovered anorexics (Kaye et al. 1987a). Dynorphin concentrations in the CSF are normal in both emaciated and recovered AN patients (Lesem et al. 1991, Kaye 1996). In plasma, β-EP levels have been reported to be higher (Brambilla et al. 1991, Tepper et al. 1992) or lower than normal (Baranowska 1990). Because endogenous opioids exert an inhibitory action on hypothalamic luteinizing hormone-releasing hormone (LHRH), the increased central OLA activity of underweight anorexics may contribute to dysfunction in the hypothalamo-pituitary-gonadal axis and to amenorrhoea in these patients.

In BN patients, β-EP concentrations in the CSF either do not differ from those of controls (Lesem et al. 1991) or are lower than normal (Brewerton et al. 1992), and dynorphyn levels do not differ from those of controls (Brewerton et al. 1992). β-EP CSF concentrations are inversely correlated with depressive scores (Brewerton et al. 1992). In plasma, β-EP concentrations are lower in some studies (Waller et al. 1986) and higher in others (Fullerton et val. 1986)

No data are available on opioid peptide secretion in BED.

Galanin (GA)

Galanin, an orexigenic 29-aminoacid peptide, originally isolated from the intestine, has been found in high concentrations in the adrenal medulla, the respiratory tract, the vascular adrenergic neurons, the enteric nervous system and particularly in the CNS. In humans, GA immunoreactive neurons can be found in the hypothalamus and substantia innominata and galaninergic fibers are widely distributed throughout the hypothalamus, hippocampus, amygdala, septum and cortex. In the basal forebrain, including the medial septal nucleus, the nucleus of the diagonal band and the nucleus basalis, and in the hippocampus this peptide is colocalized within acetylcholine neurons, with norepinephrine and epinephrine in the hypothalamic arcuate and paraventricular nucleus, and with ACTH in the pituitary (Berrettini et al. 1988, Ceresini et al. 1997, Kageyama et al. 2006). In the hypothalamus GA inhibits vasopressin secretion and peripherally aldosterone secretion thus suggesting a role of GA in regulating water metabolism, water consumption and water seeking behaviour (Brewer et al. 2005). Owing to its neuronal coexistence with serotonin, noradrenalin and acetylcholine, GA also plays a part in motivation for reinforcement of reward processes (Brewer et al. 2005), improves learning and memory task performance, does not produce perseveration but decreases reinforced strength, and is thought to exert an antidepressant effect by acting on 5HT presynaptic autoreceptors (Echevarria et al. 2005, Ogren et al. 2006). In addition, the peptide influences pain processing, sexual behaviour, fear-related behaviour, and seizure activity (Ogren et al. 2006). Galanin exerts endocrine effects by stimulating the secretion of luteinizing hormone (LH), growth hormone, prolactin, insulin and corticosterone, and by inhibiting corticotropin secretion (Akabayashi et al. 1994, Mazzocchi et al. 1998, Elsaesser et al. 2001). In the paraventricular nucleus GA stimulates feeding behaviour, preferentially increasing fat ingestion, and decreases energy expenditure (Tempel et al. 1998). The important role this peptide plays in regulating food intake occurs via activation of NPY neurons in the dorsomedial hypothalamic nucleus (Kuramochi et al. 2006) and/or inhibition of orexin in the lateral hypothalamus (Kageyama et al. 2006).

GA concentrations in CSF and plasma are normal in restricted anorexics and in anorexics weight-restored for more than 6 months, but are below than normal in anorexics weight-restored for more than 1 year. Plasma GA levels are related to Body Mass Index (BMI) values, although this has not always been confirmed (Berrettini et al. 1988, Invitti et al. 1995, Baranowska et al. 1997-2000).

No alterations in plasma GA secretion have been reported in BN patients (Berrettini et al. 1988, Pirke et al. 1993).

No data are available on GA secretion in BED.

Vasopressin (VP)

Vasopressin, a nonapeptide secreted in the hypothalamus in the cell bodies of the supraoptic nucleus, is transported within the axons to the posterior lobe of the pituitary in association with neurophysin I, enters in the systemic circulation where it controls free water clearence by the kidney (Martin and Reichlin 1987). VP secretion is co-expressed with oxytocin secretion in the supraoptic nucleus and in the median eminence (Merighi 2002). The secretion of the peptide is regulated by hypothalamic osmolarity and stress, including biochemically stressful food restriction (Wideman et al. 1997). VP exerts complex behavioural effects, specifically it enhances memory function (De Wied 1965) and influences development and tolerance to opioids (Demitrack et al. 1992). Directly or indirectly vasopressin induces carbohydrate preference and consumption (Wideman et al. 1997).

In anorexic patients VP concentrations in plasma and CSF have been reported not to differ from those of controls or to be raised, and osmoregulation of the peptide is impaired, either with a secretory deficiency or with an erratic and osmotically uncontrolled release of the hormone, which does not return to normal for up to 18 month after recovery (Gold et al. 1983).

In bulimic patients, basal VP plasma concentrations are normal, while responsiveness of the peptide to hypertonic saline infusion and to insulin-induced hypoglycemia is reduced (Demitrack et al, 1992, Chiodera et al. 1993). VP response to osmotic stimuli in cerebrospinal fluid is higher than normal in bulimics (Demitrack et al. 1992).

No data are available on VP secretion in BED.

Neuropeptides Inhibiting Hunger and Food Intake and Stimulating Satiety

α-Melanocyte Stimulating Hormone (α-MSH)

The anorexigenic α-melanocyte stimulating hormone is a 13-aminoacid peptide produced by proteolytic cleavage of the precursor proopiomelanocortin (POMC). In experimental animals and humans the peptide inhibits food intake, chiefly by blocking fat ingestion, and controls energy homeostasis and expenditure, balancing substrate oxidation, by acting on the melanocortin receptors (MRC) M-3 and M-4 in the arcuate nucleus, ventromedial and dorsal hypothalamus and paraventricular nucleus (Cone 1999, Lindbloom et al. 2000, Butler 2006). Produced chiefly in the pituitary and to a lesser extent in the arcuate nucleus and the nucleus of the solitary tract of the brain, the thalamus and the parafornical area whence it projects into the paraventricular nucleus and the lateral hypothalamus, α-MSH is believed to act through inhibitory-related pathways as a downstream mediator of the effect of leptin (Murphy et al. 1998), in turn inhibiting leptin secretion in adipocytes (Norman et al. 2002). α-MSH exerts its action via five different G protein coupled receptors of which MC3R and MC4R are abundantly expressed in the CNS. Melanocortins have a broad spectrum of physiological activities, including immunomodulatory effects and the control of host responses, with α-MSH being increased in inflammatory diseases. When centrally administered, α-MSH exerts an inhibitory effect on interleukin-1β-induced anxiety (Cragnolini et al. 2006).

In obese experimental animals and in some obese humans genetic alterations in the melanocortin receptors and α-MSH secretion have been detected (Lindbloom et al. 2000, List and Habener 2003). In obese humans, α-MSH circulating concentrations are increased, and its administration reduces body fat (Katsuki et al. 2000, Fehm et al. 2001).

In AN patients, α-MSH levels are not different from those of normal controls (Moriya et al. 2006).

No data are available for α-MSH secretion in BN and BED patients.

Brain-Derived Neurotrophic Factor (BDNF)

Brain-derived neurotrophic factor, a member of the neurotropin family, is a homodimeric protein that acts as a survival and differentiating factor for neuronal subpopulations in prenatal stages. The peptide is highly expressed in the ventromedial hypothalamus and moderately in the paraventricular nucleus (Sahu 2004). In the adult brain, BDNF has several functions, it induces changes in synaptic plasticity, in neurosurvival development, and in neurotransmitter-neuropeptide production, it plays a key role in learning, memory and behaviour (Lommatzsch et al. 2005). Neurodegenerative disorders such as Alzheimer and Parkinson Diseases are associated with decreased BDNF levels in the brain (Bariohay et al. 2005).

The peptide has recently been implicated as an anorexigenic factor and as a downstream effector of melanocortin signalling in the dorsal vagal complex (Bariohay et al. 2005), where its effects seem to be modulated by leptin and cholecystokinin (Bariohay et al. 2005).

In emaciated anorexics and in bulimics BDNF baseline serum levels are significantly reduced (more so in anorexics than in bulimics) while they are normal in patients affected by Binge-Eating Disorder (Nakazato at al 2003, Monteleone et al. 2004-2005c). BDNF levels do not change after partial weight recovery in anorexic patients (Nakazato et al. 2006). An association between BDNF gene polymorphism and restricted AN or BN has been demonstrated (Ribases et al. 2003-2004, Koizumi et al. 2004). In active AN, a strong positive correlation between BDNF levels and EDI-2 scores, B.I.T.E. scores and depression scores has been reported, suggesting a contributing role in the pathophysiology of AN and BN (Nakazato et al. 2006).

As for the pathophysiological significance of decreased circulating BDNF in ED patients one hypothesis posit that, since BDNF exerts a satiety effect, its reduction may represent an adaptive phenomenon that counteracts reduced calorie intake by increasing hunger. In addition, it has been demonstrated that mice with reduced hypothalamic BDNF expression exhibit increased locomotor activity (Kernie *et al.* 2000). It could be argued, therefore, that reduced BDNF levels in AN and BN women sustain increased physical activity of these patients.

Corticotropin-Releasing Hormone (CRH)

Corticotropin-releasing hormone, a 41-aminoacide neuropeptide, is produced in the hypothalamus, the central nucleus of the amygdala, the locus ceruleus, the nucleus of the tractus solitarious, the brain stem and is widely distributed in other brain areas where it exerts one of its primary role in regulating stress reaction and adaptation. Injected intracerebroventricularly, CRH elicits a constellation of behavioural effects, in particular inducing anxiety and depressed mood, and physiological and endocrinological changes like those observed in the stress reactions (Merali et al. 1997, Hillebrand et al. 2002). Peripherally, it stimulates proopiomelanocortin, ACTH, cortisol and adrenal androgens secretion (Gold et al. 1984). Increased by food intake, CRH reduces hunger and food consumption, and provides conditioned taste aversion learning, by acting through α1-adrenergic and dopaminergic receptors and by reducing hypothalamic noradrenaline and dopamine synthesis (Merali e al. 1997, De Pedro et al. 1998, Benoit et al. 2000). CRH exerts other behavioural effects which are typical of AN, like increased locomotor activity and decreased libido and sexual activity(Gold et al. 1984).

CRH has been observed to be altered in several psychiatric disorders, in particular depressive and anxiety disorders, with increased stimulation of the hypothalamo-pituitary-adrenal activity and increased secretion of ACTH, cortisol and DHEA (Monteleone et al 2001).

CRH concentrations in the CSF are higher than normal in anorexic patients and less frequently so in bulimics (Gerner and Gwirtsman 1981, Gwirtsman et al. 1983, Hotta et al. 1986). CRH concentrations in the CSF are especially increased in underweight anorexics during the active phase of the disease and also after short-term recovery (Hotta et al. 1986, Kaye et al. 1987b). By acting directly on the CNS, increased CRH secretion in AN and BN may stimulate anxiety and *depressed* mood, and by long-lasting stimulation of peripheral cortisol secretion it may impair cognitive aspects by acting at the hippocampal and cortical levels (Gold et al. 1984).

No data are available on CRH secretion in BED.

Thyrotropin-Releasing Hormone (TRH)

Thyrotropin-releasing hormone, a 3-aminoacid peptide and a key regulator of the hypothalamo-pituitary thyroid axis, plays an important role in energy homeostasis by setting the basal metabolic rate to stimuli. This peptide is secreted in the cerebral cortex, the hypothalamic ventromedial and dorsomedial nuclei, the arcuate nucleus, the paraventricular nucleus, the floor of the third ventricle, the septal area, the basal ganglia, the lower brain stem and the spinal cord (Grant et al. 1984). Considered an endogenous modulator in the central nervous system, TRH stimulates locomotor activity, body temperature and sympathetic outflow, induces arousal, acts on cognitive function and pain perception. In experimental animals and humans TRH is involved in the regulation of feeding behaviour, by decreasing food intake in the short-term, but it does not seem to modify eating behavior when repeatedly administered (Nemeroff et al. 1984, Choi et al. 2002).

In AN patients, TRH concentration in the CSF is decreased (Lesem et al. 1994), and the alteration persists in weight-recovered subjects. Lower than normal TRH secretion may be due to a decreased secretion of the TRH-stimulating leptin (Hillebrand et al. 2002). Diminished secretion of TRH, resulting in reduced stimulation of thyrotropin, thyroxine and triiodothyronine, may be responsible for the lower energy expenditure and the lower metabolic processes in all of the body's cells, as well as for depressed mood in AN patients (Hillebrand et al. 2002).

TRH concentrations are normal in BN, as are the peripherally dependent TSH, T-4 and T-3 secretions (Devlin et al. 1990).

No data are available on TRH secretion in BED.

Neurotensin (NT)

Neurotensin is a tridecapeptide secreted in the CNS, chiefly in the hypothalamic arcuate, the paraventricular and the dorsomedial nuclei, and in the gastrointestinal tract. Within the wide range of its effects, NT acts on the secretion of pituitary and peripheral hormones, to lower TSH and T-4 secretion, to raise corticosterone secretion and increase glucose, gluconeogenesis and glycogenolysis. Moreover, the peptide acts on the digestive tract stimulating the secretion of insulin and glucagon, on the cardiovascular system and on body temperature and pain (Boules et al. 2000, Hillebrand et al. 2002). In experimental animals and humans, NT is released during meals, modulates ingestive behaviour at the level of the CNS, restricting food intake, and appears to participate also to the regulation of digestion (Boules et al. 2000, Hillebrand et al. 2002).

In acutely ill AN patients and after refeeding NT blood levels are normal (Pirke et al. 1993).

NT levels are normal in BN patients, both in the active phase of the disease and after recovery (Nemeroff et al. 1989).

No data are available on NT levels in BED.

Somatostatin (SRIF)

Somatostatin is a tetradecapeptide first isolated in the hypothalamus, in particular in the paraventricular nucleus, but also ubiquitously throughout the central and peripheral nervous systems, the endocrine pancreas and the gastrointestinal tract. Acting on the CNS, SRIF produces a characteristic spectrum of behaviour, including arousal and stereotypes, and influences analgesia, muscle control and respiratory coordination (Beal et al. 1986). Moreover, the peptide exerts a variety of suppressive actions on the secretion of secretin, pancreozymin, gastrin, pancreatic bicarbonate, gastric acid, pepsin, insulin, glucagon. It works in concert with growth hormone-releasing hormone to regulate growth hormone secretion.

The effects of SRIF on food intake are yet not fully defined, since the peptide seems to possess a dual effect, at low doses increasing and at high doses decreasing food intake, body weight gain and gut motility (Scalera and Tarozzi 1998; Cummings et al. 1998). SRIF secretion is sensitive to protein deficiency and declines with decreases of essential aminoacids in the diet. The peptide seems to act in a circuitry that includes the anterior piriform cortex (APC), a paleocortical region in close anatomical proximity to the neocortical area responsible for taste, which connects with the amygdala, hypothalamus, nucleus of the solitary tract, and parabranchial nucleus, all areas involved in the physical and emotional regulation of food intake (Cummings et al. 1998). Food deprivation leads to significant increases in SRIF secretion (Ishikawa et al. 1997).

In AN patients, CSF and plasma baseline levels of the peptide have been reported to be either normal, or lower or higher than normal to return to normal after refeeding (Gerner and Yamada 1982a, Pirke et al. 1993, Baranowska et al. 2000). In the acute phase of the disease, however, the sensitivity of somatotrop cells to exogenous SRIF is preserved (Gianotti et al. 1999). After a test meal, SRIF is higher then normal in anorexics in the acute phase of the disease, and returns to normal after refeeding (Pirke et al. 1993).

In BN patients, SRIF concentrations in the CSF seem to be normal and then to increase at recovery from the disease (Kaye et al. 1988).

No data are available on SRIF secretion in BED.

Oxytocin (OX)

Oxytocin is an anorexigenic nonapeptide secreted together with neurophisin II from hypothalamic magnocellular neurons located in the paraventricular, supraoptic and accessory nuclei, projecting into the posterior pituitary, whence the hormone is released into the general circulation. OX is also secreted by paraventricular parvicellular neurons projecting into other hypothalamic and extrahypothalamic brain areas. OX binding sites have been found in the hippocampus, the amygdala, the hypothalamic paraventricular nucleus and the thalamic nuclei (Diatz Cabiale et al. 2000). Various peptides seem to be colocalized with OX, including CRH, colecystokinine 1-8 and metenkephalin (Demitrack and Gold 1988). OX is thought to modulate α-2-adrenoceptor-induced food intake, by acting as a satiating factor (Lokrantz et al. 1997).

OX has been recently shown to influence cognition, displaying a disruptive effect on recall of newly stored informations, and on tolerance and adaptation to opioids. This peptide influences a variety of behaviours, as well as physiological and endocrine functions, including female sexual function, cardiovascular effects, digestion and metabolism, locomotor behaviour, pain threshold, thermoregulation. It was also suggested that decreased OX together with increased VP could enhance the retention of cognitive distortion of the aversive consequence of eating (Demitrack and Gold 1988, Demitrack et al. 1990). OX contributes to the control of meal size by acting on gastric emptying and motility. OX and neurophysin II are elevated in the CSF of schizophrenic patients but not in unipolar and bipolar patients.

In emaciated anorexics, OX concentrations in the CSFare reduced , and the response to challenging stimuli is impaired (Demitrack et al. 1990, Chiodera et al. 1991).

No data are available for OX secretion in BN and BED.

PERIPHERAL PEPTIDES STIMULATING HUNGER, SATIETY AND FOOD INTAKE

Ghrelin

Ghrelin, an endogenous 28-aminoacid peptide, was initially characterized as a ligand for the growth hormone secretagogue receptor, primarily produced by endocrine cells in the stomach. The physiological relevance of such a localization remained unclear until it was found that the hormone plays an important role in the regulation of food intake and metabolism. Currently, ghrelin is considered as a "hunger hormone" that signals the brain the need to initiate food consumption (Cummings et al. 2001). Altered ghrelin physiology may therefore be involved in the onset and/or maintenance of disrupted eating behaviors in AN and BN individuals.

Fasting circulating levels of ghrelin have been consistently reported to be raised in underweight patients with AN, particularly in those with bingeing-purging (BP) AN as compared to those with restricted (R) AN, suggesting that binge-purging behavior may have some influence on circulating ghrelin (Cuntz et al. 2001, Hrsek et al. 2003, Tanaka et al. 2003a,b, Tolle et al. 2003, Misra et al. 2004, Otto et al. 2004, Soriano-Guillen et al. 2004, Misra et al. 2005b, Troisi et al. 2005). These findings were not replicated by Otto et al. (2004), who found no significant difference in fasting plasma ghrelin concentrations between AN-R and AN-BP patients, or by Troisi et al. (2005), who detected opposite results, with higher levels of plasma ghrelin in AN-R subjects than in AN-BP patients. The elevated ghrelin concentrations in underweight anorexics tend to normalize with recovery of body weight (BW), and the decline in circulating ghrelin seems to parallel the progressive increase in BW during weight restoration treatments (Cuntz et al. 2001, Soriano-Guillen et al. 2004, Tanaka et al. 2004, Otto et al. 2005, Janas-Kozik et al. 2006, Nakahara et al. 2006). This supports the view that the elevated ghrelin production in symptomatic anorexics is a state-dependent phenomenon that resolves with the restoration of normal eating habits. Nedvidkova et al. (2003) found that the food-induced suppression of circulating ghrelin in underweight anorexic patients was almost completely absent, suggesting that in AN, because of the chronic food restriction and the subsequent adaptation to starvation, a single meal is unable to suppress the drive to eat to counteract the subjects's negative energy balance. The suppressant effect of oral glucose administration on plasma ghrelin was significantly blunted in AN-R women, and normal but delayed in those with AN-BP, although both patient groups had increased baseline ghrelin concentrations (Tanaka et al. 2003c). Two other studies, with different experimental designs, reported that in AN elevated preprandial levels of ghrelin, although suppressed by food ingestion in percentages similar to those of normal subjects, remained significantly higher than in controls (Misra et al. 2004, Stock et al. 2005). Otto et al. (2005) found that postprandial ghrelin release in AN patients did not differ from that of healthy subjects; moreover, although morning ghrelin levels progressively declined with the recovery of BW, ghrelin response to food ingestion was not influenced by weight restoration. Differences in the clinical characteristics of patients'samples, the type, composition and total calories of test meals, and the timing of blood collection may partially explain the discrepancies between different studies.

A major technical issue in ghrelin studies is the rapid degradation of circulating ghrelin into inactive fragments. It is commonly accepted that concentrations of octanoylated ghrelin (active form) accounts for less than 10% of circulating ghrelin levels, which include acylated and desacylated fragments (inactive forms). Since the studies cited above measured total ghrelin concentrations without differentiating between active and inactive ghrelin, increases in total ghrelin plasma levels in AN were not representative of increased active ghrelin production. When this aspect was taken into account, conflicting results emerged. In fact, Nakai et al. (2003) showed that plasma active ghrelin was increased in underweight anorexics and normally suppressed after oral glucose administration. Hotta et al. (2004), instead, found that total ghrelin was elevated in underweight anorexics and did not decrease after glucose infusion whereas active ghrelin was reduced and showed normal glucose suppression, suggesting that decreased active ghrelin in AN results from the reduced availability of gastric secreting cells in an atrophic stomach while an increase in non-active ghrelin may be the result of reduced renal clearance of degraded forms of the peptide.

In BN it was initially reported that fasting ghrelin was increased in relatively small samples of symptomatic patients (Tanaka et al. 2002, Tanaka et al. 2003b). This increase was evident in patients with frequent binge-purging episodes, but not in non-purging ones, supporting the idea that binge/purge cycles with vomiting as opposed to binge eating episodes can influence fasting plasma ghrelin. However, subsequent studies on relatively small patients' samples, detected no significant difference in plasma ghrelin concentrations between bingeing-purging bulimics and healthy controls (Nakazato et al. 2004, Troisi et al. 2005). Monteleone et al. (2005a) measured fasting ghrelin concentrations in a relatively large sample of 56 subjects with BN-BP and 51 healthy controls, without detecting any significant difference in circulating ghrelin levels between the two groups. Moreover, no significant correlation was found between plasma ghrelin and severity of the binging/vomiting behavior.

Ghrelin responses to a macronutrient balanced meal and a fat-rich meal have been reported to be blunted in symptomatic binge/purge bulimics as compared to healthy controls (Monteleone et al. 2003a, Kojima et al. 2005, Monteleone et al. 2005b). The suppression of circulating ghrelin after food ingestion may denote a compensatory activation of a peripheral signal that promotes the termination of food ingestion. In this vein, it is noteworthy that some, but not all, experimental human studies have suggested diminished satiety responses to meals in BN patients (Owen et al. 1985, Walsh et al. 1989, Kissileff et al. 1996), which could be mediated, at least in part, by the impaired food-induced responses of ghrelin.

No data are available for ghrelin secretion in BED.

Peripheral Peptides Inhibiting Hunger and Food Intake Stimulating Satiety

Leptin

Leptin, a hormone encoded by the *ob* gene, is a 146-aminoacid protein secreted predominantly by white adipocytes (Zhang et al. 1994). It acts as a factor that informs the CNS on the amount of energy stored in the body's adipose tissue and behaves as a hunger suppressant signal involved in both the long-term and short-term regulation of energy balance (Blundell et al. 2001). Circulating leptin is also affected by dietary nutrients, since diets with different fat and carbohydrate contents differently affect leptin production (Coleman and Herrmann 1999, Havel 1999). Leptin is also crucial for many processes such as inflammation, angiogenesis, endocrine regulation, immune function and, most importantly, for reproduction (Bouloumie et al. 1998, Fantuzzi and Faggioni 2000, Moschos et al. 2002). In fact, leptin behaves as a critical factor linking adipose stores and hypothalamic centers that control gonadal function. Specifically, leptin stimulates the gonadal axis. Insufficient nutrient intake leads to decreased leptin production, resulting in the inhibition of reproductive function.

It has been consistently reported that plasma and CSF levels of leptin are markedly lower than normal in underweight anorexics (Ferron et al. 1997, Hebebrand et al. 1997, Mantzoros et al. 1997, Monteleone et al. 2000a, Monteleone et al. 2002a,b, Holtkamp et al. 2003a,b, Lob et al. 2003), but are still significantly and positively correlated to the patient's BMI and body fat mass (Grinspoon et al. 1996, Ferron et al. 1997, Eckert et al. 1998, Monteleone et al. 2000a, Holtkamp et al. 2003a). This indicates that, even at extreme low body weight (BW) and low body fat, circulating leptin still acts in AN as an endogenous signal of energy stores. The circadian rhythm of leptin is disrupted in AN patients, with attenuation or abolition of the physiological nocturnal surge (Balligand et al. 1998). Alterations in leptin resolve with the recovery of BW (Hebebrand et al. 1997, Holtkamp et al. 2003a, Lob et al. 2003, Haas et al. 2005) and both plasma and CSF levels of the hormone in long-term recovered anorexics appear to be similar to those of controls (Gendall et al. 1999). Moreover, longitudinal studies have shown that during refeeding treatments, leptin concentrations progressively increase as the BW is regained and in those cases with a too rapid weight restoration they reach values disproportionately higher than normal (Hebebrand et al. 1997, Mantzoros et al. 1997, Holtkamp et al. 2003a, Lob et al. 2003). It has been argued that hyperleptinemia in AN patients during refeeding may be a major factor in the patient’s difficulty to reach or maintain target-weight. Since circulating levels of the Ob-Re are drastically elevated in underweight anorectic women (Krizova et al. 2002, Monteleone et al. 2002a), assuming that these changes reflect an up-regulation of leptin receptors on the cell surfaces, a too rapid increase in circulating leptin in AN patients might theoretically result in a potentiation of leptin-induced appetite suppression and energy expenditure, thus counteracting the therapeutic process. Preliminary evidence indeed suggests that hyperleptinaemia is associated with an elevated risk of renewed weight loss (Holtkamp et al. 2004). However, it has been recently reported that in anorexic patients, whose caloric intake during refeeding was adapted daily to keep circulating leptin within the reference range for the patient's current BW, the outcome was not improved (Lob et al. 2003). Therefore, the role of leptin in BW recovery in AN needs to be further explored. It should be pointed out that relative hyperleptinaemia is not a universal finding in studies on AN patients who have regained weight (Djurovic et al. 2004, Popovic et al. 2004), magnitude and rate of weight gain in addition to timing of blood samples after weight gain possibly explaining these differences.

Since leptin exerts several stimulatory effects on neuroendocrine function, reduced secretion of leptin in AN patients has been held responsible for a lifetime history of amenorrhea and subnormal serum levels of luteinizing hormone (LH) (Kopp et al. 1997, Audi et al. 1998, Ballauff et al. 1999, Holtkamp et al. 2003c, Brambilla et al. 2003a). Low leptin secretion is also held potentially responsible for low hypothalamo-pituitary-thyroid function, increased hypothalamo-pituitary-adrenal function and osteopenia in AN (Wauters et al. 2000, Brambilla and Monteleone 2003b, Misra et al. 2005a).

Excessive physical activity has been reported in 31% to 80% of AN patients. In addition, an inverse correlation between food intake and rigorous physical activity has been shown during the acute weight loss phase suggesting a relationship between energy balance and activity levels (Davis et al. 1997, Exner et al. 2000, Hebebrand et al. 2003). It has been also demonstrated that, leptin concentrations in AN inpatients were not only inversely correlated with levels of physical exercise but also predicted physical activity, supporting the idea that hypoleptinemia may be an important factor to sustain an excessive physical activity in women with AN (Holtkamp et al. 2003b).

In normal weight subjects with BN, circulating leptin has been reported to be either decreased, normal or increased (Ferron et al. 1997, Monteleone et al. 2000a-b, Jimerson et al. 2000, Brewerton et al. 2000, Nakai et al. 2000) perehaps owing to heterogeneity of patient samples, since bulimic women exhibit plasma leptin levels ranging from anorexic-like to normal (Monteleone et al. 2002b). In a group of BN patients with a significantly longer duration of the illness and a significantly higher number of daily binge/vomiting episodes than those with normal leptin secretion, hyposecretion of leptin was observed in the absence of significant modifications in BW and BMI (Jimerson et al. 2000, Monteleone et al. 2002b). It therefore seems that binge eating, particularly when chronic and severe, impairs leptin production. However, no decrease in circulating leptin has been detected in patients with BED (Karhunen et al. 1997, Monteleone et al. 2000a, Adami et al. 2002), who binge more often than bulimic individuals, but do not engage in compensatory behaviors. In bulimic patients a positive correlation between circulating leptin and BMI was detected (Monteleone et al. 2000a, Monteleone et al. 2002b), suggesting that leptin still functions as an endogenous signal of BW differences. Moreover, in bulimics with hypoleptinemia, the fall in circulating leptin in response to acute fasting is almost completely blunted, whereas its response to short-term normal refeeding, although similar in percentage to that of controls, is not sufficient to restore normal blood levels of the hormone (Monteleone et al. 2000b). It therefore seems that the role of leptin as a peripheral signal of available energy stores is preserved in BN, whereas, at least in those patients with anorexic-like leptin concentrations, its function as an index of acute changes in energy balance is lost. This may have important pathogenic implications. In fact, since leptin behaves as a hunger suppressant signal (Blundell et al. 2001), lower leptin values after refeeding together with alterations in other key regulatory systems of hunger and satiety occurring in people with BN, such as the impaired meal-induced ghrelin suppression (Monteleone et al. 2003a) and the blunted post-prandial rise in cholecystokinin (Geracioti and Liddle 1988), could contribute to the binge-eating behavior in BN patients.

Insulin

Insulin, a 51-aminoacid anorexigenic peptide secreted by the pancreas, taken up in the CNS from the bloodstream where it circulates in proportion to body adiposity thus providing a dose-dependent feedback signal to the CNS, intervenes in food intake and body weight regulation. Insulin acts at binding sites in the hypothalamic paraventricular and arcuate nuclei. Insulin deficiency increases NPYmRNA and decreases POMC hypothalamic biosynthesis in the hypothalamus. Insulin blood levels results largely from peripheral insulin sensitivity related partly to body fat stores and fat distribution, specifically to intra-abdominal fat (Porte et al. 2002). This peptide influences catabolic pathways as well. Insulin secretion is elevated during and after food intake, and its major function is to eliminate absorbed fuel from the blood by stimulating the storage of these fuels in tissue depots. The anorexigenic effect of insulin is not due to food aversion, and it is mostly directed against fat consumption, possibly through the inhibition of galanin secretion which is stimulated by fat introduction (Van Dyk et al. 1997)

In AN patients, insulin secretion has been reported to be normal or decreased during the active phase of the disease to return to normal after recovery, and correlates inversely with the degree of weight loss. Insulin secretion after glucose stimulation is prolonged. However, insulin sensitivity is augmented owing to an increased number of specific receptors, which contrasts with the old theory of insulin resistance in AN. These alterations resolve to normal after refeeding and weight gain (Kalucy et al. 1976, Beumont and Russell 1982, Alderdice et al. 1985, Kobayashi et al. 1988, Weingarten et al. 1988, Fujii et al. 1989, Schreiber et al. 1991, Invitti et al. 1995, Wacschlicht- Rodbard et al. 1997, Dostalova et al. 2006a).

No signs of altered insulin and glucose metabolism have been observed in BN (Weingarten et al. 1988, Schreiber et al. 1991).

No data are available for insulin secretion in BED.

Cholecystokinin (CCK)

Cholecystokinin, an anorexigenic 39-aminopeptide, reduces hunger by inducing satiety. Chiefly secreted peripherally in the gastrointestinal tract but also centrally by nearly all brain regions, the peptide acts through CCKa receptors located on vagal afferent fibers connecting the vagus nerve with the brainstem complex, and through activation of pyloric CCK receptors (Moran 2000). CCK plays a role in several digestive functions, like contraction and emptying of the gallbladder, stimulation of pancreatic enzymes, delay of gastric emptying, reduction of gastric secretion, regulation of insulin secretion (Baranowska et al. 2000). CCK acts synergistically with leptin to inhibit food intake (Wang et al. 2000) and its secretion is mainly stimulated by fat feeding (Jacob et al. 2000). CCK also acts as an anxiogenic and fear-stimulating substance (Bradwejn et al. 1998, Tsutsumi et al. 1999), increases cognitive processes in normal young but not in older humans (Dodt et al. 1996), and decreases short-term memory (Shlik et al. 1998). It is noteworthy that CCK is co-expressed with dopamine in dopamine-secreting cells of the brain cortex and that it modulates brain dopamine function and dopamine-related behaviour, particularly cognitive function and locomotor activity (Yoshikawa et al. 1994, Feifel et al. 1995).

The genetic –36C/T polymorphism of CCK is similar in anorexics, bulimics and controls. In emaciated anorexics, baseline CCK CSF and plasma levels are normal (Gerner and Yamada 1982a, Abell et al. 1987, Harty et al. 1991, Geracioti et al. 1992) or higher than normal (Tamai et al. 1993, Fujimoto et al. 1997, Cuntz et al. 2000), but return to normal at recovery. Decreased levels of CCK-8 have been found in peripheral blood mononuclear cells, a peripheral compartment that mimics the secretion and regulation of neurotransmitters, neuropeptides and neurohormones in central neurons (Brambilla et al. 1995a). After a fat-reach meal CCK is released more rapidly in restricted anorexics than in normal controls and is higher than normal, with normalization of the response in weight-recovered patients (Harty et al. 1991, Phillipp et al. 1991, Fujimoto et al. 1997), although a normal postprandial rise in CCK has also been reported (Geracioti et al. 1992, Pirke et al. 1994). A glucose load is followed by lower than normal CCK secretion, which returns to normal at recovery (Tamai et al. 1993).

In bulimic patients, CCK basal plasma levels have been reported to be normal (Geracioti and Liddle 1988) or lower than normal (Lydiard et al. 1993) and lower than normal also in the CSF and in polymorphonucleate blood mononuclear cells (Brambilla et al. 1995b). A single intravenous injection of CCK is able to suppress binges in BN (Mitchell et al. 1985). A normal or diminished CCK release after a test-meal has been reported (Geracioti et al. 1988, Phillipp et al. 1991, Brewerton et al. 1997, Walsh 2002). Decreased CCK values do not correlate to either BMI or frequency of binging/vomiting episodes, but are significantly related to anxiety, hostility, aggression and impairments of interpersonal sensitivity (Lydiard et al. 1993).

No data are available for CCK secretion in BED.

Peptide YY (PYY)

Peptide YY belongs to the PP-fold peptide family together with neuropeptide Y and pancreatic polypeptide PP. PYY is a 36-aminoacid peptide secreted by the endocrine L-cells of the gut (ileum, colon and rectum) and also by the nerve fibers of the enteric nervous system, but it is also found in the cervical spinal cord, medulla oblongata, pons and hypothalamus (Ferrier et al. 2002). Like all PP-fold peptides, PYY binds to a family of 5 G-protein coupled receptors named the "Y-receptors family". PYY itself is an orexigenic peptide, but its truncated form PYY 3-36 is released postprandially, induces satiety, and is a short-term regulator of feeding behavior. PYY inhibits feeding by acting on the paraventricular and arcuate nuclei of the hypothalamus; it delays gastric emptying, inhibits gastric motility, stimulates duodenum and ileum motility, and exerts inhibitory or stimulatory colonic motility, stimulates absorption-secretion of water and electrolytes and gastric and pancreatic secretions. Its release is stimulated by the presence of nutrients in the intestine, specifically by short-chain fatty acids. The peptide is low during fasting, increases postprandially and acts as a satiety signal regulating the termination of individual meals (Kaye et al. 1990, Ferrier et al. 2002, Riediger et al. 2004, Monteleone et al. 2005b).

Studies on PYY_{3-36} secretion in people with ED are still scanty. In malnourished AN girls, baseline PYY_{3-36} levels were reported to be normal (Stock *et al.* 2005) or increased (Misra *et al.* 2006; Nakahara *et al.* 2006) while the PYY_{3-36} response to energy intake was reported to be time-delayed (Stock et al. 2005) or increased (Nakahara *et al.* 2006). After a partial recovery of BW, the PYY_{3-36} response to test meal was improved, but not completely restored (Nakahara *et al.* 2006).

Initial studies on BN reported normal CSF and plasma levels of PYY in both symptomatic and 1-year recovered bulimic patients (Berrettini et al 1988; Kaye et al 1990). Moreover, Kaye et al. (1990) measured plasma concentrations of PYY in 5 bulimic women during episodes of binge/vomiting and in 6 healthy women, who experimentally binged without vomiting. In the latter, plasma PYY concentrations significantly rose after meals and remained elevated for the subsequent 2 hours; in the former, circulating PYY increased after the first binge and remained elevated for the duration of bingeing and vomiting; bulimic patients exhibited a slightly higher post-prandial peak value of PYY than did healthy volunteers. Recently, two independent research groups found a blunted PYY_{3-36} response to food ingestion in symptomatic bulimic women together with a decreased response of ghrelin (Monteleone *et al.* 2005b; Kojima *et al.* 2005). Moreover, both studies showed a negative correlation between meal-induced PPY increase and ghrelin decrease, suggesting a negative interaction between PYY_{3-36} and ghrelin. Suppression of circulating ghrelin and increased plasma PYY_{3-36} after food ingestion may denote a compensatory activation of peripheral signals that promote termination of food ingestion. Hence, in symptomatic bulimics, the blunted responses of circulating ghrelin and PYY_{3-36} to a test meal would denote the occurrence in these subjects of an impaired suppression of the drive to eat following a meal, which might play a role in their increased food consumption and binge eating. Further studies are needed to clarify the state or trait-dependent nature of these alterations in BN.

No data are available for BED on PYY secretion.

Bombesin (BN)

Bombesin, a tetradecapeptide, comprises a large family of structurally related peptides with a wide array of physiological actions and behavioural effects. This peptide is widely distributed in the gastrointestinal tract and in the CNS, in particular in the hypothalamic paraventricular nucleus. Bombesin suppresses food intake and induces satiety in experimental animals and humans (Ladenheim et al. 2002). The peptide modulates exocrine and endocrine processes, activates the sympathetic nervous system, regulates smooth muscle contraction, metabolism, homeostasis, thermoregulation and behaviour.

In AN patients, CSF concentrations of bombesin are normal (Gerner and Yamada 1982a).

No data are available for bombesin secretion in BN and BED patients.

Pancreatic Polypeptide (PP)

Pancreatic polypeptide, a 36-aminoacid peptide of the NPY and PYY family, is released from the pancreas and the gastrointestinal mucosa after food ingestion in experimental animals and in humans, but it is also widely distributed in central and peripheral neurons and in the adrenal gland (Morley et al. 1987, Berntson et al. 1993). PP is thought to be a circulating satiety factor. Moreover, the peptide modulates learning and memory and controls various neuroendocrine functions (Nussdorfer and Gottardo 1997)

In restricted and bingeing-purging anorexic and in bulimic patients plasma PP levels are not different from those of controls. However, after a fat-rich meal PP plasma concentrations are frankly higher than normal in patients with restricted AN and BN but not in those with bingeing-purging AN (Fujimoto et al. 1997).

No data are available for PP secretion in BED.

Gastrin-Releasing Peptide (GRP) and Neuromedin B

Gastrin-releasing peptide and Neuromedin B are bombesin-like peptides, and potent centrally acting anorexigenic agents (Wada et al. 1998). GRP is secreted by the enterochromaffin cells of the gastrointestinal mucosa, the bronchopulmonary mucosa, the ovaries, and the prostate. GRP exerts an hypothalamo-pituitary-adrenal stimulating activity in experimental animals, and has an autocrine role as a growth factor (Watanabe and Orth 1988, Strewler 1998).

In patients with long-term recovered AN, CSF GRP concentrations have been found to be normal, while levels are significantly lower in BN than in AN patients or in controls (Frank et al. 2000). Persistent GRP abnormalities after recovery from BN suggests that this alteration may be trait-related and contribute to the hyperphagia of bulimia.

No data are available for GRP secretion in BED.

Vasoactive Intestinal Peptide (VIP)

Vasoactive intestinal peptide, an anorexigenic 14-aminoacid peptide and member of the gastroenteropancreatic peptide family, is secreted by the gastric antrum and intestine mainly after a high fat concentration in the duodenum. The peptide is also produced in the cerebral cortex, suprachiasmatic nucleus, amygdala, periaqueductal gray, spinal cord, heart, lung, pituitary, eyes, kidney, skin, middle ear, neutrophils and chromaffin granules of the adrenal medulla (Said 1984). It coexists with acetylcholine, and catecholamines. The peptide exerts an hypnogenic effect, and acts on neuroendocrine cells by stimulating the secretion of prolactin, the hypothalamo-pituitary-adrenal axis, LHRH and by inhibiting the secretion of SRIF and CCK. VIP may be stimulated by bicarbonate and CCK release (Said 1984, Hill et al. 1986, Baranowska et al. 2000). The secretion of the peptide increases significantly postprandially

In AN, baseline plasma values of VIP have been reported to be normal (Harty et al. 1991) or higher then normal, without correlating with the patient's BMI (Baranowska et al. 2000). After a liquid mixed meals VIP concentrations are normal (Harty et al. 1991).

No data are available for VIP secretion in BN and BED patients.

Gastrin

Gastrin, a 12-aminoacid peptide that stimulates satiety, is released from the gut, from the gastric antrum in particular, after food ingestion. This peptide belongs to the gastropancreatic polypeptide family, and in particular to the CCK-gastrin group. The CCK family of peptides is widely distributed throughout the digestive system, in the central nervous system and in the peripheral nervous system. Gastrin plays a role similar to that of CCK in the contraction and emptying of the gallbladder, stimulation of pancreatic enzyme secretion, delay of gastric emptying, diminution of gastric acid secretion, and regulation of insulin secretion (Furuse et al. 1999, Baranowska et al. 2000). Moreover, it suppresses drinking, stimulates locomotor activity and activates the hypothalamo-pituitary-adrenal function.

Gastrin plasma levels are lower than normal in AN patients (Baranowska et al. 2000) or relatively higher (Pirke et al. 1993). After a test meal gastrin concentration does not differ between AN patients and controls (Pirke et al. 1993).

No data are available for gastrin secretion in BN and BED patients.

Resistin

Resistin, a satiety-inducing 114-aminoacid polypeptide containing a 20-aminoacid signal sequence, is secreted by adipocytes in white adipose tissue. In experimental animals resistin and the resistin gene are both expressed in the hypothalamus, specifically in the arcuate nucleus, the cortex, the pituitary, and the circulating mononuclear cells, while is barely detectable in adipocytes (Wiesner et al. 2004). Resistin is not a unique molecule but is a member of a small family of homologous cystein-rich proteins (Steiger and Haring 2005). Because administration of the substance impairs insulin action and worsens insulin sensitivity, and because anti-resistin antibody partially restores insulin sensitivity in obese experimental animals, the peptide is considered a potential factor in regulating insulin sensitivity (Vidal-Puig and O'Rahilly 2001). The peptide acts also on skeletal muscle myocytes and hepatocytes (Hellstrom et al. 2004). The regulatory mechanism of resistin in adipogenesis has not been fully studied. Resistin mRNA is barely detectable during fasting and dramatically increases when animals and humans are fed with a high carbohydrate diet. Resistin RmNA expression is also regulated by other factors including free fatty acids. A correlation between adipose tissue resistin expression and BMI has not yet been found (Villena et al. 2002, Steiger and Haring 2005)

In a study by Housova et al. (2005) plasma resistin levels in patients with AN were not found to be significantly different form those in controls or in BN patients and showed no significant relation with BMI or body fat content. To the contrary, significantly decreased plasma resistin levels were detected in AN patients by Dostalova et al. (2006a,b), who suggested that low plasma resistin levels in AN were probably related to a defective mononuclear-macrophage number and/or function, because resistin, besides being produced in adipocytes in humans, is highly secreted by macrophages in the bone marrow (Patel et al. 2003). Interestingly, an in vivo microdialysis experiment detected increased levels of resistin in the subcutaneous adipose tissue of undernourished AN patients (Dostalova et al. 2006b). An explenation for these opposite changes of resistin in the blood and adipose tissue of AN individuals is lacking.

No data ara available for resistin secretion in BN and BED patients.

Adiponectin (AD)

Adiponectin, a member of the complement C19 factor family, displays a structural homology with the Tumor Necrosis Factor (TNF) family. This recently investigated peptide comprises 244 aminoacid and is an adipose-tissue-derived protein (adipocytokine), peripherally exclusively secreted by differentiated adipocytes and centrally expressed in the brain and the pituitary, where it is implicated in the central regulation of energy expenditure and body weight (Hellstrom et al. 2004, Steiger and Hering 2005). The effect of AD on hunger and satiety has not been fully elucidated.

AD secretion is barely detectable in fasting animals, but increases when the animals are refed with a high carbohydrate diet. Moreover, AD secretion is modulated by the degree of adiposity in experimental animals and humans, is relatively constant throughout the day and is unaffected by food intake. In humans, a negative correlation between AD plasma levels and BMI has been reported though not always confirmed (Villena et al. 2002, Kontogianni et al. 2004, Steiger and Hering 2005). In obese subjects AD and adiponectin gene transcription levels are lower than normal, while weight reduction is accompanied by an increase in plasma AD concentrations, possibly representing a key factor in the development of the metabolic syndrome (Stefan and Stumvoll 2002, Kern et al. 2003, Staiger and Haring 2005). AD increases insulin sensitivity and improve glucose tolerance, control body weight, regulates lipid homeostasis and prevents atherosclerosis. AD is also decreased in insulin resistance, in hyperinsulinemia and coronary artery disease (Steiger and Hering 2005). Diminished AD production is thought to be involved in the development of obesity, diabetes and insulin resistance (Havel 2002).

In underweight AN patients elevated levels of circulating AD were reported by some authors (Delporte et al. 2003; Pannaciulli_et al. 2003), whereas Iwahashi et al. (2003) observed no difference in AD concentrations between AN patients and healthy women, and Tagami et al. (2004) even found decreased levels in malnourished AN individuals. Increased AD levels in AN patients were recently confirmed by three independent research groups (Housova et al. 2005; Bosy-Westphal et al. 2005; Dostalova et al. 2006b). Moreover, AD was found to be strictly correlated to the patient's nutrional status, based on an inverse correlation between hormone concentrations and both BMI and percent body fat mass (Housova et al. 2005; Dostalova et al. 2006b). Less severely malnourished patients with binge/purging AN had a relatively modest increase in circulating AD, whereas a more prominent rise was found in severely malnourished restrictive AN individuals (Housova et al. 2005). In a sample of AN patients undergoing BW recovery, elevated pretreatment AD levels only minimally declined after weight gain, suggesting that factors other than changes in body fat mass are involved in the dysregulated AD production in AN. The physiologic relevance of high adiponectin levels in AN is unclear. Two main hypotheses have been put forward. The one suggests that, since intracerobreventricular administration of AD in mice decreases BW (Qi et al. 2004), hyperadiponectinemia could be a contributing etiopathogenetic factor to AN (Housova et al. 2005). Alternatively, elevated circulating AD concentrations could represent a compensatory mechanism for the increased insulin sensitivity in AN patients, since a negative correlaction between plasma AD and insulin was found in these patients (Dostalova et al. 2006a-b).

In BN patients, one study reported increased AD concentrations that resulted positively correlated with the severity of binge/purging (Monteleone *et al.* 2003b), whereas another study (Tagami *et al.* 2004) detected decreased circulating levels of the adipokine. Housova et al. (2005) found normal AD concentrations in symptomatic bulimic women. Differences in the patient's samples, assay methods and time of the day of blood collection might have been responsible for the discrepancies among the studies.

No data are available for AD secretion in BED.

Central and Peripheral Peptides Modulating Hunger and Satiety not yet Studied in Eating Disorders

Other central and peripheral peptides have been shown to influence eating habits in experimental animals and humans. Although the details of their secretion have not yet been examined in AN and BN, the data reported here will demonstrate the importance of these peptides in regulating hunger and satiety and the need to examine them in relation to ED.

Orexigenic Peptides

Melanocyte–concentrating hormone (MCH), a cyclic 19-aminoacid peptide and one of the most potent orexigenic factors, is secreted in the CNS in the lateral hypothalamus, the paraventricular and arcuate nuclei, the perifornical area and the zona incerta, and connects to several brain areas throughout the brain and to the pituitary. MCH is a functional melanocortin antagonist and exerts various endocrine action; specifically, it stimulates hypothalamo-pituitary-adrenal function and secretion of LHRH. MCHmRNA levels increases during fasting (Sone et al. 2000, Chaffer and Morris 2002, Sahu 2004).

Orexins A and B, two novel appetite-stimulating neuropeptides with 33 and 28 aminoacids respectively, besides playing a role in food intake regulation, influence also energy homeostasis and arousal. They are produced centrally in the lateral hypothalamic area and peripherally in the gastrointestinal submucosal and myenteric plexuses, the endocrine cells of the intestinal mucosa and pancreatic islets and the vagal neurones. In turns, orexin-containing neurons project into the hypothalamic arcuate and paraventricular nuclei. Orexins inhibit gastrointestinal motility, modulates insulin and glucagon release from the endocrine pancreas, stimulate arousal, locomotor activity, behavioural aspects, especially inducing stereotypies, increases blood pressure, heart rate, pain, sympathetic and neuroendocrine functions, in particular the hypothalamo-pituitary-adrenal secretion, and induce narcolepsy. Plasma concentrations of orexins increase during fasting (Kirchgessner 2002, Lopez et al. 2002, Hellstrom et al. 2004).

Anorexigenic Peptides

Cocaine-amphetamine regulated transcript (CART), a 116-aminoacid peptide with potent anorexigenic effects, is expressed in the hypothalamic arcuate, paraventricular, dorsomedial and lateral nuclei, in extrahypothalamic areas, the median eminence and the adrenal medulla. CART and CART mRNA are colocalized with proopiomelanocortin, TRH, GA, VP, OX in all the above mentioned hypothalamic areas. CART inhibits food intake, decreases body weight, stimulates energy metabolism, activates the hypothalamo-pituitary-adrenal axis, increases anxiety, is increased during fasting and low in obesity (Hillebrand et al. 2002, Stanek 2006).

Urocortin 1, a 40-aminoacids peptide and a member of the CRH family, exerts a very strong anorexic effect. Urocortin 1 is expressed centrally in the hypothalamic arcuate, lateral and paraventricular nuclei, the Edingher-Westphal nucleus, the lateral superior olive, the supraoptic and motor nuclei of the brain stem, and peripherally in the stomach, colon and endocrine organs. From the hypothalamic nuclei urocortin 1-producing neurons project into the nucleus of the tractus solitarius and the dorsomotor nucleus of the vagus in the brain stem (Hillebrand et al. 2002)

Amylin, a 37-aminoacid polypeptide synthesized in pancreatic cells and co-released with insulin in response to food intake, has been shown to elicit a strong anorexic effect, mediated by the area postrema of the hindbrain, partly by dopaminergic and histaminergic neurons (Mollet et al. 2003).

Glucagon-like peptides (GLP)1 and 2 are produced and secreted by endocrine L-cells in the mucosa of ileum and colon, primarily in response to carbohydrate and fat ingestion. They play an important role in the "ideal brake mechanism", i.e. in the slowing down of stomach and intestine motility and acid secretion after food ingestion, which is probably the mechanism by which GLP-1 exerts its short-term appetite-reducing effect. GLP-1 exerts insulinotropic and glucagonogastric actions. In experimental animals and humans, GLP-1 inhibits food and water intake by acting on the paraventricular nucleus of the hypothalamus, while GLP-2 produces the same effects only in animals (De Graaf et al. 2004, Hellstrom et al. 2004).

Neuromedin U is a 174-aminoacid protein widely expressed in the CNS, particularly in the hypothalamus. The peptide is a potent anorexigenic factor, stimulates energy expenditure, increases locomotor activity, elevates body temperature and heat production. In experimental animals it is increased in adiposity and in obesity, in association with increased leptin, insulin, lipids and glycemia. In human obesity NMU gene variants have been reported to occur, in association with an autosomal dominant inheritance (Hainerova et al. 2006).

CONCLUSIONS

The data of the literature on central and peripheral peptides regulating hunger, satiety and food choice suggest that these substances are part of complex apparently independent, redundant systems, each one possibly just superimposing its effects on the other without a definite design. However, preliminary observations indicate that at least some of these peptides are part of circuitries in which each element, with its own specific and different effect in terms of time of secretion, response to specific stimuli, type of responses, duration and consistency of effects, site of action, concurs toward a final regulatory action on eating behavior. A typical example of such a circuit is the stimulation by peripherally secreted leptin and inhibition from ghrelin of hypothalamic NPY secretion, which inhibits melanocortins secretion, which, in turn, inhibits NPY secretion, with AGRP and MCH intervening in this functional circle (Broberger and Hokfelt 2001). Another circuit is represented by GA secretion which inhibits secretion of leptin in adipose tissue and central orexins, while it stimulates NPY; or by the fat-stimulated CCK which stimulates PYY which inhibits POMC peptides. Other circuits are emerging and will perhaps shed light on an apparently extremely complex network of peptides and neurotransmitters involved in the same function, with the secretion of each peptide and the sensitivity of its receptors prevailing according to the body's needs.

A second point of interest is the site of and consequently the time of action of neuropeptides. In fact, some function as hormones influencing distant neuronal activities, like preoptic nucleus-secreted vasopressin that acts on hippocampal memory, or hypothalamic CRH that acts on amygdala-related emotions, while others act locally in a paracrine fashion, like CCK negatively acting on dopamine secretion in cortical neurons. This obviously requires different times of responses, the former needing time for the neuropeptide to reach distant neurons, and the latter being immediate, and, as a consequence, lasting differently too. It should also be kept in mind that some neuropeptides are co-expressed by the same neurons. Typical of this phenomenon is the secretions of NPY and AGRP and that of POMC and CART which are co-expressed in the same neurons in the arcuate nucleus, while in the lateral hypothalamus orexins and dynorphin are similarly co-expressed. This phenomenon may underlie a positive or negative paracrine effect of one peptide on another.

Coming to neuropeptides and peripheral peptides secretion and functions in ED, data are still too scanty to draw definite conclusions, specifically to which comes first, peptide alterations being a pathogenetic cause of ED or the ED psychopathology leading to impaired nutrition and subsequent peptide damage. Although this last hypothesis is currently considered to be the most probable, it has been demonstrated that alterations of some peptides persist or appear long after recovery from ED. A typical example of this are the normal levels of GA during the active phase of AN, which rise to higher than normal more than 1 year after recovery (Berrettini et al. 1988), the still osmotically uncontrolled VA secretion in 1 year-recovered anorexics (Gold et al. 1983), the reduced TRH secretion in long-recovered anorexics (Lesem et al. 1994), the persistently low gastrin-releasing peptide blood concentrations in BN long after recovery (Frank et al. 2000). The persistence or the occurrence of a peptide pathology long after recovery from ED have suggested that these alterations migh be trait rather than state dependent phenomena, and so may precede the onset of the diseases as one of their causal factors. Even so, it cannot be ruled out that the persistence or occurrence of peptide impairments long after recovery may be the expression of alterations occurring during the disease that are severe enough to induce a permanent damage, or that they might depend on still not well defined persistent external interfering factors.

As mentioned in the introduction, neuropeptide and peripheral peptide alterations may be relevant for the development of specific psychopathological aspects of ED. And as outlined above, it has been demonstrated that alterations in NPY, CRH, TRH, GA and opioid peptides might modulate anxiety, mood and aggressiveness, that opioid peptides and GA positively, and α-MSH negatively, modulate hedonic and rewarding psychological aspects, that GA, VP, and CCK improve cognitive processes, particularly learning and memory. This would suggest that peptide alterations may underlie some of the psychopathological aspects of ED, although these connections have not yet been sufficiently studied and need to be clarified both as mechanisms of action and in regard to their relevance in the onset, course and therapeutic approach in ED. Investigation into the connections between the secretion of central-peripheral peptides and hormones which results in clinically evident pathological symptoms could be of importance to explain the occurrence of amenorrhoea, impaired statural growth, abnormal water metabolism, hypothyroidism and all the other hormonal dysfunctions which occur in ED. It is clear that physical impairments in ED not only result from the nutritional alterations of the eating disorders but may be the consequence of the peptide derangement responsible for the concurrent physical and psychological damages of ED.

Studies of the biological alterations underlying ED are still a question of controversy. Further investigations may reveal that the central and peripheral peptide impairments represent the chore of the coexistence of psychological and physical alterations.

REFERENCES

Abell TL, Malagelada JR, Brown ML, Camilleri M, Go VLW, Aspiroz F, Callaway CW, Kao PC, Zinsmeister AR, Huse DM. (1987) Gastric electromechanical and neurohormonal function in anorexia nervosa. *Gastroenterol., 93*: 958-965

Abizaid A, Walzer CD, Woodside B. (1997) Changes in neuropeptide Y immunoreactivity in the arcuate nucleus during and after food restriction in lactating rats. *Brain Res., 761*: 306-312

Adami GF, Campostano A, Cella F, Scopinaro N. (2002) Serum leptin concentration in obese patients with binge eating disorder. *Int. J. Obesity, 26*: 1125-1128.

Akabayashi A, Zaia CTBV, Koenig JI, Gabriel SM, Silva I, Leibowitz SF. (1994) Diurnal rhythm of galanin-like immunoreactivity in the paraventricular and suprachiasmatic nuclei and other hypothalamic areas. *Peptides, 15*: 1437-1444

Alderdice JT, Dinsmore WW, Buchanan KD, Adams C. (1985) Gastrointestinal hormones in Anorexia Nervosa. *J. Psychiat. Res., 19*: 207-213

Audi L, Mantzoros CS, Vidal-Puig A, Vargas D, Gussinye M, Carrascosa A. (1998) Leptin in relation to resumption of menses in women with anorexia nervosa. *Mol. Psychiatry, 3*: 544-547.

Ballauff A, Ziegler A, Emons G, Sturm G, Blum WF, Remschmidt H, Hebebrand J. (1999) Serum leptin and gonadotropin levels in patients with anorexia nervosa during weight gain. *Mol. Psychiatry, 4*: 71-75.

Balligand JL, Brichard SM, Brichard V, Desager JP, Lambert M. (1998) Hypoleptinemia in patients with anorexia nervosa: loss of circadian rhythm and unresponsiveness to short-term refeeding. *Eur. J. Endocrinol., 138*: 415-420.

Baranowska B. (1990) Are disturbances in opioid and adrenergic systems involved in the hormonal dysfunction of anorexia nervosa?. *Psychoneuroendocrinol., 15*: 371-379

Baranowska B, Radzikowska M, Wasilewska-Dziubinska E, Plonowski A, Roguski K. (1997) Neuropeptide Y, galanin and leptin release in obese women and in women with anorexia nervosa. *Metabolism, 46*: 1384-1389

Baranowska B, Radzikowska M, Wasilewska-Dziubinska E, Roguski K, Boroviec M. (2000) Disturbed release of gastrointestinal peptides in anorexia nervosa. *Diabetes, Obesity, Metab., 2*: 99-103

Bariohay B, Lebrun B, Moyse E, Jean A. (2005) Brain-derived neurotrophic factor plays a role as an anorexigenic factor in the dorsal vagal complex. *Endocrinol., 146*: 5612-5620

Beal MF, Uhl G, Mazurek MF, Kowall N, Martin JB. (1986) Somatostatin: alterations in the central nervous system in neurological diseases. In Martin J, Barchas JD (eds) *Neuropeptides and psychiatric diseases.* Raven Press, New York, pp. 215-257

Benoit SC, Thiele TE, Heinrichs SC, Rushing PA, Blake KA, Steeley RJ. (2000) Comparison of central administration of corticotropin-releasing hormone and urocortin on food intake, conditioned taste aversion, and c-Fos expression. *Peptides, 21*: 345-351

Berntson GG, Zipf WB, O'Dorisio TM, Hoffman JA, Chance RE. (1993) Pancreatic polypeptide infusions reduce food intake in Prader-Willi syndrome. *Peptides, 14*: 497-503

Berrettini WH, Kaye WH, Sunderland T, May C., Gwirtsman HE, Mellow A, Albright A. (1988) Galanin immunoreactivity in human CSF: studies in eating disorders and Alzheimer's disease. N*europsychobiol., 19*: 64-68

Beumont PJV, Russell J. (1982) Anorexia Nervosa. In Beaumont PJV, Burrows GD (eds) *Handbook of Psychiatry and Endocrinology,* Elsevier, Amsterdan. pp. 63-79

Blundell JE, Goodson S, Halford JCG. (2001) Regulation of appetite: role of leptin in signalling systems for drive and satiety. *Int. J. Ob., 25 (Suppl 1)*: 529-534.

Bosy-Westphal A, Brabant G, Haas V, Onur S, Paul T, Nutzinger D, Klein H, Haner M, Muller M. (2005) Determinants of plasma adiponectin levels in patients with anorexia nervosa examined before and after weight gain. *Eur. J. Nutr., 44*: 355-359.

Boules M, Cusack B, Zhao L, Fauq A, McCormick DJ, Richelson E. (2000) A novel neurotensin peptide analog given extracranially decreases food intake and weight in rodents. *Brain Research, 865*: 35-44

Bouloumie A, Drexler HC, Lafontan M, Busse R. (1998) Leptin, the product of ob gene, promotes angiogenesis. *Circ. Res., 83*: 1059-1066.

Bradwejn J, LeGrand JM, Koszycki D, Bates JHT, Bourin M. (1998) Effects of cholecystokinin tetrapeptide on respiratory function in healthy volunteers. *Am. J. Psychiatry, 155*: 280-282

Brambilla F, Ferrari E, Petraglia F, Facchinetti G, Catalano M, Genazzani AR. (1991) Peripheral opioid secretory pattern in anorexia nervosa. *Psychiatry Res., 39*: 115-127

Brambilla F, Brunetta M, Peirone A, Perna G, Sacerdote P, Manfredi B. (1995a) T-lymphocyte cholecystokinin-8 and β-endorphin concentrations in eating disorders. I. Anorexia Nervosa. *Psychiatry Res., 29*: 43-50

Brambilla F, Brunetta M, Draisci A, Peirone A, Perna G, Sacerdote P, Manfredi B, Panerai A. (1995b) T-lymphocyte concentrations of cholecystokinin-8 and beta-endorphin in eating disorders. II Bulimia Nervosa. *Psychiatry Res., 59*: 51-56

Brambilla F, Monteleone P, Bortolotti F, Dalle Grave R, Todisco P, Favaro A, Santonastaso P, Ramacciotti C, Paoli R, Maj M. (2003a). Persistent amenorrhoea in weight-recovered anorexics: psychological and biological aspects. *Psychiatry Res., 118*: 249-257.

Brambilla F, Monteleone P. (2003b) Physical complications and physiological aberrations in eating disorders: a review. In: Maj M, Halmi K, Lopez-Ibor JJ, Sartorius N (eds.) *Eating Disorders.* Wiley & Sons Ltd, Chichester, England, 139-192.

Brewer A, Echevarria DJ, Langel U, Robinson JK. (2005) Assessment of functional role for galanin in the CNS. *Neuropeptides, 39*: 323-326

Brewerton TD, Lydiard RB, Laraia MT, Shook JE. (1992) CSF-β-endorphin and dynorphin in Bulimia Nervosa. *Am. J. Psychiatry, 149*: 1086-1090

Brewerton TD, Lydiard RB, Beinfeld MC, Laraia M, Stuart G, Ballenger JC. (1997) CSF CCK-8 in Bulimia Nervosa. *Biol. Psychiatry, 31*: 90A-91A

Brewerton TD, Lesem MD, Kennedy A, Garvey WT. (2000) Reduced plasma leptin concentrations in bulimia nervosa. *Psychoneuroendocrinol., 25*: 649-658.

Broberger C, Hokfelt T. (2001) Hypothalamic and vagal neuropeptide circuitries regulating food intake. *Physiol.Behav.,* 74: 669-682

Butler AA. (2006) The melanocortin system and energy balance. *Peptides 27*: 281-290

Ceresini G, Fabbro A, Freddi M, Pisolini G, Reali N, Broglio G, Valenti G. (1997) Evaluation of circulating galanin levels after exercise-induced pituitary hormone secretion in man. *Metabolism, 46*: 282-286

Chaffer CL, Morris MJ. (2002) The feeding response to melanin-concentrating hormone is attenuated by antagonism of the NPY-Y1-receptor in the rat. *Endocrinol., 143*: 191-197

Chiodera P, Volpi R, Capretti L, Marchesi C, D'Amato L, De Ferri A, Bianconi L, Coiro V. (1991) Effect of estrogens or insulin-induced hypoglycemia on plasma oxytocin levels in bulimia and anorexia nervosa. *Metabolism, 40*: 1226-1230

Chiodera P, Volpi R, Marchesi C, Caffarra P, De Ferri A, Capretti L, Speroni G, D'Amato L, Coiro V. (1993) Reduction in the arginine vasopressin responses to metoclopramide and insulin-induced hypoglycemia in normal weight bulimic women. *Neuroendocrinol. 57*: 907-911

Choi YH, Hartzell D, Azain MJ, Baile CA. (2002) TRH decreases food intake and increase water intake and body temperature in rats. *Physiol. Behav., 77*: 1-4

Cleary J, Semotuk M, Levine AS. (1994) Effects of neuropeptide Y on short-term memory. *Brain Res., 653*: 210-214

Coleman RA, Herrmann TS. (1999) Nutritional regulation of leptin in humans. *Diabetologia, 42*: 639-646.

Cone RD. (1999) The central melanocortin system and energy homeostasis. *TEM, 10*: 211-216

Cragnolini AB, Schioth HB, Scimonelli TN. (2006) Anxiety-like behavior induced by IL-1β is modulated by α-MSH trough central melanocortin-4 receptor. *Peptides, 27*: 1451-1456

Cummings SL, Truong BG, Gietzen DW. (1998) Neuropeptide Y and somatostatin in the anterior piriform cortex alter intake of aminoacid-deficient diets. *Peptides, 19*: 527-535

Cummings DE, Purnell JQ, Frayo RS, Schmidova K, Wisse BE, Weigle DS. (2001) A preprandial rise in plasma ghrelin levels suggests a role in meal initiation in humans. *Diabetes, 50*: 1714-1719.

Cuntz U, Fruhauf E, Lehnert P, Otto B, Eiepl R, Fichter MM. (2000) The role of CCK-8-S in anorexia nervosa at admission and during weight gain. A longitudinal study in 9 controls and 35 patients with Anorexia Nervosa. *Eeating Disorders Research Society congress Prien,Germany* 9-12/11/2000 abstract

Cuntz BO, Fruehauf E, Wawarta R, Folwaczny C, Riepl RL, Heiman ML, Lehnert P, Fichter M, Tschop M. (2001) Weight gain decreases elevated plasma ghrelin concentrations of patients with anorexia nervosa. *Eur. J. Endocrinol., 145*: R5-R9.

Davis C, Katzam DK, Kaptein S, Kirsh C, Brewer H, Kalmbach K, Olmsted MP, Woodside DB, Kaplan AS. (1997) The prevalence of high-level exercise in the eating disorders: etiological complications. *Compr. Psychiatry, 38*: 321-326.

De Graaf C, Blom WAM, Smeets PAM, Stafleu A, Hendriks HFJ. (2004) Biomarkers of satiation and satiety. *Am. J. Clin. Nutr., 79*: 946-961

Delporte ML, Brichard SM, Hermans MP, Beguin C, Lambert M. (2003) Hyperadiponectinemia in anorexia nervosa. *Clin. Endocrinol. (Oxf), 58*: 22-29

Demitrack MA, Gold PW (1988). Oxytocin: neurobiologic considerations and their implications for affective illness. *Progr. Neuro-Psychopharmacol. Biol. Psychiat., 12*: S32-S51

Demitrack MA, Lesem MD, Listwak SJ, Brandt HA, Jimerson DC, Gold PW. (1990) CSF oxytocin in anorexia nervosa and bulimia nervosa: clinical and pathophysiologic considerations *Am. J. Psychiatry, 147*: 882-886

Demitrack MA, Kalogeras KT, Altemus M, Pigott TA, Listvwak SJ, Gold PW. (1992) Plasma and cerebrospinal fluid measures of Arginine Vasopressin secretion in patients with Bulimia Nervosa and in healthy subjects. *J. Clin. Endocrinol. Metab., 74*: 1277-1284

De Pedro N, Delgado MJ, Pinillos ML, Alonso-Bedade M. (1998) α1-adrenergic and dopaminergic receptors are involved in the anorectic effect of corticotropin-releasing factor in goldfish. *Life Science, 63:* 180-188

Devlin MJ, Walsh B, Kraft JG, Heymsfield SB, Pi-Sunyer FX, Dantzic S. (1990) Metabolic abnormalities in bulimia nervosa. *Arch. Gen. Psychiatry, 47*: 144-148

De Wied D. (1965) The influences of the posterior and intermediate lobe of the pituitary and pituitary peptides on the maintenance of a conditioned avoidance response in rats. *Int. J. Neuropharmacol., 4*: 157-167

Diatz-Cabiale Z, Narvaez JA, Petersson M, Uvnas-Moberg K, Fuxe K. (2000) Oxytocin/alpha2-adrenoceptor interactions in feeding responses. *Neuroendocrinol., 71*: 209-218

Djurovic M, Pekic S, Petakov M, Damjanovic S, Doknic M, Dieguez C, Casanueva FF, Popovic V. (2004) Gonadotropin response to clomiphene and plasma leptin levels in weight recovered but amenorrhoeic patients with anorexia nervosa. *J. Endocrinol. Invest., 27*: 523-527.

Dodt C, Sarnighausen HE, Pietrowsky R, Fehm HL, Born J. (1996) Ceruletide improves event-related potential indicators of cognitive processing in young but not in elderly humans. *J. Clin. Psychopharmacol., 16*: 440-445

Dostalova I, Smitka K, Papezova H, Kvasnickova H, Nedvidkova J. (2006a) Increased insulin sensitivity in patients with anorexia nervosa: the role of adipocytokines. *Physiol. Res.,* in press

Dostalova I, Kunesova M, Duskova J, Papezova H, Nedvidkova J. (2006b) Adipose tissue resistin levels in patients with anorexia nervosa. *Nutrition, 22*: 977-983

Echevarria DJ, Brewer A, Bushell G, Manuzon H, Langel U, Robinson JK. (2005) Galanin and perseveration. *Brain Res., 1041*: 143-148

Eckert ED, Pomeroy C, Raymond N, Kohler PF, Thuras P, Bowers CY.(1998) Leptin in anorexia nervosa. *J. Clin. Endocrinol. Metab., 83*: 791-795.

Elsaesser F, Aldag P, Stimper B. (2001) Stimulation of porcine pituitary luteinizing hormone release by galanin: putative auto/paracrine regulation. *Neuroendocrinol., 74*: 288-299

Exner C, Hebebrand J, Remschmidt H, Wewetzer C, Ziegler A, Herpertz S, Schweiger U, Blum WF, Preibisch G, Heldmaier G, Klingenspor M. (2000) Leptin suppresses semi-starvation induced hyperactivity in rats: implications for anorexia nervosa. *Mol. Psychiatry, 5*: 476-481

Eyny YS, Horvitz JC. (1998) Opposing roles of D1 and D2 receptors in appetitive conditioning. *J. Neuroscience, 23*. 1584-1587

Fantuzzi G, Faggioni R. (2000) Leptin in the regulation of immunity, inflammation, and haematopoiesis. *J. Leukoc. Biol., 68*: 437-446.

Fehm HL, Smolnik R, Kern W, McGregor GP, Bickel U, Born J. (2001) The melanocortin melanocyte-stimulating hormone/adrenocorticotropin4-10 decreases body fat in humans. *J. Clin. Endocrinol. Metab., 86*: 1144-1148

Feifel D, Reza T, Robeck S. (1995) Antipsychotic potential of CCK-based treatments: an assessment using the prepulse inhibition model of psychosis. *Neuropsychopharmacol., 20*: 141-149

Ferrier L, Segain JP, Bonnet C, Cherbut C, Lehur PA, Jarry A, Galmiche JP, Blottiere HM. (2002) Functional mapping of NPY/PYY receptors in rat and human gastrointestinal tract. *Peptides, 23*: 1765-1771

Ferron F, Considine RV, Peino R, Lado IG, Dieguez C, Casanueva FF. (1997) Serum leptin concentrations in patients with anorexia nervosa, bulimia nervosa and non-specific eating disorders correlate with the body mass index but are independent of the respective disease. *Clin. Endocrinol., 46*: 289-293.

Frank GK, Kaye W, Landenheim EE, McConaha C. (2000) Reduced gastrin releasing peptide in CSF after recovery from bulimia nervosa. *Eating Disorders Research Society Congress*, Prien, Germany, abstract.

Fujii S, Tamai H, Kumai M, Takaichi Y, Nakagawa T, Aoki TT. (1989) Impaired glucagon secretion to insulin-induced hypoglycemia in anorexia nervosa. *Acta Endocrinol., 120*: 610-615

Fujimoto S, Inui A, Kiyota N, Seki W, Koide K, Takamiya S, Uemoto M, Nakajima Y, Baba S, Kasuga M. (1997) Increased cholecystokinin and pancreatic polypeptide responses to a fat-rich meal in patients with restrictive but not bulimic Anorexia Nervosa. *Biol. Psychiatry, 41*: 1068-1070

Fullerton DT, Swift WJ, Getto CJ, Carlson IH.(1986) Plasma immunoreactive beta-endorphin in bulimics. *Psychol. Med., 16*: 56-63

Furuse M, Ao R, Bungo T, Ando R, Shimojo M, Masuda Y, Saito N. (1999) Central gastrin inhibits feeding behavior and food passage in neonatal chiks. *Life Sciences, 65:* 305-311

Gendall KA, Kaye WH, Altemus M, McConaha CW, La Via MC. (1999) Leptin, neuropeptide Y, and peptide YY in long term recovered eating disorder patients. *Biol. Psychiatry, 46*: 292-299

Geracioti TD jr, Liddle RA. (1988) Impaired cholecystokinin secretion in bulimia nervosa. *New Engl. J. Med., 319*: 683-688.

Geracioti TD, Liddle RA, Altemus M, Demitrack MA, Gold PW. (1992) Regulation of appetite and cholecystokinin secretion in Anorexia Nervosa. *Am. J. Psychiatry, 149*: 958-961

Gerner RH, Gwirtsman HE. (1981) Abnormalities of the dexamethasone suppression test and urinary MHPG in anorexia nervosa. *Am. J. Psychiatry, 138*: 650-654

Gerner RH, Yamada T. (1982a) Altered neuropeptide concentrations in cerebrospinal fluid of psychiatric patients. *Brain Res., 238* : 298-302

Gerner RH, Sharp S. (1982b) β-endorphin immunoreactivity in normal, schizophrenic, depressed, manic, and anorexic subjects. *Brain Res., 237*: 244-247

Gianotti L, Rolla M, Arvat E, Belliti D, Valetta MR, Federghini M, Ghigo E, Muller EE. (1999) Effects of somatostatin infusion on the somatotropic responsiveness to growth hormone-releasing hormone in patients with Anorexia Nervosa. *Biol. Psychiatry, 45*: 334-339

Gold PW, Kaye W, Robertson GL, Ebert M. (1983) Abnormalities in plasma and cerebrospinal-fluid arginine vasopressin in women with anorexia nervosa. *New Engl. J. Med,. 30*8: 1117-1123

Gold PW, Chrousos G, Kellner C, Post R, Roy A, Augerinos P, Schulte H, Oldfield E, Loriaux DL. (1984) Psychiatric implications of basic and clinical studies with corticotropin-releasing factor. *Am. J.Psychiatry, 141*: 619-627

Grant LD, Bissette G, Nemeroff CB. (1984) Distribution of peptides in the central nervous system. In Nemeroff CB, Dunn AJ (eds) *Peptide hormones and behavior.* MTPpress ltd., Lancaster, England, pp. 77-98

Grinspoon S, Gulick T, Askari H, Landt M, Lee K, Anderson E, Ma Z, Vignati L, Bowsher R, Herzog D, Klibanski A. (1996) Serum leptin levels in women with anorexia nervosa. *J. Clin. Endocrinol. Metab., 81*: 3861-3863.

Gwirtsman HE, Roy P, Yager RH. (1983) Neuroendocrine abnormalities in bulimia. *Am. J. Psychiatry, 140*: 559-563

Haas V, Onur S, Paul T, Nutzinger DO, Bosy-Westphal A, Hauer M, Brabant G, Klein H, Muller MJ. (2005) Leptin and body weight regulation in patients with anorexia nervosa before and during weight recovery. *Am. J. Clin. Nutr., 81*: 889-896.

Hainerova I, Torekov SS, Ek J, Finkova M, Borch-Johnsen K, Jorgensen T, Madsen OD, Lebl J, Hansen T, Pedersen O. (2006) Association between neuromedin U gene variants and overweight and obesity. *J. Clin. Endocrinol. Metab., 91*: 5057-5063

Harrold JA, Williams G, Widdowson PS. (1999) Changes in hypothalamic agouti-related protein (AGRP) but not α-MSH or pro-opiomelanocortin concentrations in dietary-obese and food-restricted rats. *Biochem. Biophys. Res. Commun., 258*: 574-577

Harty RF, Pearson PH, Solomon TE, McGuigan JE (1991) Cholecystokinin, vasoactive intestinal peptide and peptide histidine methionine responses to feeding in anorexia nervosa. *Regul. Pept., 36*: 141-150

Havel PJ, Townsend R, Chaump L, Teff K. (1999) High-fat meals reduce 24-h circulating leptin concentrations in women. *Diabetes., 48*: 334-341.

Havel PJ. (2002) Control of energy homeostasis and insulin action by adipocyte hormones: leptin acylation stimulating protein, and adiponectin. *Curr. Opin. Lipidol., 13: 51-59*

Hebebrand J, Blum WF, Barth N, Coners H, Englaro P, Juul A, Ziegler A, Warnke A, Rascher W, Remschmidt H. (1997) Leptin levels in patients with anorexia nervosa are reduced in the acute stage and elevated upon short-term weight restoration. *Mol. Psychiatry, 2*: 330-334.

Hebebrand J, Exner C, Hebebrand K, Holtkamp C, Casper RC, Remschmidt H, Herpertz-Dahlmann B, Klingenspor M. (2003) Hyperactivity in patients with anorexia nervosa and in semistarved rats: evidence for a pivotal role of hypoleptinemia. *Physiol. Behav., 79*: 25-37.

Hellstrom PM, Geliebter A, Naslund E, Schmidt PT, Yahav EK, Hashim SA, Yeomans MR. (2004) Peripheral and central signals in the control of eating in normal, obese and binge-eating human subjects. *Brit. J. Nutrition, 92 Suppl. 1*: S47-S57

Hill P, Thijssan JH, Garbaczewski L, Koppeschaar HP, De Waard F. (1986) VIP and prolactin release in response to meals. *Scand J. Gastroenterol., 21*: 958-960

Hillebrand JJG, De Wied D, Adan RAH. (2002) Neuropeptides, food intake and body weight regulation: a hypothalamic focus. *Peptides, 23*: 2283-2306

Holtkamp K, Hebebrand J, Mika C, Grzella I, Heer M, Heussen N, Herpertz-Dahlmann B. (2003a) The effect of therapeutically induced weight gain on plasma leptin levels in patients with anorexia nervosa. *J. Psychiat. Res., 37*: 165-169.

Holtkamp K, Herpertz-Dahlmann B., Mika C, Heer M, Heussen N, Fichter M, Herpertz S, Senf W, Blum WF, Schweiger U, Warnke A, Ballauff A, Remschmidt H, Hebebrand J. (2003b) Elevated physical activity and low leptin levels co-occur in patients with anorexia nervosa. *J. Clin. Endocrinol. Metab., 88*: 5169-5174.

Holtkamp K, Mika C, Grzella I, Heer M, Pak H, Hebebrand J, Herpertz-Dahlmann B. (2003c) Reproductive function during weight gain in anorexia nervosa. Leptin represents a metabolic gate to gonadotropin secretion. *J Neural Transm, 110*: 427-435.

Holtkamp K, Hebebrand J, Mika C, Heer M, Heussen N, Herpertz-Dahlmann B. (2004) High serum leptin levels subsequent to weight gain predict renewed weight loss in patients with anorexia nervosa. *Psychoneuroendocrinol.. 29*: 791-797.

Hotta M, Shibasaki T, Masuda A, Imaki T, Demura H, Ling N, Shizume K. (1986) The responses of plasma adrenocorticotropin and cortisol to corticotropin-releasing hormone (CRH) and cerebrospinal fluid immunoreactivity CRH in anorexia nervosa patients . *J. Clin. Endocrinol. Metab., 62*: 319-324

Hotta M, Ohwada R, Katakami H, Shibasaki T, Hizuka N, Takano K. (2004) Plasma levels of intact and degraded ghrelin and their responses to glucose infusion in anorexia nervosa. *J. Clin. Endocrinol. Metab., 89*: 5707-5712.

Housova J, Anderlova K, Krizova J, Haluzikova D, Kremen J, Kumstyrova T, Papezova H, Haluzik M. (2005) Serum adiponectin and resistin concentrations in patients with restrictive and binge/purge form of anorexia nervosa and bulimia nervosa. *J. Clin. Endocrinol. Metab., 90*: 1366-1370.

Hrsek M, Rosicka M, Papezova H, Krizova J, Kotrikova E, Haluzk M, Justova V, Lacinova Z, Jarkovska Z. (2003) Plasma ghrelin levels and malnutrition: a comparison of two etiologies. *Eat. Weight Dis., 8*: 207-211.

Invitti C, Brunanti A, Pasqualinotto L, Dubini A, Bendinelli P, Maroni P, Cavagnini F. (1995) Plasma galanin concentrations in obese, normal weight and anorectic women. *Int. J. Obesity, 19*: 347-349

Ishikawa M, Mizobuchi M, Takahashi H, Bando H, Saito S. (1997) Somatostatin release as measured by in vivo microdialysis: circadian variation and effect of prolonged food deprivation. *Brain Res., 749*: 226-231

Iwahashi H, Funahashi T, Kurokawa N, Sayama K, Fukuda E, Okita K, Imagawa A, Yamagata P, Shimamura I, Miyagawa JI, Mabsuzawa Y. (2003) Plasma adiponectin levels in women with anorexia nervosa. *Horm. Metab. Res., 35*: 537-540

Jacob S, Mosenthin R, Zabielski R, Rippe C, Sorhede Winzel M, Gacsalyi U, Laubitz D, Grzesiuk E, Pierzynowski SG. (2000) Fat infused intraduodenally affects the postprandial secretion of the exocrine pancreas and the plasma concentrations of cholecystokinin but not of peptide YY in growing pigs. *J. Nutr., 130*: 2450-2455

Janas-Kozik M, Krupka-Matuszczyk I, Malinowska-Kolodziej I, Lewin-Kowalik J. (2006) Total ghrelin plasma level in patients with the restrictive type of anorexia nervosa. *Regul Pept.*, in press

Jimenez Vasquez PA, Overstreet DH, Mathé AA. (2000) Neuropeptide Y in male and female brains of flinders sensitive line, a rat model of depression. Effects of electroconvulsive stimuli. *J. Psychiat. Res., 34*: 405-412

Jimerson DC, Mantzoros C, Wolfe BE, Metzger ED. (2000) Decreased serum leptin in bulimia nervosa. *J. Clin. Endocrinol. Metab., 85*: 4511-4514.

Kageyama H, Kita T, Toshinai K, Guan JL, Date Y, Takenoya F, Kato S, Matsumoto H, Ohtaki T. (2006) Galanin-like peptide promotes feeding behaviour via activation of orexinergic neurones in the rat lateral hypothalamus. *J. Neuroendocrinol., 18*. 33-41

Kalucy RS, Crisp AH, Chard T, McNeilly A, Chen CN, Lacey JH. (1976) Nocturnal hormonal profiles in massive obesity, anorexia nervosa, and normal females. *J. Psychosom. Res., 20*: 594-604

Karhunen L, Haffner S, Lappalainen R, Turpeinen A, Miettinen H, Uusitupa M. (1997) Serum leptin and short-term regulation of eating in obese women. *Clin. Sci., 92*: 573-578.

Katsuki A, Sumida Y, Murashima S, Furuta M, Araki-Sasaki R, Tsuchihashi K, Hori Y, Yano Y, Adachi Y. (2000) Elevated plasma levels of α-melanocyte stimulating hormone (α-MSH) are correlated with insulin resistance in obese men. *Int. J. Obesity, 24*: 1260-1264

Kaye WH, Pickar D, Naber D, Ebert MH. (1982) Cerebrospinal fluid opioid activity in anorexia nervosa. *Am. J. Psychiatry, 139*: 643-645

Kaye WH, Berrettini WH, Gwirtsman HE, Chretien M, Gold PW, George DT, Jimerson DC, Ebert MH. (1987a) Reduced cerebrospinal fluid levels of immunoreactive pro-opiomelanocortin related peptides (including β-endorphin) in anorexia nervosa. *Life Sciences, 41*: 2147-2155

Kaye WH, Gwirtsman HE, George DT, Ebert MH, Jimerson DC, Tomai TP, Chrousos GP, Gold PW. (1987b) Elevated cerebrospinal fluid levels of immunoreactive corticotropin-releasing hormone in anorexia nervosa: relation to state of nutrition, adrenal function , and intensity of depression. *J. Clin. Endocrinol. Metab., 64*:203-208

Kaye WH, Rubinow D, Gwirtsman HE, George DT, Jimerson DC, Gold PW. (1988) CSF somatostatin in anorexia nervosa and bulimia: relationship to the hypothalamic-pituitary-adrenal cortical axis. *Psychoneuroendocrinol., 13:* 265-272

Kaye WH, Berrettini W, Gwirtsman H, George DT. (1990) Altered cerebrospinal fluid neuropeptide Y and peptide YY immunoreactivity in Anorexia and Bulimia Nervosa. *Arch. Gen. Psychiatry, 47*: 548-556

Kaye WH. (1996) Neuropeptide abnormalities in Anorexia Nervosa. *Psychiat. Res.,* 62: 65-74

Kaye W, Frank G, Klump K. (2003) Psychophysiology of obsessional behaviors in Anorexia and Bulimia Nervosa. In Bellodi L, Brambilla F (eds) *Eating Disorders and Obsessive Compulsive Disorder: an etiopathogenetic link?.* Centro Scientifico Editore, Turin, Italy pp. 57-76

Kelley AE, Balski VP, Haber SN, Steininger TL, Will MJ, Zhang M. (2002) Opioid modulation of taste hedonics within the ventral striatum. *Physiol. Behav., 76*: 365-377

Kelley AE, Baldo BA, Pratt WE, Will MJ. (2005) Corticostriatal-hypothalamic circuitry and food motivation: integration of energy, action and reward. *Physiol. Behav., 86*: 773-795

Kern PA, Di Gregorio GB, Lu T, Rassouli N, Ranganathan G. (2003) Adiponectin expression from human adipose tissue. Relation to obesity, insulin resistance, and Tumor Necrosis Factor-α expression. *Diabetes, 52*: 1779-1785

Kernie SG, Liebl DJ, Parada LF. (2000) BDNF regulates eating behavior and locomotor activity in mice. *EMBO J, 19*: 1290-1300.

Kirchgessner AL. (2002) Orexins in the brain-gut axis. *Endocr. Rev., 23*: 1-15

Kissileff HR, Wentzlaff TH, Guss JL, Walsh BT, Devlin MJ, Thornton JC. (1996) A direct measure of satiety disturbance in patients with bulimia nervosa. *Physiol. Behav., 60*: 1077-1085.

Kobayashi N, Tamai H, Uehata S, Komaki G, Mori K, Matsubayashi S, Nakagawa T. (1988) Pancreatic abnormalities in patients with eating disorders. *Psychosom. Med., 50*: 607-614.

Koizumi H, Hashimoto K, Itoh K, Nakazato M, Shimizu E, Ohgake S, Koike K, Okamura N, Natsushita S, Suzuki K, Murayama M, Iguchi S, Iyo M. (2004) Association between the brain-derived neurotrophic factor 196G/A polymorphism and eating disorders. *Am. J. Med. Genetic, 127B:* 125-127

Kojima S, Nakahara T, Nagai N, Muranaga T, Tanaka M, Yasuhara D, Masuda A, Date Y, Ueno H, Nakazato M, Naruo T. (2005) Altered ghrelin and peptide YY responses to meals in bulimia nervosa. *Clin. Endocrinol. (Oxf)*, *62*: 74-78.

Kontogianni MD, Dafni UG, Routsias JG, Skopouli FN. (2004) Blood level leptin and adiponectin as possible mediators of the relation between fat mass and BMD in premenopausal women. *J. Bone Min. Res., 19:* 546-551

Kopp W, Blum WF, von Prittwitz S, Ziegler A, Lubbert H, Emons G, Herzog W, Herpertz S, Deter HC, Remschmidt H, Hebebrand J. (1997) Low leptin levels predict amenorrhea in underweight and eating disordered females. *Mol. Psychiatry, 2*: 335-340.

Krizova J, Papezova H, Haluzikova D, Kremen J, Kumstyrova T, Papezova H, Haluzik M. (2002) Soluble leptin receptor levels in patients with anorexia nervosa. *Endocr. Res., 28*: 199-205.

Kuramoki M, Onaka T, Kohno D, Kato S, Yada T. (2006) Galanin-like peptide stimulates food intake via activation of neuropeptide Y neurons in the hypothalamic dorsomedial nucleus of the rat. *Endocrinol., 147*: 1744-1752

Ladenheim EE, Hampton LL, Whitney AC, White WO, Battey JF, Moran TH. (2002) Disruption in feeding and body weight control in gastrin-releasing peptide receptor deficient mice. *J. Endocrinol. 174:* 273-281

Lesem MD, Berrettini WH, Kaye WH, Jimerson DC. (1991) Measurement of CSF dynorphin A 1-8 immunoreactivity in anorexia nervosa and normal weight bulimia. *Biol. Psychiatry, 29*: 244-252

Lesem MD, Kaye WH, Bissette G, Jimerson DC, Nemeroff CF. (1994) Cerebrospinal fluid TRH immunoreactivity in anorexia nervosa. *Biol. Psychiatry, 35*: 48-53

Li AJ, Ritter S. (2004) Glucoprivation increases expression of neuropeptide Y mRNA in hindbrain neurons that innervate the hypothalamus. *Europ. J. Neuroscience 19*: 2147-2154

Lindbloom J, Schioth HB, Watanobe H, Suda T, Wikberg JES, Bergstrom L. (2000) Downregulation of melanocortin receptors in brain areas involved in food intake and reward mechanisms in obese (OLETF) rats. *Brain Res. 852*: 180-185

List JF, Habener JF (2003) Defective melanocortin 4 receptors in hyperphagia and morbid obesity. *New Engl. J. Med. 348*: 1160-1163

Lob S, Pickel J, Bidlingmaier M, Schaaf L, Backmund H, Gerlinghoff M, Stalla GK. (2003) Serum leptin monitoring in anorectic patients during refeeding therapy. *Exp. Clin. Endocrinol. Diabetes, 111*: 278-282.

Lokrantz CM, Uvnas-Moberg K, Kaplan JM. (1997) Effects of central oxytocin administration on intraoral intake of glucose in deprived and nondeprived rats. *Physiol. Behav. 62*: 347-352

Lommatzsch M, Zingler D, Schuhbaeck K, Schloetcke K, Zingler C, Schuff-Werner P, Virchow JC. (2005) The impact of age, weight and gender on BDNF levels in human platelets and plasma. *Neurobiol. Aging, 26*: 115-123

Lopez M, Seoane LM, Garcia MC, Dieguez C, Senaris R. (2002) Neuropeptide Y, but not agouti-related peptide or melanin-concentrating hormone is a target peptide for orexin-A feeding actions in the rat hypothalamus. *Neuroendocrinol., 75*: 34-44

Lydiard RB, Brewerton TD, Fossey MD, Laraia MT, Stuart G, Beinfeld MC, Ballenger JC. (1993) CSF cholecystokinin octapeptide in patients with Bulimia Nervosa and in normal comparison subjects. *Am. J. Psychiatry, 150*: 1099-1101

Mantzoros C, Flier JS, Lesem MD, Brewerton TD, Jimerson DC. (1997) Cerebrospinal fluid leptin in anorexia nervosa: correlation with nutritional status and potential role in resistance to weight gain. *J. Clin. Endocrinol.Metab., 82*: 1845-1851.

Martin JB, Reichlin S.(1987) The neurohypophysis: physiology and disorders of secretion. In Davis FA (ed) *Clinical Neuroendocrinology, Edition 2.*, Davis, Philadelphia, USA, pp. 67-109

Mazzocchi G, Rebuffat P, Gottardo G, Malendowicz LK, Nussdorfer GG. (1998) Galanin stimulates glucocorticoid secretion in rats through a receptor-dependent activation of the adenylate cyclase/protein kinase A dependent signaling pathway. *Peptides, 19*: 891-895

Merighi A. (2002) Costorage and coexistence of neuropeptides in the mammallian CNS. *Progr. Neurobiol., 66*: 161-190

Merali Z, McIntosh J, Kent P, Michaud D, Anisman H. (1997) Adversive and appetitive events evoke the release of corticotropin-releasing hormone and bombesin-like peptide at the central nucleus of the amygdala. *J. Neuroscience, 18*: 4758-4766

Misra M, Miller K, Herzog D, Ramaswamy K, Aggarwal A, Almzan C, Neubauer G, Breu J, Klibanski A. (2004) Growth hormone and ghrelin responses to an oral glucose load in adolescent girls with anorexia nervosa and controls. *J. Clin. Endocrinol. Metab., 89*: 1605-1612.

Misra M, Miller KK, Kuo K, Griffin K, Stewart V, Hunter E, Herzog DB, Klibanski A. (2005a) Secretory dynamics of leptin in adolescent girls with anorexia nervosa and healthy adolescents. *Am. J. Physiol. Endocrinol. Metab., 289*: 373-381.

Misra M, Miller KK, Stewart V, Hunter E, Kuo K, Herzog DB, Klibanski A. (2005b) Ghrelin and bone metabolism in adolescent girls with anorexia nervosa and healthy adolescents. *J. Clin. Endocrinol. Metab., 90*: 5082-5087.

Misra M, Miller KK, Tsai P, Gallagher K, Lin A, Lee N, Herzog DB, Klibanski A. (2006) Elevated peptide YY levels in adolescent girls with anorexia nervosa. *J. Clin. Endocrinol. Metab., 91*: 1027-1033.

Mitchell JE, Pyle RL, Eckert ED. (1985) Bulimia. In Hales RE, Frances A (eds) *American Psychiatric Association Annual Review.* Am. Psychiat. Press, Washington, pp. 464-480

MizunoTM, Makimura H, Silverstein J, Roberts JL, Lopingco T, Mobbs CV. (1999) Fasting regulates hypothalamic neuropeptide Y, agouti-related peptide and proopiomelanocortin in diabetic mice independent of changes in leptin or insulin. *Endocrinol., 140*: 4551-4557

Mollet A, Meyer S, Grabler V, Gilg S, Scharrer E, Lutz TA. (2003) Endogenous amylin contributes to the anorectic effects of cholecystokinin and bombesin. *Peptides, 24*: 91-98

Monteleone P, Di Lieto A, Tortorella A, Longobardi N, Maj M. (2000a) Circulating leptin in patients with anorexia nervosa, bulimia nervosa or binge-eating disorder: relationship to body weight, eating patterns, psychopathology and endocrine changes. *Psychiatry Res., 94*: 121-129.

Monteleone P, Bortolotti F, Fabrazzo M, La Rocca A, Fuschino A, Maj M. (2000b) Plasma leptin response to acute fasting and refeeding in untreated women with bulimia nervosa. *J. Clin. Endocrinol. Metab., 85*: 2499-2503.

Monteleone P, Luisi M, Colurcio B, Casarosa E, Monteleone P, Ioime R, Genazzani AR, Maj M. (2001) Plasma levels of neuroactive steroids are increased in untreated women with anorexia nervosa or bulimia nervosa. *Psychosom. Med.,63*: 62-68

Monteleone P, Fabrazzo M, Tortorella A, Fuschino A, Maj M. (2002a) Opposite modifications in circulating leptin and soluble leptin receptor across the eating disorder spectrum. *Mol. Psychiatry, 7*: 641-646.

Monteleone P, Martiadis V, Colurcio B, Maj M. (2002b) Leptin secretion is related to chronicity and severity of the illness in bulimia nervosa. *Psychosom. Med., 64*: 874-879.

Monteleone P, Martiadis V, Fabrazzo M, Serritella C, Maj M. (2003a) Ghrelin and leptin responses to food ingestion in bulimia nervosa: implications for binge-eating and compensatory behaviours. *Psychol. Med., 33*: 1387-1394.

Monteleone P, Fabrazzo M, Martiadis V, Fuschino A, Serritella C, Milici N, Maj M. (2003b) Opposite changes in circulating adiponectin in women with Bulimia Nervosa or Binge Eating Disorder. *J. Clin. Endocrinol. Metab., 88*: 5387-5391

Monteleone P, Tortorella A, Martiadis V, Serritella C., Fuschino A, Maj M. (2004) Opposite changes in the serum brain-derived neurotrophic factor in anorexia nervosa and obesity. *Psychosom. Med., 66*: 744-748

Monteleone P, Fabrazzo M, Tortorella A, Martiadis V, Serritella C, Maj M. (2005a) Circulating ghrelin is decreased in non-obese and obese women with binge eating disorder as well as in obese non-binge eating women, but not in patients with bulimia nervosa. *Psychoneuroendocrinol., 30*: 243-250.

Monteleone P, Martiadis V, Rigamonti AE, Fabrazzo M, Giordani C, Muller EE, Maj M. (2005b) Investigation of peptide YY and ghrelin responses to a test meal in bulimia nervosa. *Biol. Psychiatry, 57*: 926-931.

Monteleone P, Fabrazzo M, Martiadis V, Serritella C, Pannuto M, Maj M. (2005c) Circulating brain-derived neurotropic factor is decreased in women with anorexia nervosa and bulimia nervosa but not in women with binge-eating disorder: relationship to comorbid depression, psychopathology and hormonal variables. *Psychol. Med.,* 35: 897-905

Moran TH (2000) Cholecystokinin and satiety: current perspective. *Nutrition, 16*: 858-865

Moriya J. Takimoto Y, Yoshiuchi K, Shimosawa T, Akabayashi A. (2006) Plasma agouti-related protein levels in women with anorexia nervosa. *Psychoneuroendocrinol., 31*: 1057-1061

Morley JE, Levine AS, Gosnell BA, Kneip J, Grace M. (1987) Effect of neuropeptide Y on ingestive behaviors in the rat. *Am. J.Physiol., 252*: R599-R609

Morley JE, Blundell JE. (1988) The neurobiological basis of eating disorders: some formulations. *Biol. Psychiatr., 23*: 53-78

Moschos S, Chan JL, Mantzoros CS. (2002) Leptin and reproduction: a review. *Fertil Steril., 77*: 433-444.

Murphy B, Nunes CN, Ronan JJ, Harper CM, Beall MJ, Hanaway M, Fairhurst AM, Van der Ploeg LHT, MacIntyre TN. (1998) Melanocortin mediated inhibition of feeding behavior in rats. *Neuroendocrinol.,_32*: 491-497

Nakahara T, Kojima S, Tanaka M, Yasuhara D, Harada T, Sagiyama KI, Muranaga T, Nagai N H, Nakazato M, Naruo T. (2005) Altered ghrelin and peptide YY responses to meals in bulimia nervosa. *Clin. Endocrinol. (Oxf), 62*: 74-78.

Nakahara T, Kojima S, Tanaka M, Yasuhara D, Harada T, Sagiyama KI, Muranaga T, Nagai N, Nakazato M, Nozoe SI, Naruo T, Inui A. (2006) Incomplete restoration of the secretion of ghrelin and PYY compared to insulin after food ingestion following weight gain in anorexia nervosa. *J. Psychiat. Res.,* (in press).

Nakai Y, Hamagaki S, Takagi R, Taniguchi A, Kurimoto F. (2000) Plasma concentrations of tumor necrosis factor-alpha (TNF-alpha) and soluble TNF receptors in patients with bulimia nervosa. *Clin. Endocrinol., 53*: 383-388.

Nakai Y, Hosoda H, Nin K, Ooya C, Hayashi H, Akamizu T, Kangawa K. (2003) Plasma levels of active form of ghrelin during oral glucose tolerance test in patients with anorexia nervosa. *Eur. J. Endocrinol., 149*: R1-R3.

Nakazato M, Hashimoto K, Shimizu E, Kumakiri C, Koizumi H, Okamura N, Mitsumori M, Komatsu N, Iyo M. (2003) Decreased levels of serum brain-derived neurotrophic factor in female patients with eating disorders. *Biol. Psychiatry, 54*:485-490

Nakazato M, Hashimoto K, Shiina A, Koizumi H, Mitsumoti M, Imai M, Shimizu E, Iyo M. (2004) No changes in serum ghrelin levels in female patients with bulimia nervosa. *Prog. Neuro-Psychopharmacol. Biol. Psychiatry, 28*: 1181-1184.

Nakazato M, Hashimoto K, Yoshimura K, Hashimoto T, Shimizu E, Kumakiri C, Masaomi IYO . (2006) No change between the serum brain-derived neurotrophic factor in female patients with anorexia nervosa before and after partial weight recovery. *Progr. NeuroPsychopharmacol. Biol. Psychiatry, 30*: 1117-1121

Nedvidkova J, Krykorkova I, Bartak V, Papezova H, Gold PW, Alesci S, Pacak K. (2003) Loss of meal-induced decrease in plasma ghrelin levels in patients with anorexia nervosa. *J. Clin. Endocrinol. Metab., 88*: 1678-1682.

Nemeroff CB, Bissette G, Manberg PJ, Luttinger D, Prange AJ, (1984) Effects of hypothalamic peptides on the central nervous system. In Nemeroff CB, Dunn AJ (eds) *Peptides, hormones and behavior,* MTP press ltd, Lancaster England, pp. 217-272

Nemeroff CB, Bissette G, Widerlow E, Beckman H, Gerner R, Manberg PJ, Lindstrom L, Prange AJ, Gattaz WF. (1989) Neurotensin-like immunoreactivity in cerebrospinal fluid of patients with schizophrenia, depression, anorexia nervosa, bulimia and premenstrual syndrome. *J. Neuropsychiary, 1*: 16-20

Norman D, Isidori AM, Frajese V, Caprio M, Chew SL, Grossman AB, Clark AJ, Besser GM. Fabbri A. (2002) ACTH and α-MSH inhibit leptin expression and secretion in 3T3-L1 adypocytes: model for a central-peripheral melanocortin-leptin pathway. *Mol. Cell. Endocrinol., 200*: 99-109

Nussdorfer GG, Gottardo G. (1997) Neuropeptide-Y family of peptides in the autocrine-paracrine regulation of adrenocortical function. *Horm. Metab. Res., 30*: 368-373

Ogren SE, Kuteeva E, Hokfelt T, Kehr J. (2006) Galanin receptor antagonists. A potential novel pharmacological treatment for mood disorders. *CNS Drugs, 20*: 633-654

Olson GA, Olson RD, Kastin AJ. (1997) Endogenous opiates. *Peptides,* 18 : 1651-1688

Otto B, Tschop M, Cuntz U. (2004) Similar fasting ghrelin levels in binge eating/purging anorexia nervosa and restrictive anorexia nervosa. *Psychoneuroendocrinol.,, 29*: 692-693.

Otto B, Tschop M, Fruhauf E, Heldwein W, Fichter M, Otto C, Cuntz U. (2005) Postprandial ghrelin release in anorectic patients before and after weight gain. *Psychoneuroendocrinol., 30*: 577-581.

Owen WP, Halmi KA, Gibbs J, Smith GP. (1985) Satiety responses in eating disorders. *J. Psychiat. Res., 19*: 279-284.

Pannaciulli N, Vettor R, Milan G, Granzotto M, Catucci A, Federspil G. (2003) Anorexia Nervosa is characterized by increased adiponectin plasma levels and reduced non oxidative glucose metabolism. *J. Clin. Endocrinol. Metab., 88*: 1748-1752

Patel L, Buckels AC, Kinghorn IJ, Murdock PR, Holbrook JD, Plumpton C, Macphee C, Smith SA. (2003) Resistin is expressed in human macrophages and directly regulated by PPARγ activators. *Biochem. Biophys.Res. Commun., 300*: 472-476

Phillipp E, Pirke KM, Kellner MB, Krieg JC. (1991) Disturbed cholecystokinin secretion in patients with eating disorders. *Life Sciences, 48*: 2443-2450

Pirke KM, Philipp E, Friess E, Kellner B, Wilkens T, Krieg JC, Fichter MM. (1993) The role of gastrointestinal hormone secretion in Eating Disorders. *Adv. Biosciences, 60*: 75-79

Pirke KM, Kellner MB, Friess E, Krieg JC, Fichter MM. (1994) Satiety and cholecistokinin. *Int. J. Eat. Disord., 15*: 63-69

Popovic V, Djurovic M, Cetkovic A, Vojvodic D, Pekic S, Spremovic S, Petakov M, Damjanovic S, Milic N, Dieguez C, Casanueva FF. (2004) Inhibin B: a potential marker of gonadal activity in patients with anorexia nervosa during weight recovery. *J. Clin. Endocrinol. Metab., 89*: 1838-1843.

Porte D, Baskin DG, Schwartz MW. (2002) Leptin and insulin action in the central nervous system. *Nutr. Rev., 60*: S20-S29

Primeaux SD, Wilson SP, Cusick MC, York DA, Wilson MA. (2005) Effects of altered amygdalar neuropeptide Y expression on anxiety-related behaviors. *Neuropsychopharmacol., 30*: 1589-1597

Qi Y, Takahashi N, Hileman SM, Patel HR, Berg AH, Pajvani UB, Scherer PE, Ahima RS (2004) Adiponectin acts in the brain to decrease body weight. *Nat. Med., 10*: 524-529

Ribases M, Gratacos M, Amengol L, De Cid R, Badia A, Jimenez L, Solano R, Vallejo J, Fernandez-Aranda F, Estivil X. (2003) Met66 in the brain-derived neurotrophic factor (BDNF) precursor is associated with anorexia nervosa restricting type. *Mol. Psychiatry, 8*: 745-751

Ribases M, Gratacos M, Fernandez Arana F, Bellodi L, Boni C, Anderluh M, Cavallini MC, Cellini E, Di Bella E, Erzegovesi S, Foulon C, Gabrovsek M, Gorwood P, Hebebrand J, Hinney A, Hu X, Karwantz A, Kipman A, Komel R, Nacmias B, Remschmidt H, Ricca V, Sorbi S, Tomori M, Wagner G, Treasure J, Coffier DA, Estivil X. (2004) Association of BDNF with anorexia nervosa, bulimia and age of onset of weight loss in six european population. *Hum. Mol. Genet., 13*: 1205-1212

Riediger T, Bothe C, Beckskei C, Lutz TA. (2004) Peptide YY directly inhibits ghrelin-activated neurons of the arcuate nucleus and reverses fasting-induced c-Fos expression. *Neuroendocrinol., 79*: 317-326

Sahu A. (2004) Minireview: a hypothalamic role in energy balance with special emphasis on leptine. *Endocrinol., 145*: 2613-2620

Said SI. (1984) Vasoactive Intestinal Polypeptide (VIP): current status. *Peptides, 5*: 143:150

Saydyk TJ, Schober DA, Smiley DL, Gehlert DR. (2004) Neuropeptide Y-Y2 receptors mediate anxiety in the amigdala. *Pharmacol. Biochem. Behav., 71*: 419-423

Scalera G, Tarozzi G. (1998) Somatostatin administration modifies food intake, body weight, and gut motility in rat. *Peptides, 19*: 991-997

Schreiber W, Schweiger U, Werner D, Brunner G, Tuschl RJ, Laessle RG, Krieg JC, Fichter MM, Pirke KM. (1991) Circadian pattern of large neutral aminoacids, glucose, insulin, and food intake in Anorexia Nervosa and Bulimia Nervosa. *Metabolism, 40*: 503-507

Schwartz MW. (2006) Central Nervous System regulation of food intake. *Obesity, 14*: 1S-8S

Shen CP, Wu KK, Shearman LP, Camac HR, Tota MR,Fong TM, Van der Ploeg LHT. (2002) Plasma agouti-related protein level: a possible correlation with fasted and fed states in humans and rats. *J. Neuroendocrinol,. 14*: 607-610

Shlik J, Koszycki D, Bradwein J. (1998) Decrease in short-term memory function induced by CCK-4 in healthy volunteers. *Peptides, 19*: 969-975

Sone M, Takahashi K, Murakami O, Totsun K, Arihara Z, Satho F, Ito H, Mouri T. (2000) Binding sites for melanin-concentrating hormone in the human brain. *Peptides, 21*: 245-250

Soriano-Guillen L, Barrios V, Campos-Barros A, Argente J. (2004) Ghrelin levels in obesity and anorexia nervosa: effect of weight reduction or recuperation. *J. Pediatr., 144*: 36-42

Stanek LM. (2006) Cocaine- and amphetamine related transcript (CART) and anxiety. *Peptides, 27*: 2005-2011

Stanley BG, Daniel DR, Chin AS, Leibowitz SF. (1985) Paraventricular nucleus injections of peptide YY and neuropeptide Y preferentially enhance carbohydrate ingestion. *Peptides, 6*: 1205-1211

Stefan N, Stumvoll M (2002) Adiponectin – its role in metabolism and beyond. *Horm. Metab. Res., 34*: 469-474

Steiger H, Haring NU. (2005) Adipocytokines: fat-derived humoral mediators of metabolic homeostasis. *Exp. Clin. Endocrinol. Diabetes, 113*: 67-79

Stock S, Leichner P, Wong AC, Ghatei MA, Kieffer TJ, Bloom SR, Chanoine JP. (2005) Ghrelin, peptide YY, glucose-dependent insulinotropic polypeptide, and hunger responses to a mixed meal in anorexic, obese, and control female adolescents. *J. Clin. Endocrinol. Metab., 90*: 2161-2168.

Strewler GJ. (1998) Humoral manifestations of malignancy. InWilson D, Foster DW, Kronenberg HM, Larsen PR (eds) *Williams textbook of endocrinology.* WB Saunders co. Philadelphia, pp. 1693-1710

Tagami T, Satoh N, Usui T, Yamada K, Shimutsu A, Kuzuya H. (2004) Adiponectin in anorexia nervosa and bulimia nervosa. *J. Clin. Endocrinol. Metab., 89*: 1833-1837

Tamai H, Takemura J, Kobayashi N, Matsubayashi S, Matsukura S, Nakagawa T. (1993) Changes in plasma cholecystokinin concentrations after oral glucose tolerance test in anorexia nervosa before and after therapy. *Metabolism, 45*: 581-584

Tanaka M, Naruo T, Muranaga T, Yasuhara D, Shiiya T, Nakazato M, Matsukura S, Nozoe S. (2002) Increased fasting plasma ghrelin levels in patients with bulimia nervosa. *Eur. J. Endocrinol., 146*: 1-3.

Tanaka M, Naruo T, Yasuhara D, Tatebe Y, Nagai N, Shiiya T, Nakazato M, Matsukura S, Nozoe S. (2003a) Fasting plasma ghrelin levels in subtypes of anorexia nervosa. P*ychoneuroendocrinol,. 28*: 829-835.

Tanaka M, Narau T, Nagai N, Kuroki N, Shiiya T, Nakazato M, Matsukura S, Nozoe S. (2003b) Habitual binge/purge behavior influences circulating ghrelin levels in eating disorders. *J. Psychiat. Res., 37*: 17-22.

Tanaka M, Tatebe Y, Nakahara T, Yasuhara D, Sagiyama K, Muranaga T, Ueno H, Nakazato M, Nozoe S, Naruo T. (2003c) Eating pattern and the effect of oral glucose on ghrelin and insulin secretion in patients with anorexia nervosa. *Clin. Endocrinol., 59*: 574-579.

Tanaka M, Nakahara T, Kojima S, Nakano T, Muranaga T, Nagai N, Ueno H, Nakazato M, Nozoe S, Naruo T. (2004) Effect of nutritional rehabilitation on circulating ghrelin and growth hormone levels in patients with anorexia nervosa. *Reg. Peptides, 122*: 163-168.

Tempel DL, Leibowitz KJ, Leibowitz SF. (1998) Effects of PVN galanin on macronutrient selection. *Peptides, 9*: 309-314

Tepper R, Weixman A, Apter A, Tyano S, Beyth Y. (1992) Elevated plasma immunoreactive β-endorphin in anorexia nervosa. *Clin. Neuropharm., 15*: 387-391

Tolle V, Kadem M, Bluet-Pajot MT, Frere D, Foulon C, Bossu C, Dardennes R, Mounier C, Zizzari P, Lang F, Epelbaum J, Estour B. (2003) Balance in ghrelin and leptin plasma levels in anorexia nervosa patients and constitutionally thin women. *J. Clin. Endocrinol. Metab., 88*: 109-116.

Troisi A, Di Lorenzo G, Lega I, Tesauro M, Bertoli A, Leo R, Intorno M, Pecchioli C, Rizza S, Turriziani M, Lauro R, Siracusano A. (2005) Plasma ghrelin in anorexia, bulimia, and binge-eating disorder: relations with eating patterns and circulating concentrations of cortisol and thyroid hormones. *Neuroendocrinol., 81*: 259-266.

Tsutsumi T, Akyioshi K, Isogawa K, Kohno Y, Hikichi T, Nagayama H. (1999) Suppression of conditioned fear by administration of CCK8 receptor antagonist PD135158. *Neuropeptides, 33*: 483-486

Van Dyk G, De Groote C, Chavrez M, van der Werf Y, Steffens AB, Strubbe JH. (1997) Insulin in the arcuate nucleus of the hypothalamus reduces fat consumption in rats. *Brain Res., 777*: 147-152

Vidal Puig A, O'Rahilly S. (2001) Resistin: a new link between obesity and insulin resistance? *Clin. Endocrinol., 55*: 437-438

Villena JA, Kim KH, Sul HS. (2002) Pre-1 and ADSF/resistin: two secreted factors inhibiting adipose tissue development. *Horm. Metab. Res., 34*: 664-670

Vink H, Hinney A, van Elburg AA, Van Goozen SH, Sandkuijl LA, Sinke RJ, Herpertz Dahlman BN, Hebebrand J, Remschmidt H. (2001) Association between an agouti-related protein gene polymorphism and anorexia nervosa. *Mol. Psychiatry., 6*: 325-328

Wachschlict-Rodbard H, Gross HA, Rodbard D, Ebert MH, Roth J. (1997) Increased insulin binding to erythrocytes in anorexia nervosa: restoration to normal with refeeding. *N. Engl. J. Med., 300*: 882-887

Wada K, Wada E, Watase K, Yamada K, Ohki-Hamazaki H. (1998) Bombesin, obesity, and social behavior. *Mol. Psychiatry, 3*: 204-206

Waller DA, Kiser RS, Hardy BW, Fuchs I, Feigenbaum LP, Uauy R. (1986) Eating behavior and plasma beta-endorphin in bulimia. *Am. J. Clin. Nutr., 44*: 20-23

Walsh BT, Kissileff HR, Cassidy SM, Dantzic S. (1989) Eating behavior of women with bulimia. *Arch. Gen. Psychiatry, 46*: 54-58

Walsh BT. (2002) Psychobiology of eating behavior in bulimia. *Int. J. Neuropsychopharmacol.. 5:* S40-S43

Wang J, Dourmashkin JT, Yun R, Leibowitz SF. (1999) Rapid changes in hypothalamic neuropeptide Y produced by carbohydrate-reach meals that enhance corticosterone and glucose levels. *Brain Res., 848:* 124-136

Wang L, Barachina MD, Martinez V, Wei JY, Taché Y. (2000) Synergistic interaction between CCK and leptin to regulate food intake. *Regulatory Peptides, 92*: 79-85

Watanabe T, Orth DN. (1988) Effects of several in vitro systems on the potencies of putative adrecorticotropin secretagogues on rat anterior pituitary cells. *Endocrinol., 12*: 2299-2309

Wauters M, Considine RV, Van Gaal LF. (2000) Human leptin: from an adipocyte hormone to an endocrine mediator. *Eur. J. Endocrinol., 143*: 293-311.

Weingarten HP, Hendler R, Rodin J. (1988) Metabolism and endocrine secretion in response to a test meal in normal-weight bulimic women. *Psychosom. Med., 50:* 273-285

Wellman PJ. (2000) Norepinephrine and the control of food intake. *Nutrition 16*: 837-842

Wideman CH, Murphy HM, Tumwesigye S. (1997) Role of vasopressin in long-term preferences of sucrose and polycose under ad-libitum and food-restricted conditions. *Appetite, 28*: 189-200

Wiesner G, Morash BA, Ur E, Wilkinson M. (2004) Food restriction regulates adipose-specific cytokines in pituitary gland but not in the hypothalamus. *J. Endocrinol., 180*: R1-R6

Will MJ, Franzblau EB, Kelley AE. (2003) Nucleus accumbens μ-opioids regulate intake of high-fat diet via activation of a distributed brain network. *J. Neuroscience, 23*: 2882-2888

Wirth MM, Giraudo SQ. (2000) Agouti-related protein in the hypothalamic paraventricular nucleus: effect of feeding. *Peptides 21*: 1369-1375

Xiao E, Xia-Zhang L, Vulliemoz NR, Ferin M, Wardlaw SL. (2003) Agouti-related protein stimulates the hypothalamic-pituitary-adrenal (HPA) axis and enhances the HPA response to interleukin-1 in the primate. *Endocrinol., 144*: 1736-1741

Yoshikawa T, Kunishima C, Yamada K, Nabeshima T, Shibuya H, Toru M. (1994) Effect of a single injection of psychoactive drugs on CCK mRNA in rat brain. *Peptides, 15*: 471-473

Zhang Y, Proenca R, Maffei M, Barone M, Leopold L, Friedman JM. (1994) Positional cloning of the mouse obese gene and its human homologue. *Nature, 372*: 425-432.

In: Psychoneuroendocrinology Research Trends
Editor: Martina T. Czerbska, pp. 213-257
ISBN: 978-1-60021-665-7

Chapter VII

EFFECT OF PSYCHOSOCIAL INTERVENTIONS ON PSYCHOENUROENDOCRINE OUTCOMES IN CANCER PATIENTS: WHERE DO WE GO FROM HERE?

Linda E. Carlson[1,2,3,*] and Sheila N. Garland[1,3]
[1]Department of Psychology, University of Calgary, Canada;
[2]Department of Oncology, University of Calgary, Canada;
[3]Department of Psychosocial Resources, Tom Baker Cancer Centre, Canada.

ABSTRACT

A significant literature exists on the role of psychosocial factors in cancer initiation and progression, and effects of psychosocial interventions on eventual survival, but research investigating the effects of psychosocial interventions on psychoneuroendocrine and psychoneuroimmune outcomes in cancer patients is rare. There is some evidence that stress-reduction interventions may affect cortisol secretion profiles and aspects of cellular immunity, but the clinical significance of any observed effects is not known. Questions that require additional investigation concern: 1) the significance of various endocrine and immune outcome measures for predicting disease outcome in cancer patients (i.e. disease recurrence and/or survival); 2) the optimal timing of psychosocial interventions to affect biological outcomes (i.e. pre- or post-surgery, chemotherapy); 3) the type and stage of cancers that are potentially most responsive to psychosocial interventions (e.g. early vs. late stage, tumour type), and; 4) consideration of other factors that may be mediating any biological changes seen as a result of psychosocial interventions (i.e. health behaviours). The research in these areas will be reviewed and fruitful directions for future research outlined.

* Correspondence concerning this article should be addressed to Linda E. Carlson, Ph. D., C. Psych. Alberta Cancer Board - Holy Cross Site, Department of Psychosocial Resources, 2202 2nd St. S.W. Calgary, Alberta, Canada T2S 3C1. Phone: (403) 355-3209; Fax: (403) 355-3206; E-mail: l.carlson@ucalgary.ca.

Abbreviations

ACTH-adrenocorticotropic hormone; ANS-autonomic nervous system; AP-1-activator protein 1; BDI-Beck Depression Inventory; BEST-breast expressive-supportive therapy; BMI-body mass index; CD3, CD4, CD8-specific t-cell cluster of differentiation; CES-D-Center for Epidemiological Studies-Depression; CI-confidence interval; CMI-cell mediated immunity; CNS-central nervous system; Con-A-canavalin A; CRF-corticotrophin-releasing factor; DHEAS-dehydroepiandosterone sulphate; EORTC-QLQ-30- European Organization for the Treatment of Cancer-Quality of Life Questionnaire-30; E-estrogen; EP-epinephrine; ER-estrogen receptor; FACT-B-Functional Assessment of Cancer Therapy-Breast; HADS-Hospital Anxiety and Depression Scale; HA-hazard ratio; HPA-hypothalamic-pituitary-adrenal; HPG-hypothalamic-pituitary-gonadal; HPV-human papillomavirus; HRT-hormone replacement therapy; IES-Impact of Events Scale; IFN-γ-Interferon-gamma; IgA-immunoglobulin-A; IGF-insulin-like growth factor; IgG-immunoglobulin-G; LGL-large granular lymphocyte; IgM-immunoglobulin-M; IL-1α-interluekin-1 alpha; IL-1β-interleukin-1 beta; IL-2-interleukin 2; IL-4-interleukin 4; IL-10-interleukin 10; MAC-Mental Adjustment to Cancer; MLR-mixed lymphocyte responsiveness; MMPI-Minnesota Multiphasic Personality Inventory; NK cells-natural killer cells; NE-norepinephrine; NFkB- Nuclear Factor kappa B; PBL-peripheral blood lymphocytes; PHA-phytohemagglutinin; PNE-psychoneuroendocrine; PNI-psychoneuroimmune; PNS-parasympathetic nervous system; POMS-Profile of Mood States; PSA-prostate-specific antigen; PSS-Perceived Stress Scale; RCT-randomized controlled trial; SEGT-supportive expressive group therapy; SHBG-sex hormone-binding growth factor; sIgA-salivary immunoglobulin A; SNS-sympathetic nervous system; SOSI-Symptoms of Stress Inventory; STAI-State-Trait Anxiety Inventory; STD-sexually transmitted disease; T-testosterone; VEG-F-vascular endothelial growth factor; WBC-white blood cell.

Overview

The study of interactions between psychological and social factors and eventual disease initiation and progression in psychosocial oncology has been a hot-button issue for several decades. Controversy began with the introduction of the "Type C" personality style, purported to lead to higher incidence of cancer diseases [1,2]. The "Type C" personality was characterized by emotional repression, agreeableness, passivity and patience combined with a helpless/hopeless coping style that deferred to authority – the type of person often referred to as a "good patient" [1]. The theory holds that this type of person often unconsciously represses or purposively suppresses their emotions and turns them inward, resulting in pathophysiological internal changes that could contribute to the development of cancer. The psychoneuroimmunological (PNI) or psychoneuroendocrinological (PNE) routes by which this process was thought to occur were relatively unspecified. However, a good deal of research over the last two decades has explored these questions with increasing rigour. Another large body of literature exists detailing the associations between chronic stress and health outcomes as diverse as colds, heart disease, asthma, allergies, rheumatoid arthritis and wound healing [3-7] – the question now is - does the same apply to cancer?

The underlying rationale for this type of research is based on an understanding of the stress response and its physiological concomitants – the research on stress is also illustrative of how other psychosocial factors such as depression and social support may affect disease. Essentially, psychosocial processes which result in the perception of psychological or physical threat trigger a cascade of events in the central nervous system (CNS) and periphery that result in stress responses in the autonomic nervous system (ANS) and through the hypothalamic-pituitary-adrenal (HPA) axis. HPA responses are mediated by production of corticotrophin-releasing factor (CRF) and arginine vasopression in the hypothalamus. These releasing factors activate the pituitary gland to secrete hormones such as adrenocorticotrophic hormone (ACTH), enkephalins and endorphins. ACTH then triggers the release of glucocorticoids (cortisol in humans) from the adrenal glands, situated atop the kidneys. Glucocorticoids have a whole host of effects throughout the body, many of which are necessary for regulatory functions, but can also cause downregulation of immune function when exposure is prolonged or chronic [8]. Often in disease states such as cancer or depression, HPA feedback is dysregulated, which can result in chronic hyper-arousal of the HPA axis, or eventual collapse and exhaustion [9]. ANS responses to stress are mediated primarily through the sympathetic nervous system (SNS) through release of catecholamines (epinephrine (EP) and norepinepherine (NE)) both from sympathetic nerve endings and through the medulla of the adrenal glands. In short time frames under acute stressors, elevations in cortisol and catecholamines are adaptive, but if they become chronically elevated as a consequence of exposure to chronic stress, many physiological systems can be negatively affected, resulting in increased risk for viral infections such as common colds [4], increased risk for cardiac disease [6], and slower wound healing [7]. There is also a growing body of evidence, primarily from animal research, that chronic alterations in neuroendocrine feedback dynamics and balance can alter an array of physiological parameters related to cancer tumour development [10]. This research will be reviewed later in this chapter.

A general biopsychosocial or biobehavioral model of cancer development and progression can be helpful to tie together the different inputs and outcomes often measured in PNI and PNE research. Such a model has been proposed by a number of researchers [11-13] – Figure 1 is our adaptation showing in simple terms our understanding of interactions among a number of key elements, many of which will be discussed in this chapter. It includes the influence of psychological, social and biological background variables such as temperament, social support, and heredity, as well as health behaviours such as diet, exercise and smoking, on health outcomes. These factors interact with the experience of life stress, which can in turn influence neuroendocrine, immune and central nervous system processes that effect disease course and recovery. Psychosocial interventions can act to affect psychological processes as indicated, and change health behaviours, both of which may impact biological processes. These in turn affect disease processes and subsequent morbidity and mortality. This model will be used as a guide to understand the areas in the field that have been relatively well researched, and highlight areas and associations where little empirical understanding exists.

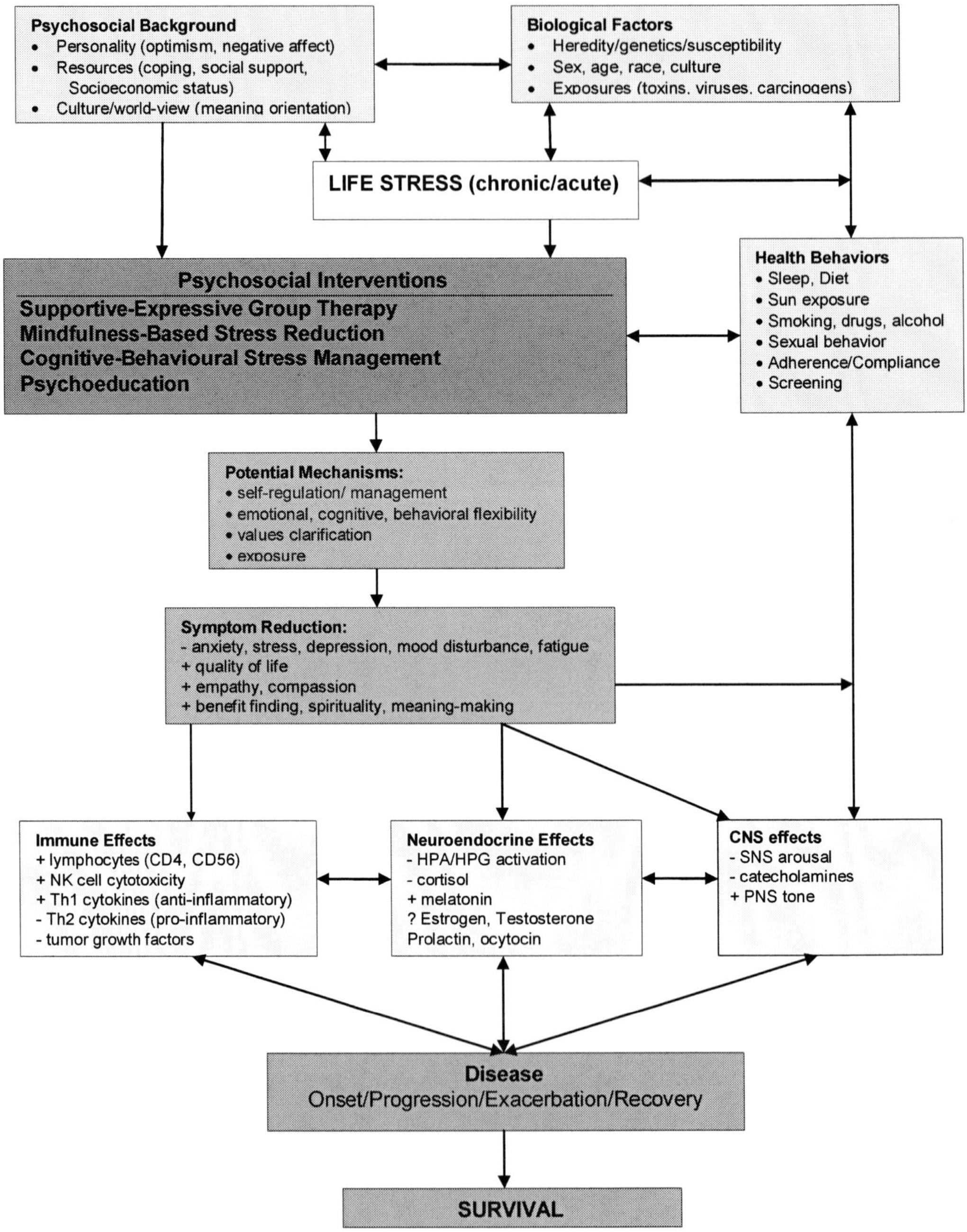

Figure 1. Biobehavioral Model of Cancer Progression.

Returning to the clinical study of these potential effects in humans, however, it is important to consider how such relationships can accurately be investigated. Several types of research methodologies have been applied to studying relationships between psychological and social factors and cancer initiation and progression. The question of disease etiology has been addressed primarily through long-term prospective (or retrospective) epidemiological cohort studies that have tracked large groups of initially healthy individuals over time to determine if social, personality or psychological variables predict later cancer incidence in the small proportion who do go on to develop cancers. Very rarely have biomarkers been collected in these studies that might also be associated with disease initiation, so little can be said about potential pathways to disease that any psychosocial factors may affect.

Other methodology has been applied to determine the effects of psychosocial factors on disease progression, recurrence and survival in people once they have been diagnosed with cancer. Because the time-frame is shorter and sample sizes are smaller, this research is substantially less expensive, time consuming and difficult to conduct than large-scale prospective cohort studies with initially healthy populations, and hence there is a much larger body of research in this area. Again, however, the pathways by which psychosocial factors may influence disease outcome are rarely investigated, and research methodology in general has not been of very good quality. However, these studies do provide some tantalizing findings that suggest psychosocial factors may be influencing relevant endocrine and immune markers that may be important in cancer progression [14-16].

Other researchers have begun to investigate the effects of manipulation of psychosocial factors, through different supportive and stress reductive interventions, on outcomes as diverse as quality of life, psychological well-being and symptomatology, endocrine and immune measures, and eventual survival. This work is well known and the survival literature has been extensively reviewed on numerous occasions, with equivocal results [17-21]. There have been a small number of reviews of studies looking at immune and endocrine changes as a result of psychosocial interventions in cancer [22,23], and much speculation about potential pathways [23,24]. Hence, the picture is far from complete.

This chapter will review studies of the above two types (predicting cancer incidence and progression from psychosocial factors, and effects of interventions on survival and psychological, immune and endocrine outcomes). This will not be an exhaustive review but carried out from the perspective of providing a general overview of the field. This chapter will then address several outstanding, relatively unresearched and unresolved areas of inquiry that may help to direct the thinking of the field regarding the larger questions. These include addressing the specific usefulness of various endocrine and immune measures for predicting disease outcome in cancer patients (i.e. disease recurrence and/or survival). This question in many cases can only be addressed indirectly through noting associations between psychosocial and immune or endocrine measures and linking this with any existing research documenting pathways between certain immune or endocrine measures and disease status.

Other more specific questions of interest for those considering the implementation of psychosocial interventions designed to alter immune/endocrine function or survival include the optimal timing of psychosocial interventions to affect biological outcomes (i.e. pre- or post-surgery, chemotherapy), the type and stage of cancers that are potentially most biologically responsive to psychosocial interventions (e.g. early vs. late stage, solid vs. hematological vs. virally-mediated tumours), and consideration of other factors that may be mediating any biological changes seen as a result of psychosocial interventions (i.e. health behaviours).

The quality and quantity of research available to be drawn upon to address each of these issues varies, and hence this review will also vary in terms of the level of detail and certainty of the conclusions that can be put forth.

Finally, before reviewing the area of psychosocial effects on cancer, the research should be put into perspective: although all the risk factors for different cancers are not definitively known, and because cancer is a term which encompasses over 200 different diseases, it is difficult to determine the relative weight associated with various risk factors. Information on specific risk factors for the major types of cancer is readily available, but it has proven more difficult to obtain actual percentage values in terms of the risk carried for the various factors. One estimate of this data suggests that 30-32% of the variability in incidence for all cancers is related to tobacco, 30-35% to diet, 10% to viruses and infection and so on (including alcohol, sexual factors, heredity, and occupational and environmental exposures), without the inclusion of any psychosocial factors and approximately 5% left unexplained [25]. Hence, the maximum amount of variance in the incidence of cancers that could be accounted for by psychosocial factors would be up to but not likely much beyond five percent. This is not insignificant, and about on par with the effects of alcohol consumption overall. Of course, these values are different for different forms of cancer; for example smoking accounts for approximately 85% of lung cancer cases but is not a risk factor for colon cancer or melanoma. Hence, it may be the case that psychosocial factors play a larger role in some cancers and are inconsequential in others. This possibility has only begun to be investigated.

Psychosocial Factors and Disease Initiation

The research in this area spans 30 years and has been reviewed on several occasions [9,26-33]. The most recent and rigorous review included only studies that used true prospective longitudinal designs, and excluded all those that were cross-sectional, comparisons between groups or semi-prospective, as these study designs do not allow conclusions about causality – 38 studies of this type were excluded [30]. Seventy studies were included in the review, but only 25 of these looked specifically at cancer initiation (while the others focussed on disease progression – reviewed below). Psychosocial factors were divided into several commonly-studied categories: stressful life events; bereavement and other losses; social relations (i.e social support); negative emotions (distress, psychiatric diagnoses); repression of emotions; and personality. Another critical review of studies looking only at disease initiation focussed on similar factors but assessed a broader base of studies, including retrospective cohort studies and case-controlled studies [28]. Several studies published after these most recent reviews also addressed issues of the association between perceived stress and breast cancer incidence, in particular [34-37].

In terms of major life events and bereavement, a meta-analytic review by Petticrew et al [38] of 15 studies looking at only breast cancer patients found that patients reported adverse life events more than twice as often as control subjects. However, they included studied of low quality including group comparisons and semi-prospective studies. Two more rigorous reviews conclude that large population-based studies using independently collected registry data failed to support the link between major life events and increased risk for cancer [28,30]. Garssen [30] did suggest, however, that negative life events in combination with other risk factors, such as social support or hopelessness, could interact to potentially effect cancer initiation. Bereavement too was concluded not to have a clear-cut effect on cancer incidence

[28], but perhaps a stronger effect on death from cancer [30]. This conclusion is based on several large-scale studies which looked at the effects of death of a spouse in the UK [39], widowhood [40] and death of a child [41] in Norway, and loss of a son through war in Israel [42] – findings were suggestive of associations but effects were small or inconclusive in most cases. In the Levav study [42], there were no effects on incidence of all cancers together, but the odds of being diagnosed with lymphoma for bereaved parents compared to the rest of the population was 1.5 times higher, and for melanoma it was 1.7 times higher.

Methodologically, it is important to note that these studies use major life events as surrogate markers for the experience of stress, but none of the studies took into account how the individuals actually perceived these events and subsequently reacted to them. For example, widowhood may be perceived very differently by a young woman with small children and a happy marriage, compared to an older woman with an unsatisfactory relationship. Hence, the subjective stressfulness of these life events rather than their mere presence or absence may be more important in determining psychophysiological responses, but this was not measured in these studies. Newer studies not covered in the most recent review have looked at specific reports of subjective life stress, and found further conflicting results. In a report from the Nurses' Health Study of a subsample of 665 women, subjective stress associated with caregiving did not predict incidence of breast cancer, but women who spent a lot of time caregiving had lower circulating levels of estradiol, which may be significant as major risk factors for breast cancer are related to estrogen exposure – in fact lower estrogen levels may be a protective factor against the growth of breast neoplasms [35]. Hence, in this case, the stress of caregiving may have the opposite effect, of lowering estradiol levels and decreasing cancer risk. Another report from the Nurses' Health Study cohort found that job stress was not related to development of breast cancer in 37, 562 women followed for up to 8 years [43]. Hence, the role of major life events and even subjective stress is likely minor in terms of the development of cancers, but the research is not conclusive.

The research evidence for the role of social support is stronger for disease progression than for cancer incidence, with fewer studies in the area of social support and incidence – some reviews of cancer incidence don't even consider this factor (e.g. 28;33). Of those that were conducted and considered of high methodological quality, one failed to find associations [44] but one found that women with few friends who experienced feelings of social isolation were at higher risk for developing cancer [45]. Issues around how to measure social support and the differential effects of structural measures of support (i.e. number of people in the social network) and perceived emotional support (i.e. satisfaction with the network) continue to make interpretation of the data difficult [30].

Negative emotions associated with mood disturbance or more serious psychiatric illness such as major depressive disorder have also been investigated, and again the results are equivocal. Specifically looking at depression, Dalton [28] found of 10 major prospective studies and two retrospective studies that the support for a link between depression and cancer incidence was strongest for specific types of cancers (i.e. breast; lung) and that the duration of depression was also an important, yet not often assessed factor. In studies that did look at the duration of depression through repeated measurements over a period of years, those who suffered more chronic and stable depression were more likely to be diagnosed with cancer [28]. There was also a moderating effect of smoking, such that smokers who were also depressed were more likely to get cancer than similarly heavy smokers who were not depressed. Garsson [30] expressed the most support for the strength of the evidence associating depression with cancer initiation, but two earlier reviewers [32,33] were less convinced. McKenna [33] failed to find a convincing link between depression or anxiety and breast cancer initiation, and McGee [32] reported a small but marginally statistically significant association, which amounted to an almost negligible effect.

It is important to note, also, that the measures of depression were quite varied and not necessarily the most up-to-date instruments that are known to be valid and reliable in this context. Summarizing the evidence, Garssen [30] noted that of nine studies looking at formal psychiatric diagnosis and incidence of cancer, four studies found a relationship as predicted, whereas five did not. Similarly, of 14 studies on measures of depressive affect and disease initiation, six studies found no effect, seven found an association between depression and cancer incidence, and one study even found depression to be a protective factor against cancer incidence. An additional newer study adds to our knowledge in this area: in a large sample of 81,612 women from the Nurses' Health Study in the USA, depressive symptoms were associated with increased risk for colorectal cancer 4-8 years later, in that those women in the highest quartile of depressive symptoms were at higher risk than those with the lowest levels [46]. Hence, the research is mixed, but there are indicators that chronic levels of depression may be associated with higher risk for cancer initiation, particularly in the case of lung cancer.

Finally, returning to the initial question of the effect of certain personality characteristics on cancer initiation, the cancer-prone or "Type C" personality (described as cooperative, unassertive, patient, suppressing negative emotions and accepting external authority) was investigated as a causal factor in several studies. Because no one theory of personality is universally accepted, many different methods for measuring personality have been utilized in this research. Results in this area have again been varied, with two of four high-quality studies finding positive associations. However, the measures of this construct have varied considerably. The most well-designed study found that men classified as "acting out/emotional" were less likely to develop cancer than those classified as normal, healthy/sensitive, loners, or having a personality characteristic of interpersonal conflict [47]. There was little direct support for a role of the entire Type C personality cluster as defined previously in cancer initiation [28,30].

Emotional repression is a characteristic of the Type C personality described as a tendency to consciously suppress or unconsciously repress negative emotions, often accompanied with denial that difficult emotions even exist. It has been studied independently of other aspects of the personality style. Again, this is a difficult construct to measure, and has been operationally defined in some studies by combining a low score on a measure of anxiety with a high score on a measure of defensiveness. Hence, the typical highly repressive person would report low levels of anxiety and have a tendency to refuse to acknowledge any emotional difficulties. There are also specific measures for emotional repression that have been developed. In terms of the initiation of cancer, the role of repression is considered questionable, as only two high quality studies support the association, while three others refute it [28,30]. One study that did support an association found that "anti-emotionality" was associated with a 19% increase in breast cancer diagnoses in a group of over 9,000 women in the Netherlands [48].

In summary, it is clear from this overview that there are a tremendous number of conflicting results regarding the role of various psychosocial factors in cancer initiation. For each factor reviewed, there were both supporting and refuting studies published. None can be ruled out as important, and the degree of support for the different factors, from strongest to weakest, can be summarized as: chronic depression, social support, emotional repression, major life events and Type C personality. The issue of how these conditions or characteristics could lead to cancer development will be addressed in subsequent sections. First however, we will consider issues around psychosocial factors and disease progression and survival, in order to better understand the nuances of this research.

Psychosocial Factors and Disease Progression/Survival

Research in this area mirrors in some ways the results of studies on psychosocial factors and cancer initiation; however in some cases the effects are much stronger when looking at disease progression. This may have as much to do with research methodology as actual effects, but it does inspire greater confidence in the results. Reviews have also been conducted in this area [49,50] and some of the reviews of factors related to cancer incidence also studied progression and survival [9,30,31,51]. Factors considered include bereavement and stressful life events, perceived stress, social support, depression, hopelessness/helplessness, emotional repression, and also the coping strategies one uses to deal with cancer and its treatments.

Considering the effect of stressful life events on disease outcome, only one study of major life events found a negative relationship between higher total number of life events and shorter disease-free interval and survival [52], where five other studies found no associations (reviewed in Garssen 2004 [30]). Specifically regarding bereavement studies, mixed results have been reported, with large population-based studies showing conflicting results. For example, in a study of 84,000 participants, no effect of losing a partner was found in women, but men were more likely to die from lung cancer and "other cancers", but not from stomach cancer – this effect was highest in younger men (aged 35-64 years) [53]. However, another study of 20,000 people found no increase in death from cancer in people who had lost a spouse [54]. Garssen concluded a weak effect exists for the impact of loss events on death from cancer, based on two studies of child-loss and one bereavement study, but the effect seems stronger for men than women [30].

Twelve studies reviewed by Garssen [30] investigated the relationships between social support and disease progression. In seven reports experiencing social support, having confidantes and a sufficient network of relatives and friends were related to a longer disease-free interval and longer survival. An example of a well-conducted study in this area is Maunsell et al (1995) [55] who identified the presence of different types confidantes in 224 newly diagnosed breast cancer patients. The larger number of categories in which women had close confidantes (i.e. spouse, children, friends, colleagues, etc), the higher the survival rate 7 years later. Fifty-six percent of women without a confidante had survived compared to 72% of those with one or more confidantes – a dose-response relationship was also seen, in that having more confidantes improved the survival rate.

There were also five negative reports reviewed that found no associations between measures of social support and cancer progression. Reasons for these discrepancies may include the specific ways that social support has been measured – structural versus functional. It may be the case that the quality of the support sytem is more important than its size, as indicated in one study which found no relationships between structural measures of support and survival, but did find associations between the subjective judgment of experienced social support and mortality [56]. A more recent report of 2,835 women who were diagnosed with breast cancer from the Nurses' Health Study in the USA found that women who were socially isolated before the diagnosis of cancer had a 66% increased risk of mortality from any cause, and twice the chance of dying from breast cancer [57]. Women without close relatives, friends or children also had elevated risks of death from breast cancer. Added to the data reviewed by Garssen in 2004 [30], this recent study suggests that women who feel less isolated and have a larger support network may have better outcomes.

The case of depression is yet another in which the evidence for a relationship with survival may be stronger than with incidence, with a good deal of research having been conducted in this area. Garssen [30] summarized that of 33 well-conducted studies in this area, six found that negative emotions predicted a more favorable disease outcome, 11 found a negative relationship, and 16 failed to find a relationship at all. When looking at studies that considered only psychiatric diagnoses, a similar pattern is seen, with two studies finding that having a psychiatric diagnosis was a favourable prognostic indicator, three studies finding the opposite, and no relationship in an additional two.

A well-conducted study published in 2003 and not included in earlier reviews followed similar methodology [58]. Measures were collected on a range of emotional and cognitive factors in the early postdiagnostic period and at 4-month intervals up to 15 months after diagnosis. These were used to predict survival time up to 10 years among 205 cancer patients heterogeneous in disease site, status, and progression. Depressive symptomology was the most consistent psychological predictor of shortened survival time using both baseline and repeated measures, after controlling for several known demographic and medical risk factors. Although controversy remains, most reviewers state that the evidence for an effect of depression on disease course is still inconclusive [23,30,50] although one major review concluded more strongly that depression likely plays a role in cancer survival [9]. These authours noted also that emotional overcontrol or repression of emotions likely adds to the deleterious effect of the depression itself.

Indeed, considering more specifically emotional repression and survival, five studies found that "repressors" had shorter disease-free intervals and shorter survival time. However, three others found no associations. Nonetheless, the role of emotional repression finds more support in terms of cancer progression than for initial cancer etiology and is described as one of the more "promising factors" for further investigation [30].

The final area that was evaluated as a factor in disease progression is coping style and adjustment to illness. Several hypothesized coping styles loosely related to the Type-C personality have been investigated. These include having a "fighting spirit", denial or minimization of the impact of the illness, stoic acceptance or fatalism, and a helpless/hopeless or pessimistic coping style. This latter type of response to cancer is also linked with depression, as some of the symptoms of depression are hopelessness and pessimism. The most support is found for a negative role of this factor in disease progression, with six of 10 studies finding patients with more hopelessness and pessimism had shorter survival times. For the other factors, little support was found for the importance of fighting spirit (1/6 studies), denial (0/5 studies), or fatalism (1/5 studies) [30].

Hence, overall a poor prognostic profile emerges of a person prone to feelings of helplessness, hopelessness and pessimism, with a tendency to repress emotions, and with the perception of little social support. The question then arises of what the physiological consequences of having a predominance of these psychological characteristics may be, and how this profile might promote cancer development. The question also arises of whether psychosocial interventions can, though altering these psychological characteristics, change this internal state and arrest cancer promoting processes already set in motion. The latter question regarding survival will be addressed next.

Psychosocial Interventions and Survival

This type of research investigates more specifically whether it is possible to manipulate some of the factors identified in the research reviewed previously that seem important for disease outcome, and subsequently change the course of the illness. This is an ambitious undertaking that many researchers would consider the final test for the role of psychosocial factors in disease processes. Again, this work has been reviewed on numerous occasions [17-21].

A recent review used the technique of meta-analysis to quantify the relative effect of psychosocial interventions on survival across all studies in the area that met criteria for good methodological quality [21]. There were 14 studies included in the review; interventions included components such as supportive therapy, emotional expression, cognitive behavioural therapy, psychoeducation, coping skills training, stress management, mental imagery, psychotherapy and hypnotherapy. Most were in formats of weekly group meetings from 6 weeks to one year in duration with follow-up periods ranging from 9-months to 20 years. When hazard ratios (HR) were calculated for each study and aggregated, the average HR was 0.85 (with a confidence interval (CI) of 0.65-1.11). In meta-analytic terminology, a HR of 1.0 means there is no advantage of the intervention over the control condition on survival – both the control patients and those who got the intervention lived about the same length of time. Values below 1.0 favour the interventions, over 1.0 favour the control group. Only if the value is below 1.0 and the confidence interval is entirely above or below 1.0 can the results be considered statistically significant. Hence, even though the HR is below one, the range of the CI includes 1.0, so one cannot conclude with certainty that the interventions prolonged survival. When interventions were analyzed separately by group vs. individual delivery, there was no effect of group programs on survival (HR=0.97, CI=0.73-1.27). However, individually-delivered interventions did have a positive effect on survival time (HR=0.55, CI=0.43-0.70). What this value means is that the odds of dying first were 35%

greater for patients in no-treatment groups compared to those receiving the individual interventions.

As a well-known example of research in this area, consider the case of the Spiegel at al Supportive Expressive Group Therapy (SEGT) study [59,60]. SEGT is a form of professionally-led group psychosocial intervention that evolved specifically to address the support needs of seriously ill medical patients. Two key interrelated goals of SEGT are to build social bonds and to facilitate the expression of emotion. Thus, the group creates a context for the expression, containment, and processing of current distress. This enables patients and their families to proactively address foreseeable challenges, to marshal appropriate resources and to make the most of whatever life remains [59]. Given that lack of social support and emotional repression were two of the strongest predictive factors for poor disease outcomes, this type of approach makes intuitive sense.

The Spiegel study of SEGT, published in 1989, reported on 10-year follow-up of women with metastatic breast cancer randomly assigned to weekly SEGT meetings over the course of an entire year, or care-as-usual, which in the late 1970s included no formal psychosocial support. They found that after 10 years, the women in the treatment group lived an average of 18-months longer than those assigned to the control condition [60]. This caused a great deal of interest internationally and a large replication study was carried out in Canada, known as the Breast Expressive-Supportive Therapy pan-Canadian trial (BEST; [61]). However, that study failed to find any survival advantage for the SEGT group, although it was successful in alleviating pain and suffering, as well as enriching the lives of patients diagnosed with metastatic cancer.

Another long-awaited replication by the original group at Stanford released the results in an abstract format in 2006 [62]: 125 women with metastatic (n = 122) or locally recurrent (n = 3) breast cancer were randomly assigned to the SEGT condition (n = 64), or to the control condition (n = 61) receiving only educational materials for a minimum of 1 year. No overall statistically significant effect of treatment on survival was found for treatment compared to control patients, but there was a statistically significant site by treatment interaction such that estrogen receptor (ER) negative participants randomized to treatment survived longer (median = 29.8 months) than ER negative controls (median = 9.3 months), while the ER positive participants showed no treatment effect (this was a secondary post-hoc unplanned analysis and hence the results must be considered cautiously).. Hence there was no overall effect of the intervention on survival, except perhaps for those whose cancers were *not* responsive to estrogens These cancers also typically have a poorer prognosis than ER positive cancers. If this result were bourne out, it can be speculated that because ER positive cancers respond well to hormonal therapies which may also have effects on HPA functioning, that any putative effect of psychosocial support on survival among ER positive women has been superseded. However, outcome among ER negative women is less affected by hormonal treatments, perhaps leaving room for the impact of other treatments including psychosocial interventions. This study illustrates well the ongoing controversy about the possible life-extending properties of psychosocial interventions and hints at some of the potential neuroendocrine moderators of potential survival effects of psychosocial interventions.

Psychosocial Interventions and Immune and Endocrine Outcomes

There is surprisingly little known about the effects of psychosocial interventions on immune and endocrine outcomes in cancer patients. A small number of narrative reviews speculated on these associations in the 1990s [22,23], but little of a conclusive nature has been discussed. Given the dearth of information, we conducted a systematic review of studies in this area by searching every combination of the following search terms in Medline, Pubmed and PsychInfo: cancer, neoplasm, psychosocial, intervention, treatment, outcomes, immune, endocrine, psychoneuroimmunology, psychoneuroendocrinology, PNI, PNE, biological, cortisol, natural killer, lymphocyte as well as searching the reference lists of selected articles in a snowball technique. Details of the sample, methods, interventions, and outcomes of each study are presented in Table 1. Papers were not vetted or rated on their methodological quality, as relatively few (17 studies) were identified, so all are included in the review. Hence the list includes study designs ranging from pre-post observational studies to larger randomized controlled trials (RCTs). Sample sizes vary from a low of 13 [63] to a high of 227 participants [64]. Of the 17 studies, 11 were RCTs with one or more comparison groups [63,65-67] [64,68-72], another 2 had nonrandomized assignment to one or more comparison groups [73,74], and 4 were pre-post assessments of one intervention only without a control group [75-78]. This distribution is not optimal, as the only study design that allows inferences of causality between participation in interventions and outcome measures is the RCT, but almost two-thirds of the studies did utilize this design. The other study designs can provide provocative preliminary evidence for promising avenues of future research.

Intervention types also varied considerably across studies, as did outcome measures. Interventions ranged from existential psychotherapy [67] to cognitive-behavioural stress management [70], mindfulness-based stress reduction [77,78] (an intervention involving intensive training in meditation and yoga), support groups [66], music therapy [76] and biofeedback [63]. Outcomes, as well, included a range of measures, mostly cortisol (measured both in saliva and blood), and indices of immune functioning (white blood cells, lymphocytes, NK cells, proliferative responses, immunoglobluins, cytokines). A broad range of psychological outcome measures were also used to assess an array of constructs, including stress, depression, anxiety, quality of life, mood, coping strategies, personality, loneliness, optimism and benefit finding. Studies primarily looked at women with breast cancer (13/17) with 2 studies investigating prostate cancer [63-71,73,77-79], 2 with mixed participants [75,76] and one each in malignant melanoma [72] and ovarian cancer [74]. In addition, the stage of cancer also was varied, as was the timing of the intervention. Most participants were diagnosed with non-metastatic disease (stages I=III) in 12/17 studies [63,64,66-68,70-73,77-79], two studies investigated metastatic cancer patients only [65,75], and three included patients with any stage of cancer [69,74,76]. Finally, in terms of timing, the majority of studies were conducted after the completion of surgery and adjuvant treatment (8/17) [63,66,67,71,75-78], followed by during the post-surgical phase (5) [72] [64,70,73,79], during treatment (3) [65] [68,74] and one was conducted pre-surgery [69].

Table 1.

	Year	Reference	Sample and Methods	Intervention	Biological Outcomes	Psychological Outcomes
1.	1989	Length of survival and lymphocyte percentage in women with mammary cancer as a function of psychotherapy [65]	100 women with metastatic breast cancer were randomly assigned to concurrent psychotherapy and chemotherapy (25 in each group: chemo, no chemo, therapy, no therapy). The groups were matched on age, social background, cancer stage, and medical treatments received.	Patients in the therapy group were randomly assigned to 1 of 3 therapies (creative novation therapy, depth psychotherapy, and relaxation therapy with desensitization). Total psychotherapy time was 30 hours. The chemotherapy group received Doxorubicin and 1 of 3 other combination of agents (cyclophosphamide, fluoruracil and vincristine, cyclophosphamide and prednisolone). Chemotherapy was given in 3-4 week cycles and was repeated 4-9 times	Leucocyte concentration and lymphocyte percentage was measured in the women in the chemotherapy group. Measurements were taken prior to chemo, two weeks after treatment initiation, and one month after for all consecutive cycles resulting in 7 samples. *Patients with no psychotherapy showed a steady decrease in lymphocyte production. Patients in therapy showed an initial decrease but then demonstrated an increase past levels at treatment initiation.*	Psychological outcomes not included
2.	1990	A structured psychiatric intervention for cancer patients: II. Changes over time in immunological measures [72]	61 patients (28 mean and 33 women, mean age = 42) with stage I or II malignant melanoma were randomly assigned to a standard care (n=35) or intervention (n=26) group. Immunological and psychological assessments were taken at baseline, 6 weeks, and 6 months. All patients had completed surgery	A short term structured psychiatric group, 9 hours in duration was provided for the treatment group. The intervention was delivered to groups of 7-10 patients for 11/2 hours for 6 weeks. The intervention consisted of health education, enhancement of problem solving skills, stress management and relaxation, and psychological support	Immune outcomes included NK cells, LGLs, major T-cell subsets, CD8 and CD4 T-cells. Groups did not differ at baseline on immunological variables. *At 6 weeks, the treatment group had higher levels of LGLs. At 6 months, the treatment group remained higher in LGLs but also had higher levels of NK cells. The majority of patients in the treatment group had increases in the percent of LDLs, NK cell and NK function and decreased in CD4 t-cells compared to the control group*	The POMS and the Dealing with Illness-Coping Inventory were used to assess affective state and coping. Coping and affective state were improved after the intervention but no significant correlations were found between affective state and immune changes when groups were analyzed independently.

Table 1. (Continued)

	Year	Reference	Sample and Methods	Intervention	Biological Outcomes	Psychological Outcomes
3.	1993	Immunological responses of breast cancer patients to behavioral interventions [63]	13 stage I breast cancer patients were enrolled in the study, mean age of 45 years. 7 patients were randomly assigned to immediate treatment and 6 patients to delayed treatment (6 months into the study). All patients had undergone surgery and completed adjuvant treatment.	The intervention was a 9 week sequence of relaxation training, guided imagery and biofeedback training in group format. The first 3 weeks were focused on relaxation and guided imagery with biofeedback being introduced in week 4. Participants were given relaxation tapes and were asked to practice twice daily. After completion of the 9 week intervention, monthly brush up sessions were held.	Blood samples were provided at baseline, throughout treatment and during 3 month follow up for both groups. Immune assays performed included NK cell activity, Con-A, MLR, IL-2, plasma IgA, IgG and IgM, total WBC and PBL. Baseline measures of psychophysiological stress response was taken to evaluate biofeedback training effects. *Between group comparisons demonstrated significant differences in PBL, WBC, Con-A, and MLR. Within group comparisons demonstrated significant effects for PBL, Con-A, MLR, and IgM.*	Psychological tests were administered at baseline and following training. Measurements included the MMPI, Millon Behavioral Health Inventory, Sarason Social Support Scale, Rotter Locus of control, the Affects Balance Scale, and the MAC Scale. *There were no significant changes demonstrated on the psychological measures from pre- to post-training.*
4.	1994	Effects of behavioral interventions on plasma cortisol and lymphocytes in breast cancer patients: An exploratory study [73]	24 women with stage I and II breast cancer. All patients had surgery and were not undergoing additional treatment. 14 patients volunteered to participate in the treatment and 10 chose to wait and serve as controls (non-randomized)	The intervention consisted of 2 hour session, once a week for 10 weeks. The intervention consisted of relaxation techniques based on autogentic principles and guided imagery, health education, development of stress and illness coping skills	Blood samples were obtained to determine plasma concentration of cortisol and WBC before and after the 2nd and 10th sessions. *There was a significant reduction in cortisol after the 2nd session and were further decreased after the 10th session, totaling a 23% decrease in cortisol levels. No significant differences were seen in numbers of leukocytes and monocytes. Lymphocyte numbers increased overall and specifically after the second session, but not after the 10th session.* There were no correlations between biological outcomes.	A German questionnaire assessed coping with severe bodily disease. Personality factors were controlled for using the Freiburger Personality Inventory. *There were no significant differences in personality factors or coping strategies*

Table 1. (Continued)

	Year	Reference	Sample and Methods	Intervention	Biological Outcomes	Psychological Outcomes
5.	1997	Coping, life attitudes, and immune responses to imagery and group support after breast cancer treatment [66]	47 women who completed treatment for stage 1-3 breast cancer with a mean age of 46. Patients were randomly assigned to a standard care control group (n=15), standard care plus six weekly 1 hour support session (n=16), or imagery/relaxation sessions (n=16).	Themes for both groups were of giving and receiving support and preparing for the future. The support group focused on the exploration of feelings, returning to normal, the impact of relationships, fears of recurrence, loss and saying goodbye. The imagery group focused on relaxation, setting goals and finding purpose, the impact of beliefs on health, coping with fears, giving and receiving support and concluding.	NK cytotoxicity, cytokines (IL-1α, IL-1β, IL-2 and IFN-γ), and beta endorphins were assessed. *Quality of imagery and frequency of practice was associated with increased NK activity and improved psychical quality of life. No significant differences were found between the groups on any of the immune measures. Neopterin decreased and IFN-γ increased for all the women.*	Assessment instruments included the FACT-B, POMS-Brief, Ways of Coping with Cancer, and the Duke-UNC Functional Social Support Questionnaire. *Coping skills improved for the support and imagery groups. The treatment groups also sought more support from others.* No significant differences were found between treatment groups
6.	1997	Effectiveness of a short-term group psychotherapy program on endocrine and immune function in breast cancer patients: An exploratory study [67]	The sample consisted of women with early stage breast cancer and distant or lymph metastases (mean age = 59). Patients were randomly assigned to an intervention group (11 patients) or a control group (12 patients). A healthy age matched control group of 15 women (mean age = 56) was included for comparison	Patients were randomly assigned to either the existential-experiential group psychotherapy or a wait list control group. The group was semi-structured and focused on fears of death, limitations of freedom, existential isolation, relationships with family, relatives and the medical professions, autonomy, helplessness and dependence and the meaning of life. Sessions were held once a week for 2.5 hours.	*Patients provided 2 blood samples before and after the intervention or at concurrent time periods for the control group. Biological measures included cortisol, ACTH, peripheral blood cells, NK cell activity and proliferative responses. Women with breast cancer had significantly higher cortisol levels than healthy matched women and lower levels of CD3 and CD4 cells. When baseline levels were adjusted, women in the treatment group had lower levels of prolactin and cortisol and lower levels of NK cells, CD4 and CD8 cells.*	*Measures were completed before and after program participation and included the BDI, the STAI, and the POMS. No correlations were found for psychological and endocrine outcomes. Trait anxiety was positively related to changes in CD4 levels. BDI scores were posiively related to percentage of NK cells. Mood disturbance was related to proliferative responses*

Table 1. (Continued)

	Year	Reference	Sample and Methods	Intervention	Biological Outcomes	Psychological Outcomes
7.	1997	Immune effects of relaxation during chemotherapy for ovarian cancer [74]	22 patients (Mean age= 57) receiving 4 cycles postoperative chemotherapy for stage 1-4 ovarian cancer. 12 patients were in the relaxation group, 10 patients were in the control group. Allocation was based on hospital ward and non-random.	Intervention was designed to train patients how to use cue controlled progressive relaxation skills rapidly and apply them when needed. Audiotapes were provided for practice and patients were asked to practice twice a day. Patients received 3 relaxation training sessions with a clinical psychologist. The relaxation sessions were delivered prior to initiation of treatment, during treatment and prior to the second course of treatment	Venous blood sampling was performed 2 days before chemotherapy and on the morning of their 3rd or 4th cycle. Routine hematological date was assessed (white blood cells, lymphocytes, granulocytes, monocytes, Con-A and NK). Groups were not different at baseline. *After intervention, the relaxation group had higher white blood cell counts and more lymphocytes.* Nonsignificant trends were observed for monocyte levels and proliferative responses to Con-A. NK levels were similar between groups.	The STAI was completed immediately before each blood sampling. No results were reported for psychological outcomes.
8.	1997	Phase II study of psychotherapeutic intervention in advanced cancer [75]	35 patients (16 males, 19 females) enrolled in the study with a mean age of 55 years. The group was heterogeneous but the primary tumor site was colorectal. No control group.	Patients were offered 12, 1.5-2 hour sessions once a week of individual counselling. They also attended twice monthly group meetings lasting 2.5 hours. Psychotherapy was experiential-existentially based and focused on feelings, needs, aims and potential of the individual.	NK cell cytotoxicity and activity was measured prior to the intervention, at 6 weeks and 12 weeks of the intervention and at 6 and 12 months following intake. *5 patients experienced an arrest in tumour growth. No effect of treatment on NK cell activity was observed. No significant relationships were found between NK and psychological outcomes.*	*Patients completed the Purpose in Life Test, Zung's Depression Scale, the Loneliness Inventory and the Cancer Locus of Control scale at the same time as blood sampling. No changes were observed on measures of loneliness, depression or locus of control. Purpose in life increased significantly.*

Table 1. (Continued)

	Year	Reference	Sample and Methods	Intervention	Biological Outcomes	Psychological Outcomes
9.	2000	Cognitive-behavioral stress management reduces serum cortisol by enhancing benefit finding among women being treated for early stage breast cancer [70]	34 women (mean age = 46) with stage I or II breast cancer recruited within 8 weeks post surgery. 24 women were randomized to the intervention group and 10 to a waitlist control group.	Cognitive behavioral stress management was delivered in 2 hour classes over the course of 10 weeks. Class material consisted of progressive muscle relaxation, meditation, abdominal breathing, guided imagery, cognitive restructuring and techniques to improve coping, assertiveness, anger management and the use of social support.	Blood samples were provided pre and post intervention or control period to assess cortisol levels. There were no significant differences between groups at baseline. *Significant decreases in cortisol were produced for the intervention group compared to the control. Cortisol reductions were significantly associated with improvements in benefit finding.*	Perceived positive contribution for women experiencing breast cancer were measured by the Benefit Finding Scale. Distress was measured by the POMS. *Significant improvements were observed in benefit finding but not in distress levels after program participation.*
10.	2000	A presurgical psychosocial intervention for breast cancer patients: psychological distress and the immune response [69]	41 women (mean age=56). All women were awaiting either surgery or surgery plus adjuvant therapy at enrolment. Patients were randomly assigned to standard care or standard care plus a 2 session psychosocial intervention. Assessments and blood draws were performed before surgery and prior to intervention, before surgery and following intervention, and 1 week following surgery.	The intervention was delivered by two clinical psychologists to individuals or small groups of 2-3. Duration was 90 minutes. The intervention had 3 objectives: psychoeducation about somatic and emotional responses to stress, problems solving skills for use in crisis, relaxation techniques and psychosocial support	NK cell activity and IFN-γ were measured. No significant time or interaction effects were observed in NK cell activity. *A significant interaction was seen for IFN-γ levels, whereby the levels of IFN-γ in the control group decreased steadily and the levels in the therapy group remained constant.* This was no longer significant when baseline measures were included as a covariate	Patients completed the CES-D scale; the Differential Emotions Scale-IV, the IES and the Life Orientation Test. *The therapy group experienced an increase on measures of overall interest and enjoyment and a decrease in sadness compared to the control group.* Optimism increased regardless of group

Table 1. (Continued)

	Year	Reference	Sample and Methods	Intervention	Biological Outcomes	Psychological Outcomes
11.	2001	Effects of stress management on testosterone levels in women with early stage breast cancer [79]	34 women (mean age = 46) with stage I or II breast cancer recruited within 8 weeks post surgery. 24 women were randomized to the intervention group and 10 to a waitlist control group.	Cognitive behavioral stress management was delivered in 2 hour classes over the course of 10 weeks. Class material consisted of progressive muscle relaxation, meditation, abdominal breathing, guided imagery, cognitive restructuring and techniques to improve coping, assertiveness, anger management and the use of social support.	Blood samples were provided pre and post intervention or control period to assess total testosterone levels. There were no significant differences between groups at baseline. *Post intervention, there were significant reductions in free testosterone and total testosterone. Reduction in testosterone were positively correlated to increases in benefit finding.*	Perceived positive contribution for women experiencing breast cancer were measured by the Benefit Finding Scale
12.	2001	A pilot study into the therapeutic effects of music therapy at a cancer help center [76]	29 patients (21 women and 8 men) participated in a residential 1 week course at the care center with an average age of 49 years (no control group). Cancer type and stage was mixed but breast cancer was predominant. All were in the process or had completed treatment Physiological and psychological assessments were conducted before and after each session.	The intervention was completed over 3 days. Day 1 included listening to 1 hour of music in a relaxed state. Day 2 included active playing of instruments for 1.5 hours in a group format. Day 3 included a 1 hour focus group meeting to discuss, compare and contract their experiences in both the listening and playing sessions.	Two salivary measures were examined: Salivary cortisol and sIgA. *Listening to music resulted in increased levels of sIgA compared to levels before music listening session. Cortisol levels decreased significantly from Day 1 to Day 2, but were not significantly affected by either music treatment.* No significant correlations were found between the physiological and psychological measures	The University of Wales Institute of Science and Technology (UWIST) Mood Adjective Checklist provided information on hedonic tone (well-being), tense arousal and energetic arousal. *Listening to music produced significant decreases in tense arousal and increases in hedonic tone. Active playing of music produced a significant increase in energetic arousal.*

Table 1. (Continued)

	Year	Reference	Sample and Methods	Intervention	Biological Outcomes	Psychological Outcomes
13.	2003	Mindfulness-based stress reduction in relation to quality of life, mood, symptoms of stress, and immune parameters in breast and prostate cancer outpatients [77]	49 women with breast and 10 men with prostate cancer. The sample had a mean age of 55 years and were diagnosed with early stage cancer a median of 1.1 years prior. Treatment was completed a median of 6 months prior (excluding hormonal). The sample was mostly comprised of patients with stage II cancer. No control group.	Mindfulness-based Stress Reduction was delivered over 8 weeks with weekly 90 minute group sessions and 1 3 hour silent retreat. Primary components of the program are: mindfulness meditation and yoga practice, psychoeducation regading stress and the stress response and group support.	Biological measurements included all leukocyte subclasses and intracellular cytokines. *Increases in eosinophils and IL-4 production in T-cells were observed as well as decreases in monocytes, T-cell production of IFN-γ, and NK cell production of IL-10.* There were no significant relationships between psychological and immunological measures	Psychological outcomes were measured with the EORTC-QLQ-30, POMS, and the SOSI. *Improvements were observed in quality of life, mood disturbance (13% reduction) and symptoms of stress (19.3% reduction).*
14.	2004	Psychological, behavioral, and immune changes after a psychological intervention: A clinical trial [64]	227 women (mean age = 51) with stage II or III breast cancer who can completed surgery but were awaiting adjuvant therapy were accrued. 113 were randomly assigned to the assessment only group and 114 were assigned to the intervention group. At 4 month assessment, the sample sizes were 91 in the assessment group and 107 in the intervention group.	The intervention was provided to small groups of 8-12 patients and led by 2 clinical psychologists. Groups met for 1.5 hours for 18 sessions. Therapy content included stress management, emotional distress and social adjustment, health behaviors and adherence to treatment.	Immune assays investigated were T lymphocytes, T-cell subsets and NK cells, NK cell cytotoxicity and blastogetnic response to PHA and Con-A. No significant group effects were found for T-lymphocyte counts. *A significant effects was reported for Con-A and PHA induced T-cell blastogenesis in the intervention group.* NK cell count and cell lysis were not significant	Psychological measures were the IES, the POMS, the Social Network Index, the Perceived Social Support Scale, the Food Habits Questionnaire and the Seven Day exercise Recall. *Patients in the intervention group showed a reduction in anxiety and for patients with higher cancer stress, the group produced reduction in mood disturbance and fatigue.* The intervention group also perceived more support

Table 1. (Continued)

	Year	Reference	Sample and Methods	Intervention	Biological Outcomes	Psychological Outcomes
						whereas the assessment only group reported a reduction in support. Health behaviors (diet and smoking) improved for the intervention group only
15.	2004	A pilot randomized trial assessing the effects of autogenic training in early stage cancer patients in relation to psychological status and immune system responses [71]	31 women with early stage breast cancer post surgery and radiotherapy, not undergoing further treatment were randomized to autogentic relaxation training or a wait list control	Autogenic relaxation training was taught in groups and focused on heaviness of limbs, warmth of limbs, calm regular heart beat, easy breathing, abdominal warmth and cooling of the forehead. Measurements were taken before and after the intervention.	CD4, CD8, B cells, NK cells, neutrophils and monocytes were examined. *There was a statistically significant difference in CD8 and NK cells after the intervention.* Caution in interpretation is warranted due to small sample size.	Anxiety and depression were measured using the HADS. *Intervention and control groups were significantly different on measures of anxiety and depression after the intervention*
16.	2004	Mindfulness-based stress reduction in relation to quality of life, mood, symptoms of stress and levels of cortisol, dehydroepiandrosterone sulphate (DHEAS) and melatonin in breast and prostate cancer outpatients [78]	49 women with breast and 10 men with pros-tate cancer. The sample had a mean age of 55 years and were diagnosed with early stage cancer a median of 1.1 years prior. Treatment was completed a median of 6 months prior (excluding hormonal). The sample was mostly comprised of patients with stage II cancer. No control group.	Mindfulness-based Stress Reduction was delivered over 8 weeks with weekly 90 minute group sessions and 1 3 hour silent retreat. Primary components of the program are: mindfulness meditation and yoga practice, psychoeducation regading stress and the stress response and group support.	A blood sample was taken for DHEAS testing and saliva samples for melatonin and cortisol. Cortisol samples were provided at 8am, 2pm, and 8pm. Melatonin samples were provided at 2pm. Patients with abnormal cortisol profiles at baseline also had higher stress scores. *No changes were observed in mean cortisol levels, but patients with higher initial cortisol levels decreased over time on mean cortisol levels as well as morning, afternoon and evening levels.* No significant changes were seen in DHEAS or melatonin levels	Psychological outcomes were measured with the EORTC-QLQ-30, POMS, and the SOSI. Psychological data is presented in the paper by Carlson, Speca, Patel and Goodey (2003) [77]

Table 1. (Continued)

	Year	Reference	Sample and Methods	Intervention	Biological Outcomes	Psychological Outcomes
17.	2006	A randomized controlled trial of psychological interventions using the psychophysiological framework for Chinese breast cancer patients [68]	76 participants (mean age = 49) were randomly assigned to a body-mind-spirit (BMS; n=27), supportive-expressive (SE; n=16), social support self help group (SS; n=16), or a no intervention group (n=17). Psychological and physiological measurements were administered at baseline, 4 months and 8 months post intervention.	The BMS group and integrated western psychotherapeutic elements with eastern philosophy and Chinese health practices. It focused on the normalization of traumatic experiences, letting go of attachments, forgiveness and self love and social support and commitment to others (15 total hours). The SE group was designed to build social support, deal with concerns of death, dying and body image, reorder life priorities, improve relationships, communication and symptom control (16 total hours). The SS group was unstructured and members decided topics (15 total hours).	The physiological outcome measure was salivary cortisol. Samples were taken at 5 times thoughout the day (waking, 45 min after waking, 12pm, 5pm, and 9pm). *Only the BMS group showed a significant reduction in total salivary cortisol level after 8 months. Mean cortisol concentration was significantly related to positive social support in that higher support was associated with lower cortisol at 8 months.*	psychological outcome measures included the General Health Questionnaire, the PSS, the Mini MAC Scale, the Courtauld Emotional Control Scale, and the Yale Social Support Index. *After 4 months, a significant reduction in health concerns, emotionality, negative emotion, and positive support was reported for the BMS group. The SS group showed and increase in negative emotion after 8 months. The control group showed a reduction in positive support after 4 months*

What can be concluded from this broad array of study designs, samples, interventions and outcome measures? Overall, 6 studies found some effect of the intervention on measures of cortisol [67,68,70,73,76,78], 14 reported positive effects on immune function [63-67,69,71-74,76-79], and one study reported no effect on immune function [75]. It might not be expected that biological changes would occur in the absence of psychological changes; hence it's important to note that 13 studies found positive effects of the intervention on psychological outcomes [64,66-72,75-78] but 2 studies did not [63,73]. The numbers don't always add up to 17 as some studies included both cortisol and immune measures, and other did not report psychological outcomes at all. It is also important to take into account publication bias, as most papers with negative results either aren't submitted for publication or are rejected at peer review. Also, most studies conducted multiple comparisons and had multiple outcome measures – hence, the finding of one of two positive outcomes inflates the error rate and perhaps exaggerates the effects of these interventions. If the percent positive outcomes were to be assessed on the basis of how many comparisons of all those conducted were statistically significant, the number would be far lower. However, this data does serve as "proof of principle" that it is possible to affect changes, particularly in measures of immune function and cortisol levels, through psychosocial intervention.

As mentioned previously, the quality of the studies varied considerably, with some of quite low quality, so perhaps it is useful to review one or two of the best studies in detail, to illustrate the potential effects. The largest study, conducted by Andersen et al (2004) [64] randomly assigned 227 women with stage II or III breast cancer who had completed surgery to either usual care or group therapy consisting of techniques for stress management, social support, emotional and social adjustment to cancer, enhancing health behaviours and adherence to treatment. After four months during which participants received chemotherapy, retention was good (80-93%) and the intervention group demonstrated greater t-cell proliferative responses to the mitogens ConA and PHA. Patients in the intervention also had greater reductions in anxiety, improvements in social support, and better health behaviors such as sleep and diet.

Similarly, a recent study reported a sample of 76 patients who were randomly assigned to a body-mind-spirit group, supportive-expressive therapy or a social support self-help group [68]. This type of study pitting various psychosocial interventions against one another creates a very hard test of specificity, as it is quite difficult to demonstrate the superiority of one group over another. Nonetheless, the authours found that only the body-mind-spirit group showed reductions in total salivary cortisol after 8-months, and higher social support was related to lower cortisol concentrations. Participants in the body-mind-spirit group also had decreases in health concerns, fewer negative emotions and felt more positive support than the other two groups.

To summarize, then, this body of work has demonstrated the potential for psychosocial interventions to affect a wide array of outcomes such as increasing NK cell cytotoxicity and decreasing salivary cortisol levels. Research in this area has focused mostly on immune and cortisol outcomes, and mostly on women with breast cancer. The potential significance of these changes will be addressed in the following section.

Endocrine/Immune Measures and Disease Progression/Survival

Much of the motivation behind investigating the neuroendocrine and immune impacts of psychosocial interventions in the previous section is based on the assumption that the intermediary outcome measures used are meaningful. Hence, the studies reviewed document associations between manipulating psychosocial factors and consequent changes in immune and endocrine measures. However, whether or not these outcomes have much or anything to do with disease processes has not been well explored. This question can be addressed by looking at both the animal and human research into the psychobiology of cancer, which has been reviewed by a number of investigators [9,10,13,14,22-24,80,81].

Researchers have looked at several potential pathways between immune and endocrine function, some of which can be altered by psychosocial factors, and disease progression in both humans and animals. Simply speaking, at least five interacting systems are involved in these relationships: 1) psychological factors (i.e. stress; depression); 2) CNS factors including HPA and SNS reactivity and regulation; 3) endocrine factors, including HPG reactivity of estrogens and testosterone and circadian rhythms of melatonin; 4) immune factors including immunosurveillance; and 5) tumour-related factors in the tumour microenvironment itself, including processes such as angiogenesis and programmed cell death (apoptosis). These systems clearly interact, so breaking them down into separate categories is an artificial heuristic that may help with understanding, but it is quickly obvious that they interact with one another in a myriad of ways. The first factor, psychological processes, has been addressed in earlier sections; the remainder of this section will selectively look at interactions between the other systems and disease progression.

HPA Axis

A growing body of research has investigated the stress-related concept of allostatic load, and a large body of evidence has associated excessive release of cortisol with suppression of the immune system (for reviews see [11,82,83]). Cortisol is largely responsible for the downregulation of immune function as a result of stress. Its hypersecretion also results in depressed mood [84,85]. Cortisol levels are typically highest in the morning, and decrease during the day, resulting in the downward sloping profile characteristic of most healthy individuals. However, cortisol levels have been reported to be elevated and overall diurnal profiles flatter in breast cancer patients compared to control women [86-88]. This supportive data stems primarily from women with metastatic, rather than earlier stage, cancers. For example, abnormal patterns of cortisol secretion have been reported in up to 75% of a sample of metastatic breast and ovarian cancer patients [89]. Further, the slope of the rate of change of cortisol levels measured four times a day for three consecutive days was associated with survival time in a group of 104 women with metastatic breast cancer. Those women who displayed less variation in salivary cortisol levels, expressed as a flatter slope and indicating a lack of normal diurnal cortisol variation, experienced earlier mortality over a 7-year follow-up period [15].

When these patients were split at the median cortisol slope for descriptive purposes, 77% of those with flat rhythms had died after surviving an average of 3.2 years. In contrast, 60% of the patients with relatively steep rhythms had died, with an average survival of 4.5 years. Hence, the women with steeper slopes survived more than 1 year longer on average. This relationship held even when other prognostic medical variables were taken into account, such as markers of disease status (e.g. location of metastases, estrogen receptor status), medical treatment (e.g. chemotherapy drugs) and psychosocial variables (e.g. stress levels and marital satisfaction) [15]. The authors speculate that these abnormal circadian rhythms of cortisol secretion represent compromised HPA axis functioning, which may be responsible for earlier mortality. Indeed other studies have reported circadian abnormalities in the secretion of 12 hormones in women at high risk for developing occurrences of breast cancer [90], as well as associations with later stages of cancer development and other prognostic indicators such as poorer performance status and more metastatic involvement [89,91]. In addition to affecting hormones, potentially through cell receptors and cell signalling pathways, excess cortisol may directly facilitate tumour growth through metabolic pathways, such that normal cells in which cortisol inhibits glucose uptake may become resistant to this effect when they mutate into cancer cells. Hence, glucocorticoids may suppress energy uptake in healthy cells but facilitate the ability of cancer cells to preferentially utilize energy, providing a metabolic advantage [92].

HPA axis dysregulation also has direct effects on immune function, both in terms of cellular and innate immunity. Immune defences against tumours are very complicated and include specific mechanisms such as tumour cell targeting by cytotoxic and helper T-cells and B-cell mediated cell lysis, and also natural immunity channels including cell death through NK cells, macrophages and ganulocytes. Abnormal cortisol rhythms were associated with reduced NK cell number and cytotoxicity in women with metastatic breast cancer [15], and decreased NK activity has been associated with tumour progression in animals [81] and humans [93]. Hence, excess cortisol in the blood stream of cancer patients due to allostatic load and HPA axis dysregulation can affect both numbers and activity levels of lymphocytes, macrophages and granulocytes [10].

SNS Factors

There is preliminary evidence suggesting that autonomic control may sometimes be impaired among BC patients [94]. Women at high familiar risk for breast cancer showed a greater catecholamine response to laboratory stressors than healthy women with normal risk levels [95], and they also had higher urinary levels of epinephrine (EP) during the work day [96]. During sleep there is usually a reduction of sympathetic nervous system (SNS) activity and an increase in parasympathetic (PNS) function, but the quality of sleep contributes to these autonomic changes. Deep-wave sleep is characterized by markedly reduced SNS activity, in that both norepinephrine (NE) and EP levels decline [97]. Hence, sleep quality can impact production of SNS catecholamines, or vice versa (higher SNS arousal can negatively impact sleep quality). If SNS activity is not reduced sufficiently during sleep, other systems may also suffer dysregulation. Hence, sleep is an important variable in maintaining SNS regulation and potentially contributing to cancer outcomes as well.

Influences of stress-related SNS products on *in vivo* tumour growth is illustrated by a wealth of studies demonstrating the negative effects of various types of stressors on tumour growth and metastasis in animal models. Typically animals may be injected with tumour cell lines, subjected to stress, and tumour growth quantified. For example, immobilization stress led to increased incidence of tumours and more growth in rats injected with the carcinogen diethylnitrosamine [98]. Other physical, social and psychological stressors including swimming stress, surgical stress, social confrontation and hypothermia all lead to increased lung metastasis from injected breast cancer cell lines [81,99,100]. In a chemical model simulating stressor effects, animals were subjected to elevated levels of injected β-adrageneric agonists, which simulate the effect of SNS activation, on tumour growth. Several animal studies have found β-adrageneric agonists cause increases in lung and mammary tumour metastases [101]. Conversely, pre-treatment of animals with β-adrageneric antagonists to block SNS activity also blocked effects of behavioural stress on lung tumour metastases previously observed [102]. There is also evidence that stress in animals may compromise mechanisms of DNA repair [10], thought to be important for recovery of cells from DNA damage due environmental or therapeutic exposure to radiation. Hence, products of the SNS elevated during the stress response or when circadian systems are dysregulated have direct facilitative effects on tumour growth and metastasis. The molecular basis for this effect is not fully understood but being actively researched.

Hormonal Factors

Estrogen (E)

Endocrine factors altered through stress and other psychological processes are extremely important for cancer development and progression, particularly in the case of cancers such as breast and prostate. Researchers have reported circadian abnormalities in the secretion of 12 hormones in women at high risk for developing occurrences of breast cancer [103] , as well as associations with later stages of cancer development and other prognostic indicators such as poorer performance status and more metastatic involvement [89,91]. Although estrogen is essential for normal mammary development and ductal growth, it also plays a role in the development and progression of breast cancer, and increased exposure to estrogens and increases in estrogen receptor expression in mammary epithelial cells increases risk. Primary risk factors for breast cancer include measures of lifelong E exposure, such as early menarche, fewer or no children, later age of parity, later menopause and obesity (for a review see Velie, 2005) [104]. Typically, women are tested at the time of diagnosis for E receptors and tumours with a high concentration of E receptors are referred to as positive (ER-positive) whereas those with a low or non-existent concentration of receptors are negative (ER-negative). Generally, ER-positive tumours respond best to hormonal treatments like Tamoxifen, which preferentially bind to ERs on breast cancer cells, taking the place of the naturally occurring estrogen and inhibiting the expression of estrogen-regulated genes that would normally induce cell proliferation [105].

Estrogen biosynthesis is catalyzed by the aromatase enzyme (aromatase cytochrome P450), which converts androgens to estrogens through the process of aromatization. Aromatase levels increase with age and BMI. The increase with rising BMI is thought to be due to tumour-adipocyte interactions [106]. Adipocytes are endocrine cells making up the bulk of the human breast (epithelial cells account for only 10% of the volume), and their numbers increase with increasing body weight and fat concentrations. They secrete various cytokines, polypeptide and hormone-like molecules, such as TNF-alpha and IL-6, which stimulate the production of aromatase [106]. Hence, more aromatase available for conversion of T to E may account for higher levels of E in obese women, as aromatase (and not T) is usually the rate-limiting factor.

Recall also that psychosocial interventions have been observed to alter levels of these aromatase-producing cytokines [78] which may result in less aromatase and subsequent less conversion of T to E in women with breast cancer participating in these interventions. No studies to our awareness have directly measured E level changes as a result of participation in psychosocial interventions. However it is well known that IL-6, TNF-alpha and other proinflammatory cytokines are elevated during periods of stress, and also in depression [107]. The cytokine pattern of depression is in fact very similar to that of cancer, leading researchers to suggest a high degree of similarity between the pathobiochemistry and immunology of cancer, cancer pain and depression [108]. Cancer treatment incorporating immune therapy with IL-2 and/or IFN-alpha is associated with depressive symptoms in a large proportion of patients [109,110]. This effect seems to be mediated by the activation of the cytokine network, including IL-6 [111], which is also elevated in depressed patients [112-114]. It makes sense, then, that if levels of pro-inflammatory cytokines can be decreased through the treatment of stress and depression, this may also affect levels of bioavailable E, especially in obese women, which is important for the development and progression of many breast cancers.

Testosterone

There is an interesting controversy around the role of testosterone in the development and progression of prostate cancer. Although the common assumption is that testosterone (T) plays a role in prostate cancer development due to the observation that castration (either surgical or biochemical) is an effective treatment for prostate cancer disease [115,116], a review of 34 studies investigating T levels in individuals newly diagnosed with prostate cancer compared to those without is inconclusive. That is, there is no clear-cut evidence that men who get prostate cancer have higher circulating levels of testosterone than those free of the disease [116]. This non-association between T levels and prostate cancer incidence has been shown in clinical trials of T supplementation, in longitudinal population-based studies, and in men who receive exogenous T treatment for hypogonadism. These findings have lead researchers to propose an indirect link between T and prostate cancer, perhaps through the acquisition of multiple non-specific sexually-transmitted diseases (STDs) caused by higher levels of sexual activity in men with higher T in their youth. This may promote transformation of prostate cells and damage to Leydig cells in the testis, which may contribute to the pathogenesis of prostate cancer, but not be observed as higher T levels around the time of diagnosis [116].

Not all researchers agree, as another meta-analysis found a 2.34 fold increase in the risk of prostate cancer in men in the highest quartile of T production, but they also found the same level of risk associated with higher serum IGF-1, and conversely, higher levels of sex hormone-binding globulin were associated with lower risk. Estrogens and DHT did not seem to play a role in prostate cancer risk [117]. The question of whether T levels can be altered through stress or psychosocial interventions is open. It is well known that T can affect mood states, and in aging there is a moderate decline of total T and a more significant decline of bioavailable T [118]. Elderly men who are most depressed also have the lowest T values, and T replacement in men with low T levels often results in improved mood. Indeed, even in hypogonadal men with clinical depression, T administration in some cases alleviated the depression [118]. Whether treatment of depression through psychosocial interventions may influence T production (and potentially prostate cancer progression) is not known, but one study did find that a combination treatment of diet, stress reduction, meditation and yoga resulted in arrest or regression, in some cases, of prostate-specific antigen (PSA) levels, the primary marker of disease activity in prostate cancer [119]. The mechanism for any such action is currently unknown.

Melatonin

The pineal hormone melatonin has been implicated in the treatment of many types of cancers and other diseases [120,121]. Proposed mechanisms of action include its effects as a free radical scavenger, an antioxidant, as well as an immunomodulatory agent and through the promotion of apoptosis of cancer cells in animal and human models [122]. In both *in vitro* and *in vivo* investigations, melatonin protected healthy cells from radiation-induced and chemotherapeutic drug-induced toxicity [123,124]. In humans, a series of clinical trials using melatonin in conjunction with standard treatment found superior survival response in patients with advanced cancer receiving adjuvant melatonin therapy [125], and higher tolerance of standard chemotherapy regimes [125,126]. A review of the animal and human literature concluded that converging evidence supports large transnational research-based clinical trials of melatonin therapy for a wide variety of cancers [122]. For years reports have indicated that women with breast cancer have suppressed or absent nocturnal melatonin peaks [127]. Epidemiological studies have also shown increased risk of breast cancer in women who work night shifts [128,129]. One biologically plausible explanation for this association is that these women have blunted melatonin secretion rhythms, and lack the nocturnal melatonin peak that is associated with normal sleep cycles.

We have evaluated the associations between sleep, stress, urinary melatonin and catecholamines in women recovering from breast cancer and matched healthy controls, and found higher levels of depressive symptoms, anxiety, fatigue, confusion, cardiopulmonary symptoms of stress and sleep disturbance in breast cancer patients than the comparison women; however, despite these disturbances there were no group differences on any of the biomarkers, including salivary cortisol, urinary catecholamines and melatonin, with the exception of higher dopamine levels in the control participants [130]. The women with breast cancer in our group were diagnosed with stages I-III breast cancer, primarily stage II, with no metastatic spread. This may account for the failure to find differences in any of the endocrine measures. However, the issue of the potential role of melatonin in disease development, circadian rhythm dysregulations and mood and stress remains a promising area of research [91,122].

Immune Factors

Cell-mediated immunity is the most studied outcome area in the intervention studies reviewed above, but the question arises as to the importance of the measures shown to be responsive to psychosocial interventions – most commonly increases in the number of lymphocytes, including NK cells, [64,67,72,73] NK cell cytotoxicity [72] and cytokine production [69,77].

Even though NK cells have been the target outcome measure in a number of psychosocial intervention trials and some changes have been observed, they have rarely been correlated with disease outcome (i.e. Fawzy et al, 1991 [72]). Many authours are skeptical of a role for immune factors in cancer progression (e.g. [23], including mainstream cancer biologists). Points raised by Garssen and Goodkin [23] against an important role of immune factors in cancer development include observations that: 1) spontaneously arising cancers in mice (in contrast to those induced by experimental methods) provoke little or no immune response; 2) tumour cells can easily mutate and avoid being targeted by cytotoxic T cells by constantly changing their antigens; 3) tumour cells can directly suppress the efficacy of various immune cells; and 4) due to the fast proliferation of some tumour cells the immune response is not capable of limiting their growth.

However, decreased NK activity has been associated with tumour progression in some studies of animals [81] and humans [93] and abnormal cortisol rhythms were associated with reduced NK cell number and cytotoxicity in women with metastatic breast cancer [15]. A persuasive argument has been made by Ben-Eliyahu for a role of cell-mediated immunity in the promotion of metastatic spread [81]. He speculates that the postoperative period when immunosupression occurs as a result of the surgical assault is a vulnerable period, in that dislodged tumour cells are most likely to travel through the bloodstream and metastasize to other parts of the body, if not controlled through immunomodulatory mechanisms. Mechanisms thought to be important for this process include cell-mediated immunity (CMI) though cytotoxic T lymphocytes, NK cells, NKT cells, tissue macrophages, dendritic cells and helper T cells. Research has shown that cell-mediated immunosuppression following surgery or stress coincides with periods of compromised resistance to metastasis of a cancer cell line known to be NK-sensitive [131]

Hence Ben-Eliyahu (2003) [81] makes an argument for the role of surgery in promoting metastatic spread through suppression of CMI based on the following points: 1) In human models, general anaesthesia has been shown to suppress aspects of human CMI, such as NK activity and helper T-cell ratios, but local or regional anaesthesia is not immunosuppressive. When given in conjunction, local anaesthesia blunts the SNS and HPA reactivity to surgery and attenuates the degree of immunosuppression. In animals local anaesthesia when added to general, results in less suppression of CMI and less metastatic spread; 2) Blood transfusions suppress CMI (decreased NKCC and T-cell blastogenesis) and are also associated with poorer prognosis, independent of other complications and risk factors; 3) In the past, colorectal cancer was treated with two successive surgeries; when these were reduced to one, lower rates of metastatic spread and better survival was seen, in both humans and animal models; 4) minimally invasive surgeries are less immunosuppressive and also reduce the promotion of metastatic spread, in both animals and humans. Hence, an argument can be made for an important role of CMI in preventing metastatic spread after surgery, but little is known about what specific level of various immune cells or function is required to prevent such spread. It may also be the case that beyond the post-surgical period immune factors are relatively *un-important* in the progression of tumour growth, which is supported by angiogenesis and the failure of apoptosis.

Growth Factors and Angiogenesis

Effects of stress on the tumour microenvironment including effects on the growth of a blood supply to tumours have been investigated quite extensively in recent years. Associations between elevated stress, lower social support and elevated levels of IL-6, a proinflammatory cytokine, and vascular endothelial growth factor (VEG-F), both promoters of angiogenesis in the tumour site, have been documented in ovarian cancer patients as well as *in vitro* cell cultures [13,132-134]. Immobilization stress increased tumour burden and enhanced angiogenesis and production of VEG-F within the tumour cell – angiogenesis is critical for tumour mass to increase beyond a certain size as the tumour cannot nourish itself without this development of its own blood supply. VEG-F also stimulates endothelial cell migration and proliferation and interferes with the development of T cells and the functional maturation of dendritic cells, which suggests additional negative effects on anti-tumour immune responses [10].

There are also effects of stress on viral oncogenesis, with many studies demonstrating accelerated growth of virally-induced tumours in animals who have been subjected to stressors [10]. Antoni et al [10] reviewed the neuroendocrine influences on various virally-related tumours and drew associations between HPA dysregulations and liver as well as cervical and head and neck cancers caused in part by Hepatitis B and C and human papilomavirus (HPV) viral exposure. They also drew associations between ANS reactivity and leukemia/lymphoma and Kaposi sarcomas cased in part by T-cell lymphotropic virus and Kaposi sarcoma-associated herpes virus, respectively [10], based on a review of primarily *in vitro* studies.

In summary, there are many potential avenues that stress and other psychological factors can "get in the body" and affect a whole host of processes important in cancer development and progression. Although the role of immunosurveillance is not a promising area of study for many cancer biologists, it may play an important role in prevention of metastatic spread of cancer cells following surgery. Other substances such as elevated epinephrine and cortisol may be cancer-promoting through several different mechanisms such as suppression of CMI. Elevated estrogens as a result of stress may also play a role in the pathogenesis of breast cancer, and testosterone may be important indirectly for its ability to aromatize to estrogens which can feed breast cancer cells, or by directly impacting prostate cancer development.

OTHER ISSUES FOR INTERVENTION RESEARCH

Timing of Interventions

Little research has been done to investigate the importance of the timing of psychosocial intervention delivery, but there is reason to believe that the impact may be different depending on where in the process of diagnosis, treatment and recovery interventions are applied. Keeping in mind the research on the effects of surgery on CMI, it seems that post-surgery might be an opportune time to apply psychosocial interventions likely to enhance cellular immunity. Levels of distress are also high at this time, as evidenced not only by large surveys of distress levels [78,125], but also by studies of anxiety around the time of biopsy [136,137]. Women assessed at that time, both pre- and post-biopsy also showed changes in immune function. Lymphocytes at pre biopsy showed a lower expression of NFkB and Ap-1, transcription factors that regulate lymphocyte function, compared to post-biopsy. At post biopsy NFkB activity increased threefold and AP-1 activity nearly doubled. Post biopsy, the transcription factors were similar to those of the healthy controls [137]. In another similar study, women undergoing breast biopsy (later confirmed either malignant or benign) and a control group were assessed on levels of T-lymphocytes, NK cells, helper lymphocytes, cytotoxic lymphocytes, circulating monocytes and cytokines [136]. NK cell activity was reduced pre-biopsy and remained so for 3 months after results for the benign group of women. NK cell activity remained reduced for the women in the malignant group. IFN-γ was reduced in the malignant and nonmalignant groups pre and post biopsy but returned to normal levels for both groups at 3 months. This data illustrates that the time around biopsy and initial surgery is not only stressful, but also may be a time of biological vulnerability.

Only one study we are aware of investigated the effects of delayed versus early psychosocial interventions for women with early stage breast cancer [138], but they did not assess any immune or endocrine markers. All participants had received surgery no longer than 4 months prior to study participation. Participants were randomized to 1 of 4 conditions: early or late start and experiential existential group psychotherapy or a support group. Groups were closed and limited to 6-10 women. Thirty-three women were assigned to the early start: 19 in psychotherapy and 14 in social support, and 34 women were assigned to the delayed start condition: 16 in psychotherapy and18 in social support. In both interventions, women met for 2.5 hours for 12 weeks and included follow up groups at 1 and 2 months post-completion. Women who were assigned to the late intervention appeared more distressed at follow up than did the early starters, after controlling for baseline differences. Overall there were improvements in distress, body image, social interaction, and recreational activities.

Hence it may be possible, through carefully choosing the timing of interventions, to maximize the effects of psychosocial treatments to have the most beneficial effects on both reduction of psychological symptoms, and biological processes potentially important for disease course. Whether concomitant changes occurred in important parameters related to CMI as a result of these interventions, either soon after surgery or later, remains to be investigated in further research.

Type and Stage of Cancer

The issue of the importance of the specific type of cancer for determining psychological and biological responses to interventions has been addressed only infrequently in the psychosocial literature, but it is likely quite a critical point, as the pathophysiology of cancers varies greatly. Unfortunately, most of the research in these areas has been conducted on breast cancer populations, and most in early stage women after treatment. Hence it is difficult to directly compare effects of interventions on different types of cancer patients at a variety of stages of disease progression.

However, in one review on the etiology and progression of cancer in relation to psychosocial characteristics, effects were compared among different types and stages of cancers. The proportion of studies that found a link between psychosocial factors and disease progression was 73% in breast cancer, 75% in malignant melanoma, 67% in haematological malignancies, and an astounding 100% in lung cancers [30]. This is interesting as some researchers have suggested that psychosocial factors are more likely to be important in diseases with more favourable prognoses, where the tumour burden is perhaps less and more amenable to respond to changes in immune and endocrine function caused by psychosocial interventions [50]. The result that all studies included in this review found an association between lung cancer incidence and psychosocial factors can partly be explained by the behaviour of smoking, as not all studies controlled for the correlation between depression and smoking. Looking just at those that did statistically control for smoking, two-thirds failed to find a relationship between depression and lung cancer initiation, but one-third still upheld the relationship. However, if smoking is controlled for, there appears to be no stronger association between psychosocial factors and disease initiation in lung cancer than other commonly studied types. The authors of this review conclude that there is little convincing evidence that psychosocial factors are more important in the initiation of some types of cancers more than others.

Another perspective on this issue is put forth by Antoni et al [24], who suggest that virally-mediated tumours are likely to be more responsive to psychosocial influences than solid epithelial tumours. This is based on their work in HIV and AIDS-related conditions where they have shown significant effects of cognitive-behavioral stress management on numbers and function of important T-cell subtypes and NK cells. They speculate that cervical cancer, a causal factor of which is HPV infection, may be more responsive to psychosocial interventions than solid tumours with different pathogenesis such as breast cancers. This possibility has not yet been tested empirically.

The impact of the stage of tumour development on the effect that psychosocial interventions can have on disease progression has also been considered in the literature, but not extensively evaluated. The literature in the area of breast cancer is the only area extensive enough within which to directly address this question. In one review, 64% of 14 studies looking at nonmetastatic breast cancer found relationships between psychosocial factors and disease initiation or progression, and this was even higher in metastatic breast cancer at 83% [30]. However, some reviewers have concluded that interventions for treating cancer patients might best be applied to earlier-stage diseases, in the hopes of intervening earlier in the disease process before the tumour burden becomes overwhelming [50]. This issue awaits resolution with further research.

Health Behaviours

Included in the biopsychosocial model of disease etiology and progression are other factors with potential to influence disease outcomes, including health behaviours such as sleep, exercise, and diet. There is evidence linking each of these to cancer initiation and progression. In terms of sleep, in the general North American population 1/3 of adults experience intermittent insomnia, and 10% suffer chronic insomnia [139,140]. The prevalence of chronic insomnia is much higher in cancer, with anywhere from 30-50% of patients reporting sleep difficulties [141,142] that often persist well into the post-treatment period. In metastatic breast cancer patients, 63% reported serious sleeping problems [143]. Individuals with insomnia have been characterized by increases SNS activity, increased 24-hour metabolic rate, and elevated cortisol and NE. A general state of physiological hyperarousal is considered to be symptomatic of these poor sleepers, and insufficient sleep duration has recently been shown to be associated with all-cause mortality [144]. Hence, it is very important to consider the effects of sleep quality on disease outcome, and the impact improving sleep during psychosocial interventions such as stress reduction could have on overall outcomes.

Exercise has emerged in the literature of late as a risk factor for both cancer incidence and progression [145,146]. Recent epidemiological research analyzed the relation between physical activity and breast cancer incidence between 1990 and 2002 among 90,509 French women between 40 and 65 years of age [147]. A linear decrease in risk of breast cancer was observed with increasing amounts of moderate and vigorous recreational activities. Compared with women who reported no recreational activities, those with more than five weekly hours of vigorous exercise had a relative risk of 0.62 (CI=0.49-0.78). This decrease was still observed among women who were overweight, had no children, had a family history of breast cancer, or used hormone replacement therapy [147]. After diagnosis, exercise rehabilitation programs have been successful in improving quality of life and reducing all-cause mortality [148,149], and recent observational evidence suggests that moderate levels of physical activity may reduce the risk of death from breast cancer after diagnosis [150]. This study was based on responses from 2987 nurses in the Nurses' Health Study who were diagnosed with stage I, II, or III breast cancer. Women who engaged in higher levels of physical activity (the equivalent of walking at average pace of 2 to 2.9 mph for 1 hour, 3 times/week, at minimum) were less likely to die from breast cancer than those who exercised less. The benefit of physical activity was particularly apparent among women with hormone-responsive tumours (ER positive). This is the opposite of the recent study of supportive-expressive therapy which found effects only for women who were ER negative [62]. The potential interactions between ER status and psychosocial interventions have yet to be investigated in great detail.

Another study examined colorectal cancer death rates and exercise, and found similar results: Increasing levels of exercise after diagnosis of nonmetastatic colorectal cancer reduced cancer-specific mortality ($p = .008$) and overall mortality ($p = .003$) [151]. In this case, prediagnosis physical activity was not predictive of mortality. Women who increased their activity (when comparing prediagnosis to postdiagnosis values) had a hazard ratio of 0.48 for colorectal cancer deaths and a hazard ratio of 0.51 for any-cause death, compared with those with no change in activity. Hence, there is little question that physical activity can be an important factor in improving disease outcomes, and changes in this variable associated with decreased depression and stress should be addressed in future research.

Finally, diet is known as one of the most important risk factors for the development of a host of cancers, and thought to account for about 30-35% of overall risk [25]. In general, plant-based diets high in fruits and vegetables and low in red meat and saturated fats are considered to be the healthiest in terms of avoiding or recovering from cancer. Specifically, according to Key et al., (2004) [152] fruits and vegetables probably reduce the risk for cancers of the oral cavity, oesophagus, stomach and colorectum; preserved meat and red meat probably increase the risk for colorectal cancer; salt preserved foods and high salt intake probably increase the risk for stomach cancer; and very hot drinks and foods probably increase the risk for cancers of the oral cavity, pharynx and oesophagus.

Recommendations regarding diet in several other reviews are similar, but focussing more on the role of obesity and alcohol consumption. Indeed, as reviewed previously, obesity increases breast cancer risk in postmenopausal women by around 30%, probably by increasing serum concentrations of estradiol. Moderate alcohol intake increases breast cancer risk by about 7% per alcoholic drink per day, perhaps also by increasing estrogen levels. Populations with high fat intake generally have higher rates of breast cancer, but studies of individual women have not confirmed an association of high fat diets with breast cancer risk. Speculation is that nutrition might affect breast cancer risk by altering levels of growth factors such as insulin-like growth factor (IGF)-I [153].

In colorectal and other gastric cancers, consumption of fruits and vegetables appeared to have a modest role in prevention [154]. In contrast, the roles of alcohol consumption and overweight on risk of gastrointestinal cancer are more clear and similar to breast cancer: overweight and obesity are important risk factors for adenocarcinoma (but not squamous carcinoma) of the esophagus and colorectal cancer (particularly in men). Alcohol consumption is a risk factor for squamous carcinoma (but not adenocarcinoma) of the esophagus, gastric cancer and colorectal cancer [154].

One research group wished to investigate the interactions among diet, exercise and markers of prostate cancer risk including measures of insulin, free T, E, and IGF as well as sex hormone-binding globulin (SHBG), a positive prognostic marker [155]. They found that in men who undertook a low-fat diet and/or exercise program all these factors changed in the predicted directions, and these changes impacted prostate cancer cell lines in vitro to reduce cell growth and induce apoptosis (programmed cell death), primarily through increasing tumour cell p53 proteins [155]. This is a compelling example of the potential for dietary factors to directly affect processes known to be important for cancer development and progression.

In summary, it is possible that psychosocial interventions that often include discussion of health behaviours and extol the virtues of taking control of lifestyle factors may also result in changes in these important behaviours, not to mention stopping smoking, which accounts for about 30% of the risk for all cancers combined. Stress reduction interventions often report better sleep in participants [156], which may be a pathway though which changes in immune and endocrine function are mediated. Some interventions explicitly recommend or include exercise, and the literature on specific exercise effects in cancer patients is growing, although it was not reviewed in this paper [145,146].

Conclusion

In conclusion, this chapter has addressed associations between psychosocial variables and both disease initiation and progression, showing that the evidence is strongest for a role of social support, emotional repression, hopelessness and depression as potentially important factors in disease progression. As well, the evidence that psychosocial interventions can effect survival was reviewed, with the conclusion that there may be effects, but if so they are quite weak. The literature on the effects of psychosocial interventions on immune and endocrine outcomes was reviewed in greater detail, with the conclusion that a wide variety of psychosocial programs may effect aspects of cell-mediated immunity and stress hormone production. Potential mechanisms of action between immune and endocrine variables and cancer progression were summarized, including HPA and SNS mechanisms, the role of various hormones, immunosuppression and other mechanisms such as HPA promotion of angiogenesis and virally-induced tumour growth. All pathways show some promise but perhaps the most hope for future research finding links between psychosocial interventions and important factors in disease pathogenesis lies with the effect of stress on the regulation of circadian rhythms, growth factors such as VEG-F, and hormones such as melatonin, although very little research has been conducted in these areas to date. Finally, specific factors such as the timing of interventions, the types of cancers targeted and the stage of disease progression are important to consider, as are other factors that may affect outcomes such as sleep, exercise, diet and smoking. Application of a comprehensive biopsychosocial model of disease progression will help to guide future research efforts.

Acknowledgements

Dr. Linda E. Carlson is supported by a New Investigator Award from the Canadian Institutes of Health Research and holds the Enbridge Endowed Research Chair in Psychosocial Oncology . Sheila Garland is funded by the University of Calgary and the Social Sciences and Humanities Research Council of Canada. Thanks to Jennifer Lowden for help with managing references.

References

[1] Temoshok L: Personality, coping style, emotion and cancer: towards an integrative model. *Cancer Surv* 1987; 6(3):545-567.

[2] Greer S, Watson M: Towards a psychobiological model of cancer: psychological considerations. *Soc Sci Med* 1985; 20(8):773-777.
[3] Chen E, Hanson MD, Paterson LQ, Griffin MJ, Walker HA, Miller GE: Socioeconomic status and inflammatory processes in childhood asthma: the role of psychological stress. *J Allergy Clin Immunol* 2006; 117(5):1014-1020.
[4] Cohen S, Doyle WJ, Skoner DP: Psychological stress, cytokine production, and severity of upper respiratory illness. *Psychosom Med* 1999; 61(2):175-180.
[5] Daruna JD: *Introduction to psychoneuroimmonology*. London, UK, Elsevier Academic Press, 2004.
[6] Miller GE, Cohen S, Ritchey AK: Chronic psychological stress and the regulation of pro-inflammatory cytokines: A glucocorticoid-resistance model. *Health Psychol* 2002; 21(6):531-541.
[7] Rozlog LA, Kiecolt-Glaser JK, Marucha PT, Sheridan JF, Glaser R: Stress and immunity: implications for viral disease and wound healing. *J Periodontol* 1999; 70(7):786-792.
[8] Segerstrom SC, Miller GE: Psychological stress and the human immune system: a meta-analytic study of 30 years of inquiry. *Psychol Bull* 2004; 130(4):601-630.
[9] Spiegel D, Giese-Davis J: Depression and cancer: mechanisms and disease progression. *Biol Psychiatry* 2003; 54(3):269-282.
[10] Antoni MH, Lutgendorf SK, Cole SW, Dhabhar FS, Sephton SE, McDonald PG, Stefanek M, Sood AK: The influence of bio-behavioural factors on tumour biology: pathways and mechanisms. *Nat Rev Cancer* 2006; 6(3):240-248.
[11] Andersen BL, Kiecolt-Glaser JK, Glaser R: A biobehavioral model of cancer stress and disease course. *Am Psychol* 1994; 49(5):389-404.
[12] Cameron LD, Leventhal H: *The self-regulation of health and behavior*. New York, Routledge, 2003.
[13] Lutgendorf SK, Costanzo ES: Psychoneuroimmunology and health psychology: an integrative model. *Brain Behav Immun* 2003; 17(4):225-232.
[14] Sephton S, Spiegel D: Circadian disruption in cancer: A neuroendocrine-immune pathway from stress to disease? *Brain, Behavior, & Immunity* 2003; 17(5):321-328.
[15] Sephton SE, Sapolsky RM, Kraemer HC, Spiegel D: Diurnal cortisol rhythm as a predictor of breast cancer survival. *Journal of the National Cancer Institute* 2000; 92(12):944-1000.
[16] Spiegel D, Sephton SE: Psychoneuroimmune and endocrine pathways in cancer: effects of stress and support. [Review] [184 refs]. *Seminars in Clinical Neuropsychiatry 6(4)*:252-65, 2001.
[17] Classen C, Koopman C, Hales R, Spiegel D: Acute stress disorder as a predictor of posttraumatic stress symptoms. *Am J Psychiatry* 1998; 155(5):620-624.
[18] Newell SA, Sanson-Fisher RW, Savolainen NJ: Systematic review of psychological therapies for cancer patients: overview and recommendations for future research. *J Natl Cancer Inst* 2002; 94(8):558-584.
[19] Ross L, Boesen EH, Dalton SO, Johansen C: Mind and cancer: does psychosocial intervention improve survival and psychological well-being? [Review] [69 refs]. *European Journal of Cancer 38(11)*:1447-57, 2002.

[20] Meyer TJ, Mark MM: Effects of psychosocial interventions with adult cancer patients: a meta-analysis of randomized experiments. *Health Psychol* 1995; 14(2):101-108.

[21] Smedslund G, Ringdal GI: Meta-analysis of the effects of psychosocial interventions on survival time in cancer patients. *J Psychosom Res* 2004; 57(2):123-131.

[22] van der Pompe G, Antoni M, Visser A, Garssen B: Adjustment to breast cancer: the psychobiological effects of psychosocial interventions. [Review] [77 refs]. *Patient Education & Counseling 28(2)*:209-19, 1996.

[23] Garssen B, Goodkin K: On the role of immunological factors as mediators between psychosocial factors and cancer progression. *Psychiatry Res* 1999; 85(1):51-61.

[24] Antoni MH: Psychoneuroendocrinology and psychoneuroimmunology of cancer: Plausible mechanisms worth pursuing? *Brain Behav Immun* 2003; 17 Suppl 1:S84-S91.

[25] Dollinger M, Resenbaum EH, Cable G: *Everyone's Guide to Cancer Therapy: How Cancer is Diagnosed*, Treated, and Managed Day to Day. ed 3, Kansas City, Andrews McMeel Publishing, 1997.

[26] Butow PN, Hiller JE, Price MA, Thackway SV, Kricker A, Tennant CC: Epidemiological evidence for a relationship between life events, coping style, and personality factors in the development of breast cancer. *J Psychosom Res* 2000; 49(3):169-181.

[27] Price MA, Tennant CC, Butow PN, Smith RC, Kennedy SJ, Kossoff MB, Dunn SM: The role of psychosocial factors in the development of breast carcinoma: Part II. Life event stressors, social support, defense style, and emotional control and their interactions. *Cancer* 91(4):686-97, 2001.

[28] Dalton SO, Boesen EH, Ross L, Schapiro IR, Johansen C: Mind and cancer. Do psychological factors cause cancer? *Eur J Cancer* 2002; 38(10):1313-1323.

[29] Edelman S, Kidman A: Mind and cancer: Is there a relationship? A review of evidence. *Australian Psycologist* 1997; 32(2):79-85.

[30] Garssen B: Psychological factors and cancer development: evidence after 30 years of research. [Review] [81 refs]. *Clinical Psychology Review* 24(3):315-38, 2004.

[31] Geyer S: The role of social and psychosocial factors in the development and course of cancer. *Wien Klin Wochenschr* 2000; 112(23):986-994.

[32] McGee R, Williams S, Elwood M: Are life events related to the onset of breast cancer. *Psychol Med* 1996; 26(3):441-447.

[33] McKenna MC, Zevon MA, Corn B, Rounds J: Psychosocial factors and the development of breast cancer: a meta-analysis. *Health Psychol* 1999; 18(5):520-531.

[34] Duijts SF, Zeegers MP, Borne BV: The association between stressful life events and breast cancer risk: a meta-analysis. *Int J Cancer* 2003; 107(6):1023-1029.

[35] Kroenke CH, Hankinson SE, Schernhammer ES, Colditz GA, Kawachi I, Holmes MD: Caregiving stress, endogenous sex steroid hormone levels, and breast cancer incidence. *Am J Epidemiol* 2004; 159(11):1019-1027.

[36] Helgesson O, Cabrera C, Lapidus L, Bengtsson C, Lissner L: Self-reported stress levels predict subsequent breast cancer in a cohort of Swedish women. *Eur J Cancer Prev* 2003; 12(5):377-381.

[37] Nielsen NR, Zhang ZF, Kristensen TS, Netterstrom B, Schnohr P, Gronbaek M: Self reported stress and risk of breast cancer: prospective cohort study. *BMJ* 2005; 331(7516):548.

[38] Petticrew M, Fraser J, Regan MF: Adverse life-events and risk of breast caner: A meta-analysis. *British Journal of Health Psychology* 1999; 4:1-17.

[39] Jones DR, Goldblatt PO, Leon DA: Bereavement and cancer: some data on deaths of spouses from the longitudinal study of Office of Population Censuses and Surveys. *Br Med J (Clin Res Ed)* 1984; 289(6443):461-464.

[40] Kvikstad A, Vatten LJ, Tretli S, Kvinnsland S: Death of a husband or marital divorce related to risk of breast cancer in middle-aged women. A nested case-control study among Norwegian women born 1935-1954. *Eur J Cancer* 1994; 30A(4):473-477.

[41] Kvikstad A, Vatten LJ: Risk and prognosis of cancer in middle-aged women who have experienced the death of a child. *Int J Cancer* 1996; 67(2):165-169.

[42] Levav I, Kohn R, Iscovich J, Abramson JH, Tsai WY, Vigdorovich D: Cancer incidence and survival following bereavement. *Am J Public Health* 2000; 90(10):1601-1607.

[43] Schernhammer ES, Hankinson SE, Rosner B, Kroenke CH, Willett WC, Colditz GA, Kawachi I: Job stress and breast cancer risk: the nurses' health study. *Am J Epidemiol* 2004; 160(11):1079-1086.

[44] Bleiker EM, van der Ploeg HM: Psychosocial factors in the etiology of breast cancer: review of a popular link. *Patient Educ Couns* 1999; 37(3):201-214.

[45] Reynolds P, Kaplan GA: Social connections and risk for cancer: prospective evidence from the Alameda County Study. *Behav Med* 1990; 16(3):101-110.

[46] Kroenke CH, Bennett GG, Fuchs C, Giovannucci E, Kawachi I, Schernhammer E, Holmes MD, Kubzansky LD: Depressive symptoms and prospective incidence of colorectal cancer in women. *Am J Epidemiol* 2005; 162(9):839-848.

[47] Shaffer JW, Graves PL, Swank RT, Pearson TA: Clustering of personality traits in youth and the subsequent development of cancer among physicians. *J Behav Med* 1987; 10(5):441-447.

[48] Bleiker EM, van der Ploeg HM, Hendriks JH, Ader HJ: Personality factors and breast cancer development: a prospective longitudinal study. *J Natl Cancer Inst* 1996; 88(20):1478-1482.

[49] Petticrew M, Bell R, Hunter D: Influence of psychological coping on survival and recurrence in people with cancer: systematic review. *BMJ* 2002; 325(7372):1066.

[50] Cwikel JG, Behar LC, Zabora JR: Psychosocial factors that affect the survival of adult cancer patients: A review of research. *Journal of Psychosocial Oncology* 1997; 15(3/4):1-34.

[51] Fox BH: The role of psychological factors in cancer incidence and prognosis. *Oncology (Williston Park)* 1995; 9(3):245-253.

[52] Forsen A: Psychosocial stress as a risk for breast cancer. *Psychother Psychosom* 1991; 55(2-4):176-185.

[53] Martikainen P, Valkonen T: Mortality after the death of a spouse: rates and causes of death in a large Finnish cohort. *Am J Public Health* 1996; 86(8):1087-1093.

[54] Jones DR, Goldblatt PO: Cancer mortality following widow(er)hood: Some further results from the office of population censuses and surveys longitudinal study. *Stress Medicine* 1986; 2:129-229.

[55] Maunsell E, Brisson J, Deschenes L: Social support and survival among women with breast cancer. *Cancer* 1995; 76(4):631-637.

[56] Reynolds P, Boyd PT, Blacklow RS, Jackson JS, Greenberg RS, Austin DF, Chen VW, Edwards BK: The relationship between social ties and survival among black and white breast cancer patients. National Cancer Institute Black/White Cancer Survival Study Group. *Cancer Epidemiol Biomarkers Prev* 1994; 3(3):253-259.

[57] Kroenke CH, Kubzansky LD, Schernhammer ES, Holmes MD, Kawachi I: Social networks, social support, and survival after breast cancer diagnosis. *J Clin Oncol* 2006; 24(7):1105-1111.

[58] Brown KW, Levy AR, Rosberger Z, Edgar L: Psychological distress and cancer survival: a follow-up 10 years after diagnosis. *Psychosom Med* 2003; 65(4):636-643.

[59] Spiegel D, Classen C: *Group psychotherapy for cancer patients: A research-based handbook of psycosocial care.* New York, Basic Books, 2000.

[60] Spiegel D, Bloom JR, Kraemer HC, Gottheil E: Effect of psychosocial treatment on survival of patients with metastatic breast cancer. *Lancet* 1989; 2(8668):888-891.

[61] Goodwin PJ, Leszcz M, Ennis M, Koopmans J, Vincent L, Guther H, Drysdale E, Hundleby M, Chochinov HM, Navarro M, Speca M, Hunter J: The effect of group psychosocial support on survival in metastatic breast cancer. *N Engl J Med* 2001; 345(24):1719-1726.

[62] Spiegel D, Butler LD, Giese-Davis J, Koopman C, Miller E, DiMiceli S, Classen C, Fobair P, Carlson RW, Kraemer H: Effects of Supportive-Expressive Group Therapy on Survival of Patients with Metastatic Breast Cancer: A Randomized Clinical Intervention Trial. *Breast Cancer Res Treat* 2006; 100(Supplement 1):S240.

[63] Gruber BL, Hersh SP, Hall NR, Waletzky LR, Kunz JF, Carpenter JK, Kverno KS, Weiss SM: Immunological responses of breast cancer patients to behavioral interventions. *Biofeedback Self Regul* 1993; 18(1):1-22.

[64] Andersen BL, Farrar WB, Golden-Kreutz DM, Glaser R, Emery CF, Crespin TR, Shapiro CL, Carson WE, III: Psychological, behavioral, and immune changes after a psychological intervention: a clinical trial. *J Clin Oncol* 2004; 22(17):3570-3580.

[65] Grossarth-Maticek R, Eysenck HJ: Length of survival and lymphocyte percentage in women with mammary cancer as a function of psychotherapy. *Psychol Rep* 1989; 65(1):315-321.

[66] Richardson MA, Post-White J, Grimm EA, Moye LA, Singletary SE, Justice B: Coping, life attitudes, and immune responses to imagery and group support after breast cancer treatment. *Altern Ther Health Med* 1997; 3(5):62-70.

[67] van der Pompe G, Duivenvoorden HJ, Antoni MH, Visser A, Heijnen CJ: Effectiveness of a short-term group psychotherapy program on endocrine and immune function in breast cancer patients: An exploratory study. *J Psychosom Res* 1997; 42(5):453-466.

[68] Chan CL, Ho RT, Lee PW, Cheng JY, Leung PP, Foo W, Chow LW, Sham JS, Spiegel D: A randomized controlled trial of psychosocial interventions using the

psychophysiological framework for Chinese breast cancer patients. *J Psychosoc Oncol* 2006; 24(1):3-26.

[69] Larson MR, Duberstein PR, Talbot NL, Caldwell C, Moynihan JA: A presurgical psychosocial intervention for breast cancer patients. psychological distress and the immune response. *J Psychosom Res* 2000; 48(2):187-194.

[70] Cruess DG, Antoni MH, McGregor BA, Kilbourn KM, Boyers AE, Alferi SM, Carver CS, Kumar M: Cognitive-behavioral stress management reduces serum cortisol by enhancing benefit finding among women being treated for early stage breast cancer. *Psychosom Med* 2000; 62(3):304-308.

[71] Hidderley M, Holt M: A pilot randomized trial assessing the effects of autogenic training in early stage cancer patients in relation to psychological status and immune system responses. *European Journal of Oncology Nursing* 2004; 8(1):61-65.

[72] Fawzy FI, Kemeny ME, Fawzy NW, Elashoff R, Morton D, Cousins N, Fahey JL: A structured psychiatric intervention for cancer patients. II. Changes over time in immunological measures. *Arch Gen Psychiatry* 1990; 47(8):729-735.

[73] Schedlowski M, Jung C, Schimanski G, Tewes U, Schmoll H: Effects of behavioral interventions on plasma cortisol and lymphocytes in breast cancer patients: An exploratory study. *Psychooncology* 1994; 3:181-187.

[74] Lekander M, Furst CJ, Rotstein S, Hursti TJ, Fredrikson M: Immune effects of relaxation during chemotherapy for ovarian cancer. *Psychother Psychosom* 1997; 66(4):185-191.

[75] de Vries MJ, Schilder JN, Mulder CL, Vrancken AM, Remie M, Garssen B: Phase II study of psychotherapeutic intervention in advanced cancer. *Psychooncology* 1997; 6(2):129-137.

[76] Burns SJ, Harbuz MS, Hucklebridge F, Bunt L: A pilot study into the therapeutic effects of music therapy at a cancer help center. *Altern Ther Health Med* 2001; 7(1):48-56.

[77] Carlson LE, Speca M, Patel KD, Goodey E: Mindfulness-based stress reduction in relation to quality of life, mood, symptoms of stress, and immune parameters in breast and prostate cancer outpatients. *Psychosom Med* 2003; 65(4):571-581.

[78] Carlson LE, Speca M, Patel KD, Goodey E: Mindfulness-based stress reduction in relation to quality of life, mood, symptoms of stress and levels of cortisol, dehydroepiandrosterone sulfate (DHEAS) and melatonin in breast and prostate cancer outpatients. *Psychoneuroendocrinology* 2004; 29(4):448-474.

[79] Cruess DG, Antoni M, Kumar M, McGregor BA, Alferi SM, Boyers AE, Carver CS, Kilshourn L: Effects of stress management on testosterone levels in women with early stage breast cancer. *International Journal of Behavioral Medicine* 2001; 8:194-207.

[80] Bovbjerg DH: Conditioning, cancer, and immune regulation. *Brain Behav Immun* 2003; 17 Suppl 1:S58-S61.

[81] Ben-Eliyahu S: The promotion of tumor metastasis by surgery and stress: immunological basis and implications for psychoneuroimmunology. *Brain Behav Immun* 2003; 17 Suppl 1:S27-S36.

[82] Spiegel D, Sephton SE, Terr AI, Stites DP: Effects of psychosocial treatment in prolonging cancer survival may be mediated by neuroimmune pathways. *Ann N Y Acad Sci* 1998; 840(674):683.

[83] Cohen S, Williamson GM: Stress and infectious disease in humans. *Psychol Bull* 1991; 109(1):5-24.

[84] Sikes CRA, Lasley BJ: Cognitive sequelae of hypothalamic-pituitary-adrenal (HPA) dysregulation in depression. *Biological Psychiatry* 1989; 25:148A-149A.

[85] Wolkowitz OM: Prospective controlled studies of the behavioral and biological effects of exogenous corticosteroids. *Psychoneuroendocrinology* 1994; 3(19):233-255.

[86] Abercrombie HC, Giese-Davis J, Sephton S, Epel ES, Turner-Cobb JM, Spiegel D: Flattened cortisol rhythms in metastatic breast cancer patients. *Psychoneuroendocrinology* 2004; 29(8):1082-1092.

[87] van der Pompe G, Antoni MH, Heijnen CJ: Elevated basal cortisol levels and attenuated ACTH and cortisol responses to a behavioral challenge in women with metastatic breast cancer. *Psychoneuroendocrinology* 1996; 21(4):361-374.

[88] Porter LS, Mishel M, Neelon V, Belyea M, Pisano E, Soo MS: Cortisol levels and responses to mammography screening in breast cancer survivors: a pilot study. *Psychosom Med* 2003; 65(5):842-848.

[89] Touitou Y, Bogdan A, Levi F, Benavides M, Auzeby A: Disruption of the circadian patterns of serum cortisol in breast and ovarian cancer patients: relationships with tumour marker antigens. *British Journal of Cancer* 74(8):1248-52, 1996.

[90] Ticher A, Haus E, Ron IG, Sackett-Lundeen L, Ashkenazi IE: The pattern of hormonal circadian time structure (acrophase) as an assessor of breast-cancer risk. *International Journal of Cancer* 1996; 65(5):591-593.

[91] Mormont MC, Levi F: Circadian-system alterations during cancer processes: a review. *Int J Cancer* 1997; 70(2):241-247.

[92] Turner-Cobb J, Sephton SE, Spiegel D: Psychosocial effects on immune function and disease progression in cancer:human studies.; in: Ader R, Pelten D, Cohen N, (eds): *Psychoneuroimmunology.* San Diego, Academin Press, 2001, pp 565-582.

[93] Levy SM, Herberman RB, Lippman M, D'Angelo T, Lee J: Immunological and psychosocial predictors of disease recurrence in patients with early-stage breast cancer. *Behav Med* 1991; 17(2):67-75.

[94] Bettermann H, Kroz M, Girke M, Heckmann C: Heart rate dynamics and cardiorespiratory coordination in diabetic and breast cancer patients. *Clin Physiol* 2001; 21(4):411-420.

[95] Gold SM, Zakowski SG, Valdimarsdottir HB, Bovbjerg DH: Stronger endocrine responses after brief psychological stress in women at familial risk of breast cancer. *Psychoneuroendocrinology* 2003; 28(4):584-593.

[96] James GD, Berge-Landry HH, Valdimarsdottir HB, Montgomery GH, Bovbjerg DH: Urinary catecholamine levels in daily life are elevated in women at familial risk of breast cancer. *Psychoneuroendocrinology* 2004; 29(7):831-838.

[97] Linsell CR, Lightman SL, Mullen PE, Brown MJ, Causon RC: Circadian rhythms of epinephrine and norepinephrine in man. *J Clin Endocrinol Metab* 1985; 60(6):1210-1215.

[98] Laconi E, Tomasi C, Curreli F, Diana S, Laconi S, Serra G, Collu M, Pani P: Early exposure to restraint stress enhances chemical carcinogenesis in rat liver. *Cancer Lett* 2000; 161(2):215-220.

[99] Ben-Eliyahu S, Shakhar G, Page GG, Stefanski V, Shakhar K: Suppression of NK cell activity and of resistance to metastasis by stress: a role for adrenal catecholamines and beta-adrenoceptors. *Neuroimmunomodulation* 2000; 8(3):154-164.

[100] Page GG, Ben-Eliyahu S: A role for NK cells in greater susceptibility of young rats to metastatic formation. *Dev Comp Immunol* 1999; 23(1):87-96.

[101] Ben-Eliyahu S, Yirmiya R, Liebeskind JC, Taylor AN, Gale RP: Stress increases metastatic spread of a mammary tumor in rats: evidence for mediation by the immune system. *Brain, Behavior, & Immunity* 1991; 5(2):193-205.

[102] Melamed R, Rosenne E, Shakhar K, Schwartz Y, Abudarham N, Ben-Eliyahu S: Marginating pulmonary-NK activity and resistance to experimental tumor metastasis: suppression by surgery and the prophylactic use of a beta-adrenergic antagonist and a prostaglandin synthesis inhibitor. *Brain Behav Immun* 2005; 19(2):114-126.

[103] Ticher A, Haus E, Ron IG, Sackett-Lundeen L, Ashkenazi IE: The pattern of hormonal circadian time structure (acrophase) as an assessor of breast-cancer risk. *International Journal of Cancer* 1996; 65(5):591-593.

[104] Velie EM, Nechuta S, Osuch JR: Lifetime reproductive and anthropometric risk factors for breast cancer in postmenopausal women. *Breast Dis* 2005; 24:17-35.

[105] Jaiyesimi IA, Buzdar AU, Decker DA, Hortobagyi GN: Use of tamoxifen for breast cancer: twenty-eight years later. *J Clin Oncol* 1995; 13(2):513-529.

[106] Lorincz AM, Sukumar S: Molecular links between obesity and breast cancer. *Endocr Relat Cancer* 2006; 13(2):279-292.

[107] Reiche EM, Morimoto HK, Nunes SM: Stress and depression-induced immune dysfunction: implications for the development and progression of cancer. *Int Rev Psychiatry* 2005; 17(6):515-527.

[108] Holden RJ, Pakula IS, Mooney PA: An immunological model connecting the pathogenesis of stress, depression and carcinoma. *Med Hypotheses* 1998; 51(4):309-314.

[109] Hauser P, Soler R, Reed S, Kane R, Gulati M, Khosla J, Kling MA, Valentine AD, Meyers CA: Prophylactic treatment of depression induced by interferon-alpha. *Psychosomatics* 2000; 41(5):439-441.

[110] Capuron L, Ravaud A, Dantzer R: Early depressive symptoms in cancer patients receiving interleukin 2 and/or interferon alfa-2b therapy. *J Clin Oncol* 2000; 18(10):2143-2151.

[111] Tilg H, Atkins MB, Dinarello CA, Mier JW: Induction of circulating interleukin 10 by interleukin 1 and interleukin 2, but not interleukin 6 immunotherapy. *Cytokine* 1995; 7(7):734-739.

[112] Maes M, Bosmans E, De JR, Kenis G, Vandoolaeghe E, Neels H: Increased serum IL-6 and IL-1 receptor antagonist concentrations in major depression and treatment resistant depression. *Cytokine* 1997; 9(11):853-858.

[113] Seidel A, Arolt V, Hunstiger M, Rink L, Behnisch A, Kirchner H: Cytokine production and serum proteins in depression. *Scand J Immunol* 1995; 41(6):534-538.

[114] Sluzewska A, Rybakowski J, Bosmans E, Sobieska M, Berghmans R, Maes M, Wiktorowicz K: Indicators of immune activation in major depression. *Psychiatry Res* 1996; 64(3):161-167.

[115] Morgentaler A: Testosterone replacement therapy and prostate risks: where's the beef? *Can J Urol* 2006; 13 Suppl 1:40-43.

[116] Slater S, Oliver RT: Testosterone: its role in development of prostate cancer and potential risk from use as hormone replacement therapy. *Drugs Aging* 2000; 17(6):431-439.

[117] Shaneyfelt T, Husein R, Bubley G, Mantzoros CS: Hormonal predictors of prostate cancer: a meta-analysis. *J Clin Oncol* 2000; 18(4):847-853.

[118] Margolese HC: The male menopause and mood: testosterone decline and depression in the aging male--is there a link? *J Geriatr Psychiatry Neurol* 2000; 13(2):93-101.

[119] Saxe GA, Hebert JR, Carmody JF, Kabat-Zinn J, Rosenzweig PH, Jarzobski D, Reed GW, Blute RD: Can diet in conjunction with stress reduction affect the rate of increase in prostate specific antigen after biochemical recurrence of prostate cancer? *J Urol* 2001; 166(6):2202-2207.

[120] Saez MC, Barriga C, Garcia JJ, Rodriguez AB, Masot J, Duran E, Ortega E: Melatonin increases the survival time of animals with untreated mammary tumours: neuroendocrine stabilization. *Mol Cell Biochem* 2005; 278(1-2):15-20.

[121] Bubenik GA, Blask DE, Brown GM, Maestroni GJ, Pang SF, Reiter RJ, Viswanathan M, Zisapel N: Prospects of the clinical utilization of melatonin. *Biol Signals Recept* 1998; 7(4):195-219.

[122] Vijayalaxmi, Thomas CR, Jr., Reiter RJ, Herman TS: Melatonin: from basic research to cancer treatment clinics. *J Clin Oncol* 2002; 20(10):2575-2601.

[123] Vijayalaxmi, Meltz ML, Reiter RJ, Herman TS: Melatonin and protection from genetic damage in blood and bone marrow: whole-body irradiation studies in mice. *Journal of Pineal Research* 1999; 27(4):221-225.

[124] Vijayalaxmi, Reiter RJ, Meltz ML, Herman TS: Melatonin: possible mechanisms involved in its 'radioprotective' effect. *Mutation Research* 1998; 404(1-2):187-189.

[125] Lissoni P, Barni S, Mandala M, Ardizzoia A, Paolorossi F, Vaghi M, Longarini R, Malugani F, Tancini G: Decreased toxicity and increased efficacy of cancer chemotherapy using the pineal hormone melatonin in metastatic solid tumour patients with poor clinical status. *Eur J Cancer* 1999; 35(12):1688-1692.

[126] Lissoni P, Cazzaniga M, Tancini G, Scardino E, Musci R, Barni S, Maffezzini M, Meroni T, Rocco F, Conti A, Maestroni G: Reversal of clinical resistance to LHRH analogue in metastatic prostate cancer by the pineal hormone melatonin: efficacy of LHRH analogue plus melatonin in patients progressing on LHRH analogue alone. *Eur Urol* 1997; 31(2):178-181.

[127] Tamarkin L, Danforth D, Lichter A, DeMoss E, Cohen M, Chabner B, Lippman M: Decreased nocturnal plasma melatonin peak in patients with estrogen receptor positive breast cancer. *Science* 1982; 216(4549):1003-1005.

[128] Schernhammer ES, Laden F, Speizer FE, Willett WC, Hunter DJ, Kawachi I, Colditz GA: Rotating night shifts and risk of breast cancer in women participating in the nurses' health study. *J Natl Cancer Inst* 2001; 93(20):1563-1568.

[129] Davis S, Mirick DK, Stevens RG: Night shift work, light at night, and risk of breast cancer. *J Natl Cancer Inst* 2001; 93(20):1557-1562.

[130] Carlson LE, Campbell TS, Garland SG, Grossman P: Associations among salivary cortisol, melatonin, catecholamines, sleep quality and stress in women with breast cancer and healthy controls. *J Behav Med.* In press.

[131] Ben-Eliyahu S, Page GG, Yirmiya R, Shakhar G: Evidence that stress and surgical interventions promote tumor development by suppressing natural killer cell activity. *Int J Cancer* 1999; 80(6):880-888.

[132] Costanzo ES, Lutgendorf SK, Sood AK, Anderson B, Sorosky J, Lubaroff DM: Psychosocial factors and interleukin-6 among women with advanced ovarian cancer. *Cancer* 2005; 104(2):305-313.

[133] Thaker PH, Han LY, Kamat AA, Arevalo JM, Takahashi R, Lu C, Jennings NB, rmaiz-Pena G, Bankson JA, Ravoori M, Merritt WM, Lin YG, Mangala LS, Kim TJ, Coleman RL, Landen CN, Li Y, Felix E, Sanguino AM, Newman RA, Lloyd M, Gershenson DM, Kundra V, Lopez-Berestein G, Lutgendorf SK, Cole SW, Sood AK: Chronic stress promotes tumor growth and angiogenesis in a mouse model of ovarian carcinoma. *Nat Med* 2006; 12(8):939-944.

[134] Lutgendorf SK, Logan H, Costanzo E, Lubaroff D: Effects of acute stress, relaxation, and a neurogenic inflammatory stimulus on interleukin-6 in humans. *Brain Behav Immun* 2004; 18(1):55-64.

[135] Zabora J, BrintzenhofeSzoc K, Curbow B, Hooker C, Piantadosi S: The prevalence of psychological distress by cancer site. *Psycho-Oncology 10(1)*:19-28, 2001;-Feb.

[136] Witek-Janusek L, Gabram S, Mathews HL: Psychologic stress, reduced NK cell activity, and cytokine dysregulation in women experiencing diagnostic breast biopsy. *Psychoneuroendocrinology* 2006.

[137] Nagabhushan M, Mathews HL, Witek-Janusek L: Aberrant nuclear expression of AP-1 and NFkappaB in lymphocytes of women stressed by the experience of breast biopsy. *Brain Behav Immun* 2001; 15(1):78-84.

[138] Vos PJ, Visser AP, Garssen B, Duivenvoorden HJ, de Haes HC: Effects of delayed psychosocial interventions versus early psychosocial interventions for women with early stage breast cancer. *Patient Educ Couns* 2006; 60(2):212-219.

[139] Hossain JL, Shapiro CM: The prevalence, cost implications, and management of sleep disorders: an overview. *Sleep Breath* 2002; 6(2):85-102.

[140] Kushida CA, Nichols DA, Simon RD, Young T, Grauke JH, Britzmann JB, Hyde PR, Dement WC: Symptom-Based Prevalence of Sleep Disorders in an Adult Primary Care Population. *Sleep Breath* 2000; 4(1):9-14.

[141] Savard J, Simard S, Blanchet J, Ivers H, Morin CM: Prevalence, clinical characteristics, and risk factors for insomnia in the context of breast cancer. *Sleep* 2001; 24(5):583-590.

[142] Savard J, Morin CM: Insomnia in the context of cancer: a review of a neglected problem. *J Clin Oncol* 2001; 19(3):895-908.

[143] Koopman C, Butler LD, Classen C, Giese-Davis J, Morrow GR, Westendorf J, Banerjee T, Spiegel D: Traumatic stress symptoms among women with recently diagnosed primary breast cancer. *J Trauma Stress* 2002; 15(4):277-287.

[144] Kripke DF, Garfinkel L, Wingard DL, Klauber MR, Marler MR: Mortality associated with sleep duration and insomnia. *Arch Gen Psychiatry* 2002; 59(2):131-136.

[145] Courneya KS: Exercise in cancer survivors: an overview of research. *Med Sci Sports Exerc* 2006; 35:1846-1852.

[146] Courneya KS: Exercise interventions during cancer treatment: biopsychosocial outcomes. *Exerc Sport Sci Rev* 2001; 29(2):60-64.

[147] Tehard B, Friedenreich CM, Oppert JM, Clavel-Chapelon F: Effect of physical activity on women at increased risk of breast cancer: results from the E3N cohort study. *Cancer Epidemiol Biomarkers Prev* 2006; 15(1):57-64.

[148] Farrell SW, Braun L, Barlow CE, Cheng YJ, Blair SN: The relation of body mass index, cardiorespiratory fitness, and all-cause mortality in women. *Obes Res* 2002; 10(6):417-423.

[149] McNeely ML, Campbell KL, Rowe BH, Klassen TP, Mackey JR, Courneya KS: Effects of exercise on breast cancer patients and survivors: a systematic review and meta-analysis. *CMAJ* 2006; 175(1):34-41.

[150] Holmes MD, Chen WY, Feskanich D, Kroenke CH, Colditz GA: Physical activity and survival after breast cancer diagnosis. *JAMA* 2005; 293(20):2479-2486.

[151] Meyerhardt JA, Giovannucci EL, Holmes MD, Chan AT, Chan JA, Colditz GA, Fuchs CS: Physical activity and survival after colorectal cancer diagnosis. *J Clin Oncol* 2006; 24(22):3527-3534.

[152] Key TJ, Schatzkin A, Willett WC, Allen NE, Spencer EA, Travis RC: Diet, nutrition and the prevention of cancer. *Public Health Nutr* 2004; 7(1A):187-200.

[153] Key TJ, Allen NE, Spencer EA, Travis RC: Nutrition and breast cancer. *Breast* 2003; 12(6):412-416.

[154] van den Brandt PA, Goldbohm RA: Nutrition in the prevention of gastrointestinal cancer. *Best Pract Res Clin Gastroenterol* 2006; 20(3):589-603.

[155] Barnard RJ, Aronson WJ: Preclinical models relevant to diet, exercise, and cancer risk. *Recent Results Cancer Res* 2005; 166:47-61.

[156] Carlson LE, Garland SN: Impact of mindfulness-based stress reduction (MBSR) on sleep, mood, stress and fatigue symptoms in cancer outpatients. *International Journal of Behavioral Medicine* 2005; 12(4):278-285.

In: Psychoneuroendocrinology Research Trends
Editor: Martina T. Czerbska, pp. 259-292 ISBN: 978-1-60021-665-7

Chapter VIII

INVESTIGATING ADOLESCENCE AS A SENSITIVE PERIOD FOR STRESS RESPONSE PROGRAMMING

Lisa D. Wright, Kimberly E. Hébert and Tara S. Perrot-Sinal
Psychology Department / Neuroscience Institute, Dalhousie University, 1355 Oxford Street, Life Sciences Centre, Halifax, Nova Scotia, Canada B3H 4J1.

ABSTRACT

Aspects of the hypothalamic-pituitary-adrenal (HPA) stress response system are programmed by experiential factors in rats during the early postnatal period. Important developmental events occur later on, during adolescence, in frontal brain regions that are critical for modulating HPA activity in the adult. However, it is currently unknown to what extent these alterations are sensitive to environmental conditions. In particular, the mesocorticolimbic dopamine system undergoes major restructuring in frontal regions during adolescence, including regions critically involved in regulating HPA output. The overall purpose of the work presented in this chapter is to begin evaluating the potential for environmental conditions to program adolescent development of the stress response system toward a context-specific optimum. In order to manipulate environmental conditions during this period, we have developed a novel, ecologically-relevant adolescent stressor paradigm involving repeated presentation of cat odour (periadolescent predator odour; PPO). This model was used in three separate objectives. In Objective 1 we examined long-term behavioural outcomes in the adult, in comparison with the commonly used maternal separation (MS) paradigm and a sham-separated control group. The purpose of Objective 2 was to examine adolescent corticosterone responses to the cat odour stressor, and in Objective 3 we investigated alterations in adult levels of D1 and D2 dopamine receptors in stress-responsive medial prefrontal cortex (mPFC) subregions. In this study, significant adolescent cort responses to cat odour were measurable in females but not males (Objective 2), and this corresponded with long-term alterations in the behavioural phenotype of PPO-exposed females but not males. Specifically, adult females who had been exposed to PPO exhibited a phenotype characterized by increases in measures of generalized anxiety (Objective 1). In addition, levels of D2 receptors were lower in mPFC subregions of both male and female adults who had been exposed to PPO. These findings support the contention that adolescence is a sensitive period for stress response programming, and a need for extensive future work is indicated.

INTRODUCTION

Environmental Programming and Critical Periods

Features of the environment relevant for survival, such as resource availability and predation threat, play significant roles in guiding development, and therefore, in modulating expression of adult behaviour. Flexible developmental systems have evolved as a survival strategy allowing species to maximize fitness over a range of suitable habitats. Each potential habitat presents a unique set of advantages and challenges. Thus, behaviours that are adaptive in one habitat, with its unique challenges, may not maximize fitness in another. Evolutionary theory predicts that physiological systems with a lengthy period of developmental flexibility have the capacity to act as environmental gauges. The stress response repertoire represents such a system and is particularly amenable to the study of developmental programming.

To date, most of the research examining developmental programming of stress responding has been conducted using manipulations administered prenatally or during the early postnatal period. For example, exposing rodents to early life stressors, such as repeated maternal separation, impacts the development of stress response systems at neural, endocrine, and behavioural levels [1]. In contrast, relatively few studies have examined effects of manipulations administered later, during potentially critical periods of adolescent development. This is despite the fact that environmental demands experienced during adolescence (re: habitat with its balance of advantages and challenges) will be a more reliable predictor of the adult environment. Environmental modulation of adolescent development should therefore steer programming toward optimizing fitness for prevailing environmental demands. Such modulation is likely to affect neural systems that are 'under construction' at that time. Figure 1 represents a schematic overview of the theoretical framework within which our hypotheses for this chapter are based.

Adolescence as a categorical developmental stage is a loose construct, because it actually consists of overlapping neuroendocrine cascades involving adrenal and gonadal hormones and major restructuring of neuronal architecture in frontal regions [2,3]. While these cascades interact once initiated, each has a relatively independent onset and temporal progression [2]. In mammalian species, there are several systematic alterations in behaviour that occur across adolescent development, coinciding with late maturation of the prefrontal cortex (PFC) [4]. These include increases in risk-taking, novelty-seeking, and social behaviours [2], as well as emergence of defensive-type behavioural responses to threat [5], and various other alterations in stress responding [6,5]. In addition, recent findings using impulsive decision-making paradigms indicate that adolescents evaluate rewards differently than adults [2,8]. Mesocortical dopamine (DA) is critically involved in regulating all of these behaviours;

therefore, developmental alterations in PFC affecting components of the DA system might underlie the behavioural changes characteristic of adolescents.

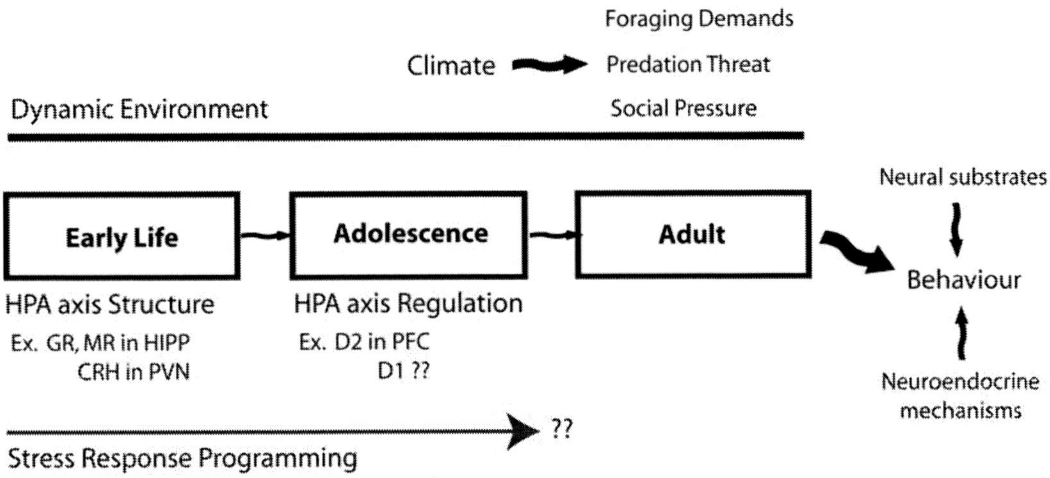

Figure 1. A hypothetical model that provides a framework in which to study stress response programming. Across generations, the environment of most animals is dynamic, as major factors such as climate affect critical variables within species-specific ecological niches. For an adult animal (ie. one that has gained independence) to maximize fitness, behavioural output in response to such changing demands needs to be appropriate. One way in which behavioural responses can be tailored to the present ecological demands is to program the underlying neural and neuroendocrine mechanisms after birth. In other words, physiological systems, such as stress responding, contain a certain level of flexibility that is afforded by epigenetic modulation of components of the system. In the rat, there is ample evidence for a critical period for hypothalamic-pituitary-adrenal (HPA) axis development early in life. For example, certain structural aspects of the HPA axis, such as levels of glucorcorticoid receptors (GR) and mineralocorticoid receptors (MR) in hippocampus (HIPP) and corticotrophic releasing hormone (CRH) levels in paraventricular nucleus of hypothalamus (PVN) are programmed during the first two weeks of life in the rat. Here, we hypothesize that adolescence represents a sensitive period for stress response programming. In particular, we propose that the dopamine (DA) system, which regulates HPA axis function, continues to be programmed throughout adolescence. As we argue in this chapter, adolescence represents the point at which an animal attains independence and has to finally face the environment alone. It is also a time of massive neuroendocrine and neural change, which has consequences for adaptive behaviour. In this chapter, we demonstrate long-lasting changes in D2 receptors in prefrontal cortex (PFC) in animals exposed to repeated adolescent predation threat. We expect that, as we and others further investigate programming of the DA system, we will find programming of other aspects of this system during adolescence.

In humans, expansion of the frontal lobes peaks during adolescence, at which point extensive synaptic and receptor pruning processes begin to shape the mature, adult circuitry [3,9,10]. As mentioned, many biological systems have evolved flexible developmental programs, in order to support successful growth and survival in more than one possible niche. Interactions with specific environmental features play an important role in regulating this flexibility, by guiding the unfolding of genetic events, and a number of mechanisms of action have been identified and described. For example, methylation and acetylation patterns on specific regions of DNA are sensitive to environmental factors and determine how tightly compacted chromatin will be, and thus, how accessible various genes will be to transcriptional machinery. Such processes have been shown to be sensitive to maternal care during early postnatal life in rats, programming specific aspects of adult stress responding [11].

The temporal progression of cortical development and the behaviours it regulates during adolescence invites opportunity for environmental conditions to steer flexible developmental programmes toward a context-specific optimum. At this time, however, we know relatively very little of the mechanisms underlying adolescent cortical development. In this chapter, we will introduce the idea that the adolescent period may be a critical period for developmental programming of certain aspects of the adult stress response. In particular, it is our contention that the PFC DA system may be programmed during adolescence by environmental conditions, such that responding of this system in adulthood will be tuned to such conditions.

Introduction to Stress Responding

The hypothalamic-pituitary-adrenal (HPA) axis is but one of a number of systems in the body's repertoire to maintain homeostasis in response to stressor exposure. However, as the key regulator of glucocorticoid secretion, it is considered very important. Limbic and prefrontal brain regions (ex. amygdala, hippocampus, infralimbic cortex), and hypothalamic nuclei devoted to sensory processing, send projections to the paraventricular nucleus of the hypothalamus (PVN), which go on to activate the pituitary gland. Glucocorticoid secretion from the adrenal cortex is the culmination of HPA axis activation and plays a major role in ensuring a state of 'restoration' following challenges [12]. It has been proposed that cell function, mental performance, and general health are dependent on the integrated orchestration of glucocorticoids acting on two intracellular receptors: mineralocorticoid receptors (MR) and glucocorticoid receptors (GR) [13]. The major glucocorticoids, cortisol and corticosterone, bind with much higher affinity to MR than GR [14], which are co-localized in several structures including the hippocampus, amygdala and also PFC [15].

Glucocorticoids play an active role in organizing behavioural strategies during environmental perturbations [16]. Allostasis is defined as 'maintaining stability through change' and is used to describe the dynamic process by which the body responds to homeostatic threats. The concept can be extended to include 'allostatic load', the cumulative physical costs of allostasis ('wear and tear' of maintaining homeostasis over time). The frequency and severity of challenges encountered, as well as the efficiency of response systems, will therefore determine how quickly allostatic load accumulates for each individual. An inability to deal with conditions leading to increased allostatic load can culminate in pathological conditions as a result of excess release of glucocorticoids and other mediatiors of allostasis [17,18]. Imbalance of MR and GR can lead to disease states, such as unwarranted levels of anxiety [13], by decreasing negative feedback of the HPA axis, resulting in exaggerated stress responses. Glucocorticoids are catabolic, and excessive stress responding promotes development of a diverse array of disease processes, including cardiovascular, metabolic, and mental illnesses [19].

It is critical for long-lived animals to deal with changing conditions *during* a lifespan and in many species, major unpredictable perturbations initiate behavioural strategies (for example, proactive/reactive coping styles; flight/fight responses to rapid emergencies) that serve to deal with the perturbation and to avoid chronically high levels of glucocorticoids (avoid allostatic overload) [16]. The role that developmental programming plays in the initiation of such behavioural strategies in adulthood is unknown. A main objective of this chapter is to evaluate the potential for such programming to occur during the adolescent period.

Adolescence as a Critical Period for Stress Response Programming

Aspects of the stress response are programmed during early life. Childhood abuse and neglect have long-term effects on adult physical and mental health by altering the HPA axis [20,21]. In laboratory rats, the first two weeks are critical for programming systems important in adult behavioural and neural stress responding [22-26]. For example, daily short-term (3 - 15 min) separations of pups from the dam during the first 1-2 weeks of life (early handling; EH) results in blunted adrenal stress responses and reduced fear- and anxiety-related behaviour in adulthood, relative to pups left undisturbed until weaning (not handled; NH) [27,28]. This manipulation also attenuates adult DAergic responses to stress in nucleus accumbens [29].

While PFC development during adolescence is being vigorously studied, there is still relatively little conclusive information available on changes in expression levels of DA signaling markers. Nor is there information on whether or not developmental changes are sensitive to environmental factors. PFC volume increases through to adolescence, with a later peak for white matter increases than for gray matter, reflective of late axonal myelination. As in other cortical and subcortical regions, extensive synaptic pruning occurs in PFC during adolescence [2]. This is reflected by decreases in whole PFC [4] and in white matter volumes [3] towards the latter end of the adolescent period.

Presuming these pruning events are Hebbian in nature, the possibility of programming during adolescence is supported. Late myelination may aid in strengthening connections exhibiting increased activity, while those less active die away. Features of the adolescent environment, such as levels of adversity, could impact this process by influencing neuronal activity, through DAergic and/or other neurotransmitter signaling. We know that DA responses are evoked in PFC during adolescence even in response to mild psychological stressors [30], and therefore, levels of adversity in the adolescent environment could program aspects of the DA system during the adolescent period. In adulthood, the overall effect of stress-induced DA release in PFC is a reduction in neural output from this region, which can effectively augment limbic control over behavioural responses. Because DA receptors are co-localized with various other receptors in PFC, the effects of their activation on neural activity when DA is released can be conceptualized to function as a biological gearshift. DA released in response to stress kicks the PFC into high gear, for instance, via interaction with D1 receptors on inhibitory interneurons in stress-responsive subregions. This would establish a requirement for more excitatory neurotransmission (more glutamate release, for example) in order to boost activity in these circuits to pre-stress levels. While this possibility is only one simplified illustration, it demonstrates a potential mechanism by which levels of environmental adversity could modulate development of circuits that regulate PFC DA-mediated negative feedback on HPA activity. The infralimbic cortex is thought to play a major role in this effect, is sensitive to early life environmental conditions, and appears to regulate adaptation to repeated stress [31]. Given the massive and widespread, sex-specific changes that occur in DA system structure during the adolescent period, repeated stress experienced at this time might be expected to permanently alter synaptic stabilization in specific PFC subregions by altering the balance of activity between DA-responsive and non-responsive connections. This could provide a means by which PFC is programmed toward an adult phenotype best suited to cope with stressful conditions. McCormick and colleagues have shown that pubertal social stress can interact with nicotine sensitization to alter the adult corticosterone response to immobilization stress in a sex-dependent manner [32]. Furthermore, they have provided evidence that pubertal social stress enhances adult sensitivity to drugs of abuse in females only [33].

Role of Dopamine in Execution of Cognitive Tasks Mediated by the Prefrontal Cortex

The evolution of human consciousness may be viewed as a sort of capitalization on one adaptive behavioural strategy: cognition. Although responding to a stressor is partly reflexive, it often includes cognitive appraisal/evaluation of the situation. Cognitive processing allows information to be gathered from the environment so that relevant material can be integrated with internal cues, such as emotional state and level of physiological arousal, in order to adaptively guide behaviour.

Consciousness and flexibility in human behaviour have arisen alongside an extensive, but punctuated, hominid cortical expansion. Long periods of evolutionary stasis in early hominid populations (see [34]) were interrupted by massive changes in the quality of the surrounding environment. As cumulative change occurs in factors regulating climate, critical points are reached that induce major alterations in global metereological systems. A pattern has emerged of sudden flips every 100 000 years or so, between two stable climates: warm and wet (i.e. present day) and cool and dry (i.e. ice age). Rapid climate shifts are thought to have exerted pressure for selection of enhanced cognitive capacity/capabilities by imposing resource limitations and challenges not previously encountered ([35] see William H. Calvin's book "A Brain for All Seasons: Human Evolution & Abrupt Climate Change" for a synthesis of these ideas). At those points of sudden change, only individuals clever enough to adopt novel coping strategies survived and reproduced.

A number of periods of heavy selection pressure over the course of hominid evolution (i.e. following shifts into cool and dry periods or other major environmental disturbances) have therefore resulted in an exquisite refinement of the neural substratum mediating cognitive abilities: the PFC. While this refinement consists partially of a volumetric expansion of the anterior frontal lobes in the primate lineage [4], a capacious PFC is not wholly responsible for the advent of human consciousness. Modern magnetic resonance imaging (MRI) techniques allow allometric comparisons among living primate species, adding to what has been learned from paleontological data of total brain volume increases in hominids. Using a trajectory derived from nonhuman anthropoid primates, the human frontal cortex is larger than expected in relation to body size and whole brain volume [36]. However, it was later shown that humans and great apes share the same frontal to whole cortex volumetric ratio [37]. These investigations support a notion long posited by psychologists and, more recently, by neuroscientists - that refinement in functional aspects of PFC must have been required in addition to its expansion, in order to support the emergence of human conscious thought.

PFC-mediated neural processes related to higher cognition have been investigated using functional MRI studies in humans, as well as electrophysiological and/or behavioural studies in animal models (for reviews, see [38,39]). Using these techniques, specific components of cognitive functions can be isolated for study. For example, attention and working memory processes, whereby information is retained "on-line" for short-term use, are concepts intimately related to the idea of a cognitive percept. Brain function is examined during execution of tasks designed to tap these processes, thereby providing neural correlates of cognition. While the primate dorsolateral PFC is of primary importance for execution of cognitive tasks employing attention and working memory processes, the medial PFC (mPFC) plays an analogous role in rodents (see [38]). These broader areas are comprised of subregions with distinct cytoarchitecture and connectivity [40], and these accordingly play differential roles in execution of cognitive tasks. The human PFC boasts an extremely rich connectivity to other brain regions, and diverse connections among its subregions.

Work directed toward elucidating neurochemical mechanisms of action underlying goal-directed behaviour points toward an integral role for mesocortical DAergic drive [38,41-43]. Various DA receptor subtypes are present within the subregions of PFC, and downstream signal transduction events induced by DA depend on the type and location of the activated receptors.

DA receptors are G-protein coupled receptors embedded in cellular membranes. In general, they are classified as D1-type receptors, which include D1 and D5 subtypes, and D2-type receptors, which include D2, D3, and D4 subtypes. The major difference between the two types is that activation of D1-type receptors by ligand binding induces changes in membrane potential that bring a neuron closer to it's firing threshold (depolarizing effects), whereas activation of D2-type receptors move a neuron further from it's firing threshold (hyperpolarizing effects). Therefore, D1-type receptors are considered excitatory and D2-type receptors are inhibitory. D1, D2, and D4 subtypes are all present within PFC [42,44], with D1 receptors being expressed to a greater extent than D2-like on excitatory principle pyramidal neurons [38]. Both types are also expressed on γ-aminobutyric acid (GABA)ergic interneurons, which act to hyperpolarize cells onto which they synapse. Therefore, excitatory and inhibitory receptors exist on excitatory and inhibitory neurons within subregions of PFC. This complexity hints at the varied nature of DA's effects on neural activity within prefrontal circuits.

Pharmacological studies in rodents can be used to explore functional aspects of DA signaling. For instance, much information has been gleaned by locally administering DA receptor agonists and antagonists into the PFC during execution of cognitive tasks. Furthermore, the use of drugs selective for specific receptor subtypes has allowed a delineation of each subtype's involvement in different tasks. Working memory processes, for example, can be assessed in rodents using delayed response tasks in T-, Y-, or radial arm mazes. These paradigms take advantage of the role of the PFC in working memory tasks that require storage of information across a delay period.

The spontaneous alternation paradigm is such a task. Rats can be trained to move down the start arm of a T-maze and select between two distinctive but equally rewarded goal arms. Over time, rats will tend to alternate the two maze arm choices across successive trials, which is frequently interpreted as a preference for novelty [45,46]. In order to alternate maze arm choices successfully, the arm chosen on the previous trial must be recalled after an ~30 s delay period, and prefrontally guided working memory processes are primary subservients of this function [47]. There is a body of literature describing induction of a compulsive behavioural profile in rodents via sensitization to the mixed D2/D3 dopamine receptor agonist quinpirole (QP). Since various forms of compulsive behaviour can be induced in different test environments, we examined the role of the context of drug exposure in the induction of perseverative behaviour in a T-maze. Female rats exposed to the T-maze context during the sensitization procedure perseverated maze arm choices more than would be expected by chance, while males actually showed the opposite pattern and *alternated* maze arm choices if they had received prior context exposure (Figure 2). Control rats were also sensitized to QP but were returned immediately to the homecage following each dose, and these did not show proficient spontaneous alternation in a single T-maze test session. Rather, both males and females alternated arm choices at levels close to chance (Figure 2).

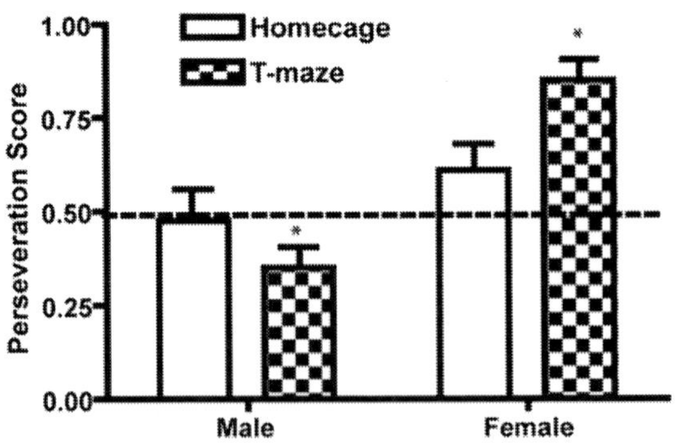

Figure 2. Over a 3-week period, groups of male and female rats were sensitized to quinpirole hydrochloride (QP), via repeated drug administration (0.5 mg/kg; s.c.) followed on each occasion by either: 1) return to the familiar homecage environment or 2) one hour of exposure to a T-maze test apparatus. Following the sensitization period, rats were tested for spontaneous alternation (or perseveration) of maze-arm choices in nine, delayed-response (~30 s), equally rewarded trials. Effects of context exposure (Homecage vs. T-maze) during drug sensitization were sexually divergent, with females demonstrating more perseveration and males more alternation than would be expected by chance (dashed line) in the T-maze task. *$p<0.05$ from chance levels.

Studies using these types of tasks involving short delay intervals (<2 min) have demonstrated an integral role for D1 receptor activation in the "on-line" maintenance of information across the delay. Moreover, the effects of D1 activation on successful responding produce an inverted-U function, such that too much or too little activation deteriorates performance [48,49]. Electrophysiological monitoring demonstrates that D1 activation modulates persistent, task-related neural firing in PFC circuits [42,50]. Principle pyramidal neurons, which form connections with inhibitory interneurons, show persistent firing across a delay period. D2 receptor activation, on the other hand, does not appear to be as important in these processes [50], although it is important for other types of working memory tasks (discussed below).

Other studies have used 30-minute delay intervals, and these have demonstrated a critical role for D1 activation in PFC for the manipulation and retrieval of task-relevant information in trained rats [38]. Information unique to each trial is likely stored in hippocampus using this longer delay paradigm [42,51], demonstrating an important role for D1 in various components of working memory processes.

Interestingly, the inverted-U function describing effects of D1 activation on task performance may apply only to some of these components, or may be shifted depending on specific task demands. Local administration of a D1 agonist into the PFC impaired performance of trained rats using short delay periods; however, the same dose enhanced performance when the delay period was extended enough to degrade baseline performance levels [38,43].

Working memory processes are intimately related to attention. Both attention and working memory are conceptualized as emergent properties of processes mediated by PFC [39]. Accordingly, in addition to influencing working memory processes, D1 receptor activity in PFC influences performance on tasks designed to tax attentional processes, while D2 receptors do not appear to have a profound impact on either. While blockade of D1 receptors in medial (m)PFC impaired attentional accuracy in rodents when baseline performance was good, performance was improved with administration of D1 agonists when baseline performance was poor [52]. Behavioural flexibility, a relatively recent concept that has been shown to rely heavily on PFC DA innervation, is also reliant on attentional processes [38]. Simple reversal learning, for example, requires alternation between two response choices in order to receive rewards. Learning to alternate in these tasks depends on the integrity of the orbital PFC [53]. The mPFC, in contrast, is tapped when rats are trained on more complex tasks requiring different aspects of attention, such as attentional set shifting.

In attentional set shifting tasks, rats are required to first base a decision on information gathered in one 'stimulus dimension' (e.g. visual cue), and then to switch the basis of the next decision to information gathered in a different dimension (e.g. tactile cue). Successful performance therefore requires both the inhibition of a previous response and the acquisition of a new attentional set in a different sensory modality. Pharmacological studies using local administration of receptor agonists and antagonists have revealed important involvement of both D1 and D2 receptor activation in successful set shifting, as well as a selective effect of D4 receptor activation in antagonizing the effects over behavioural flexibility exerted cooperatively by D1 and D2 [38]. More specifically, enhancement of D4 activity leads to persistent focus in one dimension and, thus, impaired set shifting.

The reward component of these tasks is an important consideration, as the DA system is classically known to mediate reward signaling. However, DA signaling in PFC is arguably more associated with cue salience, as opposed to reward (i.e., hedonia) *per se*. Reward predictors evoke the DA response in PFC that is normally emitted in response to the actual reward, using classical conditioning paradigms [54]. Thus, attentional processes are implicated in PFC DA-mediated judgment of the reward-relevance of a stimulus, since both are bound to DA action in PFC circuits.

Rodent behavioural tasks are also used to determine the roles of PFC DA receptors in impulsive decision-making. Tasks that require rats to choose between an immediate small reward and a larger delayed reward are used to provide an index of impulsivity, and systemic administration of DA agonists increases the preference for the larger, delayed reward [55]. Reward evaluation in this paradigm, however, does not incorporate a punishment component reflective of the 'down-side' that usually exists naturally when reward influences impulsive decision-making. Floresco and Magyar (2006) created a model to better capture this process using a conditioned punishment paradigm. In rats, a tone that predicts an electric shock creates an aversion to the tone alone. If rats are then trained to press levers for rewards, they will forego the optimal foraging strategy, in order to avoid pressing levers associated with the aversive tone. D1, D2, and D4 receptors in mPFC all contribute to this effect [38]. DA signaling in PFC is proposed to mediate a cost-benefit analysis upon presentation of rewards, as altering behaviour to attain rewards disrupts homeostasis.

In summary of this section, the PFC is a brain region that mediates higher-level cognitive processes, and does so via DA signaling. These cognitive processes involve behavioural flexibility, a feature of organisms that may be programmed by environmental conditions during sensitive developmental periods.

Prefrontal Cortex Dopaminergic Modulation of Stress Responding

DA is a key player mediating stress responding [31]. DA release in PFC has been confirmed by microdialysis in behaving animals in response to handling and other psychological stressors [56,57]. As stimuli become stress provoking via association with punishment, such as in the conditioned punishment paradigm described above, their avoidance is also mediated by DA signaling in PFC [38]. This demonstrates a role for DA signaling in response to both controllable and uncontrollable stressors. As mentioned above, enhanced DA transmission in PFC increases the tolerance of delay to attain a reward. In accordance with this, DA depletion in PFC differentially alters responses to negative events, depending on whether they are controllable or uncontrollable, with increased punished responding in two paradigms involving controllable stressors [58].

Fear- and anxiety-related behavioural output during stressful conditions has come to be understood in terms of a regulatory balance between limbic drive and cortical drive within the stress response circuitry. The hippocampus and amygdala each connect reciprocally to PFC, and there is evidence that both these limbic structures relay contextual information during stress responding. However, learned contextual associations are stored within the hippocampus when environmental cues come to predict reward (food, mates) or challenge (predators, competitors). The hippocampus is therefore involved in storing training-related information gathered during contextual fear conditioning [59], while the amygdala appears to play a more general role in memory consolidation of emotionally arousing experiences [60,61]. Fear extinction occurs when a conditioned stimulus is repeatedly presented without the original negative association, and this sort of adaptive response is mediated by PFC output to the amygdala [62]. The PFC therefore integrates its inputs dynamically during stressful conditions, in order to assess the current emotional and physical state of the organism and direct adaptive coping strategies to minimize allostatic load.

Taken together, these findings imply that DAergic signaling in PFC is very important for cognitive integration of contextual information, and thus, adaptively guiding behaviour associated with stress responding. Drawing on our earlier analogy of the biological gearshift, gearing up can be accomplished by the activation of inhibitory D2 receptors on principle pyramidal neurons, which increases the change in membrane potential required to induce neuronal firing in these cells (hyperpolarization). Alternatively, gearing up could occur via activation of excitatory D1 receptors located on inhibitory interneurons synaptically connected to the principle pyramidal neurons. Taking the analogy further, hyperpolarization of principle pyramidal neurons in either of these cases results in a requirement for increased drive, in order to transmit equal amounts of energy (neural activity).

As discussed above, there are cost-benefit health trade-offs that are inherent in stress responding. A stress response mounted to deal with a particular stressor in one environmental context may not be necessary or sufficient for successful coping within another context. Simply put, stress responding needs to be efficient and specific for the stressor encountered. Therefore, organisms with the capability to cognitively fine-tune stress responses in accordance with environmental assessment will be selected for, by avoiding unnecessary energy expenditure. Species occupying broad and varied niches would be expected to show more fine-tuning capability than species that occupy very specific niches, since environmental features would be more unpredictable.

Dopamine Signaling and Developmental Alterations of Prefrontal Cortex Across the Adolescent Period

Studies in rats demonstrate that pruning of DA receptors occurs in PFC during adolescence [63], and this is reflected by receptor subtype specific changes in mRNA expression levels across this period [64]. Changes in levels of DA receptors in other brain regions follow different developmental trajectories, depending on the region [65]. Sex of the organism is also a critical factor in some cases [66], implying potential involvement of gonadal steroids in some regions. In addition, D1 and *N*-methyl-*d*-aspartate (NMDA) receptors acquire the ability to instate persistent depolarization upon co-activation in PFC pyramidal neurons during adolescence [67]. Very little is currently known about how these events relate to developmental changes in behaviours mediated by PFC, or which aspects of these processes are sensitive to environmental modulation. There is, however, a wealth of evidence supporting environmental interaction with earlier developmental processes and related behavioural output [1,68,69].

A characteristic sex-specific profile of behavioural change across adolescence is conserved across mammalian species, implying deep evolutionary roots, despite the fact that some manifestations, such as an increased propensity to abuse drugs, may be viewed as maladaptive. This type of manifestation could be an artifact of environmental sensitivity in developing reward circuitry. There are also a number of human neuropsychiatric illnesses that first present during adolescence, and their prevalence rates and symptomatology often differ between sexes [70,71]. While many of these illnesses have a genetically basis, environmental buffering is purported to play a protective role in predisposed individuals (stress effects, environmental enrichment; [6,72]). These clues together suggest that adolescence may in fact be a sensitive period for development of PFC.

In the search for aspects of adolescent PFC development that might be constrained by environmental factors, markers of DA signaling should figure prominently, given DA's widespread role in integrating cognitive neural representations. For example, there may be changes in baseline expression levels of one or more DA receptor subtypes, the DA transporter, or catechol-O-methyltransferase (COMT), which is responsible for most DA degradation in PFC [73,74], in one or more subregions. Downstream changes in expression of mediators of DA signaling following receptor activation may also occur, since sensitivity of this and other receptor-effector systems is known to vary with factors other than receptor number [75].

Using an Ecologically Relevant Approach to Examine Developmental Programming During Adolescence

In order to examine the possibility that adverse environmental conditions experienced during adolescence can program aspects of adult stress responding, we chose to expose male and female rats to repeated predation threat (cat odour) during the periadolescent period (periadolescent predator odour (PPO)). Predation threat is a natural means to mimic adverse environmental conditions in laboratory animals. Predatory cues, such as cat odour, robustly induce well-characterized changes in defensive and non-defensive behaviours in many species of rodent [76-79], as well as in HPA axis function [76,80-82]. As such, predatory cues can be used to assess adaptive stress responding in adult animals. This model offers many advantages over other commonly used models of stress, by incorporating a purely psychological stressor that induces robust and quantifiable behavioural responses that are measurable *during* the stressor exposure [79,83]. The more commonly used restraint model, in contrast, does not offer the possibility of measuring *any* behavioural responses during exposure, nor is it a stressor to which rats would be expected to have evolved characteristic defense tactics. In the same vein, another commonly used stressor model, electric footshock, also lacks ecological validity, since most wild rats would never be confronted with this. Thus, we wouldn't expect that such stressors would produce the same effects on developmental programming.

In each of the following objectives, the PPO stressor model was administered to adolescent rats. In order to avoid repeating the methodology, the basics are outlined here, and any deviations from this procedure are pointed out in the methods for the specific objectives. Rats were exposed with cagemates to predator odour (PO) or control stimulus for 30 minutes on five occasions (PND's 41, 42, 44, 47, and 48) between PND 40 and 48, a developmental timeframe considered to represent late adolescence. During the dark phase of the light:dark cycle, rats were transported to a dark, quiet room where they were placed with the cagemate(s) into clean, clear, 60 x 27 x 22 cm Plexiglas arenas containing the odour source.

The odour source was attached to the centre of one end wall with an alligator clip located 2 cm above the floor. Clear Plexiglas lids covered the arenas during exposures. The PO stimulus used for this study was prepared fresh before each exposure, by coating 2.5 x 15 cm strips of non-antibacterial disposable cloth with cat hair and dander acquired from 2-4 female, reproductively active, domestic cats housed in the Psychology Department. Control stimuli were clean strips of cloth. Arenas were cleaned thoroughly with unscented lab soap following each use. All experimental procedures described for each objective were performed in accordance with the guidelines of the Canadian Council on Animal Care and were approved by the Dalhousie University Committee on Laboratory Animals. Data were analyzed using analysis of variance (ANOVA) with sex and experimental group as between-subject factors and test period as a within-subject factor, when applicable. Simple effects analyses and post-hoc testing were conducted where appropriate.

Objective One – Comparison of the Effects of Neonatal Maternal Separation Versus Periadolesent Predator Odour on Adult Stress-Related Behavioural Phenotypes

Specific Objective and Background

The first objective of this study was to compare the long-term effects of neonatal maternal separation (MS) and periadolescent predator odour exposure (PPO; described above) on open field and defensive behaviours in adult male and female rats.

MS paradigms (typically 2 - 4 h/day) have been used to alter stress-related environmental parameters during neonatal life, a period critical for development of certain aspects of stress response circuitry. The adult phenotype of MS animals is characterized by altered corticosterone, nucleus accumbens DA, and behavioural responses to stressors [29,84-86]. A body of literature supports the idea of exaggerated stress responding in MS animals; however, there are also recent reports describing decreased anxiety-related responses in offspring exposed to MS [87,88]. Differences in the long-term effects of MS depend on factors such as the specific stress-related behavioural measure, as well as the sex and strain of rat [87-89]. Furthermore, dissociation is often observed to exist between endocrine and behavioural effects (i.e. MS can induce long-term alterations in one without obvious change in the other). Behaviour is related to neural activity in regulatory brain circuits, which can be influenced but not controlled by changes in the hormonal milieu; thus, more work is required for understanding relationships between altered HPA activity and changes in other aspects of neural stress responding, such as mesocortical DA function. Here, we use a natural stressor model, exposure to predator odour [79], to investigate adult defensive behaviour. This will be a useful tool for decoding the neurobiological bases of MS effects on adult stress-related behaviour.

Methods

Male and female Sprague-Dawley rats were bred in a colony room in the Life Sciences Centre at Dalhousie University. Males and females were paired for five days, after which time the male was removed and females were monitored twice a day, beginning 19 days following the first day of pairing. On the day of birth, male and female offspring were assigned for exposure to one of two conditions (n = 6 / group): 1) maternal separation (MS), 2) repeated periadolescent predator odour (PPO) exposure. Two additional control groups (n = 6 / group) were tested. All pups were weaned at 23 d of age and maintained in groups of 2-3 from that point forward. Body weights of animals were measured periodically across the lifespan.

MS paradigm - Rats in the MS group received daily 3h maternal separations beginning on the day after birth (PND1) and continuing for 10 days. During a separation period, all pups within a litter were removed from the dam in the colony room, transferred to a clean cage, and transported to a dark, quiet room. Pups were left undisturbed on heating pads warmed to 32 °C. Separations took place during the dark phase of the light cycle (~1000 - 1300 h). The control group for this stressor consisted of sham-separated pups that were picked up and immediately replaced with the dam at the beginning and end of separations.

PPO paradigm: As described at the end of the Introduction section, rats in the PPO group were exposed to PO on five occasions (PND's 41, 42, 44, 47, and 48).

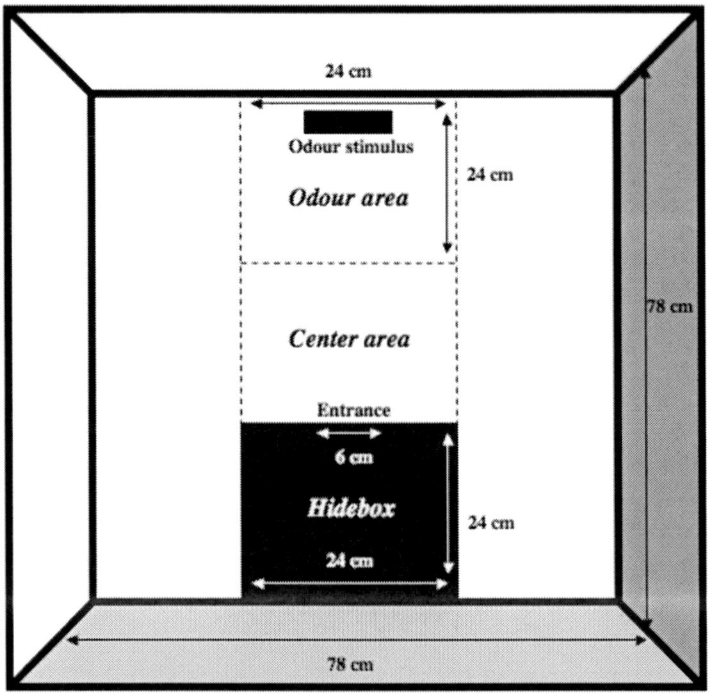

Figure 3. Behavioural test arena used for open field and predator odour testing. The arena was divided into a number of virtual regions depicted by dashed lines, and movement among these, scored as line crosses, was used to measure general activity. The odour (or control) stimulus was present only during the predator odour test.

Open-field behaviour in adulthood: Adult testing occurred when rats were between 3.5-4 months of age. Activity within an open field was measured for two days: a Novel day (rats' first exposure to this environment) and Day 2 (familiar environment 24 h later). The open field consisted of a black Plexiglas chamber (78 x 78 x 35 cm), with a bare Plexiglas floor and a clear Plexiglas lid with holes for air circulation (see Figure 3). A black Plexiglas hide box (HB; 24 x 24 x 22 cm) was in the center of one wall. Rats were placed into the center of the open field facing the HB at the beginning of a test session and allowed to remain in the chamber for 20 minutes. The chamber was washed with unscented laboratory soap, rinsed, and dried after each test. The video recorder was suspended from the ceiling above the test chamber during all testing.

Predator odour stress test in adulthood: Rats' defensive and fear-related behaviour was examined on the day following Day 2 of open-field testing. Rats were tested in a control odour condition (clean piece of cloth), followed immediately by a predator odour condition (cloth strip as described for PPO stressor), in the same test chamber that was used for open-field testing (see above for description). The order of stimulus presentation cannot be counter-balanced in this test, due to the rapid formation in rodents of conditioned aversions to a context in which a predation threat is encountered. During each condition, the cloth strip was placed in the test chamber at the opposite side from the HB. Rats were placed into the chamber facing the HB, and allowed to remain in the chamber for 10 min in each condition. The video camera was suspended above the chamber, and behaviour was recorded for each condition.

Results

There was a main effect of treatment (MS vs. PPO) on male body weight changes for two periods, one across late adolescence (PND40 - 53/54) and one from late adolescence (PND53/54) into adulthood (end point PND99 - 103; ($F(3,17)=7.397$, $p=0.002$). Results of Games-Howell post-hoc testing revealed that PPO exposed males gained less weight during the weigh periods, compared to control males ($p<0.05$). PPO males also tended to gain less weight than MS males ($p=0.06$); however, this difference did not reach significance, due to high variability in the MS males that stemmed from 2/6 males who gained drastically less weight during the late adolescent period (12% and 5% increases in body weight, relative to a group mean of 58%) but drastically more weight in adulthood (122% and 155%, relative to a group mean of 78%).

Analysis of behaviours exhibited across two 20 min open-field sessions administered on consecutive days, and a predator odour stress response test administered the following day, revealed that the MS and PPO paradigms each produced unique alterations in adult behavioural phenotypes, relative to a sham-separated control group. Furthermore, the effects of each treatment were sexually divergent, with MS exerting effects exclusively on adult behaviour in male littermates, and PPO exerting effects exclusively on females. Specifically, MS affected attention-related behavioural measures in males, while PPO affected defensive and risk assessment behaviours in females.

During open-field testing, MS males reared more frequently than sham-separated males (p=0.007; see Figure 4A). During the cat odour session of predator odour testing, MS males entered the area containing the odour source significantly more often than sham-separated males (Group X Session interaction p=0.051; see Figure 4B), despite the fact that there were no differences in the overall duration of time spent within this area. Thus, MS males made repeated short duration entries into the odour area. These results could be interpreted as suggesting that MS males were avoiding the odour source as much as sham-separated males (because of the similar total duration spent in the odour area) but were making more frequent assessments of the threat.

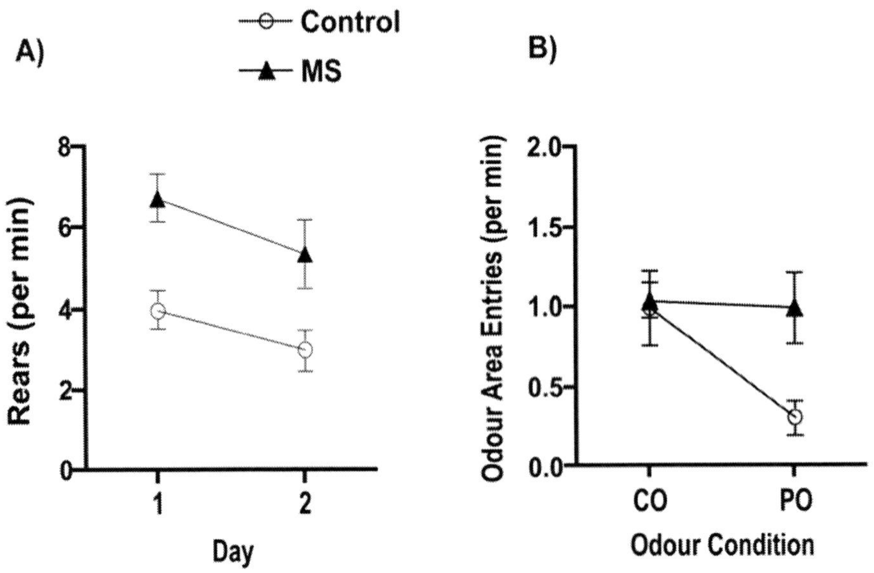

Figure 4. A. Maternally separated (MS) males showed an increased frequency of rearing during a 20 min open-field test (p=0.007). B. During the predator odour test, MS males displayed an increased rate of entry into the odour area when exposed to the cat odour stimulus, relative to a sham-separated comparison group (Control; Group X Session interaction p=0.051). However, both MS and Control rats avoided spending time within the odour area (data not shown). These behavioural alterations suggest that maternal separation influenced development of adult attention-related behaviours in males specifically, since these effects were not seen in female littermates. The increased rearing rate is indicative of increased shifts of attention during open field exploration. Increased cat odour approaches suggest failure to recall the location of the stimulus, rather than a failure to recognize the significance of the odour.

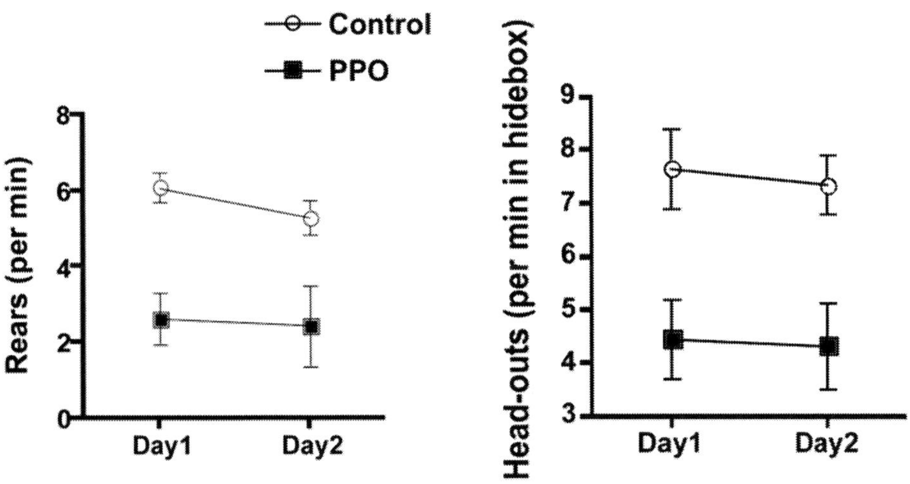

Figure 5. Across two 20 min open field sessions administered on consecutive days, adult females who had received periadolescent predator odour (PPO) exposure exhibited a phenotype that was characterized by increased levels of behavioural inhibition and vigilance. Relative to a sham-separated comparison group (Control), rates of rearing were significantly lower (A; p=0.022), while the total duration of hidebox (HB) time spent in the head-out position was significantly higher (B; p=0.001), for PPO females. These effects were not present in male littermates who had received the same PPO treatment.

PPO females, on the other hand, exhibited a phenotype characterized by increased levels of behavioural inhibition and vigilance. This was displayed in response to a novel open field (i.e., Day 1 of open-field testing) and did not habituate with a second exposure to the same apparatus (i.e., Day 2; see Figure 5). Compared with sham-separated females, PPO females exhibited lower rates of rearing (p=0.022) and spent more time engaged in a head-out position while in the HB (p=0.001), across both days of testing. Across both control and cat odour exposure conditions during predator odour testing (see Figure 6), PPO females displayed lower rates of line crossing (a measure of general activity; p=0.009), rearing (p=0.011), and entries into the odour area (p=0.029). They also spent less time rearing (p=0.036). PPO females showed increased rates of entry into the HB (p=0.027), spent more time within the HB (p=0.009), and spent more time engaged in a head-out position while in the HB (p=0.013). None of these effects were observed in male littermates who had received the same PPO treatment.

Summary

MS and PPO paradigms each had long-term effects on aspects of behaviour exhibited in the open field and during predator odour stress testing in adulthood. Three patterns emerged from the results: 1) there were differences in the types of behaviours impacted by each paradigm (attention-related behaviours affected by MS vs. defense/anxiety-related behaviours affected by PPO), 2) the results had consequences for adaptive responding to predatory threat, and 3) there were sex-specific effects. These three features will be discussed briefly for each paradigm in turn, within the context of current opinion regarding the validity of the MS model as an early life stressor and with an aim toward evaluating the usefulness of the PPO manipulation as a means to program development during later, critical periods.

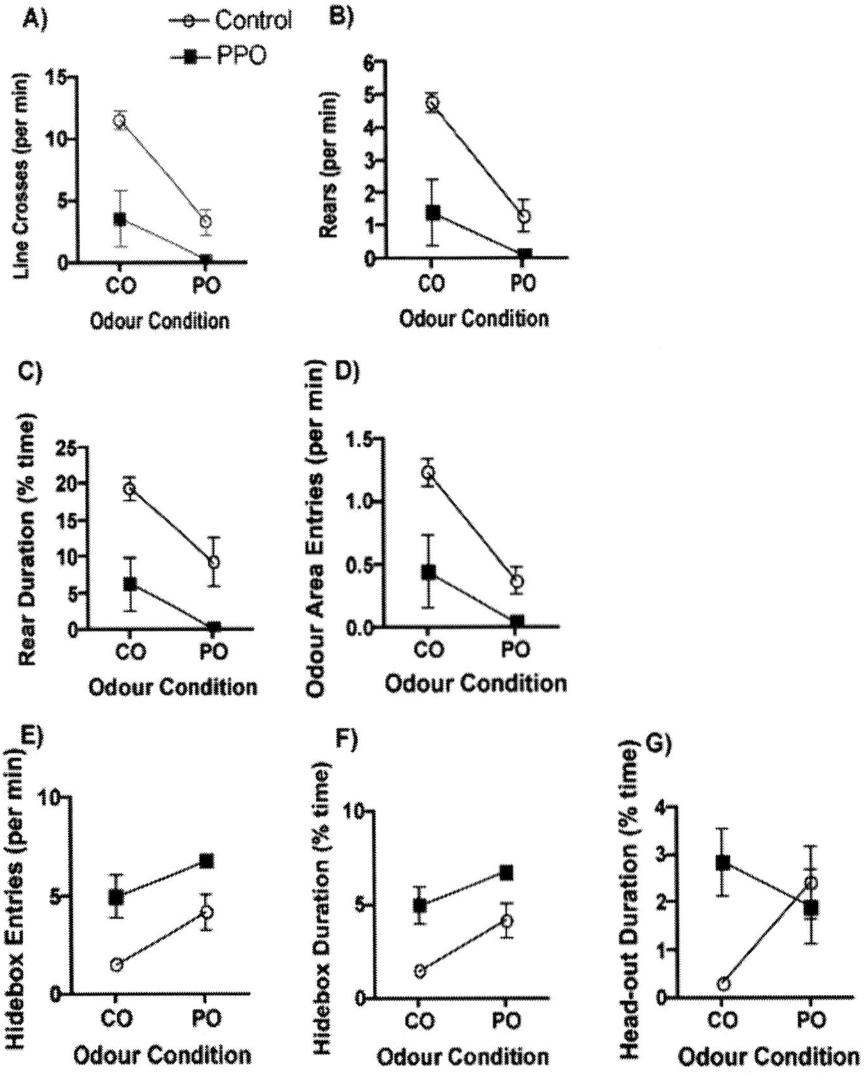

Figure 6. The response to predator odour stressor in adult females is programmed during adolescence. Adult females who had received periadolescent predator odour (PPO) exposure exhibited a phenotype characterized by increased levels of behavioural inhibition, vigilance, and risk assessment, in comparison with a sham-separated comparison group (Control). This was inferred from significant reductions in rates of line crossing (A; a general measure of activity; p=0.009), rates of rearing (B; p=0.011), total duration rearing (C; p=0.036), and rates of odour area entrance (D; p=0.029). PPO females displayed increased rates of entrance into the hidebox (E; p=0.027) and higher duration spent within the hidebox (F; p=0.009), as well as increased head-out durations (G; p=0.013). These effects were absent in male littermates who had received the same PPO treatment.

In this study, MS impacted orienting and impulsive (risk-taking) behaviours in males only. Males who had received MS treatment showed higher frequencies of rearing during open-field testing and also approached the odour source more frequently during cat odour testing, relative to control animals. It might be suggested that these behavioural alterations are related to changes in the regulation of attentional processes. MS males were not simply hyperactive, as they did not show increased movement (line crosses) during behavioural testing. They also did not show increases in rearing duration, despite the increased frequencies of this behaviour. This indicates a selective increase in interruptions of rearing behaviour. Furthermore, they did not spend more time than control animals in the vicinity of the predator odour source during stress testing, indicating that they retained an avoidance response, despite the increased number of approaches. This pattern of findings is suggestive of increased attention shifting in MS males.

Colorado *et al.* (2006) recently conducted a study comparing adolescent behaviours in MS, EH, and standard facility reared (SFR) groups. Interestingly, MS *decreased* adolescent orienting behaviour (both frequencies and durations of rearing), but increased measures of impulsivity. It is difficult to determine how these findings might relate to the adult behavioural profile produced by MS in our study, however, because several methodological issues may contribute significantly to outcomes in individual MS studies [91,92]. Colorado *et al.* (2006) began testing their animals at PND 26, which represents an early phase of adolescence. Furthermore, while their MS and EH groups each consisted of an equal number of males and females, their SFR group was solely male [90]. Sex differences were not analyzed [90], despite a plethora of evidence demonstrating such differences in effects of early environment on adult stress-related behaviour (e.g. [92,93]). This is only one example of the sorts of inconsistencies that exist throughout the vast MS/EH literature. Slotten *et al.* (2006) conducted a study examining adult behavioural outcomes in MS males and females. They found that MS males had shorter emergence latencies than nonhandled controls in a T-maze task, but MS males did not show reductions in anxiety-related behaviour on the elevated plus maze [93], suggestive again of increased impulsivity and less focused attention. Another group found *enhanced* selective attention in MS animals, as evidenced by improved avoidance learning in a latent inhibition paradigm [94]. Given such variation in the direction of behavioural effects following application of MS procedures, it might be useful to step back and first consider the *types* of behaviours that are sensitive to alteration by these manipulations during this period, rather than attributing specific outcomes to the stress of experiencing MS. In this regard, important observations in MS animals include: frequent sex differences in effects when these are examined and sensitivity within systems that will ultimately regulate attention- and impulse-related behaviours. Such alterations appear to impact adaptive behavioural responding to a cat odour stressor, since MS males in our study were more willing to engage in risky approach behaviours.

Adult behaviour in females, on the other hand, was impacted exclusively and pervasively by the PPO paradigm. Relative to control animals, PPO increased the display of defensive and risk assessment behaviours, as well as overall behavioural inhibition (reduced display of non-defensive behaviours), across all adult behavioural test conditions. This female behavioural profile suggests a PPO-induced increase in generalized anxiety in adulthood. While the long-term allostatic cost of this behavioural profile may be high, vigorous responses were mounted in the presence of the cat odour stressor in adulthood, which would be considered adaptive within a threatening context.

The pattern of findings obtained for this objective therefore supports the use of a PPO paradigm in modeling exposure to a purely psychological stressor (perception of predation threat) in the environment. The findings also suggest that PPO may interact with late developmental events in neural and endocrine stress response systems, producing an increase in generalized anxiety in adult females. We are currently examining whether these long-term behavioural effects of PPO are dependent on its administration during the adolescent period, or whether a similar effect might be obtained with repeated exposure to predator odour in early adulthood. This distinction will help clarify potential biological bases for the altered behavioural profile.

Objective 2: Determination of Corticosterone Responses to Predator Odour on the First and Last Day of a Repeated Periadolescent Exposure Paradigm

Specific Objective and Background

The second objective of this study was to determine plasma corticosterone (cort) levels in adolescent male and female rats following PPO exposure.

Cat odour stimuli, as used for each of our objectives, consistently produce defensive responses in behavioural test paradigms. The specific behavioural profile exhibited upon stressor (cat odour) exposure depends on environmental parameters. In our lab, responses of adult rats exposed within a prone environmental context are typically characterized by behavioural inhibition of non-defensive behaviours (rearing, grooming, line crosses). Provision of a protective HB allows measurement of hiding and risk assessment/vigilance behaviours, such as HB usage, and display of 'head-out' posture whereby a rat surveys the environment from this protected vantage point (see Results for Objective 1).

Adolescence is becoming generally recognized as a period critical for increased vulnerability to stressors [6]. It is also a time of neurodevelopmental alteration in stress sensitive mesocorticolimbic DA function [2]. However, work examining the neurobiological dynamics of these changes across adolescence is still limited. While it is known that regulation of endocrine responses to stressors changes across adolescence [95], only a crude characterization of these changes can be derived from the scant literature currently available. Furthermore, there is accumulating evidence demonstrating environmental sensitivity in some of these adolescent developmental processes [68,96,97], as well as interaction with early environment [98]. Much more work needs to be done examining endocrine stress responding during adolescence, using natural stressor models such as the cat odour model employed herein.

Our objective for the present study was to quantify plasma cort levels following exposure to cat odour during adolescence. Cort levels were examined in two conditions, first following acute exposure to cat odour in mid-adolescence (PND 40), which was compared with a control-odour exposed group. Exposures were then repeated for each group, as outlined for the PPO manipulation, and cort levels were examined again following a fifth exposure on PND 48, in the latter phase of adolescence.

Methods

Male and female Long-Evans rats purchased from Charles River, Canada were used as breeders to produce three litters (using similar husbandry practices as described for Objective 1), and all offspring were raised to adolescence. Until weaning, pups were left undisturbed with the dam in standard 22 x 24 x 48 cm polypropylene cages, except for once weekly cage changing, and were then housed in same-sex littermate pairs or in a group of three, if necessary. Cages were covered with wire lids and contained wood-chip bedding, and an ~5-inch piece polyvinyl-carbonate tubing for enrichment, as well as paper towel nesting material before weaning. The colony room was maintained at 20 ± 1 °C on a 12:12 reverse light:dark cycle (lights off at 0930). Food (Purina Lab Chow) and tap water were available *ad libitum*. Prior to initiation of experimental manipulations, rats were handled for six days (PND 34 to 39 inclusive) for at least 2 min per rat. This always occurred during the rats' subjective night (active period).

PPO stressor paradigm: Males and females from each litter were randomly allocated to a predator odour (PO) exposure group (n=11 males and 10 females) or a control exposure group (n=10 males and 12 females). Details of the stressor paradigm are outlined at the end of the Introduction section.

Assessment of plasma corticosterone levels: Rats were transported in their home cages from the colony room to the laboratory, and blood samples were collected from the saphenous vein of each rat into 600 µl Microtainer plasma separator tubes containing lithium heparin (Becton Dickinson, United States) within one hour of light period onset, the circadian nadir of circulating cort, on PND's 37 and 45. These samples were used to determine physiological baseline levels of cort secretion. Experimental samples were collected between 1000h and 1600h in the same fashion, ~5 min following the first (PND 40) and last (PND 48) exposure session. Immediately following blood collection, samples were centrifuged at 6000 x g for two minutes at 4 °C. Separated blood plasma was then pipetted into three 80 µl aliquots, which were frozen at -80 °C until assay. For assay using the Correlate-EIA Corticosterone Enzyme Immunoassay Kit (Assay Designs, Michigan, USA), samples were diluted 1:50 with assay buffer containing steroid displacement reagent and run in duplicate according to the manufacturer's instructions. Samples were kept chilled on ice during preparation.

Data manipulation: Cort levels determined from experimental samples are theoretically representative of circulating cort plus bound cort released by displacement reagent during assay, and these were normalized to respective group physiological baseline means. Normalized group means were then compared statistically as described in the Introduction.

Results

Data were excluded from analyses if variability calculated from duplicate samples was beyond a predetermined criterion based on the coefficient of variance. Final sample sizes ranged from 5-10 per group. Males and females were placed into the exposure context every second day between PND's 38-48, and a PO or control stimulus was introduced following a 5 min habituation period on each occasion. Responses to the first and last stimulus exposure were normalized to group baseline cort levels determined from samples taken on PND 37 and 45, respectively, at the circadian nadir of cort secretion. In the control-exposed animals, cort levels were higher following the last exposure, relative to the first (p=0.01; see Figure 7), a pattern not evident in the PO-exposed rats. Overall, cort levels were higher in females, relative to males (p=0.012; see Figure 7). Accordingly, only females showed an overall significant response to PO (p=0.028). Furthermore, there was a trend toward a Session X PO interaction for females (p=0.079), which is reflective of an initial response to PO that subsequently habituated with repeated exposures (see Figure 7). A similar trend toward a Session X PO interaction was seen in males (p=0.063), though in this case, the effect was driven primarily by the increased cort response in the control animals on the final exposure day. This increase appears to be suppressed in PO-exposed males.

Summary

Two findings of interest emerged here: 1) cort responses increased across the adolescent period in rats exposed to the control condition, but not those exposed to PPO, and 2) only adolescent females showed a significant cort response to predator odour.

With regard to the first finding, it cannot be determined from results of this study alone whether the increase in cort levels in the control animals is an expression of normal developmental change in endocrine function, or is instead related to experience with the exposure procedure. Unfortunately, there has been little work characterizing changes in HPA axis development across adolescence.

Adolescent cort responses to predator odour were observed exclusively in females, which is interesting given the sex specificity of long-term behavioural effects induced by the PPO paradigm (see Objective 1). While there is clearly an elevated cort response in PPO females upon initial (acute) cat odour exposure on PND 40, it appears to have dissipated by PND 48 (see Figure 7). We are currently investigating whether PPO also impacts adult cort responses. Sandstrom and Hart (2005) examined cort responding in male rats exposed to a social isolation stress paradigm during the third postnatal week. They found increases in plasma cort levels during the isolation period that lasted into adulthood [99].

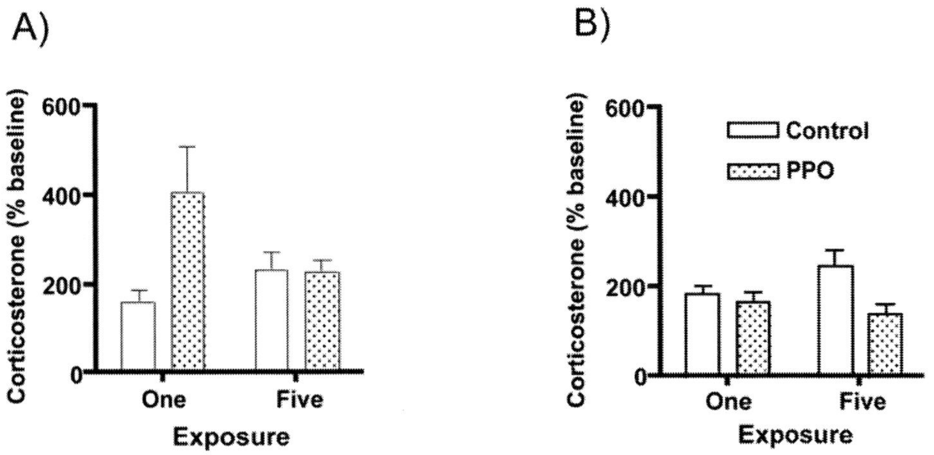

Figure 7. Corticosterone (cort) levels were assayed (see Objective 2 – Methods, for further details) in periadolescent rat blood samples taken following the first and last of five control or predator odour (PO) exposures administered across postnatal days (PND) 40 - 48. Cort levels were normalized to a physiological group mean baseline determined from samples taken on PND37 or 45, at the circadian nadir of circulating cort. Females (A) showed higher overall cort responses as compared with males (B; p=0.012). In addition, PO-exposed females showed higher responses than control-exposed females (p=0.028) and PO-exposed males (p=0.004). Only control-exposed rats (both males and females) demonstrated an increase in the cort response across the adolescent period (p=0.010).

Fos-immunoreactivity is a crude measure of neural activity that can be used to compare relative involvement of different cell populations in responding to external stimuli. Kellogg *et al.* (1998) used this technique to show that several brain regions, including cortical and medial amygdaloid nuclei, are brought 'online' during stress responding within the periadolescent period. That is, they begin to contribute input to a coordinated stress response that is dependent on the particular environmental challenge in question. For responses to immobilization, this occurs at some point between PND28 and 60 (i.e. neural populations within said nuclei begin to show Fos immunoreactivity following restraint [100]). Sims and Holberton (2000) investigated development of cort responding in wild Northern Mockingbirds by examining cort levels in seven age classes following the acute stress of capture and handling. They found increased cort responses as birds approached the age of independence. It was suggested that this effect was attributable to developmental changes at the neural level, since adrenals of young nestlings showed capacity to increase cort secretion following intrajugular injection of adrenocorticotrophic hormone (ACTH; [101]). Our study suggests that female rats have developed the capacity to mount a cort response to a cat odour stressor by PND40, whereas males do not show a similar sensitivity to the same stressor at this timepoint. Future studies will include a more detailed examination of corticosterone release to PPO during adolescence and examination of Fos immunoreactivity in various stress-responsive brain regions following PPO exposure.

Objective Three – Examination of Adult Expression of Dopamine D1 and D2 Receptors in Prefrontal Cortex Following Repeated PPO Exposure

Specific Objective and Background

The final objective of the present investigation was to examine changes in adult expression of DA receptor subtypes 1 and 2 in prefrontal cortex in male and female rats that had been exposed to PPO.

As outlined above, expression of DA receptors is dynamically altered in prefrontal and limbic brain regions across adolescent development. Patterns of change follow different developmental trajectories, depending on the brain region, with some showing evidence of sensitivity to sex or experiential factors [63,65,66]. Work examining environmental influences on these developmental processes is still very limited, however. In particular, no previous study has examined levels of D1 and D2 DA receptors in mPFC of adult rats that had been exposed to repeated adolescent stress, using a natural stressor model. Infralimbic DA is highly responsive to stressors and particularly to those that evoke strong emotional arousal [102,103]. Sullivan and Dufresne (2006) recently demonstrated involvement of the IL D1/D2 receptor system in regulating feedback inhibition of HPA axis stress responding [31]. Given the widespread involvement of PFC in stress responding, stress experienced during adolescence could interact with developmental processes that regulate DA receptor levels. An understanding of stress-related environmental impact on development of mesocorticolimbic DA function will have broad implications, given DA's widespread role in execution of cognitive processes involving attention and executive function, as well as in regulating cognitive control over stress responding.

The present experiment was undertaken to begin investigating levels of expression of D1 and D2 receptors in infralimbic cortex, following repeated exposure to a natural cat odour stressor during adolescence.

Methods

Husbandry and housing procedures were the same as those described for Objective 2.

PPO stressor paradigm: As described at the end of the Introduction section, Long Evans rats in the PPO group were exposed to PO or control on five occasions (PND's 41, 42, 44, 47, and 48). In the present objective, the PPO group consisted of 3 males and 5 females, and the control group consisted of 3 males and 6 females.

Tissue collection: Animals were decapitated in early adulthood (PND62) following complete anesthesia in a CO_2 chamber. Brains were rapidly removed, flash frozen, and stored at –80 °C until processing. Specimens were warmed to -4 °C, and frontal regions between +3.20 to +2.20 relative to Bregma were sliced on a cryostat into thick sections (200 - 400 μm) with the knife temperature set at -10 °C. Sections were mounted on glass slides chilled by dry ice, and the infralimbic and dorsal peduncular cortical subregions of PFC were micropunched out using an 18 gauge blunted needle attached to a syringe. The syringe was filled with air before taking punches for sections from each brain, which permitted the punches to then be expelled from the needle into a microtube. Tubes were placed immediately on dry ice and stored at –80 °C until further processing, and the needle was cleaned with ethanol following PFC tissue collection from each animal.

Western immunoblotting: PFC micropunches from each animal were homogenized in 40 μl chilled lysis buffer consisting of 50 mM Tris-HCl, 0.25% Na-deoxycholate, 1% w/v Triton X-100, 150 mM NaCl, 1mM EDTA, 1 mM activated Na-orthovanadate, and a protease inhibitor cocktail (1 μg/ml aprotinin, leupeptin, and pepstatin; 1 mM phenylmethylsulfonyl fluoride). This involved manual disruption with a teflon pestle, followed by 10 min of ultrasonication in an ice-cold water bath. Total protein concentrations were determined using the colorimetric method of Bradford. Western blotting methods were similar to those described previously [104]. Briefly, protein (30 μg) from each sample was loaded onto 10% SDS-polyacrylamide gels that were cast the preceding evening and stored covered by 0.1% SDS at 4 °C. Electrophoresis was conducted for 3 hours at 125 V, and separated protein was then transferred to a polyvinyl difluoride membrane (BioRad, Hercules, CA). Membranes were washed briefly in 0.1 M Tris-buffered saline containing 0.05% Triton X-100 (TTBS) and then blocked with constant agitation for one hour at room temperature in TTBS containing 4% non-fat dry milk. Membranes were then incubated with agitation in 1:1000 rabbit anti-D1 dopamine receptor (D1DR: Santa Cruz) and 1:20000 monoclonal anti-glyceraldehyde-3-phosphate dehydrogenase (Gapdh; Chemicon; used as a loading control) overnight at 4 °C. The following day, membranes were washed three times for 5 min each in TTBS and then incubated in a goat anti-rabbit and goat anti-mouse horseradish peroxidase-conjugated IgG's in TTBS with agitation for 30 min at room temperature, followed by another three 5-min rinses in TTBS and 2 rinses in TBS. Immunoreactive bands were detected using an enhanced chemiluminescence kit (New England Biolabs, Inc., Beverly, MA), and membranes were exposed using Kodak Imagestation 440. D1DR was detected as a single band at the expected molecular weight of 74 kDa and was analyzed relative to Gapdh intensity. The following day, blots were stripped and reincubated with rabbit anti-D2 dopamine receptor (D2DR: Santa Cruz) antibodies (1:1000). Blots were visualized as above, and the band corresponding to a molecular weight of 51 kDa was analyzed.

Results

While adult levels of D1 receptor expression in PFC were not affected, expression of D2 receptors in this brain region was significantly decreased following repeated periadolescent exposure to PO (p=0.019; see Figure 8).

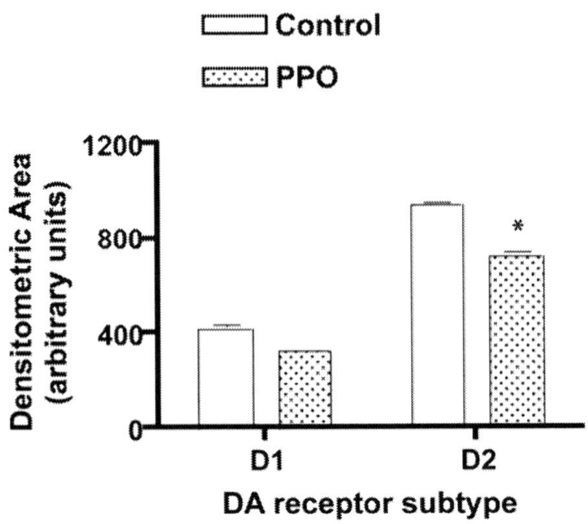

Figure 8. The periadolescent predator odour (PPO) manipulation resulted in a reduction in adult levels of D2 dopamine receptors in infralimbic/dorsopeduncular cortices of both males and females, relative to sham-separated controls (*p=0.019). In contrast, there were no differences in levels of D1 DA receptors between PPO and control animals.

Summary

As far as we know, we are the first to show a change in adult levels of DA receptors in infralimbic and dorsopeduncular cortical subregions of PFC in male and female rats that had been exposed to a natural stressor during the latter phase of adolescence. This effect of PPO was not sex-specific. A next step for this objective will be to label D2 receptors in PFC of PPO-exposed animals immunohistochemically, in order to visualize their cellular and subcellular locations within infralimbic and dorsopeduncular cortices. Signal transduction mechanisms utilized by the D2 receptor are known for specific populations of neurons (extracellular signal regulated kinase (ERK)-dependent pathway is one example). Furthermore, activation of D2 receptors on forebrain neural stem cells inhibits cell proliferation, decreasing rates of neurogenesis [105], while activation of D2 receptors on cells in other neurogenerative brain regions has been shown to increase cell differentiation and maturation, providing the basis for a growth factor-like role of DA. Thus, localization of the increased D2 receptor protein will aid in determining functional implications and relationships to PPO-induced alterations in adult behaviour.

Considering the widespread role of PFC D2 receptors in attention, working memory, and impulsivity tasks (see Introduction), and demonstration of a role for DA receptors in the infralimbic cortex in mediating negative feedback regulation of HPA output, PPO-induced long-term alterations in baseline levels of D2 receptor expression in this region and other PFC regions could have broad consequences for cognitive function in general, as well as for cognitive control of stress-related behaviour.

CONCLUSIONS

We have demonstrated some preliminary work characterizing a novel way in which to study the role of adolescent programming on adult stress-related responses. These results suggest that adolescence is a sensitive period during which features of the environment program adult defensive behaviour and the DA system in PFC.

ACKNOWLEDGEMENTS

The authors would like to acknowledge the Natural Sciences and Engineering Research Council of Canada, the Canadian Institutes of Health Research, and the Nova Scotia Health Research Foundation (NSHRF) for funding (to TSPS) and a studentship (NSHRF, to LDW) to complete this work.

REFERENCES

[1] Bock, J., M. Gruss, S. Becker, and K. Braun, Experience-induced changes of dendritic spine densities in the prefrontal and sensory cortex: correlation with developmental time windows. *Cereb Cortex*, 2005. 15(6): p. 802-8.

[2] Spear, L.P., The adolescent brain and age-related behavioral manifestations. *Neurosci Biobehav Rev*, 2000. 24(4): p. 417-63.

[3] Nagel, B.J., K.L. Medina, J. Yoshii, A.D. Schweinsburg, I. Moadab, and S.F. Tapert, Age-related changes in prefrontal white matter volume across adolescence. *Neuroreport*, 2006. 17(13): p. 1427-31.

[4] Fuster, J.M., Frontal lobe and cognitive development. *J Neurocytol*, 2002. 31(3-5): p. 373-85.

[5] Hubbard, D.T., D.C. Blanchard, M. Yang, C.M. Markham, A. Gervacio, I.L. Chun, and R.J. Blanchard, Development of defensive behavior and conditioning to cat odor in the rat. *Physiol Behav*, 2004. 80(4): p. 525-30.

[6] Charmandari, E., C. Tsigos, and G. Chrousos, Endocrinology of the stress response. *Annu Rev Physiol*, 2005. 67: p. 259-84.

[7] Laviola, G., W. Adriani, S. Morley-Fletcher, and M.L. Terranova, Peculiar response of adolescent mice to acute and chronic stress and to amphetamine: evidence of sex differences. *Behav Brain Res*, 2002. 130(1-2): p. 117-25.

[8] Volkow, N.D., What do we know about drug addiction? *Am J Psychiatry*, 2005. 162(8): p. 1401-2.

[9] Giedd, J.N., Structural magnetic resonance imaging of the adolescent brain. *Ann N Y Acad Sci*, 2004. 1021: p. 77-85.

[10] Seeman, P., Images in neuroscience. Brain development, X: pruning during development. *Am J Psychiatry*, 1999. 156(2): p. 168.

[11] Cameron, N.M., F.A. Champagne, C. Parent, E.W. Fish, K. Ozaki-Kuroda, and M.J. Meaney, The programming of individual differences in defensive responses and reproductive strategies in the rat through variations in maternal care. *Neurosci Biobehav Rev*, 2005. 29(4-5): p. 843-65.

[12] Herman, J.P., H. Figueiredo, N.K. Mueller, Y. Ulrich-Lai, M.M. Ostrander, D.C. Choi, and W.E. Cullinan, Central mechanisms of stress integration: hierarchical circuitry controlling hypothalamo-pituitary-adrenocortical responsiveness. *Front Neuroendocrinol*, 2003. 24(3): p. 151-80.

[13] De Kloet, E.R., Hormones and the stressed brain. *Ann N Y Acad Sci*, 2004. 1018: p. 1-15.

[14] Reul, J.M. and E.R. de Kloet, Two receptor systems for corticosterone in rat brain: microdistribution and differential occupation. *Endocrinology*, 1985. 117(6): p. 2505-11.

[15] Helm, K.A., J.S. Han, and M. Gallagher, Effects of cholinergic lesions produced by infusions of 192 IgG-saporin on glucocorticoid receptor mRNA expression in hippocampus and medial prefrontal cortex of the rat. *Neuroscience*, 2002. 115(3): p. 765-74.

[16] Wingfield, J.C., Control of behavioural strategies for capricious environments. *Anim Behav*, 2003. 66: p. 807-816.

[17] McEwen, B.S., Early life influences on life-long patterns of behavior and health. *Ment Retard Dev Disabil Res Rev*, 2003. 9(3): p. 149-54.

[18] McEwen, B.S., Protection and damage from acute and chronic stress: allostasis and allostatic overload and relevance to the pathophysiology of psychiatric disorders. *Ann N Y Acad Sci*, 2004. 1032: p. 1-7.

[19] Heuser, I. and C.H. Lammers, Stress and the brain. *Neurobiol Aging*, 2003. 24 Suppl 1: p. S69-76; discussion S81-2.

[20] Nemeroff, C.B., Early-Life Adversity, CRF Dysregulation, and Vulnerability to Mood and Anxiety Disorders. *Psychopharmacol Bull*, 2004. 38 Suppl 1: p. 14-20.

[21] Nemeroff, C.B. and W.W. Vale, The neurobiology of depression: inroads to treatment and new drug discovery. *J Clin Psychiatry*, 2005. 66 Suppl 7: p. 5-13.

[22] Levine, S., M. Alpert, and G.W. Lewis, Infantile experience and the maturation of the pituitary adrenal axis. *Science*, 1957. 126(3287): p. 1347.

[23] Levine, S., Infantile experience and resistance to physiological stress. *Science*, 1957. 126(3270): p. 405.

[24] Gutman, D.A. and C.B. Nemeroff, Neurobiology of early life stress: rodent studies. *Semin Clin Neuropsychiatry*, 2002. 7(2): p. 89-95.

[25] Liu, D., J. Diorio, B. Tannenbaum, C. Caldji, D. Francis, A. Freedman, S. Sharma, D. Pearson, P.M. Plotsky, and M.J. Meaney, Maternal care, hippocampal glucocorticoid receptors, and hypothalamic-pituitary-adrenal responses to stress. *Science*, 1997. 277(5332): p. 1659-62.

[26] Francis, D., J. Diorio, D. Liu, and M.J. Meaney, Nongenomic transmission across generations of maternal behavior and stress responses in the rat. *Science*, 1999. 286(5442): p. 1155-8.

[27] Levine, S., Developmental determinants of sensitivity and resistance to stress. *Psychoneuroendocrinology*, 2005.

[28] Caldji, C., J. Diorio, and M.J. Meaney, Variations in maternal care in infancy regulate the development of stress reactivity. *Biol Psychiatry*, 2000. 48(12): p. 1164-74.

[29] Brake, W.G., T.Y. Zhang, J. Diorio, M.J. Meaney, and A. Gratton, Influence of early postnatal rearing conditions on mesocorticolimbic dopamine and behavioural responses to psychostimulants and stressors in adult rats. *Eur J Neurosci*, 2004. 19(7): p. 1863-74.

[30] Elsworth, J.D., B.A. Morrow, and R.H. Roth, Prenatal cocaine exposure increases mesoprefrontal dopamine neuron responsivity to mild stress. *Synapse*, 2001. 42(2): p. 80-3.

[31] Sullivan, R.M. and M.M. Dufresne, Mesocortical dopamine and HPA axis regulation: role of laterality and early environment. *Brain Res*, 2006. 1076(1): p. 49-59.

[32] McCormick, C.M., D. Robarts, E. Gleason, and J.E. Kelsey, Stress during adolescence enhances locomotor sensitization to nicotine in adulthood in female, but not male, rats. *Horm Behav*, 2004. 46(4): p. 458-66.

[33] McCormick, C.M., D. Robarts, K. Kopeikina, and J.E. Kelsey, Long-lasting, sex- and age-specific effects of social stressors on corticosterone responses to restraint and on locomotor responses to psychostimulants in rats. *Horm Behav*, 2005. 48(1): p. 64-74.

[34] Schwartz, J.H., Race and the odd history of human paleontology. *Anat Rec B New Anat*, 2006. 289(6): p. 225-40.

[35] Schrago, C.G., On the time scale of new world primate diversification. *Am J Phys Anthropol*, 2006.

[36] Rilling, J.K. and T.R. Insel, The primate neocortex in comparative perspective using magnetic resonance imaging. *J Hum Evol*, 1999. 37(2): p. 191-223.

[37] Semendeferi, K., A. Lu, N. Schenker, and H. Damasio, Humans and great apes share a large frontal cortex. *Nat Neurosci*, 2002. 5(3): p. 272-6.

[38] Floresco, S.B. and O. Magyar, Mesocortical dopamine modulation of executive functions: beyond working memory. *Psychopharmacology (Berl)*, 2006. 188(4): p. 567-85.

[39] Postle, B.R., Working memory as an emergent property of the mind and brain. *Neuroscience*, 2006. 139(1): p. 23-38.

[40] Ongur, D. and J.L. Price, The organization of networks within the orbital and medial prefrontal cortex of rats, monkeys and humans. *Cereb Cortex*, 2000. 10(3): p. 206-19.

[41] Seamans, J.K., S.B. Floresco, and A.G. Phillips, D1 receptor modulation of hippocampal-prefrontal cortical circuits integrating spatial memory with executive functions in the rat. *J Neurosci*, 1998. 18(4): p. 1613-21.

[42] Goldman-Rakic, P.S., Regional and cellular fractionation of working memory. *Proc Natl Acad Sci U S A*, 1996. 93(24): p. 13473-80.

[43] Chudasama, Y. and T.W. Robbins, Dopaminergic modulation of visual attention and working memory in the rodent prefrontal cortex. *Neuropsychopharmacology*, 2004. 29(9): p. 1628-36.

[44] Wedzony, K., A. Chocyk, M. Mackowiak, K. Fijal, and A. Czyrak, Cortical localization of dopamine D4 receptors in the rat brain--immunocytochemical study. *J Physiol Pharmacol*, 2000. 51(2): p. 205-21.

[45] Lalonde, R., The neurobiological basis of spontaneous alternation. *Neurosci Biobehav Rev*, 2002. 26(1): p. 91-104.
[46] Montgomery, K.C., A test of two explanations of spontaneous alternation. *J Comp Physiol Psychol*, 1952. 45: p. 287-293.
[47] Sanchez-Santed, F., J.P. de Bruin, R.P. Heinsbroek, and R.W. Verwer, Spatial delayed alternation of rats in a T-maze: effects of neurotoxic lesions of the medial prefrontal cortex and of T-maze rotations. *Behav Brain Res*, 1997. 84(1-2): p. 73-9.
[48] Zahrt, J., J.R. Taylor, R.G. Mathew, and A.F. Arnsten, Supranormal stimulation of D1 dopamine receptors in the rodent prefrontal cortex impairs spatial working memory performance. *J Neurosci*, 1997. 17(21): p. 8528-35.
[49] Arnsten, A.F., Catecholamine regulation of the prefrontal cortex. *J Psychopharmacol*, 1997. 11(2): p. 151-62.
[50] Williams, G.V. and P.S. Goldman-Rakic, Modulation of memory fields by dopamine D1 receptors in prefrontal cortex. *Nature*, 1995. 376(6541): p. 572-5.
[51] Lee, I. and R.P. Kesner, Time-dependent relationship between the dorsal hippocampus and the prefrontal cortex in spatial memory. *J Neurosci*, 2003. 23(4): p. 1517-23.
[52] Granon, S., F. Passetti, K.L. Thomas, J.W. Dalley, B.J. Everitt, and T.W. Robbins, Enhanced and impaired attentional performance after infusion of D1 dopaminergic receptor agents into rat prefrontal cortex. *J Neurosci*, 2000. 20(3): p. 1208-15.
[53] McAlonan, K. and V.J. Brown, Orbital prefrontal cortex mediates reversal learning and not attentional set shifting in the rat. *Behav Brain Res*, 2003. 146(1-2): p. 97-103.
[54] Berridge, K.C., The debate over dopamine's role in reward: the case for incentive salience. *Psychopharmacology (Berl)*, 2006.
[55] van Gaalen, M.M., R. van Koten, A.N. Schoffelmeer, and L.J. Vanderschuren, Critical involvement of dopaminergic neurotransmission in impulsive decision making. *Biol Psychiatry*, 2006. 60(1): p. 66-73.
[56] Matsumoto, M., H. Togashi, A. Kaku, M. Kanno, K. Tahara, and M. Yoshioka, Cortical GABAergic regulation of dopaminergic responses to psychological stress in the rat dorsolateral striatum. *Synapse*, 2005. 56(2): p. 117-21.
[57] Pehek, E.A., C. Nocjar, B.L. Roth, T.A. Byrd, and O.S. Mabrouk, Evidence for the preferential involvement of 5-HT2A serotonin receptors in stress- and drug-induced dopamine release in the rat medial prefrontal cortex. *Neuropsychopharmacology*, 2006. 31(2): p. 265-77.
[58] Ravard, S., P. Carnoy, D. Herve, J.P. Tassin, M.H. Thiebot, and P. Soubrie, Involvement of prefrontal dopamine neurones in behavioural blockade induced by controllable vs uncontrollable negative events in rats. *Behav Brain Res*, 1990. 37(1): p. 9-18.
[59] Pentkowski, N.S., D.C. Blanchard, C. Lever, Y. Litvin, and R.J. Blanchard, Effects of lesions to the dorsal and ventral hippocampus on defensive behaviors in rats. *Eur J Neurosci*, 2006. 23(8): p. 2185-96.
[60] Malin, E.L. and J.L. McGaugh, Differential involvement of the hippocampus, anterior cingulate cortex, and basolateral amygdala in memory for context and footshock. *Proc Natl Acad Sci U S A*, 2006. 103(6): p. 1959-63.

[61] Davidson, R.J. and W. Irwin, The functional neuroanatomy of emotion and affective style. *Trends Cogn Sci*, 1999. 3(1): p. 11-21.
[62] Sotres-Bayon, F., D.E. Bush, and J.E. LeDoux, Emotional perseveration: an update on prefrontal-amygdala interactions in fear extinction. *Learn Mem*, 2004. 11(5): p. 525-35.
[63] Andersen, S.L., A.T. Thompson, M. Rutstein, J.C. Hostetter, and M.H. Teicher, Dopamine receptor pruning in prefrontal cortex during the periadolescent period in rats. *Synapse,* 2000. 37(2): p. 167-9.
[64] Weickert, C.S., M.J. Webster, P. Gondipalli, D. Rothmond, R.J. Fatula, M.M. Herman, J.E. Kleinman, and M. Akil, Postnatal alterations in dopaminergic markers in the human prefrontal cortex. *Neuroscience*, 2007. 144(3): p. 1109-19.
[65] Teicher, M.H., S.L. Andersen, and J.C. Hostetter, Jr., Evidence for dopamine receptor pruning between adolescence and adulthood in striatum but not nucleus accumbens. *Brain Res Dev Brain Res*, 1995. 89(2): p. 167-72.
[66] Andersen, S.L. and M.H. Teicher, Sex differences in dopamine receptors and their relevance to ADHD. *Neurosci Biobehav Rev*, 2000. 24(1): p. 137-41.
[67] Tseng, K.Y. and P. O'Donnell, Post-pubertal emergence of prefrontal cortical up states induced by D1-NMDA co-activation. *Cereb Cortex*, 2005. 15(1): p. 49-57.
[68] Neddens, J., F. Bagorda, A. Busche, S. Horstmann, G.H. Moll, R.R. Dawirs, and G. Teuchert-Noodt, Epigenetic factors differentially influence postnatal maturation of serotonin (5-HT) innervation in cerebral cortex of gerbils: interaction of rearing conditions and early methamphetamine challenge. *Brain Res Dev Brain Res*, 2003. 146(1-2): p. 119-30.
[69] Walker, C.D., Nutritional aspects modulating brain development and the responses to stress in early neonatal life. *Prog Neuropsychopharmacol Biol Psychiatry*, 2005. 29(8): p. 1249-63.
[70] Gater, R., M. Tansella, A. Korten, B.G. Tiemens, V.G. Mavreas, and M.O. Olatawura, Sex differences in the prevalence and detection of depressive and anxiety disorders in general health care settings: report from the World Health Organization Collaborative Study on Psychological Problems in General Health Care. *Arch Gen Psychiatry*, 1998. 55(5): p. 405-13.
[71] De Bellis, M.D. and M.S. Keshavan, Sex differences in brain maturation in maltreatment-related pediatric posttraumatic stress disorder. *Neurosci Biobehav Rev*, 2003. 27(1-2): p. 103-17.
[72] Serbin, L.A. and J. Karp, The intergenerational transfer of psychosocial risk: mediators of vulnerability and resilience. *Annu Rev Psychol*, 2004. 55: p. 333-63.
[73] Wahlstrom, D., T. White, C.J. Hooper, S. Vrshek-Schallhorn, W.S. Oetting, M.J. Brott, and M. Luciana, Variations in the Catechol O-methyltransferase Polymorphism and Prefrontally Guided Behaviors in Adolescents. *Biol Psychiatry*, 2006.
[74] Zhang, T.Y., P. Chretien, M.J. Meaney, and A. Gratton, Influence of naturally occurring variations in maternal care on prepulse inhibition of acoustic startle and the medial prefrontal cortical dopamine response to stress in adult rats. *J Neurosci*, 2005. 25(6): p. 1493-502.

[75] Kim, D.S., R.D. Palmiter, A. Cummins, and C.R. Gerfen, Reversal of supersensitive striatal dopamine D1 receptor signaling and extracellular signal-regulated kinase activity in dopamine-deficient mice. *Neuroscience,* 2006. 137(4): p. 1381-8.

[76] Blanchard, R.J., J.N. Nikulina, R.R. Sakai, C. McKittrick, B. McEwen, and D.C. Blanchard, Behavioral and endocrine change following chronic predatory stress. *Physiol Behav*, 1998. 63(4): p. 561-9.

[77] Blanchard, R.J., D.C. Blanchard, S.M. Weiss, and S. Meyer, The effects of ethanol and diazepam on reactions to predatory odors. *Pharmacol Biochem Behav*, 1990. 35(4): p. 775-80.

[78] Endler, J.A. and J. Mappes, Predator mixes and the conspicuousness of aposematic signals. *Am Nat*, 2004. 163(4): p. 532-47.

[79] Kavaliers, M. and E. Choleris, Antipredator responses and defensive behavior: ecological and ethological approaches for the neurosciences. *Neurosci Biobehav Rev*, 2001. 25(7-8): p. 577-86.

[80] File, S.E., H. Zangrossi, Jr., F.L. Sanders, and P.S. Mabbutt, Dissociation between behavioral and corticosterone responses on repeated exposures to cat odor. *Physiol Behav*, 1993. 54(6): p. 1109-11.

[81] Figueiredo, H.F., A. Bruestle, B. Bodie, C.M. Dolgas, and J.P. Herman, The medial prefrontal cortex differentially regulates stress-induced c-fos expression in the forebrain depending on type of stressor. *Eur J Neurosci*, 2003. 18(8): p. 2357-64.

[82] Perrot-Sinal, T.S., K.P. Ossenkopp, and M. Kavaliers, Brief predator odour exposure activates the HPA axis independent of locomotor changes. *Neuroreport*, 1999. 10(4): p. 775-80.

[83] Dielenberg, R.A., G.E. Hunt, and I.S. McGregor, "When a rat smells a cat": the distribution of Fos immunoreactivity in rat brain following exposure to a predatory odor. *Neuroscience*, 2001. 104(4): p. 1085-97.

[84] Caldji, C., D. Francis, S. Sharma, P.M. Plotsky, and M.J. Meaney, The effects of early rearing environment on the development of GABAA and central benzodiazepine receptor levels and novelty-induced fearfulness in the rat. *Neuropsychopharmacology*, 2000. 22(3): p. 219-29.

[85] Macri, S., G.J. Mason, and H. Wurbel, Dissociation in the effects of neonatal maternal separations on maternal care and the offspring's HPA and fear responses in rats. *Eur J Neurosci*, 2004. 20(4): p. 1017-24.

[86] Plotsky, P.M. and M.J. Meaney, Early, postnatal experience alters hypothalamic corticotropin-releasing factor (CRF) mRNA, median eminence CRF content and stress-induced release in adult rats. *Brain Res Mol Brain Res*, 1993. 18(3): p. 195-200.

[87] McIntosh, J., H. Anisman, and Z. Merali, Short- and long-periods of neonatal maternal separation differentially affect anxiety and feeding in adult rats: gender-dependent effects. *Brain Res Dev Brain Res*, 1999. 113(1-2): p. 97-106.

[88] Neumann, I.D., A. Wigger, S. Kromer, E. Frank, R. Landgraf, and O.J. Bosch, Differential effects of periodic maternal separation on adult stress coping in a rat model of extremes in trait anxiety. *Neuroscience*, 2005. 132(3): p. 867-77.

[89] Zimmerberg, B., S.H. Rackow, and K.P. George-Friedman, Sex-dependent behavioral effects of the neurosteroid allopregnanolone (3α, 5α-THP) in neonatal and adult rats after postnatal stress. *Pharmacol Biochem Behav*, 1999. 64(4): p. 717-724.

[90] Colorado, R.A., J. Shumake, N.M. Conejo, H. Gonzalez-Pardo, and F. Gonzalez-Lima, Effects of maternal separation, early handling, and standard facility rearing on orienting and impulsive behavior of adolescent rats. *Behav Processes*, 2006. 71(1): p. 51-8.

[91] Marmendal, M., C.J. Eriksson, and C. Fahlke, Early deprivation increases exploration and locomotion in adult male Wistar offspring. *Pharmacol Biochem Behav*, 2006. 85(3): p. 535-44.

[92] de Jongh, R., M.A. Geyer, B. Olivier, and L. Groenink, The effects of sex and neonatal maternal separation on fear-potentiated and light-enhanced startle. *Behav Brain Res*, 2005. 161(2): p. 190-6.

[93] Slotten, H.A., M. Kalinichev, J.J. Hagan, C.A. Marsden, and K.C. Fone, Long-lasting changes in behavioural and neuroendocrine indices in the rat following neonatal maternal separation: gender-dependent effects. *Brain Res*, 2006. 1097(1): p. 123-32.

[94] Lehmann, J., T. Stohr, and J. Feldon, Long-term effects of prenatal stress experiences and postnatal maternal separation on emotionality and attentional processes. *Behav Brain Res*, 2000. 107(1-2): p. 133-44.

[95] Romeo, R.D., R. Bellani, I.N. Karatsoreos, N. Chhua, M. Vernov, C.D. Conrad, and B.S. McEwen, Stress history and pubertal development interact to shape hypothalamic-pituitary-adrenal axis plasticity. *Endocrinology*, 2006. 147(4): p. 1664-74.

[96] Morley-Fletcher, S., M. Rea, S. Maccari, and G. Laviola, Environmental enrichment during adolescence reverses the effects of prenatal stress on play behaviour and HPA axis reactivity in rats. *Eur J Neurosci*, 2003. 18(12): p. 3367-74.

[97] Marquardt, A.R., L. Ortiz-Lemos, A.B. Lucion, and H.M. Barros, Influence of handling or aversive stimulation during rats' neonatal or adolescence periods on oral cocaine self-administration and cocaine withdrawal. *Behav Pharmacol*, 2004. 15(5-6): p. 403-12.

[98] Faure, J., J.D. Uys, L. Marais, D.J. Stein, and W.M. Daniels, Early maternal separation followed by later stressors leads to dysregulation of the HPA-axis and increases in hippocampal NGF and NT-3 levels in a rat model. *Metab Brain Dis*, 2006. 21(2-3): p. 181-88.

[99] Sandstrom, N.J. and S.R. Hart, Isolation stress during the third postnatal week alters radial arm maze performance and corticosterone levels in adulthood. *Behav Brain Res*, 2005. 156(2): p. 289-96.

[100] Kellogg, C.K., G.B. Awatramani, and D.T. Piekut, Adolescent development alters stressor-induced Fos immunoreactivity in rat brain. *Neuroscience*, 1998. 83(3): p. 681-9.

[101] Sims, C.G. and R.L. Holberton, Development of the corticosterone stress response in young northern mockingbirds (Mimus polyglottos). *Gen Comp Endocrinol*, 2000. 119(2): p. 193-201.

[102] Sullivan, R.M. and A. Gratton, Prefrontal cortical regulation of hypothalamic-pituitary-adrenal function in the rat and implications for psychopathology: side matters. *Psychoneuroendocrinology*, 2002. 27(1-2): p. 99-114.

[103] Sullivan, R.M., Hemispheric asymmetry in stress processing in rat prefrontal cortex and the role of mesocortical dopamine. *Stress*, 2004. 7(2): p. 131-43.

[104] Perrot-Sinal, T.S., A.M. Davis, and M.M. McCarthy, Developmental sex differences in glutamic acid decarboxylase (GAD(65)) and the housekeeping gene, GAPDH. *Brain Res*, 2001. 922(2): p. 201-8.

[105] Kippin, T.E., S. Kapur, and D. van der Kooy, Dopamine specifically inhibits forebrain neural stem cell proliferation, suggesting a novel effect of antipsychotic drugs. *J Neurosci*, 2005. 25(24): p. 5815-23.

In: Psychoneuroendocrinology Research Trends
Editor: Martina T. Czerbska, pp. 293-319 ISBN: 978-1-60021-665-7

Chapter IX

IMMUNE INFLUENCES ON BEHAVIOR AND ENDOCRINE ACTIVITY IN EARLY-EXPERIENCE AND MATERNAL SEPARATION PARADIGMS

Michael B. Hennessy[1,*], *Terrence Deak*[2], *Patricia A. Schiml-Webb*[1] *and Christopher J. Barnum*[2]
[1]Department of Psychology, Wright State University, Dayton, OH 45435, USA;
[2]Department of Psychology, Binghamton University, Binghamton, NY 13902, USA.

ABSTRACT

Early life stressors have long been known to have powerful immediate and lasting biobehavioral consequences. Though much has been learned about the neural and endocrine substrates of such effects, increasing evidence suggests that elements of the immune system may also play a substantial role. Neonatal exposure to lipopolysacchride (LPS) has been found to affect later behavioral and endocrine endpoints in rats and mice in ways that parallel the effects of more-traditional early experiences. In guinea pigs—a rodent model for studies of filial attachment and separation—proinflammatory factors appear to contribute to the behavioral reaction of pups during isolation in a novel environment. Evidence for this assertion includes findings that: (1) exposure of guinea pig pups to LPS produces the same constellation of passive behavioral effects as does protracted isolation from the mother in a novel environment; (2) isolation in a novel environment increases proinflammatory cytokine expression and core body temperature; and, (3) administration of anti-inflammatory agents attenuate passive responses during isolation. Further, it appears that corticotropin-releasing factor (CRF) may activate the immune system during isolation since administration of this peptide increases the passive responses in the same way as does prolonged isolation or injection of LPS, and anti-

* Correspondence concerning this article should be addressed to Michael B. Hennessy, PhD, Department of Psychology, 335 Fawcett Hall, Wright State University, Dayton, OH 35435. Phone: 937.775.2943; FAX: 937.775.3347; Email: michael.hennessy@wright.edu.

inflammatory treatment attenuates the behavioral effect of CRF. These findings will be reviewed, and their implications for the understanding of early-experience effects and the development of psychopathology will be discussed.

Introduction

Whereas the importance of the endocrine system for understanding the biology of behavior became apparent in the middle of the last century (Beach, 1948), it was only about 25 years ago that the role of the immunological system in behavioral study began to achieve widespread acceptance (Ader, 1981). Since that time there has been a growing appreciation of the ways in which immunological events can be inter-twined with psychological concepts and effects. Studies revealing neural and endocrine influences on immunological function have provided much of our understanding of the physiological and conceptual bases of psychosomatic disorders. More recently, there has been increasing attention paid to effects proceeding in the opposite direction; that is, activity of the immune system impacting neural and endocrine function. Indeed, signals from the immune system to the brain appear to occur routinely, and have even been hypothesized to constitute a sixth sensory modality, one that presumably evolved to inform the brain about pathogen exposure and bodily injury (Blalock, 2005; Maier & Watkins, 1999). These developments now are beginning to force investigators to consider events such as the encountering of bacteria as akin in crucial ways to traditionally construed psychological experience.

Examples of this interfusion of the immunological with the psychological can be seen in two broad areas of study in the realm of developmental psychobiology. These areas, often referred to as "early experience" and "maternal separation", are related and overlapping. Separation from the mother certainly is an "early experience" and common forms of early experience actually involve some, though usually brief, separation from the mother. Furthermore, both types of events have been studied primarily in relation to their influence on emotional behavior. Nonetheless, these areas have largely evolved as separate bodies of research with different histories, principal subject species, and points of emphasis. The early experience literature has primarily emphasized the lasting, even permanent, outcomes for the developing animal, while the maternal separation literature has mostly been concerned with the response of the infant during the actual period of separation and immediately thereafter. Importantly for the purposes of the present paper, the way in which immunological factors have been found to contribute to effects in the early experience and maternal separation paradigms also varies. In this chapter, we will first review recent studies illustrating how the effects of neonatal activation of the immune system parallel, and in many ways seem interchangeable with, traditionally studied early experiences. Second, we will examine some immunological influences during maternal separation. Findings that maternal separation can suppress immunological function, particularly forms of specific immunity in primates, is well established (Coe, 1993; Laudenslager, Reite, & Harbeck, 1982). The focus of the second portion of the present chapter, however, is not on immunological changes as endpoints of maternal separation, but rather as mediators of separation effects. In this regard, we will focus on ongoing work from our group suggesting that behavioral effects of maternal separation can be attributed to an activation of nonspecific, or innate, immunity by separation procedures.

A potential point of confusion for the reader in discerning between the two areas of research (early experience and maternal separation) to be reviewed is that one form of early experience involves a prolonged (i.e., several hour) period of separation from the mother, and typically is referred to as maternal separation. Therefore, for clarity in the present paper, maternal separation employed in this way (as an early experience) will be referred to as "maternal removal" whereas the term "maternal separation" or, when appropriate, "isolation" will be reserved for describing effects in the second area of research (i.e., maternal separation).

TRADITIONAL EARLY-EXPERIENCE EFFECTS

Paradigms for studying the lasting effects of early experience in rodents that are widely used today, such as "handling", first appeared in the scientific literature about 50 years ago (Levine, Chevalier, & Korchin,1956). Handling, which simply involves removing infant rats or mice from the home cage for several minutes a day during a portion of the preweaning period, was found to have a plethora of effects on adult behavior and physiology (Denenberg, 1967; Levine, 1962). No physiological system was studied more in this regard than was the stress-responsive hypothalamic-pituitary-adrenal (HPA) axis. Most often early handling was found to reduce the magnitude and/or duration of the later HPA response to various stressors (e.g., Pfeifer, Rotundo, Myers, & Denenberg, 1976; Meaney, Aitken, van Berkel, Bhatnagar, & Sapolsky, 1988; Weinberg & Levine, 1977). These findings, together with behavioral studies indicating that handling enhanced such processes as exploration and performance in some learning paradigms (DeNelsky & Denenberg, 1967; Levine, 1956), supported the conclusion that handling produced effects that were "adaptive" for the adult (Levine, 1966). This interpretation foreshadowed latter day hypotheses of early experience "programming" the infant for its likely adulthood environment.

Other studies varied parameters of the handling experience, or employed other forms of early stimulation (e.g., electric shock). It was found that the nature of the effect on HPA activity and behavior often differed with a number of these factors, including the intensity (level of electric shock (Bell & Denenberg, 1963)), timing (different days of age (Denenberg & Karas, 1960)), and duration (minutes vs hours of removal from the mother and nest (Plotsky & Meaney, 1993)) of the early experience. In some cases, for instance following exposure to prolonged maternal removal, later HPA responsiveness to stressors has been found to be increased rather than diminished (Plotsky et al, 2005).

A variety of potential mechanisms or mediators of early postnatal experience were proposed. These included a direct action of the tactile stimulation on physiological development, an action secondary to the lowering of the pup' body temperature during stimulation, an influence of the mother upon the return of her stimulated pups, and some form of "stress immunization" due to the early treatment (Russell, 1971). While in all likelihood, one variety of mediator cannot account for the full breadth of neonatal stimulation effects, evidence has accumulated over the years that the mother's treatment of the stimulated pups is a critical means by which early experience can effect outcomes, particularly in regards to the HPA axis (Hennessy, Vogt, & Levine, 1982; Liu, et al, 1997; Smotherman & Bell, 1980; Schrieber, Bell, Kufner, Villescas, 1977). That is, sensory cues from the pups (e.g., ultrasonic vocalizing) upon their return to the nest stimulates active maternal care, such as licking. The maternal stimulation appears to alter the development of neural and endocrine systems of the pups, which then accounts for many of the later effects of early experience (Meaney et al., 1996). Differences in the sensory qualities of pups produced by various forms of stimulation, and the subsequent differential responses of the mother to these differing sensory cues, probably accounts for much of the variation in effects of neonatal stimulation (Bell, Nitschke, Bell & Zachman, 1974).

Immunological Activation as Early Experience

The idea of a relation between early experience and immune function is not new (Levine & Cohen, 1959). In fact, a body of work now exists showing that traditional early experiences have an assortment of effects on later immunological activity (Shanks & Lightman, 2001). However, in the last decade evidence has begun accumulating that antigen-stimulated immunological activity might serve as an independent variable (i.e., as an early experience) affecting typical early-experience endpoints. It is these effects of early immune manipulations on the measures most often examined in early experience studies (i.e., HPA and behavioral responses) that will be the focus here.

The standard procedure for stimulating the immune system has been with injection of lipopolysaccharide (LPS), a component of the cell wall of gram negative bacteria, which elicits a robust immunological response free from contaminating effects of an actual replicating infection. Typically LPS is administered on one or more occasions during the first week of life. Adult rats administered LPS in infancy have been found to exhibit increased plasma ACTH and corticosterone responses to stressors (Shanks, Larocque, & Meaney 1995; Shanks et al., 2000; Hodgson, Knott, & Walker, 2001). These findings are the opposite of those typically observed following early handling, but correspond with results reported for rats undergoing several hours of maternal removal during the first 2 weeks of life (Plotsky et al., 2005). Traditional early experiences, including handling and prolonged maternal removal, appear to affect central (e.g., hippocampal) glucocorticoid receptor density, increasing or decreasing negative feedback and thereby levels of corticotropin-releasing factor (CRF) and arginine vasopressin (AVP) available in the median eminence for release in response to restraint (Meaney et al., 1996). Rats exposed as infants to LPS exhibit results similar to those found previously for rats exposed during infancy to prolonged maternal removal: They had reduced glucocorticoid receptor density in the hippocampus, hypothalamus, and frontal cortex; reduced suppression of the ACTH response to restraint by dexamethasone; and increased levels of CRF and AVP peptide levels and mRNA in the median eminence prior to restraint (Shanks et al., 1995).

Early LPS exposure typically has not been found to affect later basal levels of pituitary-adrenal hormones (Shanks et al., 1995; Walker et al., 2006), though one study in which blood was collected every 10 min over a 24-hr period did find an increase in basal corticosterone concentrations (Shanks et al., 2000). Likewise, traditional early manipulations generally do not affect basal levels of corticosterone, though protracted maternal removal has been observed to elevate basal HPA hormone concentrations (Rots et al., 1996).

In mice, early LPS exposure was found to produce no later change in plasma corticosterone levels in males that either were or were not placed into an enclosure with a strange male (Granger, Hood, Ikeda, Reed, & Block, 1996). However, the LPS-exposed mice had lower levels of hypothalamic CRF following the social challenge and, in one of the two strains examined, also when no challenge occurred. Furthermore, in this same strain CRF content of the LPS-exposed mice was related to behavioral effects during the social challenge: Those mice having lower CRF levels after the challenge were much less reactive to the unfamiliar animal (i.e., exhibiting fewer startles, jumps, and kicks) than were those having higher CRF levels. Reconciling the opposite direction of the effect of LPS exposure on CRF content in this study (a reduction) and that by Shanks et al (1995) with rats (an increase) is difficult because of procedural differences. In particular, Shanks et al assessed CRF only in the median eminence, whereas peptide content within the entire hypothalamus was evaluated in Granger et al's (1996) mice.

Though LPS typically does not have extensive effects on behavioral outcomes, some lasting behavioral consequences of early LPS on traditional early-experience outcomes have been observed (Granger et al., 1996; Granger, Hood, Dreschel, Sergeant, & Likos, 2001; Hood et al., 2003; Shanks et al., 2000; Spencer, Heida, & Pittman, 2005; Walker, March, & Hodgson, 2004). Early LPS administration was found to increase indicants of emotionality, including behavioral immobility (Granger et al., 1996; Granger et al., 2001) and reactivity (Granger et al., 1996; Granger et al., 2001) during a social encounter in a novel cage, reactivity following noise exposure (Shanks et al., 2000), as well as entries and time spent in the open arms of a plus maze (Walker et al., 2004, b). Early LPS also was observed to reduce the exploration of objects in an open field (Spencer et al., 2005). In contrast, early manipulations such as handling and shock quite consistently *reduce* measures of emotionality and *increase* exploration in rats and mice (Denenberg, 1967). Recall that early LPS exposure appeared to produce long-term effects on the HPA axis that were more like those of protracted maternal removal than those produced by handling. It should be emphasized that reported behavioral outcomes of prolonged maternal removal have varied considerably with differences in the particular methods employed (Lehmann & Feldon, 2000). With that caveat in mind, however, effects of procedures involving early maternal removal have mirrored those of early LPS to some degree: Notably both have been found to increase emotional responses in the elevated plus maze (Patchev et al., 1997; Walker et al., 2004 b; Wigger & Neumann, 1999).

Attempts to understand the mechanism by which early LPS initiates lasting effects have focused on the maternal mediation hypothesis: that the effect of LPS on the pup alters its sensory qualities, and consequently the mother's treatment of the pup in ways that are similar to those changes in maternal behavior that appear to mediate other early-experience effects. Two laboratories have found that injection of rat pups with LPS reduces subsequent physical contact between mothers and pups (Meaney et al., 1996; Walker, Brogan, Smith, & Hodgson, 2004). Because maternal interaction with pups is reduced as a direct consequence of prolonged maternal removal, but appears to be intensified as a result of handling (Bell,

Nitschke, Gorry, & Zachman, 1971; Meaney et al., 1996), the effects of early LPS on maternal behavior seem best to correspond with those of early prolonged maternal removal. This correspondence is at least consistent with the notion that maternal behavior could mediate effects of early LPS administration.

In mice, the influence of different maternal environments on the outcome of early LPS treatment has been directly examined. Mouse pups treated or not treated with early LPS were either fostered or not fostered to mothers of different strains (Granger et al., 2001; Hood et al., 2003). Although most outcomes of early LPS were not affected by the differences in maternal environment, some were. Granger et al (2001) found that early LPS increased social reactivity in one line of mice, but only if they had been reared by their biological mother, not if they had been raised by a foster mother of a different line. Hood et al (2003) obtained effects of early LPS exposure on the amount of exploration, and frequency of attack, of a partner in a later social confrontation that varied in a complex manner with which of three lines of foster mothers the pups were reared. The pattern of results suggested that increased maternal responsiveness ameliorated some effects of early LPS exposure. This interpretation is consistent with results reviewed above for rats suggesting that greater time out of contact with the pups is associated with larger later effects of early LPS exposure.

Studies of early immune activation are not free of methodological concerns that have plagued the traditional early-experience literature. Most noteworthy is the issue of "litter effects", which arise when experimental groups consist of multiple pups from individual litters, and each pup is assumed to represent an independent observation. The issue is more than just a concern of genetic relatedness among pups. If maternal behavior mediates an early postnatal-experience effect and, as is known (Liu et al., 1997), mothers differ in critical aspects of the care they provide, the mother's behavior is imposing litter effects. Under these conditions, the number of mothers (litters), not the number of pups, should be used as the statistical unit (Abbey & Howard, 1973) unless the litter is treated as a covariate (Zorilla, 1997), statistical models not assuming littermates to be independent observations are employed (Zorrilla, 1997), or the variation within a litter can be shown to be no smaller than the variation between litters (Denenberg, 1977). Broad appreciation of the importance of strict control of such litter effects was only gradually recognized in studies of traditional early experience. A similar, gradually increasing appreciation of the importance of controlling litter effects may be beginning to play out in studies of the effects of early immune activation.

In summary, while any conclusions must be tempered with a regard both for the diverse outcomes of seemingly similar early experiences reported in the literature, and for the issue of litter effects just described, it appears that early activation of the immune system with LPS affects many of the same later outcome measures in terms of HPA activity and behavior as has been found for traditional forms of early experience. Further, maternal behavior appears to be one mediator of effects of both traditional early experiences and early LPS exposure. In all, it seems that the effects of early immune activation bear the most resemblance to a subset of outcomes found for early prolonged maternal removal. Consequences of early immune activation, like findings in recent studies of traditional experiences, are most often interpreted in terms of the early environment programming the organism for challenges it is most likely to encounter in its future environment (e.g., Meaney et al., 1996). In all, there appears to be little to differentiate early immune activation from other forms of early experience.

Maternal Separation and Behavioral "Despair" in Monkeys

During the 1960s and 70s, another focus of developmental studies, in this case in primates rather than rats and mice, was on the response of infants during periods of maternal separation. The impetus for this body of work derived from observations of human children maintained in hospitals and other institutions without regard for the importance of the mother. In an era in which the absence of "germs" trumped the presence of a loving caretaker in minds of health care professionals, it was not uncommon for young children to be deprived of anything more than fleeting interaction with their mother for days or weeks at a time (Blum, 2002; Bowlby, Robertson, & Rosenbluth, 1952; Spitz, 1945). Eventually the distressed behavior of these children attracted the attention of psychologists and began to be systematically recorded. After a period of separation, a subset of the children became withdrawn, inactive, and, of note in the present context, even appeared physically ill (Spitz, 1946). This stage was termed "despair" as a descriptor of the children's apparent affective state. Spitz (1946) contended that the response to prolonged separation constituted a clinical entity, proposing the term "anaclitic depression". Since that time, maternal separation has remained central to discussions of the ontogeny of affective illness. Not only was the response of some individuals during the period of separation considered a form of depression, but evidence began accumulating that early stressors, often including prolonged maternal separation or other forms of disruption of the attachment bond were important risk factors for the later development of mood disorders (Bernet & Stein, 1999; Brown, Harris, & Copeland, 1977; Mancini, van Ameringen, & Macmillan, 1995). Today this notion is central to a major hypothesis concerning the development of depression (Gold, Goodwin, & Chrousos, 1988; Ladd, Owens, & Nemeroff, 1996).

Studies indicating that nonhuman primates exhibited responses to maternal separation that corresponded with those seen in children began to appear in the 1960s. Kaufman and Rosenblum (1967) reported that infant pigtail macaques (*Macaca nemestrina*) that were deprived of their mothers showed biphasic behavioral responses remarkably similar to those exhibited by separated children. After an initial period of "protest" in which infants would vocalize and apparently seek maternal contact, a portion of the infants would display behaviors and postures fitting the characterization of "despair" (Figure 1). On the basis of such observations, maternal separation in monkeys became an accepted animal model for studying depressive illness (McKinney, Moran, & Kraemer, 1984; Wilner, 1991). Furthermore, other research has demonstrated central monoaminergic, CRF, and pituitary-adrenal changes as a result of separation that suggested commonalities with physiological mechanisms underlying depression. (e.g., Coplan et al., 1996; Higley, Suomi, & Linnoila, 1991; Kraemer, Ebert, Lake, & McKinney, 1984; Lyons et al., 1999). Despite these findings, nonhuman primate separation models of depression never became widely used, and have fallen farther from favor in recent years, no doubt in part to the difficulty, expense, and ethical issues inherent in separating nonhuman primate infants from their mothers for extended periods.

Figure 1. Separated pigtail macaque infant exhibiting "despair" behavior. Reprinted with permission from Kaufman and Rosenblum (1967).

The Guinea Pig Model

Of the commonly used laboratory rodents, findings in the guinea pig (*Cavia porcellus*) best accord with those of separated primate infants. Guinea pig pups display a strong attraction to the mother and findings in guinea pigs map well onto criteria for filial attachment used in studies with nonhuman primates (Hennessy, 1997; Jäckel & Trillmich, 2003). Unlike more commonly used rats and mice, guinea pig pups are born physically mature and are fully capable of wandering off from the mother beginning moments after birth. Guinea pig mothers do not retrieve pups, and maternal behavior in general is extremely passive in this species (Hennessy & Jenkins, 1994; König, 1985). Rather it is the strong attraction, or attachment, of the pup for the mother that keeps the preweaning guinea pig pup in proximity to her. If guinea pig pups are isolated in a novel environment, they show elevated levels of pituitary-adrenal hormones and increased turnover of central monoamines not seen if the mother accompanies them to the novel surroundings (Harvey, Moore, Lucot, & Hennessy, 1994; Hennessy & Moorman, 1989; Hennessy, Tamborski, Schiml, & Lucot, 1989). Indeed, guinea pig pups show a number of similarities to nonhuman primate infants in their responses to separation (Hennessy, 2003), including a two-stage, active/passive response.

When guinea pig pups are isolated in a novel cage, they initially vocalize at high rates much as is seen in separated monkeys (Hennessy & Moorman, 1989; Mineka & Suomi, 1978). The vocalizations appear to serve as contact calls to help the pup re-establish contact with the mother. After about an hour or so of isolation, the guinea pig pup begins to settle down and starts to exhibit a constellation of passive behaviors consisting of a characteristic crouched stance, prolonged closure of the eyes, and extensive piloerection (Figure 2). These responses are linked to the absence of the attachment object: The presence of the mother in the test cage almost entirely eliminates passive behavior, whereas the presence of an unfamiliar adult female does not (Hennessy & Morris, 2005). The passive responses become prominent by three hours of isolation and persist for more than 24 hours (Hennessy, Long, Nigh, Williams, & Nolan, 1995; Hennessy et al., 2004). Thus, the two-stage, active/passive response is at least superficially similar to that described in monkeys, though it is greatly compressed in time, progressing to the second stage in hours rather than days or weeks.

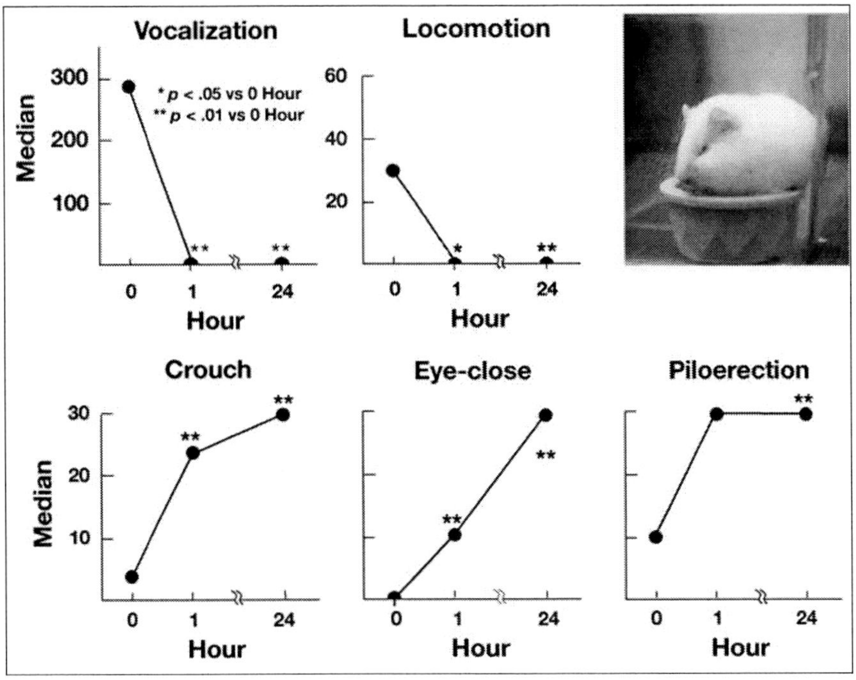

Figure 2. Median number of vocalizations and line-crossings (locomotion) and of 60-s intervals in which the crouched stance, eye-closing, and piloerection were observed in untreated guinea pig pups during 30-min observation sessions beginning immediately following pup's placement into isolation (0 hr) as well as 1 and 24 hr later. Numbers in parentheses are the interquartile ranges. Reprinted with permission from Hennessy et al (1995). The inset shows a pup exhibiting crouching, eye-closing, and piloerection during isolation.

INFLUENCES OF CRF ON BEHAVIOR DURING SEPARATION

In terms of mechanism, early studies indicated that CRF plays a role in generating the passive behaviors observed during maternal separation. If a pup is injected peripherally with CRF, returned to the home cage for an hour and then isolated, it exhibits high levels of the passive responses and low levels of the active responses from the outset of the testing period (Hennessy et al., 1995; Figure 3). This effect does not appear to be secondary to the action of glucocorticoids, ACTH, or to opioids co-released from the pituitary together with ACTH (Hennessy, Becker, & O'Neil, 1991). Further, the behavioral effect is not due to an impairment of motor activity, elicitation of behaviors incompatible with active responses, or a hypotensive effect of the CRF (Becker & Hennessy, 1993; Hennessy et al., 1995). The actions of CRF are blocked if a specific CRF-receptor antagonist is injected together with the CRF (Hennessy et al., 1995; Figure 3), indicating a CRF-receptor-mediated mechanism. Furthermore, injection of the antagonist alone both increases active behavior and reduces the number of pups exhibiting passive responses by the end of the first hour of separation, presumably by blocking the action of endogenous CRF (McInturf & Hennessy, 1996). That is, it appears that endogenous CRF acts to suppress the active, and enhance the passive, responses. For these reasons, it has been proposed that CRF is part of the mechanism that normally helps switch the pup from the initial active, to the subsequent passive, stage of responsiveness (Hennessy et al., 1999).

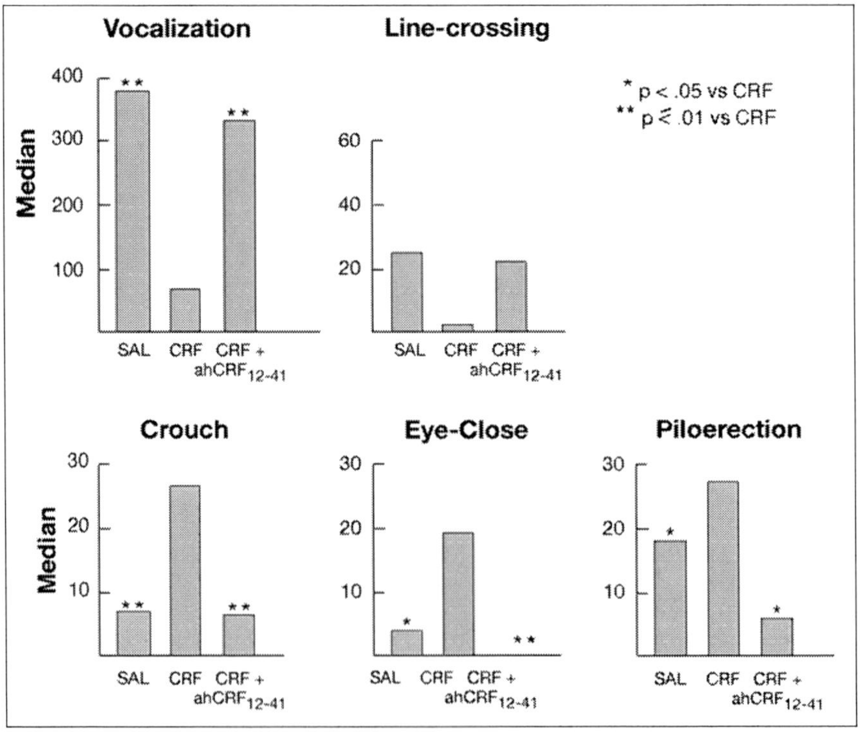

Figure 3. Median number of vocalizations and line-crossings (locomotion) and of 60-s intervals in which the crouched stance, eye-closing, and piloerection were observed during the first 30 min of isolation in pups that had received two injections of saline vehicle (SAL), saline followed by 7 μg of CRF (CRF) or 50 μg of the antagonist CRF_{12-41} followed by 7 μg of CRF (CRF + CRF_{12-41}). Numbers in parentheses are interquartile ranges. Reprinted with permission from Hennessy et al (1995).

Since these studies administered CRF and other compounds peripherally, the site of CRF's action is unclear. There is evidence from studies of intracerebroventricular (ICV) administration that central CRF can produce some of these outcomes (Hennessy et al., 1992), but the effects did not appear as powerful or consistent, and there is little to support the idea that peripherally administered CRF can cross the blood-brain barrier in appreciable quantities

(Martins, Banks, & Kastin, 1997). There are, however, peripheral sources of CRF and CRF receptors, for instance in the adrenal medulla and sympathetic nerve terminals, that may increase activity during stress (Bruhn, Engeland, Anthony, Gann, & Jackson, 1987; Krukoff, 1986; Udelsman et al., 1986). Of particular interest here, CRF is known to promote various inflammatory responses (Leu & Singh, 1992; Singh, Pang, Alexacos, Letourneau, & Theoharides, 1999; Webster et al., 1996). Proinflammatory cytokines have widespread effects on physiology and behavior that appear relevant for understanding the observed actions of CRF, as well as the behavioral effects of separation in general.

The Acute Phase Response

The acute phase response (APR) is an element of innate immunity, and a first line of defense against invading pathogens. When macrophages and other cells of the immune system encounter antigens, such as those of replicating pathogens, the cells begin to emit proinflammatory cytokines, including interleukin-1 (IL-1), IL-6, and tumor necrosis factor alpha (TNF-α). These peptide messengers not only have local effects, but ultimately result in a systemic reaction, the APR, or sickness, which maximizes host defense against infection (Bauman & Gauldie, 1994). Production of proteins by the liver switches from soluble proteins (e.g., albumin) to acute phase proteins (e.g., C-reactive protein, alpha1-acid glycoprotein) which help opsonize foreign material for rapid clearance by immune cells. One complex outcome is fever, which tends to both inhibit pathogen reproduction and promote immune activity (Hart, 1988). The proinflammatory cytokines of the APR can activate the HPA axis and also produce behavioral changes, so-called "sickness behaviors" (Dantzer, 2004; Maier & Watkins, 1998). These include reductions in goal-directed activity, such as social and sexual interaction and food and water intake, increased sleepiness, huddling, piloerection, and moving toward heat sources. These behavioral reactions serve adaptive functions such as conserving energy and promoting fever (Aubert, 1999; Dantzer, 2004; Hart, 1988). In all, the behavioral effects appear to represent a centrally mediated motivational state that ultimately promotes survival in the face of immune challenge (Aubert, 1999).

Clearly, behavioral reactions indicate CNS involvement. Because proinflammatory cytokines are peptides, understanding how a proinflammatory reaction initiated in the periphery might affect the brain would seem to pose the same issue regarding crossing the blood-brain barrier as described above for CRF. However, there is evidence for several means by which peripheral cytokines can result in central cytokine activity (see Watkins & Maier, 2005 for a recent review). Importantly, the brain can synthesize cytokines, and appears to do so in response to peripheral cytokine signals. These signals can be conveyed, for instance, neurally by means of cytokine stimulation of vagal afferents.

Stress-Induced Sickness Behaviors during Separation

The APR does not occur only in response to direct immune activation by antigens. Rather, behaviors and physiological responses characteristic of the APR also can be elicited by some, though not all, stressors (Deak et al., 1997; Deak et al., 2005; Maier & Watkins, 1998). Involvement of proinflammatory factors is indicated by the finding that components of the APR can be reversed with anti-inflammatory agents (e.g., Milligan et al., 1998). In other words, under some conditions the stress response incorporates elements of the APR. This conclusion may appear contradictory given the well known suppressive effect of the stress-responsive glucocorticoids. Indeed, the typical glucocorticoid rise produced by stressor exposure powerfully limits the degree to which stressors can induce acute phase-like changes in cytokine activity (Nguyen et al., 1998; Nguyen et al., 2000). But the relation between stress-related hormones and inflammatory factors is much more complex (e.g., Ghoshal, Wang, Sheridan, & Jacob, 1998; Hernandez et al., 2000; Ito, Takii, Matsumara, & Onozaki, 1999) and as noted above, CRF can stimulate proinflammatory activity in a variety of contexts.

Based on such considerations, we proposed that the passive behaviors exhibited by isolated guinea pig pups represent stress-induced sickness behaviors stimulated by proinflammatory cytokines, which are activated by the stressor of separation from the mother in a novel environment (Hennessy, Deak, & Schiml-Webb, 2001). Crouching, closing of the eyes, and piloerection are among the behaviors that have been associated with sickness in other contexts (Hart, 1988), and together they convey a clear impression of malaise in the isolated animal (Figure 2). Because of findings reported above suggesting that CRF helps shift the guinea pig from the initial active, to a subsequent passive, stage of responsiveness, and because of the evidence cited that CRF can induce inflammation, we proposed that CRF mediates the stress-induced cytokine response.

Subsequent investigations have provided support for this proposal. These investigations broadly can be divided into three categories: (1) comparison of the behavioral effects of separation with unequivocal sickness behaviors; (2) assessment of non-behavioral elements of the APR during separation; and, (3) attempts to reverse behavioral effects of separation and CRF injection with anti-inflammatory agents. To begin with, any claim that behavioral effects during separation represent stress-induced sickness behaviors pre-supposes that guinea pig pups do exhibit the same constellation of crouching, eye-closing, and piloerection during indisputable bouts of sickness elicited by traditional immune stimuli. LPS is probably the most commonly used means of inducing the APR and sickness behaviors in the laboratory. Therefore, guinea pig pups were administered an intraperitoneal (IP) injection of saline vehicle or 50 or 250 μg/kg of LPS (Hennessy, Deak, Schiml-Webb, Wilson, Greenlee, & McCall, 2004). After return to the home cage for 90 min, the pups were isolated in a novel cage for 60-min during which time behavior was observed. As seen in Figure 4, the saline-injected pups displayed the typical response to a brief separation of this sort; that is, they exhibited high levels of vocalizing and some crossing of grid lines into different segments of the test cage, and low levels of the passive responses. If, however, the pups had been injected with LPS, they showed a suppression of active behavior and dose-dependent elevations of crouching, eye-closing, and piloerection during the 60-min test. Thus, a compound that directly activates an APR produced the same behavioral reaction as injection with CRF, or as a several hour separation in non-treated pups. These results confirm that the behavioral reaction induced by CRF or prolonged isolation corresponds to sickness behavior in this species.

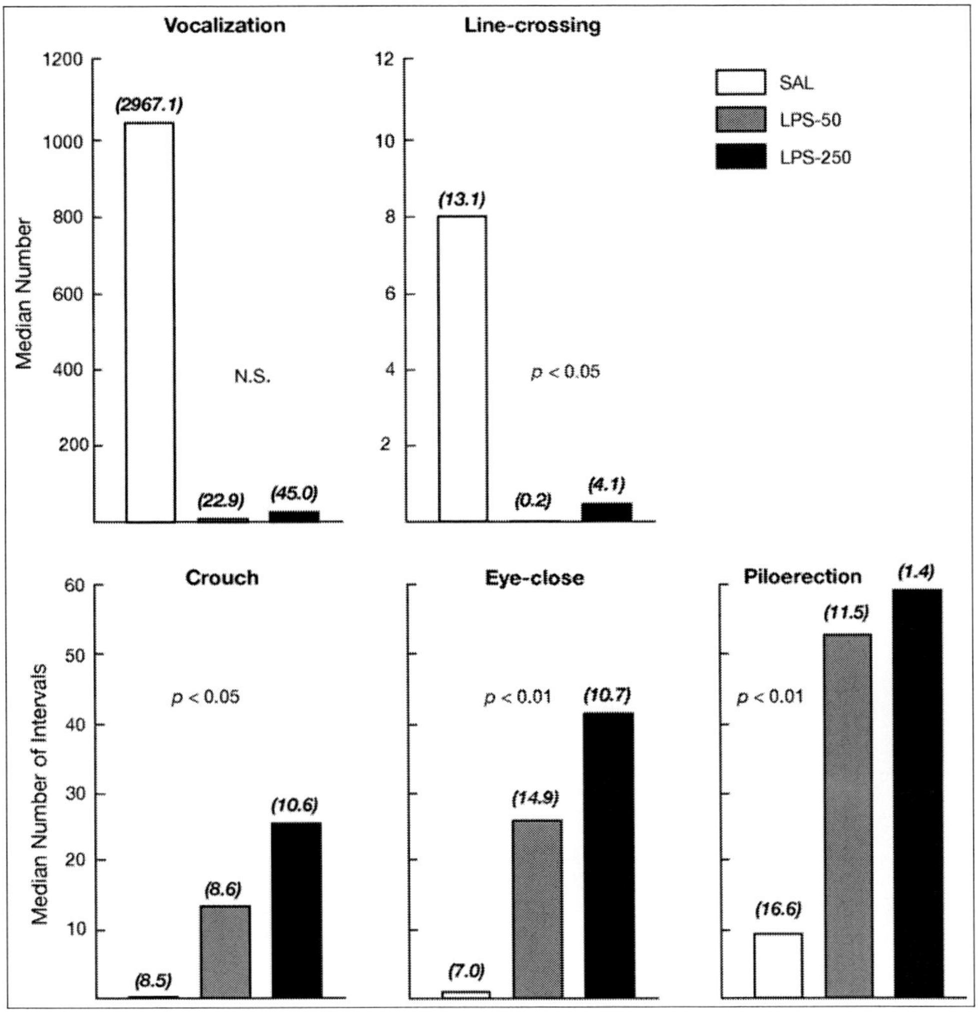

Figure 4. Median number of vocalizations and line-crossings and 60-s intervals in which crouching, eye-closing, and piloerection were observed in guinea pig pups during 60 min of isolation following injection of either saline or 50 or 250 μg/kg LPS. Numbers in parentheses are the semi-interquartile ranges. Probability levels refer to differences across injection conditions. Reprinted with permission from Hennessy et al (2004).

If separation does induce true sickness behavior, then we would expect that other aspects of an APR, such as elevated levels of at least some proinflammatory cytokines would also be apparent. We initially sought to measure IL-1β with ELISA procedures. These attempts were hampered by the lack of specific antibodies for guinea pigs, but results with a mouse ELISA kit following verification with LPS injection and Western Blot analysis, suggested that IL-1β at least was not elevated in brain or periphery in response to maternal separation (Hennessy et al., 2004). In order to provide a more-sensitive analysis of a wider array of cytokines, we have been developing protocols for using real time PCR for guinea pig cytokines. Our preliminary analysis focused on cytokine expression in the spleen because proinflammatory cytokines are highly responsive to stressors in this peripheral immune organ (Blandino, Barnum, & Deak, 2006). Early results show that levels of the proinflammatory cytokine TNF-α in spleen were significantly elevated during a 3-hr isolation period (Figure 5), with the effect appearing to grow larger as separation proceeded. This finding provides strong empirical support for the notion that separation experience can mobilize endogenous inflammatory factors that have more conventionally been associated with immune activation. Ongoing work is examining other key cytokines in both peripheral and central tissue compartments to ascertain whether our results are unique to specific structures or representative of a more global (i.e., systemic) inflammatory cascade.

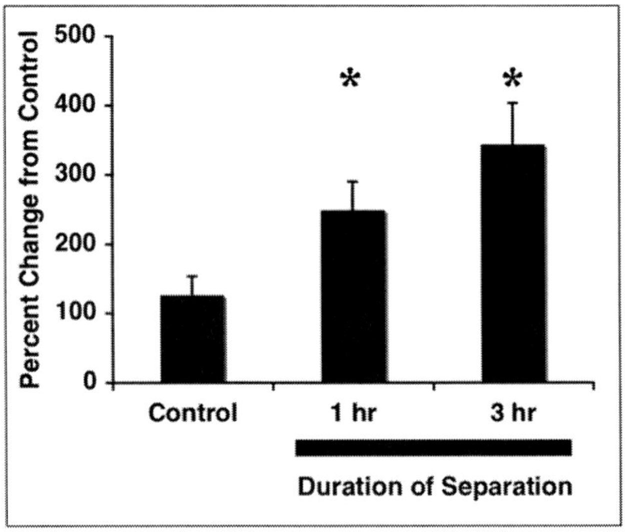

Figure 5. Guinea pig pups remained in their home cages (Control) or were isolated from the dam for 1 or 3 hr. At the appointed time, pups were killed and tissue was harvested and frozen for later analysis. RNA was extracted, purified and reverse transcribed as previously described [Deak et al 2005] for subsequent measurement of relative gene expression using real time RT-PCR. Primers for TNFα were: forward: 5'-ACGCTCACACTCAGATCAGCTTCT-3'; reverse: 5'-ACTCCAAAGTAGATCTGCCCGGAA-3'. Data were expressed relative to GAPDH (forward: 5'-TATCGTGGAAGGACTCATG-3'; reverse: 5'-GATCCACAACCGACACATT-3') using the 2^-ΔΔCt method (Livak & Schmittgen, 2001)]. Data are expressed as mean percent change relative to home cage controls (±SEM); * indicates significantly different from control animals ($p < 0.05$). Relative expression of TNF-α was increased in spleen after both 1 hr and 3 hr of maternal separation.

Another focus of ongoing work is the effect of the test procedure on body temperature. Since fever is a component of an APR, one might predict that isolation would induce a change in core body temperature. In a study using simple rectal probes, we found that pups isolated for 90 min exhibited an approximately three-quarters degree C increase in temperature relative to pups removed directly from the home cage. At 180 min, the increase was about one quarter degree and no longer significant (Hennessy et al., 2004). We are now examining core temperature with greater temporal clarity using more-sensitive procedures (i.e., abdominal telemetry probes). These studies are aimed at providing a detailed assessment of temporal changes in core temperature in pups while isolated or with their mother, and ultimately should allow us to temporally align changes in core temperature with measures of central cytokine activity.

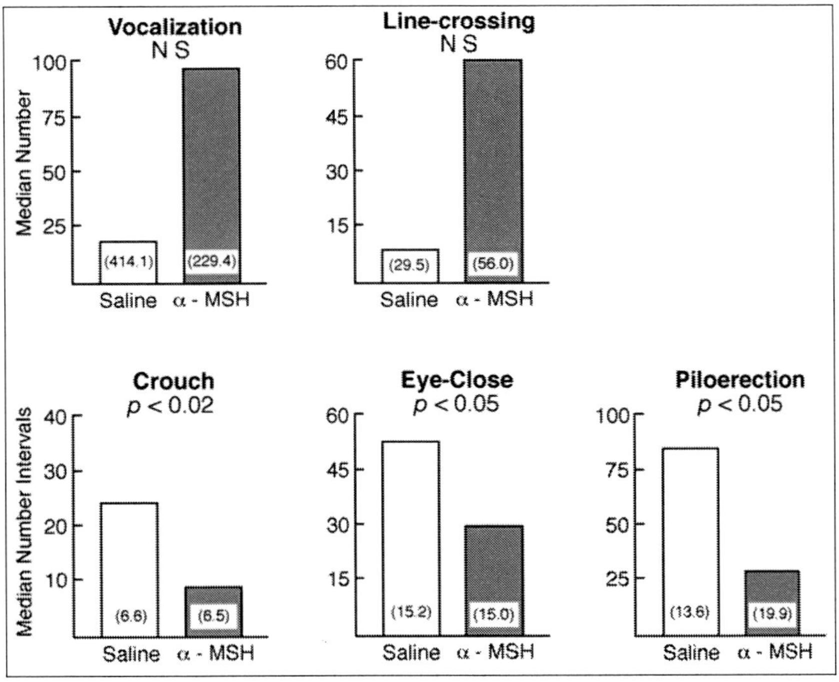

Figure 6. Median number of vocalizations and line-crossings and 60-s intervals in which guinea pig pups exhibited crouching, eye-closing, and piloerection during Min 0-30, 60-90, and 150-180 of a 180 min period of isolation following ICV infusion of either 25μg of α-MSH or saline vehicle. Numbers in parentheses are the semi-interquartile ranges. Reprinted with permission from Schiml-Webb et al (2006).

The results presented above indicate that induction of an APR results in the same passive behaviors as seen during protracted separation or following injection with CRF, and that the test procedure produces a portion of the physiological reaction seen during an APR. However, the findings do not provide direct evidence that a proinflammatory reaction actually produces the passive behaviors. To address this matter, we attempted to reverse the behavioral effects of a protracted separation with an anti-inflammatory compound. For this work, we used alpha-melanocyte-stimulating hormone (α-MSH), an endogenous peptide in the brain and periphery with a wide range of anti-inflammatory effects, including the inhibition of proinflammatory cytokines, and the induction of anti-inflammatory cytokines that counter proinflammatory effects (Lipton et al., 1999; Luger, Scholzen, Brzoska, & Bohm, 2003). Because α-MSH is a peptide that does not appear to cross the blood-brain barrier in appreciable quantities, we administered α-MSH ICV through an indwelling cannula. Pups were isolated in the test cage for 3 hr on two occasions: once 60 min following infusion of 25 μg of α-MSH and once 60 min following infusion of vehicle. We found that α-

MSH significantly reduced the number of 60-s intervals that pups displayed crouching, eye-closing, or piloerection (Schiml-Webb et al., 2006; Figure 6). As seen in the figure, instances of active behavior tended to be increased, but there was a large degree of variation, and these results were not statistically significant. A follow-up experiment administered the same dose of α-MSH via the IP route. No behavioral effect whatsoever was detected following peripheral administration.

In sum, these results suggest that central proinflammatory cytokines do mediate the crouching, eye-closing, and piloerection of guinea pig pups during prolonged separation. have other influences on emotional and motivational states (Gonzalez, Vaziri, & Subheder, 1996; Rao et al., 2003), and findings to this point in our investigation had not demonstrated that the dose and administration procedures we used did indeed have anti-inflammatory effects in guinea pig pups when challenged with an inflammatory stimulus. Accordingly, in a follow-up experiment we attempted to use the same α-MSH dose and delivery procedures to reduce the frank sickness behaviors produced by LPS. We found that IP injection of LPS (200 μg/kg) robustly increased passive behaviors of pups during a 60-min isolation period, replicating our previously published findings, and that 25 μg of ICV α-MSH significantly reduced the effect of LPS on crouching and piloerection (Hennessy, Schiml-Webb, Miller, Maken, Bullinger, & Deak, 2007). We conclude then that the procedures used by Schiml-Webb et al (2006) to attenuate passive responses during protracted separation are capable of inhibiting proinflammatory effects on behavior in guinea pig pups.

Another experiment (Hennessy et al., 2007) examined the effect of a second anti-inflammatory agent, indomethacin, which has a more specific anti-inflammatory action—that of blocking synthesis of prostaglandins, a downstream mediator of many cytokine effects. Injection of indomethacin (10 mg/kg, IP) 30 min prior to a 3-hr test significantly reduced the number of 60-s intervals in which otherwise untreated pups crouched or simultaneously exhibited all three passive responses (Figure 7). Because its anti-inflammatory effect is the only recognized means by which indomethacin would be expected to affect behavior, these findings provide strong convergent evidence that proinflammatory cytokines contribute to the passive behaviors of guinea pig pups during isolation. Moreover, the findings point to involvement of a prostaglandin-dependent pathway in the behavioral effect.

Other experiments have inquired as to whether proinflammatory cytokines mediate the effect of CRF on the passive behavior of guinea pig pups. In agreement with studies described above, pups injected subcutaneously with 10 μg of CRF 60 min prior to testing subsequently exhibited much less active behavior and much higher levels of passive responses during a 60-min isolation than did pups injected with saline. If, however, the pups were infused ICV with 25 μg of α-MSH at the same time as CRF injection, levels of each of the passive responses were reduced (Schiml-Webb, Deak, Miller, & Hennessy, submitted; Table 1). Active behaviors were not affected. These results suggest that CRF may mobilize inflammatory-related factors, thereby resulting in the expression of sickness-like behaviors.

An additional experiment asked whether the same dose of α-MSH administered peripherally would diminish the effect of CRF on behavior. As was the case for ICV administration in the previous experiment, SC α-MSH significantly reduced levels of each of the three passive responses (Table 1). Again there was no effect on active behaviors. That is, regardless of whether the α-MSH was administered centrally or peripherally, it reduced the influence of CRF on behavior. It would seem that the most parsimonious explanation for these findings is that peripheral CRF administration activates peripheral components of the APR, which then trigger central proinflammatory processes that produce the passive responses. Disrupting the proinflammatory cascade in either the peripheral or central compartments attenuates the behavioral response.

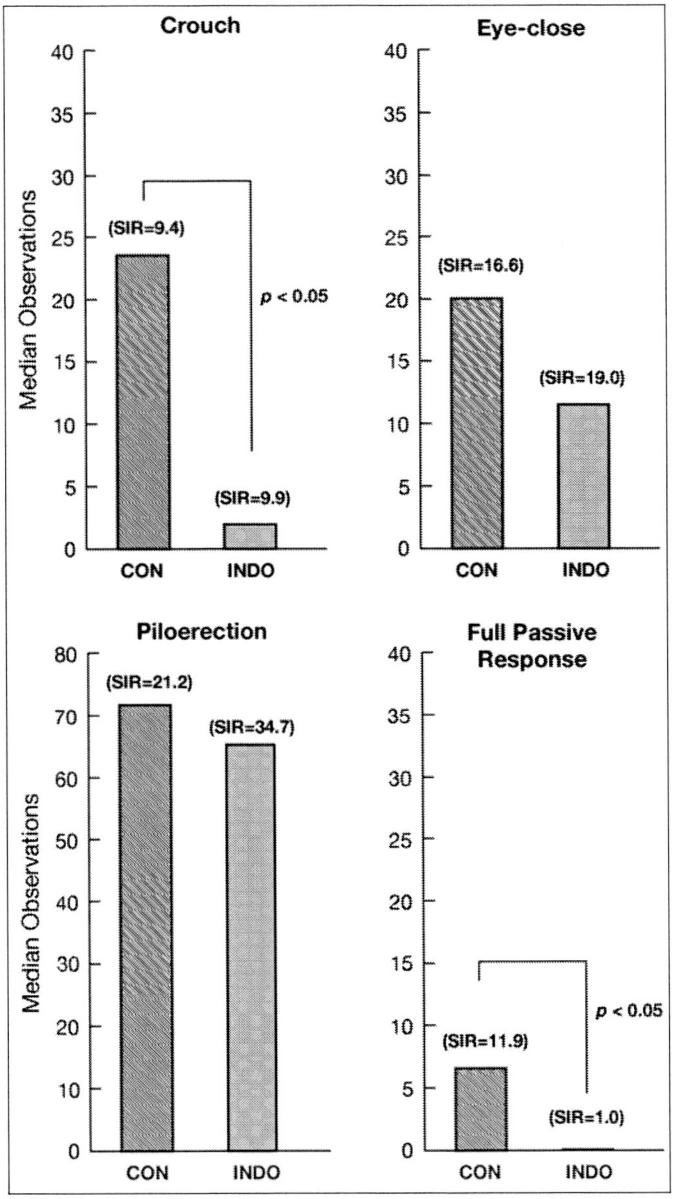

Figure 7. Median number of 60-s intervals in which guinea pigs exhibited crouching, eye-closing, piloerection, or all three of these behaviors during Min 0-30, 60-90, or 150-180 of a 180-min period of isolation. The controls (CON) were either not injected or injected with vehicle (these groups did not differ) and the INDO animals were injected with 10 mg/kg of indomethacin 30 min prior to testing. SIR indicates the semi-interquartile ranges. Reprinted with permission from Hennessy et al (2007).

Table 1. Median (and semi-interquartile range) behavioral scores for guinea pigs receiving CRF following central and peripheral administration of vehicle or α-MSH

	Vehicle	α-MSH
Central administration		
Vocalization	69.0 (72.6)	66.0 (161.0)
Line-crossing	1.0 (6.8)	4.0 (9.6)
Crouch	45.0 (11.0)	23.0 (13.1)*
Eye-close	45.0 (11.4)	23.5 (17.0)*
Piloerection	47.5 (8.6)	28.5 (22.4)*
Peripheral administration		
Vocalization	17.0 (68.9)	15.5 (44.8)
Line-crossing	0.0 (2.5)	0.0 (0.8)
Crouch	44.0 (5.3)	38.0 (9.6)*
Eye-close	32.0 (5.8)	21.0 (11.8)*
Piloerection	52.0 (9.8)	42.0 (10.9)*

*$p < 0.05$;

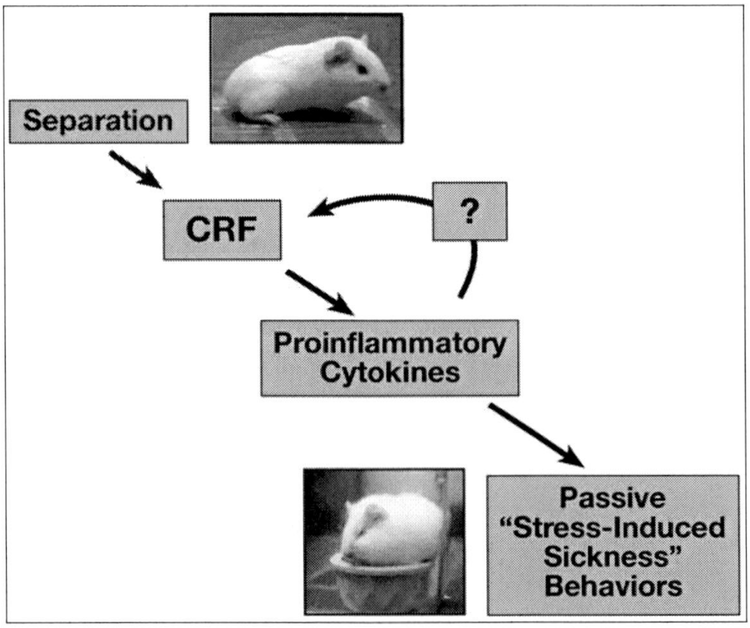

Figure 8. Schematic representation of the working model. The inset pictures are of guinea pig pups showing (bottom) and not showing (top) the passive behaviors of crouching, eye-closing, and piloerection.

Figure 8 is a simple schematic representation of our hypothesis. Separation from the maternal attachment figure in novel surroundings is posited to be a stressor that elicits release of CRF. Though we have not yet directly measured CRF during these separation procedures, the reliable elevation of plasma levels of ACTH and cortisol during such separations (e.g., Hennessy et al., 1989; Hennessy & Moorman, 1989) provides presumptive evidence for this point, as least as concerns release from the median eminence. Behaviorally, the release of CRF is proposed to induce the passive responses of crouching, eye-closing, and piloerection, while suppressing the active behaviors of vocalizing and locomotor activity. This action helps shift the animal from the initial active, to the subsequent passive, stage of responsiveness. This conclusion follows from studies cited above of the effects of administration of CRF and CRF-receptor antagonists (e.g., Hennessy et al., 1995; McInturf & Hennessy, 1996). CRF is hypothesized to induce passive responses in large part through activation of proinflammatory cytokines. The proposed cytokine effects on behavior are of the type referred to in the literature as stress-induced sickness behaviors (Maier &Watkins, 1998). Figure 8 also indicates a tentative pathway from the inflammatory reaction to the secretion of CRF. Proinflammatory cytokines, notably IL-1, appear to induce hypothalamic release of CRF (Turnbull & Rivier, 1999; Wieczorek & Dunn, 2006). Thus, this tentative pathway indicates the possibility of positive feedback in which not only does CRF induce proinflammatory cytokines, but the heightened level of proinflammatory cytokines prompts further CRF release.

The studies cited above provide overall support for the model, but they also raise additional questions that remain unresolved. Among these are the presumed proinflammatory cytokines involved, the site(s) of action (e.g., CNS or periphery) of the interaction between CRF and cytokines, and the nature and sequence of this interaction. Given the variety of proinflammatory cytokines identified, the complexity and reciprocal nature of their interactions with each other and CRF, and the presence of both cytokines and CRF in the CNS and periphery, there are a bewildering array of possibilities. As a single, though perhaps extreme, example, increased CRF release from peripheral sources in response to the isolation procedure might lead to release of cytokines in the periphery, which in turn might cause release of a different group of cytokines in the brain, which then effect behavioral change, perhaps through a CRF-mediated central mechanism. This interaction could be further promoted by positive feedback by cytokines on additional CRF.

Although such specifics of the peptide interactions seem impenetrable at the present time, the findings to date do clearly indicate that cytokines and other proinflammatory factors need to be considered when addressing the physiological mechanisms of behavior during social separation. Our subjects for this work have been guinea pigs, but the findings suggest a substitute for “despair” in conceptualizing some of the responses of primates to prolonged separation. Focusing on inflammatory factors does not discount the emotional component of separation responses, but rather may provide an alternative, objective approach to understanding one mechanism of the emotional reaction. As noted above, the responses of primates to prolonged separation have long been associated with, and are considered to provide insight about, the development of depressive illness (McKinney et al., 1984; Spitz, 1946). It is of particular interest, therefore, that recent evidence indicates that proinflammatory cytokines may contribute to the development of some forms of human depression (Schiepers, Wichers, & Maes, 2005). The findings presented here indicating a role of proinflammatory cytokines in the behavioral effects of separation may provide a means by which the link between separation and depression can further be explored.

Conclusion

The multiple, reciprocal connections between the nervous, endocrine, and immune systems have now been widely accepted for 25 years. Most of the original research into the function of these connections focused on neural and endocrine influences on the immune system and the apparent consequences of these influences for health. Recently, there has been increased attention on influences *from* the immune system to neural and endocrine activity, as well as behavior. Here we described how such influences seem to be at work in two long-established paradigms in the field of developmental psychobiology. Early immune activation appears to represent experience like any other, though proceeding mainly at the microscopic level. This activation affects the same systems, and in some of the same ways, as other forms of early experience. Its effects, like the effects of other early experiences, appear to be mediated in part by behavior of the mother, and current interpretations of both forms of early experience have focused on the role of early environmental exposure in predicting and preparing (i.e., programming) the animal for its likely adult environment.

In the case of maternal separation, immunological events appear to affect behavior in much the same manner as observed during actual physical illness, though these events seem to be triggered by the stress of the separation procedure rather than by pathogens. In all, the findings reviewed in this chapter indicate that immunological factors can have significant behavioral and endocrine influences in the context of traditional paradigms of developmental study. Further examination of such influences may provide fresh approaches for studying the relation between early life events and later normal and abnormal development.

Acknowledgements

Preparation of this chapter and previously unpublished work were supported by grant MH 068228 from the National Institute of Mental Health.

References

Abbey, H. & Howard, E. (1973). Statistical procedure in developmental studies on species with multiple offspring. *Developmental Psychobiology, 6*, 329-335.

Ader, R. (Ed., 1981). Psychoneuroimmunology. New York: Academic Press.

Aubert, A. (1999). Sickness and behaviour in animals: a motivational perspective. *Neuroscience and Biobehavioral Reviews, 23*, 1029-1036.

Baumann, H. & Gauldie, J. (1994). The acute phase response. *Immunology Today, 15*, 74-80.

Beach, F.A. (1948). Hormones and Behavior. New York: Paul B. Hoeber.

Becker, L.A. & Hennessy, M.B. (1993). Further characterization of the behavioral effects of peripherally administered corticotropin-releasing factor in guinea pigs. *Pharmacology Biochemistry and Behavior, 44*, 925-930.

Bell, R.W. & Denenberg, V.H. (1963). The interrelationships of shock and critical periods in infancy as they affect adult learning and activity. *Animal Behaviour, 11*, 21-27.

Bell, R.W., Nitschke, W., Bell, N.J., & Zachman, T.A. (1974). Early experience, ultrasonic vocalizations, and maternal responsiveness in rats. *Developmental Psychobiology, 7*, 235-242.

Bell, R.W., Nitschke, W., Gorry, T.H., & Zachman, T.A. (1971). Infantile stimulation and ultrasonic signaling: a possible mediator of early handling phenomena. *Developmental Psychobiology, 4*, 181-191.

Bernet, C.Z. & Stein, M.B. (1999). Relationship of childhood maltreatment to the onset and course of major depression in adulthood. *Depression and Anxiety, 9*, 169-174.

Blalock, J.E. (2005). The immune system as a sixth sense. *Journal of Internal Medicine, 257*, 126-138.

Blandino Jr., P., Barnum, C.J., & Deak, T. (2006). The involvement of norepinephrine and microglia in hypothalamic and splenic IL-1beta responses to stress. *Journal of Neuroimmunology, 173*, 87-95.

Blum, D. (2002). *Love at Goon Park: Harry Harlow and the Science of Affection*. New York: Berkeley Books.

Bowlby, J, Robertson, J., & Rosenbluth, D. (1952). A two-year-old goes to the hospital. *Psychoanalytical Study of the Child, 7*, 82-94.

Brown, G.W., Harris, T., & Copeland, J.R. (1977). Depression and loss. *British Journal of Psychiatry, 130*, 1-18.

Bruhn, T.O., Engeland, W.C., Anthony, E.L.P., Gann, D.S., & Jackson, I.M.D. (1987). Corticotropin-releasing factor in the adrenal medulla. *Annals of the New York Academy of Sciences, 512*, 115-128.

Coe, C.L. (1993). Psychosocial factors and immunity in nonhuman primates: a review. *Psychosomatic Medicine, 55*, 298-308.

Coplan, J.D., Andrews, M.W., Rosenblum, L.A., Owens, M.J., Friedman, S., Gorman, J.M., & Nemeroff, C.B. (1996). Persistent elevations of cerebrospinal fluid concentrations of corticotropin-releasing factor in adult nonhuman primates exposed to early-life stressors: implications for the pathophysiology of mood and anxiety disorders. *Proceedings of the National Academy of Sciences, 93*, 1619-1623.

Dantzer, R. (2004). Cytokine-induced sickness behaviour: a neuroimmune response to activation of innate immunity. *European Journal of Pharmacology, 500*, 399-411.

Deak, T., Border, K.A., McElderry, N.K., Barnum, C.J., Blandino, P., Deak, M.M., & Tammariello, S.P. (2005). Stress-induced increases in hypothalamic IL-1: a systematic comparison of multiple stressor paradigms. *Brain Research Bulletin, 64*, 541-556.

Deak, T., Meriwether, J.L., Fleshner, M., Spencer, R.L., Abouhamze, A., Moldawer, L.L., Grahn, R.E., Watkins, L.R., & Maier, S.F. (1997). Evidence that brief stress may induce the acute phase response in rats. *American Journal of Physiology, 273*, R1998-R2004.

DeNelsky, G.Y. & Denenberg, V.H. (1967). Infantile stimulation and adult exploratory behavior in the rat: effects of handling upon visual variation-seeking. *Animal Behaviour, 15*, 568-573.

Denenberg, V.H. (1967). Stimulation in infancy, emotional reactivity, and exploratory behavior. In: S.C. Glass (Ed.). *Biology and Behavior: Neurophysiology and Emotion* (pp. 161-222). New York: Russell Sage Foundation and Rockefeller Press.

Denenberg, V.H. (1977). Assessing the effects of early experience. In: R.D. Myers (Ed.). *Methods in Psychobiology, Vol. 3*. New York: Academic Press.

Denenberg, V.H. & Karas, G.G. (1960). Interactive effects of age and duration of infantile experience on adult learning. *Psychological Reports, 7*, 313-322.

Ghoshal, K., Wang, Y., Sheridan, J., & Jacob, S. (1998). Metallothionein induction in response to restraint stress. Transcriptional control, adaptation to stress, and the role of glucocorticoid. *Journal of Biological Chemistry, 273*, 27904-27910.

Gold, P.W., Goodwin, F.K., & Chrousos, G.P. (1988). Clinical and biochemical manifestations of depression: relation to the neurobiology of stress (Part 2). *New England Journal of Medicine, 319*, 413-420.

Gonzalez, M.I., Vaziri, S., & Wilson, C.A. (1996). Behavioral effects of α-MSH and MCH after central administration in the female rat. *Peptides, 17*, 171-177.

Granger, D.A., Hood, K.E., Dreschel, N.A., Sergeant, E. & Likos A. (2001). Developmental effects of early immune stress on aggressive, socially reactive, and inhibited behaviors. *Development and Psychopathology,* 13, 599-610.

Granger, D.A., Hood, K.E., Ikeda, S.C., Reed, C.L., & Block, M.L. (1996). Neonatal endotoxin exposure alters the development of social behavior and the hypothalamic-pituitary-adrenal axis in selectively bred mice. *Brain Behavior and Immunity, 10*, 249-259.

Hart, B.L. (1988). Biological basis of the behavior of sick animals. *Neuroscience and Biobehavioral Reviews, 12*, 123-137.

Harvey, A.T., Moore, H., Lucot, J.B., & Hennessy, M.B. (1994). Monoamine activity in anterior hypothalamus of guinea pig pups separated from their mothers. *Behavioral Neuroscience, 108*, 171-176.

Hennessy, M.B. (1997). Hypothalamic-pituitary-adrenal responses to brief social separation. *Neuroscience and Biobehavioral Reviews, 21*, 11-29.

Hennessy, M.B. (2003). Enduring maternal influences in a precocial rodent. *Developmental Psychobiology, 42*, 225-236.

Hennessy, M.B., Becker, L.A., & O'Neil, D.R. (1991). Peripherally administered CRF suppresses vocalizations of isolated guinea pig pups. *Physiology and Behavior, 50*, 17-22.

Hennessy, M.B., Davis, H.N., McCrea, A.E., Harvey, A.T., & Williams, M.T. (1999). Short- and long-term consequences of corticotropin-releasing factor in early development. *Annals of the New York Academy of Sciences, 897*, 76-91.

Hennessy, M.B., Deak, T., Schiml-Webb, P.A. (2001). Stress-induced sickness behaviors: An alternative hypothesis for responses during maternal separation. *Developmental Psychobiology, 39*, 76-83.

Hennessy, M.B., Deak, T., Schiml-Webb, P.A., Wilson, S.E., Greenlee, T.M., & McCall, E. (2004). Responses of guinea pig pups during isolation in a novel environment may represent stress-induced sickness behaviors. *Physiology and Behavior, 81*, 5-13.

Hennessy, M.B. & Jenkins, R. (1994). A descriptive analysis of nursing behavior in the guinea pig (*Cavia porcellus*). *Journal of Comparative Psychology, 108*, 23-28.

Hennessy, M.B., Long, S.J., Nigh, C.K., Williams, M.T., & Nolan, D.J. (1995). Effects of peripherally administered corticotropin-releasing factor (CRF) and a CRF antagonist: Does peripheral CRF activity mediate behavior of guinea pig pups during isolation? *Behavioral Neuroscience, 109*, 1137-1145.

Hennessy, M.B., Moorman, L. (1989). Factors influencing cortisol and behavioral responses to maternal separation in guinea pigs. *Behavioral Neuroscience, 103*, 378-385.

Hennessy, M.B. & Morris, A. (2005). Passive responses of young guinea pigs during exposure to a novel environment: influences of social partners and age. *Developmental Psychobiology, 46*, 86-96.

Hennessy, M.B., O'Neil, D.R., Becker, L.A., Jenkins, R. Williams, M.T., & Davis, H.N. (1992). Effects of centrally administered corticotropin-releasing factor (CRF) and α-helical CRF on the vocalizations of isolated guinea pig pups. *Pharmacology Biochemistry and Behavior, 43*, 37-43.

Hennessy, M.B., Schiml-Webb, P.A., Miller, E.E., Maken, D.S., Bullinger, K.L., & Deak, T. (2007). Anti-inflammatory agents attenuate the passive responses of guinea pig pups: evidence for stress-induced sickness behavior during maternal separation. *Psychoneuroendocrinology, 32,* 508-515.

Hennessy, M.B., Tamborski, A., Schiml, P.A., & Lucot, J.B. (1989). The influence of maternal separation on plasma concentrations of ACTH, epinephrine, and norepinephrine in guinea pig pups. *Physiology and Behavior, 45*, 1147-1152.

Hennessy, M.B., Vogt. J., & Levine, S. (1982). Strain of mother determines long-term effects of early handling: Evidence for maternal mediation. *Physiological Psychology, 10*, 153-157.

Hernandez, J., Carrasco, J., Belloso, E., Giralt, M., Bluethmann, H., Lee, D.K., Andrews, G.K., & Hidalgo, J. (2000). Metallothionein induction of restraint stress: role of glucocorticoids and IL-6. *Cytokine, 12*, 791-796.

Higley, J.D., Suomi, S.J., & Linnoila, M. (1991). CSF monoamine metabolite concentrations vary according to age, rearing, and sex, and are influenced by the stressor of social separation. *Psychopharmacology, 103*, 551-556.

Hodgson, D.M., Knott, B., & Walker, F.R. (2001). Neonatal endotoxin exposure influences HPA responsivity and impairs tumor immunity in Fischer 344 rats in adulthood. *Pediatric Research, 50*, 750-755.

Hood, K.E., Dreschel, N.A., & Granger, D.A. (2003). Maternal behavior changes after immune challenge of neonates with developmental effects on adult social behavior. *Developmental Psychobiology, 42*, 17-34.

Ito, A., Takii, T., Matsumara, T., & Onozaki, K. (1999). Augmentation of type I IL-1 receptor expression and IL-1 signaling by IL-6 and glucocorticoid in murine hepatocytes. *Journal of Immunology, 162*, 4260-4265.

Kaufman, I.C. & Rosenblum, L.A. (1967). The reaction to separation in infant monkeys: Anaclitic depression and conservation withdrawal. *Psychosomatic Medicine, 29,* 648-675.

König, B. (1985). Maternal activity budget during lactation in two species of Caviidae (*Cavia porcellus* and *Galea musteloides*). *Zeitschrift für Tierpsychologie, 68*, 215-230.

Kraemer, G.W., Ebert, M.H., Lake, C. R., & McKinney, W.T. (1984). Cerebrospinal fluid measures of neurotransmitter changes associated with pharmacological alteration of the despair response to social separation in rhesus monkeys. *Psychiatry Research, 11*, 303-315.

Jäckel, M. & Trillmich, F. (2003). Olfactory individual recognition of mothers by young guinea pigs (*Cavia porcellus*). *Ethology, 109*, 197-208.

Krukoff, T.L. (1986). Segmental distribution of corticotropin-releasing factor-like and vasoactive intestinal peptide-like immunoreactivities in presumptive sympathetic preganglionic neurons of the cat. *Brain Research, 382*, 153-157.

Ladd, C.O., Owens, M.J., & Nemeroff, C.B. (1996). Persistent changes in corticotropin-releasing factor neuronal systems induced by maternal separation. *Endocrinology, 137*, 1212-1218.

Laudenslager, M.L., Reite, M., & Harbeck, R.J. (1982). Suppressed immune response in infant monkeys associated with maternal separation. *Behavioral and Neural Biology, 36*, 40-48.

Leu, S.J. & Singh, L.K. (1992). Stimulation of interleukin-6 production by corticotrophin-releasing hormone. *Cellular Immunology, 143*, 220-227.

Lehmann, J. & Feldon, J. (2000). Long-term biobehavioral effects of maternal separation in the rat: consistent or confusing. *Reviews in the Neurosciences, 11*, 383-408.

Levine, S. (1956). A further study of infantlile handling and adult avoidance learning. *Journal of Personality*, 1956, 25, 70-80.

Levine, S. (1962). The effects of infantile experience on adult behavior. In: A.J. Bachrach (Ed.). *Experimental Foundations of Clinical Psychology* (pp.139-169). New York: Basic Books

Levine, S. (1966). Infantile stimulation and adaptation to stress. In: R. Levine (Ed.) *Endocrines and the Central Nervous System* (pp. 280-291). Baltimore: Williams and Wilkins.

Levine, S., Chevalier, J.A., & Korchin, S. J. (1956). The effects of early shock and handling on later avoidance learning. *Journal of Personality*, 1956, 475-493.

Levine, S. & Cohen, C. (1959). Differential survival to leukemia as a function of infantile stimulation in DBA/2 mice. *Proceedings of the Society for Experimental Biology and Medicine, 102*, 53-54.

Lipton, J.M., Zhao, H., Ichiyama, T., Barsh, G.S., & Catania, A. (1999). Mechanisms of antiinflammatory action of alpha-MSH peptides. In vivo and in vitro evidence. *Annals of the New York Academy of Sciences, 885,* 173-182.

Liu, D., Diorio, J., Tannenaum, B., Daldji, C., Francis, D., Freedman, A., Sharma, S., Pearson, D., Plotsky, P.M., & Meaney, M.J. (1997). Maternal care, hippocampal glucocorticoid receptors, and hypothalamic-pituitary-adrenal responses to stress. *Science, 277*, 1659-1662.

Livak, K.J. & Schmittgen, T.D.(2001). Analysis of relative gene expression data using real-time quantitative PCR and the 2(-Delta Delta C(T)) method. *Methods, 25*, 402-408.

Luger, T.A., Scholzen, T.E., Brzoska, T. Bohm, M. (2003). New insights into the functions of alpha-MSH and related peptides in the immune system. *Annals of the New York Academy of Sciences, 994*, 133-140.

Maier, S.F. & Watkins, L.R. (1998). Cytokines for psychologists: implications of bidirectional immune-to-brain communication for understanding behavior, mood, and cognition. *Psychological Review, 105*, 83-107.

Maier, S.F. & Watkins, L.R. (1999). Bidirectional communication between the brain and the immune system: implications for behaviour. *Animal Behaviour, 57*, 741-751.

Mancini, C., van Ameringen, M., & Macmillan, H. (1995). Relationship of childhood sexual and physical abuse to anxiety disorders. *Journal of Nervous and Mental Disease, 183*, 309-314.

Martins, J.M., Kastin, A.J., & Banks, W.A. (1996). Unidirectional specific and modulated brain to blood transport of corticotropin-releasing hormone. *Neuroendocrinology, 63*, 338-348.

McInturf, S.M. & Hennessy, M.B. (1996). Peripheral administration of a corticotropin-releasing factor antagonist increases the vocalizing and locomotor activity of isolated guinea pig pups. *Physiology and Behavior, 60*, 707-710.

McKinney, W.T., Moran, E., & Kraemer, G.W. (1984). Separation in nonhuman primates as a model for human depression: Neurobiological implications. In: R.M. Post & J. Ballenger (Eds.). *Neurobiology of Mood Disorders* (pp. 393-406). Baltimore: Williams and Wilkins.

Meaney, M.J., Aitken, D.H., van Berkel, C., Bhatnagar, S., & Sapolsky, R.M. Effect of neonatal handling on age-related impairments associated with the hippocampus. *Science, 239*, 766-768.

Meaney, M.J., Diorio, J., Francis, D., Widdowson, J., LaPlante, P., Caldji, C., Sharma, S., Seckl, J.R., & Plotsky, P.M. (1996). Early environmental regulation of forebrain glucocorticoid receptor gene expression: implications for adrenocortical responses to stress. *Developmental Neuroscience, 18*, 49-72.

Milligan, E.D., Nguyen, K.T., Deak, T., Hinde, J.L., Fleshner, M., Watkins, L.R., & Maier, S.F. (1998). The long term acute phase-like responses that follow acute stressor exposure are blocked by alpha-melanocyte stimulating hormone. *Brain Research, 810*, 48-58.

Mineka, S. & Suomi, S.J. (1978). Social separation in monkeys. *Psychological Bulletin, 85*, 1376-1400.

Nguyen, K.T., Deak, T., Owens, S.M., Kohno, T., Fleshner, M., Watkins, L.R., & Maier, S.F. (1998). Exposure to acute stress induces brain interleukin-1β protein in the rat. *Journal of Neuroscience, 18*, 2239-2246.

Nguyen, K.T., Deak, T., Will, M.J., Hansen, M.K., Hunsaker, B.N., Fleshner, M., Watkins, L.R., & Maier, S.F. (2000). Time course and corticosterone sensitivity of the brain, pituitary, and serum interleukin-1β protein response to acute stress. *Brain Research, 859*, 193-201.

Patchev, V.K., Montkowski, A., Rouskova, D., Koranyi, L., Holsboer, F., & Almeida, O.F.X. (1997). Neonatal treatment of rats with the neuroactive steroid tetrahydrodeoxycorticosterone (THDOC) abolishes the behavioral and neuroendocrine consequences of adverse early life events. *The Journal of Clinical Investigation, 99*, 962-966.

Pfeifer, W.D., Rotundo, R., Myers, M., & Denenberg, V.H. (1976). Stimulation in infancy: unique effects of handling. *Physiology and Behavior*, 17, 781-784.

Plotsky, P.M. & Meaney, M.J. (1993). Early, postnatal experience alters hypothalamic corticotropin-releasing factor (CRF) mRNA, median eminence CRF content and stress-induced release in adult rats. *Molecular Brain Research, 18*, 195-200.

Plotsky, P.M., Thrivikraman, K.V., Nemeroff, C.B., Caldji, C., Sharma, S., & Meaney, M.J. (2005). Long-term consequences of neonatal rearing on central corticotropin-releasing factor systems in adult male rat offspring. *Neuropsychopharmacology, 30*, 2192-2204.

Rao, T.L., Kokare, D.M., Sarkar, S., Khisti, R.T., Chopde, C.T., & Subhedar, N. (2003). GABAergic agents prevent alpha-melanocyte stimulating hormone induced anxiety and anorexia in rats. *Pharmacology, Biochemistry, and Behavior, 76*, 417-423.

Rots, N.Y., de Jong, J., Workel, J.O., Levine, S., Cools, A.R., De Kloet, E.R. (1996). Neonatal maternally deprived rats have as adults elevated basal pituitary-adrenal activity and enhanced susceptibility to apomorphine. *Journal of Neuroendocrinology, 8*, 501-506.

Russell, P.A. (1971). “Infantile stimulation” in rodents: a consideration of possible mechanisms. *Psychological Bulletin, 75*, 192-202.

Schiepers, O.J.G., Wichers, M.C., & Maes, M. (2005). Cytokines and major depression. *Progress in Neuro-Psychopharmacology and Biological Psychiatry, 29*, 201-217.

Schiml-Webb, P.A., Deak, T., Greenlee, T.M., Maken, D.S., & Hennessy, M.B. (2006). Alpha-melanocyte stimulating hormone reduces putative stress-induced sickness behaviors in isolated guinea pig pups. *Behavioural Brain Research, 168,* 326-330.

Schreiber, H.L., Bell, R.W., Kufner, M. & Villescas, R. (1977). Maternal behavior: a determinant of amphetamine toxicity in rats. *Psychopharmacology, 52*, 173-176.

Shanks, N. Larocque, S., & Meaney, M.J. (1995). Neonatal endotoxin exposure alters the development of the hypothalamic-pituitary-adrenal axis: early illness and later responsivity to stress.

Shanks, N. & Lightman, S.L. (2001). The maternal-neonatal neuro-immune interface: are there long-term implications for inflammatory or stress-related disease? *The Journal of Clinical Investigation, 108*, 1567-1573.

Shanks, N., Windle, R.J., Perks, P.A., Harbuz, M.S., Jessop, D.S., Ingram, C.D., & Lightman, S.L. (2000). Early-life exposure to endotoxin alters hypothalamic-pituitary-adrenal function and predisposition to inflammation. *Proceedings of the National Academy of Sciences, 97*, 5645-5650.

Singh, L.K., Pang, X., Alexacos, N., Letourneau, R., & Theoharides, T.C. (1999). Acute immobilization stress triggers skin mast cell degranulation via corticotropin-releasing hormone, neurotensin, and substance P: A link to neurogenic skin disorders. *Brain Behavior and Immunity, 13*, 225-239.

Smotherman, W.P. & Bell, R.W. (1980). Maternal mediation of early experience. In: R.W. Bell & W.P. Smotherman (Eds.) *Maternal Influences and Early Behavior*. New York: Spectrum Press.

Spencer, S.J., Heida, J.G., & Pittman, Q.J. (2005). Early life immune challenge—effects on behavioural indices of adult rat fear and anxiety. *Behavioural Brain Research, 164*, 231-238.

Spitz, R.A. (1945). Hospitalism: an inquiry into the genesis of psychiatric conditions in early childhood. *Psychoanalytical Study of the Child, 1*, 53-74.

Spitz, R.A. (1946). Anaclitic depression: An inquiry into the genesis of psychiatric conditions in early childhood II. *Psychoanalytical Study of the Child, 2*, 313-342.

Turnbull, A.V. & Rivier, C.L. (1999). Regulation of the hypothalamic-pituitary-adrenal axis by cytokines: actions and mechanisms of action. *Physiological Reviews, 79*, 1-71.

Udelsman, R., Harwood, J.P., Millan, M.A., Chrousos, G.P., Goldstein, D.S., Zimlichman, R., Catt, K.J., & Aguilera, G. (1986). Functional corticotropin-releasing factor receptors in the primate peripheral sympathetic nervous system. *Nature, 319*, 147-150.

Walker, F.R., Brogan, A., Smith, R., & Hodgson, D.M. (2004a). A profile of the immediate endocrine, metabolic and behavioural responses following a dual exposure to endotoxin in early life. *Physiology and Behavior, 83*, 495-504.

Walker, F.R., March, J., & Hodgson, D.M. (2004b). Endotoxin exposure in early life alters the development of anxiety-like behaviour in the Fischer 344 rat. *Behavioural Brain Research, 154*, 63-69.

Walker, F.R., Owens, J., Ali, S., & Hodgson, D.M. (2006). Individual differences in glucose homeostasis: do our early life interactions with bacteria matter? *Brain Behavior and Immunity, 20*, 401-409.

Watkins, L.R. & Maier, S.F. (2005). Immune regulation of central nervous system functions: from stress responses to pathological pain. *Journal of Internal Medicine, 257*, 139-155.

Webster, E.L., Lewis, D.B., Torpy, D.J., Zachman, E.K., Rice, K.C., & Chrousos, G.P. (1996). In vivo and in vitro characterization of antalarmin, a nonpeptide corticotropin-releasing hormone (CRH) receptor antagonist: Suppression of pituitary ACTH release and peripheral inflammation. *Endocrinology, 137*, 5747-5750.

Weinberg, J. & Levine, S. (1977). Early handling influences on behavioral and physiological responses during active avoidance. *Developmental Psychobiology, 10*, 161-169.

Wieczorek, M. & Dunn, A.J. (2006). Relationships among the behavioral, noradrenergic, and pituitary-adrenal responses to interleukin-1 and the effects of indomethacin. *Brain, Behavior, and Immunity, 20*, 477-487.

Wigger, A. & Neumann, I.D. (1999). Periodic maternal deprivation induces gender-dependent alterations in behavioral and neuroendocrine responses to emotional stress in adult rats. *Physiology and Behavior, 66*, 293-302.

Willner, P. (1991). Animal models as simulations of depression. *Trends in Pharmacological Sciences, 12*, 131-136.

Zorrilla, E.P. (1997). Multiparous species present problems (and possibilities) to developmentalists. *Developmental Psychobiology, 30*, 141-150.

In: Psychoneuroendocrinology Research Trends
Editor: Martina T. Czerbska, pp. 321-341
ISBN: 978-1-60021-665-7

Chapter X

THYROID HORMONES, SEROTONIN AND BEHAVIOR: THE ROLE OF GENOTYPE

Alexander V. Kulikov and Nina K. Popova*
Institute of Cytology and Genetics, Russian Academy of Sciences, Novosibirsk, Russia.

ABSTRACT

Crucial role of thyroid hormones (TH) in the regulation of development and function of the brain is commonly accepted. Thyroid deficit in early development results in mental retardation or cretinism. Numerous clinical observations indicate association of mental disorders in adult patients both with excess and deficit of TH. Despite an abundance of clinical data, the processes underlying the link between TH and mood disorder as well as individual sensitivity to thyroid hormone disturbances are still obscure. This essay concentrates on the role of the brain serotonin (5-HT) system in the genetically defined sensitivity to excess and deficit of TH. It describes: (1) the effect of TH alterations on such kinds of behavior as locomotion, anxiety-related, depressive-like, startle reflex, sexual and freezing (catalepsy); (2) the effect of TH excess and deficit on the key enzyme of 5-HT synthesis in the brain, tryptophan hydroxylase-2, 5-HT transporter and 5-HT receptors; (3) the interaction between TH, the 5-HT system and behavior. The review provides converging lines of evidence that: (1) both deficit and excess of TH depress sexual motivation, (2) TH deficit predisposes animal to catalepsy, (3) the effect of T4 administration on catalepsy is dependent on genotype, (4) TH produce a consistent tonic effect on the brain 5-HT_{2A} serotonin receptor. In general, these experimental findings are in good agreement with the clinical observations of implication of the 5-HT system in the pathway TH – behavior and suggest the 5-HT_{2A} receptor as the most probable molecular link between TH and mental disorders. Significant role of genetic factors in the sensitivity to T4 treatment was shown.

* Correspondence concerning this article should be addressed to Prof. Popova Nina K. Laboratory of Behavioral Neurogenomics, Institute of Cytology and Genetics, 10 Lavrentiev Avenue, Novosibirsk, 630090, Russia. Fax: 7-383-3331278; Phone: 7-383-3323101; e-mail: npopova@bionet.nsc.ru.

INTRODUCTION

Thyroid hormones (TH), thyroxine (T4) and triiodothyronine (T3), play a crucial role in the central nervous system development: at the early period of ontogenesis they stop neuronal proliferation and start differentiation, later they function as trophic factors necessary for nervous cells survival [66]. Considerable alterations in TH concentrations at the early ontogenesis result in grave brain disturbances and mental retardation.

The role of TH in the regulation of brain functions and in the mechanisms of mood disorders in adults is still obscure. During the last decade about 300 articles and 65 reviews concerning the association between thyroid dysfunction and the risk of depression were published. However, the clinical observations are controversial and evidence link a of depression with activation [43,68,76] and attenuation [9-12] of thyroid function or find no association at all [42]. Another direction of clinical investigations was focused on the application of TH along or together with antidepressants for depression treatment. It was demonstrated that high doses of T4 were able to ameliorate the mood in some depressive patients [11,12]. Adjuvant therapy with low doses of T3 sometimes augments and accelerates response to tricyclic antidepressants [42,76].

The abundance of clinical data contrasts with a shortage of consistent experimental data: there are only a few studies on TH involvement in the regulation of interneuronal interaction in the adult brain. Although, the studies of effects of experimental TH alterations on brain neurotransmitters and animal behavior are scanty, several fundamental reviews summarizing possible functions of TH in the brain and their effects on brain neurotransmitters were published [9,10,21,23,45,99].

At the present time, the brain serotonin system is considered as the most possible mediator between TH and depression [9,45], since it is implicated in the mechanism of depression and its molecular entities are the targets for many antidepressants [62]. The data on effects of experimental alterations of TH on the metabolism, transporter and receptors of serotonin were reviewed and a possible role of the brain 5-HT$_{1A}$ and 5-HT$_{2A}$ serotonin receptors as mediators between TH and depression was proposed [9].

At the same time, a considerable individual variability in the antidepressant effect of T4 was noted [10]. The variations in sensitivity to hormone appear to be dependent on genetically defined peculiarities of the brain 5-HT system. However this problem was never discussed in detail.

The experiments carried out in our laboratory during the last ten years allow us to clarify this problem. Thus, this essay concentrates on the involvement of TH in the regulation of behavior and on the role of brain 5-HT system in the determination of genetically defined sensitivity to excess and deficit of TH. It discusses: (1) the effect of TH on various kinds of animal behavior, (2) the effect of TH excess and deficit on the key enzyme of 5-HT synthesis in the brain, tryptophan hydroxylase-2, 5-HT transporter and 5-HT receptors and (3) the interaction between genotype, brain 5-HT system and TH in the regulation of behavior.

ROLE OF THYROID HORMONES IN MATURE NERVOUS SYSTEM

The most impressive proof of the importance of TH in the mammalian adult nervous system seems to be the presence of thyroid hormone homeostasis which maintains the brain concentrations of TH in relatively narrow boundaries regardless considerable variation of their blood concentration [21,22]. The homeostasis is based on the correlative changes of activities of 5'-deiodinase I (5'DI) in the liver and 5'-deiodinase II (5'-DII) in the brain in response to blood TH alterations. TH deficit inhibits 5'DI activity in the liver and T4 is converted to T3 predominantly in the brain. On the contrary, excess of TH attenuates the brain 5'DII activity and directs T4 to the liver [21].

At the same time, a little is known about the processes regulated by TH in adult brain. Dratman and Gordon [23] hypothesized that T3 is a neuromodulator or neurotransmitter. Indeed, T3 meets two main requirements for putative neurotransmitter, namely 1) it is synthesized from T4 in the central nervous system and high activity of the enzyme 5'D II converted T4 to T3 was demonstrated in the brain [21,23,30] and 2) T3 is released by a calcium-dependent mechanism [65]. Moreover, T3 can attenuate GABA-A receptor function [63]. Dratman and Gordon [23] suggest that in the adult brain T3 can function as a co-transmitter of noradrenalin.

The neurotransmitter hypothesis predicts a rapid behavioral response to acute T3 treatment. Unfortunately this attractive hypothesis is inconsistent with most experimental and clinical observations which do not show rapid behavioral alteration in reply to acute administration of moderate doses of TH. Moreover, molecular structure of this hypothetic membrane T3 receptor is also unknown. Therefore, a function of T3 as a putative brain neurotransmitter remains still unproved.

There are consistent molecular evidences only for hormonal effects of T3 mediated via the nuclear receptors, which was identified as the product of the c-erb A proto-oncogene [21]. The receptor molecule includes the hormone binding, DNA-binding and dimerisation domains. The latter domain binds two TH receptor molecules together to produce the active dimer structure. The hormone binding domain binds T3 and then the dimer complex links to specific palindromic sequence – thyroid response element (TRE) in the promoter of thyroid regulated genes (each monomer binds to a half-palindrome) and thus regulates the gene transcription [21]. Two different T3 receptors genes, namely TRα and TRβ, were distinguished. Each of these types of receptor molecules is presented in two forms (TRα_1, TRα_2 and TRβ_1, TRβ_2) resulted from alternative splicing of the corresponding receptor genes.

The only functional TH receptor in the mammalian brain is TRα_1 [21]. The receptor mediates the TH participation in the maintenance of neuronal plasticity in the mature brain similar to the way the growth factors do. Thyroid deficit produced by thyroidectomy reversibly decreased the density of dendritic spines of cortical pyramidal cells. This thyroidectomy-induced spines reduction is partially reversed with chronic T4 treatment [81].

Thus, at present the hormonal and trophic roles of TH in the regulation of neuronal plasticity seem to be the most consistent and experimentally proved.

THYROID HORMONES AND BEHAVIOR

Numerous clinical data reviewed by Joffe and Levitt [42] evidenced a link of depression with activation and attenuation of thyroid function or found no association at all. One mean of understanding this contradiction is to study the effects of experimental TH alterations on behavior of laboratory animals.

In organism TH are bound to the thyroid, various tissues or blood proteins [21]. This huge pool of the bound TH is a kind of natural buffer ensuring relative stability of free hormone concentration in blood. Another mechanism of TH homeostasis in the brain is the mentioned above cooperation between 5'DI in the liver and 5'DII in the brain described by Dratman [21]. These two mechanisms provide resistance of TH hormone concentration in the brain to acute treatment of moderate doses of hormones. Only prolonged inhibition of synthesis or chronic administration of exogenous hormones is able to alter TH level in the brain and produce some effect on the brain functions and behavior.

Thyroidectomy or chronic administration of thyroxine synthesis inhibitors, propylthiouracil (PTU) or methymazole, are widely used means of TH level reduction. A considerable decrease in T4 and T3 levels in the blood was shown three weeks after thyroidectomy [8,25,53,58,61] or PTU treatment [37,60,95]. The main disadvantage of thyroidectomy is removing along with thyroid parathyroid cells as well and, therefore, the disturbance of calcium metabolism in the organism, although the problem is partially solved by addition of 0.5% $CaCl_2$ in drinking water. Chronic treatment of T4 synthesis inhibitors, such as PTU, produced effective decrease of TH concentration in blood [37,60,95]. At the same, the drug does not seem to affect parathyroid cells. However, PTU is a strong inhibitor of 5'DI deiodinase in the liver [92] and chronic administration of the drug can disturb the cooperation between the enzymes in the liver and brain and shift the main direction of T4 metabolism from the liver to the brain. This can lead to an increase of T3 level in the brain. Sarkar and Ray [86] found three-fold increase of T3 concentration in synaptosomes in the cortex of rats treated for 3 weeks with PTU.

The main means of experimental elevation of blood TH concentration are chronic administration of T3 or T4. Some authors used s.c. [33,85,94] or i.p.[64,80,81] injection of hormones. This procedure is stressful for animal and can not be recommended for prolonged treatment. Other authors used implanted minipumps to minimize the negative effect of subcutaneous administration [25]. On contrast to i.p, s.c. injections or minipump, the administration of TH in drinking water is not stressful. It has been shown that chronic administration of supported and moderate doses of T4 in drinking water normalized blood TH levels in thyroidectomized rats [58,61] and ensured hyperthyroidism in euthyroid rats [53,60,95].

Some authors treated animals with T3 to ensure direct and rapid action of the hormone on brain and behavior. However, we have found that even moderate doses of T3 markedly increased heart rate and risk of animal mortality. So, administration of T3 should be restricted either by low doses used or by short time of treatment. On the contrary, T4 administration does not produce any serious negative effects and can be recommended for chronic administration. In our numerous experiments rats were treated with 100 μg/kg of T4 for 28 days without visible effect on the animal health [60,95].

The data on effect of TH alterations on various kinds of behavior in adult animals are summarized in the Tables 1 and 2.

Table 1. Behavioral alterations induced by adult hypothyroidism

Behavior	Animals and Treatments	Effects	References
Locomotion	Long-Evans rats, PTU	0	[29]
	Wistar and Wistar-Kyoto rats, thyroidectomy, PTU	0	[80]
	Wistar rats, methymazole, 0.02%	0	[82]
	Wistar rats, thyroidectomy	0	[8]
	GC rats, thyroidectomy	↑	[8]
	GC rats, hereditary hypothyroidism	↓	[7]
	hyt/hyt mice, hereditary hypothyroidism	↓	[4]
	Wistar rats, PTU, 0.005%	0	[95]
Anxiety-related behavior, elevated plus-maze	Wistar rats, methymazole,	↓	[82]
	Wistar and GC rats	0	[8]
Depressive-like behavior, forced swim immobility	Sprague-Dawley rats, PTU	↓	[40]
	Wistar and Wistar Kyoto, thyroidectomy, PTU	0	[80]
	Wistar, thyroidectomy	↑	[61]
	GC rats, hereditary hypothyroidism	↑	[70]
Startle reflex	Wistar rats, PTU	↓	[95]
	GC rats, hereditary hypothyroidism	↑	[73]
	Rdw rats, congenital hypothyroidism	↓	[41]
Sexual behavior	Wistar-Imamichi rats, thyroidectomy	↓	[96]
	Wistar rats, PTU	↓	[95]
Catalepsy (freezing)	GC rats, hereditary hypothyroidism	↑	[8]
	Wistar, thyroidectomy	↑	[8]
	Wistar, PTU	↑	[60, 95]

The main part of experimental studies was performed to investigate effect of TH dysfunctions on the locomotion in the open field test. Some authors disclosed an increase of locomotor activity in rats with neonatal [1,19,49,69,93] or adult [26] hypothyroidism, while others did not revealed any effect of TH deficit on locomotion of adult animals [8,29,80,82,95]. However, a tendency to decreased locomotion in the open field in thyroidectomized [8] and treated with PTU [95] Wistar rats should be noted. Significant decrease of the locomotion in hyt/hyt mice [4] and GC rats [7,8] with genetically defined hypothyroidism was shown. Paradoxically, surgical thyroidectomy produced a twofold increase of locomotion in GC rats [8].

Table 2. Effect of T3 and T4 on behavior of adult rats

Behavior	Animals and Treatments	Effects	References
Locomotion	Long-Evans, T3	0	[29]
	Wistar, T3 or T4	↑	[80]
	Wistar-Kyoto, T3 or T4	0	[80]
	Wistar, T4	0	[95]
Anxiety-related behavior, elevated plus-maze	Wistar, T4	↑	[19]
	Wistar, T3	0	[82]
Depressive-like behavior, forced swim immobility	Wistar, T3 and T4	↓	[80]
Startle reflex amplitude	Wistar, T4	0	[95]
Sexual behavior	Wistar, T4	↓	[95]
Catalepsy	Wistar, T4	0	[60, 95]
	GC, T4	↓	[60]

0 – no effect, ↑ - activate or increase, ↓ - decrease.

Effect of chronic T4 treatment on locomotion seems to be dependent on animal genotype and dose. Moderate dose of T4 (100 μg/kg) did not affect locomotion [95]. At the same time, high doses of T3 and T4 significantly increase locomotion in Wistar, but not in Wistar-Kyoto [80].

The forced swim test introduced by Porsolt et al. [75] is the main test for depressive-like behavior and for screening antidepressant drugs. Since about 90% antidepressants significantly reduce immobility time, the latter was suggested as a valuable index of "depression" [100]. The data on the association between TH and forced swim immobility are contradictory. Jefferys and Funder [40] demonstrated that hypothyroidism produced by chronic PTU treatment prevented acquisition of immobility response in Sprague-Dawley rats. Acute T4 administration an hour before pretest restored normal acquisition. Other authors found no effect of PTU or thyroidectomy on immobility response in Wistar and Wistar-Kyoto rats. However, chronic treatment with T3 and T4 significantly increased immobility in Wistar, but not in Wistar-Kyoto rats [80]. We have shown that thyroidectomy significantly increased immobility time in Wistar rats and the effect of hormone deficiency could be restored by chronic treatment with moderate doses of T4 [61]. Genetically hypothyroid GC rats were more immobile in the forced swim test compared to Wistar rats [70]. So, at present time no consistent data on the involvement of TH in regulation of depressive-like behavior are available.

The elevated plus-maze is the most used test for anxiety related behavior [18,77]. Rats usually avoid open arms in the elevated plus-maze. So, the number of entries in the open arms and time spent therein are negatively correlated with animals anxiety. No effect of thyroidectomy or PTU on the number of entries and time spent in the open arms in the elevated plus-maze in Wistar rats was found [8,95]. Chronic T4 administration did not also affect anxiety related behavior in the elevated plus-maze in Wistar rats [95].

The acoustic startle reflex is a rapid sensorimotor response elicited by a sudden and intense acoustic stimulus and mediated by a relatively simple neuronal circuit located in the lower brainstem. Neonatal hypothyroidism was shown to reduce the amplitude of startle reflex caused by hearing loss [27,28,90]. A significant reduction of startle reflex amplitude in adult Wistar rats treated with PTU was shown [95]. On the contrary, genetically defined hypothyroidism in GC rats was accompanied with a significant increase in the startle reflex amplitude [73]. Chronic T4 administration failed to affect the trait [95].

Sexual behavior is regarded as one of the most biologically essential forms of behavior. Many psychical disorders are accompanied with sexual dysfunctions. There are clinical evidences of association between sexual dysfunctions and thyroid disorders [39,101]. The disturbances in copulatory behavior in rats with congenital [41], pharmacological or thyroidectomy-induced hypothyroidism [96] were shown. Sexual arousal induced by presence of receptive female is the initial component of male sexual behavior. This trait is evaluated by the time which male spent at the partition separated the compartment with receptive female and by female-induced elevation of the testosterone level [3]. Both TH deficit and excess produced by respective chronic PTU or T4 treatments significantly reduced the time that male spent at the compartment with receptive female and blunted its testosterone response (Figure 1).

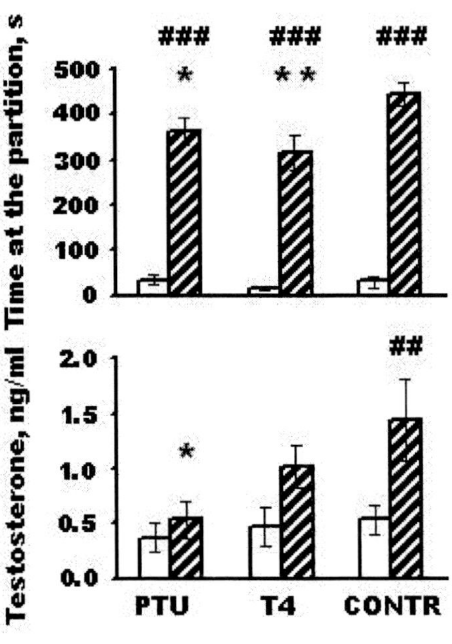

Figure 1. The effect of chronic PTU and T4 treatment on sexual arousal and testosterone level in Wistar males. Sexual arousal is evaluated by time which male spent at the partition separating receptive female (dashed bars) or empty compartment (open bars) [95]. $^{\#\#}p<0.01$, $^{\#\#\#}p<0.001$ vs empty compartment (sexual arousal) or vs resting testosterone level. $^{*}p<0.05$, $^{**}p<0.01$ vs control animals.

There are numerous and consistent evidences of the key role of TH in the regulation of hereditary catalepsy (animal hypnosis, tonic immobility). Catalepsy is a state of exaggerated immobility accompanied with a plastic muscle tonus. Animal or human in a state of catalepsy can maintain an imposed awkward posture for a long time. Catalepsy represents a natural defensive reaction against predator and it was found in all vertebrates [20,44,48]. Pronounced catalepsy in human, rats or mice is a syndrome of grave brain dysfunction [83]. In laboratory rodents catalepsy is usually induced by dopamine D_2 receptor agonists, such as haloperidol [47,83], while spontaneous freezing is rare. Prolonged selective breeding random bred Wistar rats for catalepsy resulted in the GC (Genetic Catalepsy) rat strain. About 50% of GC rats maintain a forced vertical position for at least 20 s or more, while only 5-10% of parental Wistar rats show similar immobility [7,50,51] (Figure 2).

Figure 2. Spontaneous catalepsy in a GC rat.

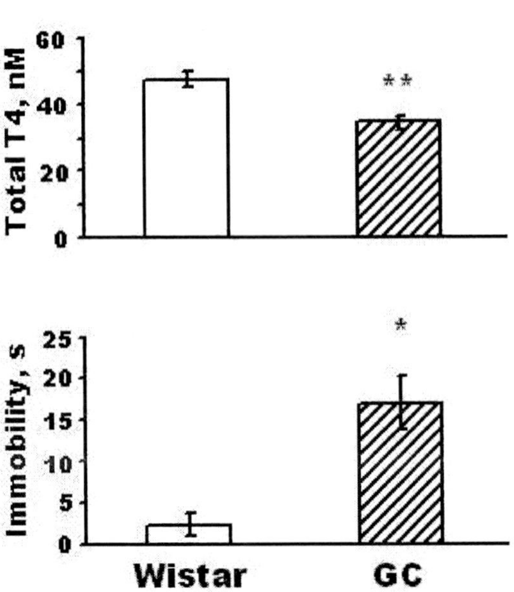

Figure 3. Blood T4 level and immobility time (s) in Wistar (open bars) and GC (dashed bars) rats [8]. *p<0.05, **p<0.01 vs Wistar rats.

A close association between catalepsy and TH was shown: 1) the selective breeding GC rat for catalepsy was accompanied with a marked decrease of their blood T4 level compared with parental Wistar rats (Figure 3); 2) chronic T4 administration reduced freezing in GC rats (Figure 4); 3) the TH deficit induced by thyroidectomy or chronic PTU treatment increased immobility time in Wistar rats (Figure 5) and 4) the thyroidectomy-induced catalepsy was prevented with supported dose of T4 (Figure 5A).

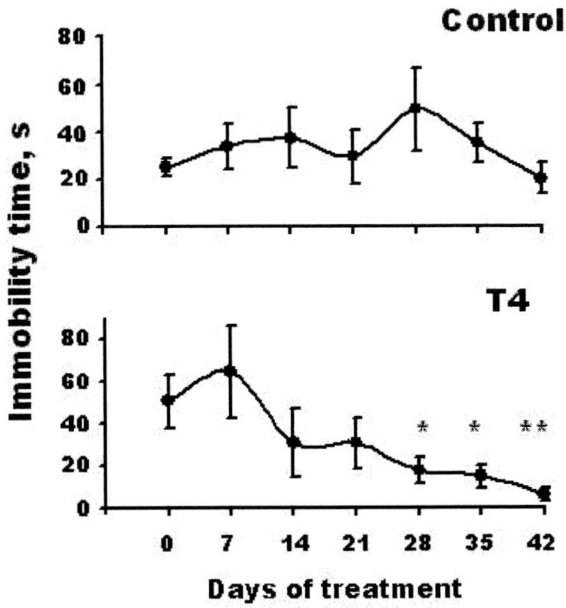

Figure 4. Dynamic of immobility time in control and T4 treated GC rats [53]. *p<0.05, **p<0.01 vs the respective initial values.

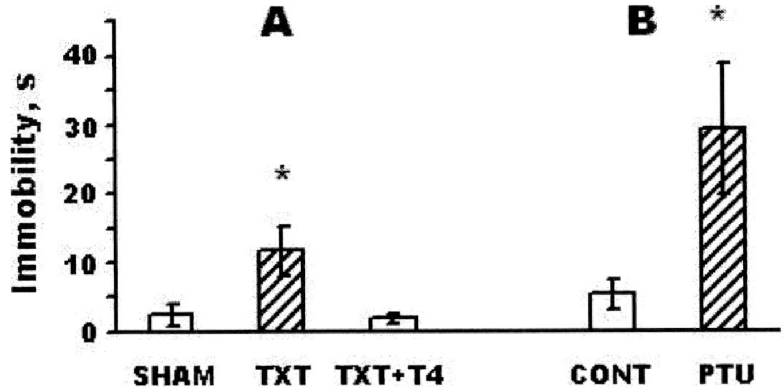

Figure 5. Effect of thyroidectomy (A) and chronic PTU treatment (B) on immobility time (s) in Wistar rats. SHAM – sham operated, TXT – thyroidectromized, TXT+T4, thyroidectomized rats treated with 15 μg/kg of T4, CONT – normal, PTU – PTU-treated rats [53,59,95]. *p<0.05, **p<0.01 vs respective control.

Therefore, these data present consistent evidences of TH involvement in the regulation of two biologically essential forms of behavior, sexual arousal and freezing. One possible explanation of close link between TH, sexual arousal and catalepsy is that TH affect the neurotransmitter participated in the regulation of these kinds of behavior. There are numerous evidences on the involvement of the brain 5-HT system in the regulation of sexual arousal [72] and catalepsy [74].

THYROID HORMONES AND THE BRAIN SEROTONIN SYSTEM

The key enzyme determined the rate of 5-HT synthesis in the brain is tryptophan hydroxylase-2 [97,98]. The main enzyme of 5-HT degradation is monoamine oxydase A (MAO A). Neurotransmitter binds to 14 subtypes of 5-HT receptors [6,103], which provide the molecular basis of the endless diverse of 5-HT regulation [87]. Among these subtypes the 5-HT_{1A} and 5-HT_{2A} are the most studied. These receptor types produced the opposite effects: 5-HT_{1A} receptor inhibits, while 5-HT_{2A} stimulates neurons [2,84]. The 5-HT_{1A} receptor located on the cell bodies of 5-HT neurons in the midbrain is presynaptic autoreceptor and being activated decreases their firing rate and the neurotransmitter release [2,36,91]. The 5-HT_{1A} receptor molecule is the direct target for buspirone-like anxiolytic drugs and is involved in the mechanism of antidepressant drugs [13,71]. The 5-HT_{2A} receptor participates in the mechanisms of some hallucinogens, neuroleptics and antidepressants [24]. 5-HT is removed from synaptic cleft with specific transporter molecule, which is considered to be the main factor of temporal and spatial regulation of 5-HT neurotransmission and the target for many antidepressant drugs [62,67].

It was hypothesized that TH affect mood via their action on the molecular entities [45]. Recently the studies on the effects of experimental thyroid disorders on the TPH-2, transporter and receptors of 5-HT in adult brain were reviewed by Bauer et al. [11]. The main problem of comparison of the results obtained using different animal genotypes, models and methods is evaluation of their relative weight or reliability.

Here, we analyze the data in the light of the results of repeated experiments carried out during the last several years. In order to understand interaction between TH and the brain 5-HT system, the effects of experimental decrease and increase of the blood TH levels on the turnover, TPH-2 activity, transporter and 5-HT_{1A} and 5-HT_{2A} receptors of 5-HT in the brain were investigated.

There were a few published results consistently indicating that neither deficit [46,58] nor excess [58,78] of TH affected TPH-2 activity. The enzyme activity in the hippocampus and midbrain of normothyroid Wistar and hypothyroid GC rats did not differ [56]. Similarly no effect of thyroidectomy or chronic T4 treatment on the density of 5-HT transporter in brain structures of Wistar [54,58] and Sprague-Dawley [94] rats was found.

The results obtained by two research groups [54,58,85] indicated absence of any effect of TH deficit and excess on 5-HT_{1A} receptor density in the brain structure of rats. Other authors [94] reported on thyroidectomy-induced increase of [^{3}H] 8-OH-DPAT binding in the cortex and hippocampus of Sprague-Dawley rats. Chronic treatment of the thyroidectomized rats with T4 failed to restore normal 5-HT_{1A} receptor density in the hippocampus [94].

The TPH-2, 5-HT_{1A} receptor and transporter control the synthesis, release and elimination of the mediator from the synaptic cleft, respectively, and, therefore, are the main factors of spatial and temporal regulation of 5-HT synapse function. No experimental evidence on involvement of TH in the regulation of the molecular entities is available. At the same time, numerous data showed the alterations in 5-HT turnover in the brain of hypo- and hyperthyroid rats compared with controls [9]. This discrepancy between the resistance of TPH-2, transporter and 5-HT_{1A} receptor to TH deviations and the sensitivity of 5-HT turnover to alterations the blood hormone level looks paradoxical and inconsistent. The paradox can be partially explained by a hypothetical effect of TH on the activity of the main enzyme of 5-HT degradation, MAO A. Although no experimental evidences on the effect of TH alterations on MAO A activity are published, this hypothesis is consistent with numerous data indicating that TH deficit, induced by thyroidectomy [34,35] or PTU [89], increased, while hyperthyroidism, produced by chronic administration of T4 [88] or T3 [35,78,79], decreased the 5-hydroxyindole acetic acid / 5-HT ratio in rat brain.

The role of TH in the 5-HT_{2A} receptor regulation is more studied. Mason et al. [64] showed that thyroidectomy did not affect [^{3}H] ketanserin binding to 5-HT_{2A} receptors in the cortex, striatum and hippocampus of Sprague-Dawley rats. We demonstrated that thyroidectomy significantly decreased [^{3}H] ketanserin binding to 5-HT_{2A} receptors in the cortex of Wistar rats. This decrease of receptor density was reversed by chronic administration of supported (15 μg/kg) and moderate (100 μg/kg) doses of T4 (Figure 6). A consistent significant attenuation of the 5-HT_{2A} receptor mRNA level in the frontal cortex of thyroidectomized Wistar rats was shown (Figure 7). These results indicate, that TH deficit is able to decrease the 5-HT_{2A} receptor activity, density and expression in rat brain. At the same time, a paradoxical activation of 5-HT_{2A} receptor in PTU-treated Wistar rats was reported (Figure 8). As it has been mentioned above, PTU inhibits 5'DI in the liver and directs the T4 stream to the brain. So, the observed receptor activation could be the result of increased T3 level in the brain of PTU-treated rats.

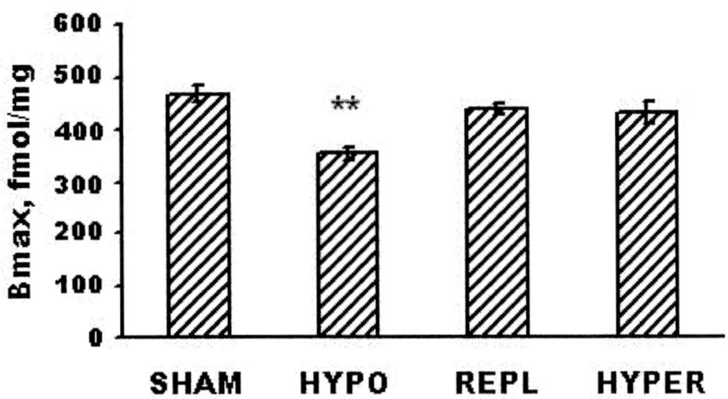

Figure 6. [^{3}H] ketanserin specific binding to 5-HT_{2A} serotonin receptors in the frontal cortex in the sham operated (SHAM), thyroidectomized (TXT) and thyroidectomized rats treated with replacing (REPL) and hyperthyroid (HYPER) doses of T4 males of Wistar rats [54,58]. Thyroidectomized rats in the REPL and HYPER groups chronically received 15 and 100 μg/kg of T4 and were eu- and hyperthyroid, respectively. $^{**}p<0.01$ vs SHAM.

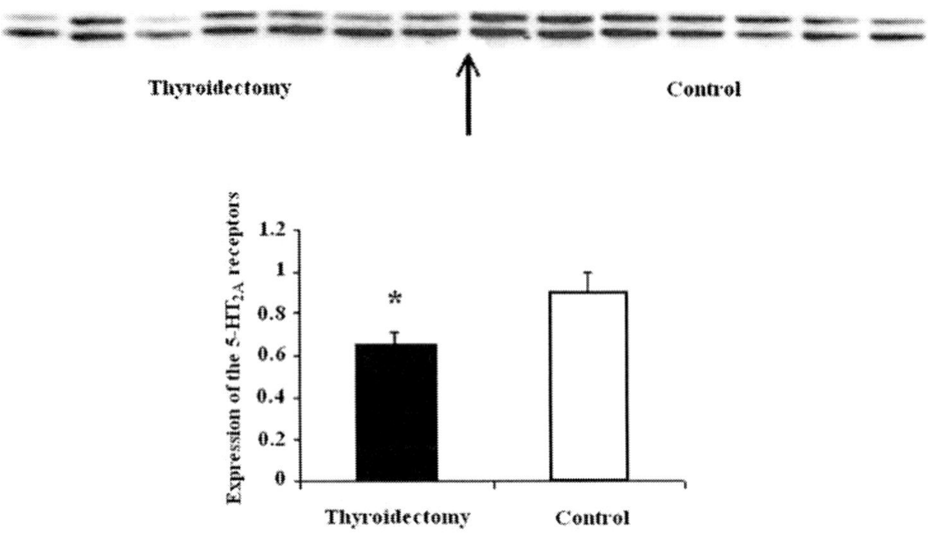

Figure 7. 5-HT_{2A} receptor mRNA level in the frontal cortex in the sham operated (SHAM) and thyroidectomized (TXT) males of Wistar rats. The mRNA level was evaluated with PT-PCR using β-actin mRNA as internal control [57]. Panel A. the PCR product bands for β-actin and 5-HT_{2A} receptor. Panel B. The 5-HT_{2A} mRNA /β-actin mRNA ratio. *$p<0.05$ vs SHAM.

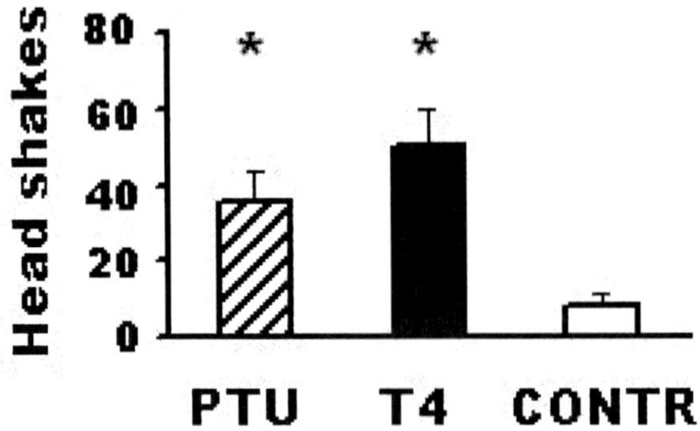

Figure 8. Effect of chronic PTU and T4 treatment on the number of DOI-induced head shakes in males of Wistar rats. Head shakes were counted for 20 min [95]. *$p<0.05$ vs control.

Chronic administration of a high dose of T3 (100 μg/kg) produced a significant reduction of 5-HT_{2A} receptor density in the cortex of mice [33] and Wistar rats [85]. At the same time no effect of chronic administration of moderate doses of T4 (100 μg/kg) on the cortical 5-HT_{2A} receptor density in Sprague-Dawley [64] and Wistar [54,58] rats was found. Chronic administration of 100 μg/kg T4 to Wistar rats also did not affect the receptor mRNA level in the frontal cortex, but significantly increased sensitivity of 5-HT_{2A} receptor to its agonist DOI.: a fivefold activation of DOI-induced head shakes in T4-treated compared to control rats was demonstrated (Figure 8). Mason et al. [64] using high doses of T4 (250-500 μg/kg) showed the increase in the number of [^{3}H] ketanserin binding sites in the striatum, hippocampus and midbrain of thyroidectomizrd Sprague-Dawley rats.

These data unambiguously indicate the involvement of TH in the tonic regulation of the 5-HT_{2A} receptor in the brain demonstrating that TH deficit attenuates, while its excess activates the receptors. At the same time, molecular mechanism of this regulation is not clear, since no TRE motif was found in the receptor promoter. It can be hypothesized that TH regulates 5-HT_{2A} receptor indirectly via alteration of the pyramidal cell plasticity. This hypothesis is based on two well established facts: 1) localization of 5-HT_{2A} receptors on the dendrites of pyramidal neurons [15,102] and 2) the involvement of TH in the regulation of the pyramidal neuron plasticity [81]. In the light of this hypothesis, the paradoxical depressive effect of T3 on the 5-HT_{2A} receptor density in the experiments of some authors [33,85] could be explained as the toxic effect the high dose of the hormone on the pyramidal neuron plasticity.

THYROID HORMONES, 5-HT_{2A} SEROTONIN RECEPTOR AND BEHAVIOR: ROLE OF GENTYPE

An important and still unresolved problem is the mechanisms of sensitivity (or vulnerability) to TH dysfunction. In the preceding part of the essay the TH involvement into tonic regulation of the brain 5-HT_{2A} serotonin receptor has been shown. At the same time, both activity and density of the receptor in the brain is under genetic control. Significant interstrain differences in 5-hydroxytryptophan-induced head twitches as well as the [^{3}H] spiperone and [^{3}H] ketanserin specific binding to the 5-HT_{2A} receptor in the frontal cortex in mice were revealed [52,74].

The 5-HT_{2A} receptor is coupled positively to phospholipase C and being activated produces neuronal depolarization [2,84]. The receptor molecule possesses two sites: one for binding of antagonists, such as ketanserin or ritanserin, and another for binding of agonists [103]. The receptor is predominately located on the dendrites of pyramidal cell in the cortex [15,102]. Relatively high density of [^{3}H] ketanserin specific binding was found also in the striatum. Stimulation of 5-HT_{2A} receptor with DOI induced specific movement activation, namely head shakes in rats and head twitches in mice [32].

The participation of 5-HT_{2A} receptors in regulation of different component of sexual behavior was demonstrated for rats and mice [14,31,72]. Tikhonova et al. [95] found a negative association between the number of head shakes and sexual arousal in males of Wistar rats. This result is consistent with the observation of other authors indicated reciprocal relations between the number of head shakes and sexual behavior expression: chronic treatment with corticosterone inhibited sexual behavior and increased head shakes in rats [31], while melatonin facilitated sexual behavior and reduced functional activity of the 5-HT_{2A} receptor [14]. Therefore, the observed dysfunction of sexual arousal in rat males produced by TH deficit or excess can be mediated by an activation of the brain 5-HT_{2A} receptors [95].

Hereditary catalepsy in GC rats is accompanied with; 1) a decrease of 5-HT_{2A} receptor mRNA level in the frontal cortex [59], 2) the deficit of the density of specific [^{3}H] ketanserin binding sites in the striatum and 3) an attenuated functional activity of the 5-HT_{2A} receptor

[74]. However, interaction between the 5-HT_{2A} receptor and catalepsy is more complex, since the receptor antagonists, ritanserin and ketanserin fail to affect catalepsy [52,74].

At the same time, the 5-HT_{2A} receptor is undoubtedly involved in the molecular mechanism of thyroidectomy-induced catalepsy in Wistar rats. Indeed, thyroidectomy increased predisposition to catalepsy [8] and decreased 5-HT_{2A} receptor density [58] and the mRNA level [57] in the cortex, while treatment of thyroidectomized rats with a supported dose of T4 prevented both catalepsy [53] and the 5-HT_{2A} receptor depression [58]. However, catalepsy induced by chronic PTU administration was accompanied with activation of the receptor [95].

The interaction between TH, 5-HT_{2A} receptor and catalepsy is undoubtedly dependent on animal genotype. TH deficit induced catalepsy in Wistar rats with normal 5-HT_{2A} receptor, but it failed to alter immobility time in GC rats with the genetically defined decrease of the receptor expression and activity. On the contrary, chronic T4 treatment attenuated catalepsy in GC, but did not alter immobility time in Wistar rats [53]. Therefore, a link between the observed difference in the vulnerability to TH dysfunction and the brain 5-HT_{2A} receptor is hypothesized.

CONCLUSION

The essay reviewed the data on the role of TH in the regulation of locomotion, anxiety-related behavior in the plus-maze test, immobility in the forced swim test, startle reflex amplitude, sexual behavior and catalepsy. Consistent experimental evidences were obtained only for the TH involvement in the regulation of sexual arousal and catalepsy. Depression of sexual arousal produced by both TH deficit and excess indicated that an euthyroide level of hormones was needed for normal sexual motivation. Interaction between TH and catalepsy was unimodal: the deficit of TH reversibly increased risk of catalepsy in Wistar rats, while exogenous T4 attenuated immobility in genetically hypothyroid GC rats.

Effects of TH alterations on the functional elements of the brain 5-HT system, tryptophan hydroxylase-2, transporter, 5-HT_{1A} and 5-HT_{2A} receptors, were analyzed. It was demonstrated that among the studied molecular entities only the 5-HT_{2A} receptor was sensitive to TH dysfunction: thyroidectomy reversibly decreased the density and expression, while T4 activated the receptor. It was suggested that the observed cataleptogenic effect of thyroidectomy in rats could be the result of a decrease of cortical 5-HT_{2A} receptor density and expression.

The presented data bring up additional neurochemical evidences for the crucial role of genotype in the sensitivity to T4. Chronic T4 treatment significantly decreased hereditary catalepsy in GC rats. At the same time, chronic hormonal treatment failed to affect freezing in "healthy" Wistar rats. These findings can explain the clinically observed differences in sensitivity to chronic T4 treatment. It was revealed that lowered blood T4 level and 5-HT_{2A} receptor dysfunction were the main factors and biological predictors of the increased sensitivity to exogenous T4 treatment.

Spontaneous catalepsy in GC rats meets the criteria of face, predictive and construct validity for a new genetic model of depression [55]. It may be hypothesized that the depressive disorders which share their molecular and genetic mechanisms with the spontaneous catalepsy in rats can be provoked by TH deficit and accompanied with low 5-HT_{2A} activity. This suggestion is in good accordance with clinical observations. Clear et al. [16,17] found a group of depressives with low blood TH level and activity of 5-HT_{2A} receptors. Chronic treatment of these patients with T4 normalized their TH concentration, receptor activity and mood.

At the same time, many authors associated depression with elevated 5-HT_{2A} receptors receptor density [5,38] and with a hyperthyroid state of the brain [42,68]. Evidently, spontaneous catalepsy in rats is not an adequate animal model for modeling this type of depressive disorder.

Thus, this essay is another attempt to construct a bridge between clinical and experimental study. It points to the 5-HT_{2A} receptor as the most probable molecular link between thyroid hormone and the specific kind of depressive disorders and to the significance of genetic factors in the sensitivity to thyroid dysfunction.

REFERENCES

[1] Akaike, M., Kato, N., Ohno, H., & Kobayashi, T. (1991). Hyperactivity and spatial maze learning impairment of adult rats with temporary neonatal hypothyroidism. *Neurotoxicol. Teratol., 13*, 317-322.

[2] Aghajanian, G.K. (1995). Electrophysiology of serotonin receptor subtypes and signal transduction pathways. In F.R. Bloom, & D.J. Kupfer (Eds.), *Psychopharmacology: The Fourth Generation of Progress* (pp. 1451-1459). N.Y.: Raven Press.

[3] Amstislavskaya, T.G., & Popova, N.K. (2004). Female-induced sexual arousal in male mice and rats: behavioral and testosterone response. *Hormones and Behavior, 46*, 544-550.

[4] Anthony, A., Adams, P.M., & Stein, S.A.(1993). The effects of congenital hypothyroidism using the hyt/hyt mouse on locomotor activity and learned behavior. *Horm Behav., 27*, 418-33.

[5] Arora, R., & Meltzer, H.Y. (1989). Serotonergic measures in the brains of suicide victims: 5-HT_2 binding sites in the frontal cortex of suicide victims and control subjects. *Am. J. Psychiat., 146*, 730-736.

[6] Barnes, N.M., & Sharp, T. (1999). A review of central 5-HT receptors and their function. *Neuropharmacol., 38*, 1083-1152.

[7] Barykina, N., Chepkasov, I., Alekhina, T., & Kolpakov, V. (1983).Breeding of Wistar rats for prediaposition to catalepsy. *Genetika, 19*, 2014-2021 (in Russian).

[8] Barykina, N.N., Chuguy, V.F., Alekhina, T.A., Kolpakov, V.G., Maksiutova, A.V., & Kulikov, A.V. (2002) Effects of thyroid hormone deficiency on behavior in rat strains with different predisposition to catalepsy. *Physiol. Behav., 75*, 733-737.

[9] Bauer, M., Heinz, A., & Whybrow, P.C. (2002). Thyroid hormones, serotonin and mood: of synergy and significance in the adult brain. *Mol. Psychiatry, 7*, 149-156.

[10] Bauer, M., London, E.D., Silverman, D.H.S., Rasgon, N., Kirchheiner, J., & Whybrow, P.C. (2003). Thyroid, brain and mood modulation in affective disorder: insights from

molecular research and functional brain imaging. *Pharmacopsychiatry*, *36(Suppl. 3)*, S215-S221.

[11] Bauer, M.S., & Whybrow, P.C. (1988). Thyroid hormones and the central nervous system in affective illness: interactions that may have clinical significance. *Integrative Psychiatry*, *6*, 75-100.

[12] Bauer, M.S., & Whybrow, P.C. (1990). Rapid cycling bipolar affective disorders. II. Yteatment of refractory rapid cycling with high-dose levothyroxine: a preliminary study. *Arch. Gen. Psychiatr.*, *47*, 435-440.

[13] Blier, P., & de Montigny, C. (1994). Current advances and trends in the treatment of depression. *Trends Pharmacol. Sci.*, *15*, 220-226.

[14] Brotto, L.A., & Gorzalka, B.B. (2000). Melatonin enhances sexual behavior in the male rat. *Physiol. Behav.*, *68*, 483-486.

[15] Burnet, P.W.J., Eastwood, S.L., Lacey, K., & Harrison, P.J. (1995). The distribution of 5-HT1A and 5-HT2A receptor mRNA in human brain. *Brain Res.*, *676*, 157-168.

[16] Cleare, A.J., McGregor, A., & O'Kean, V. (1995) Neuroendocrine evidence for an association between hypothyroidism, reduced central 5-HT activity and depression. *Clin. Endocrinol.*, *43*, 713-719.

[17] Cleare, A.J., McGregor, A., Chambers, S.M., Dawling, S., & O'Kean, V. (1996). Thyroxine replacement increases central 5-HT activity and reduces depressive symptoms in hypothyroidism. *Neuroendocrinology*, *64*: 65-69.

[18] Cruz, A.P.M., Frei, F., & Graeff, F.G. (1994). Ethopharmacological analysis of rat behavior on the elevated plus-maze. *Pharmacol. Biochem. Behav.*, *49*, 171-176.

[19] Darbra, S., Balada, F., Garau, A., Gatell, P., Sala, J., & Marti-Carbonell, M.A. (1995). Perinatal alterations of thyroid hormones and behaviour in adult rats. *Behav Brain Res.*, *68*, 159-164.

[20] Dixon, A.K. (1998). Ethological strategies for defense in animals and humans: Their role in some psychiatric disorders. *Br. J. Med. Psychol.*, *71*, 417-445.

[21] Dratman, M.B. (1993). Cerebral versus peripheral regulation and utilization of thyroid hormones. In: R.T. Joffe, & A.J. Levitt (Eds.),.*The thyroid axis and psychiatric illness* (pp. 3-94), Washington: American Psychiatric Press Inc.

[22] Dratman, M.B., Crutchfield, F.L., Gordon, J.T., & Jennings, A.S. (1983). Iodothyronine homeostasis in rat brain during hypo- and hyperthyroidism. *Am. J. Physiol.*, *245*, E185-E193.

[23] Dratman, M.B., & Gordon, J.T. (1996). Thyroid hormones as neurotransmitters. *Thyroid*, *6*, 139-147.

[24] Eison, A.S., & Mullins, U.L. (1996). Regulation of central 5-HT_{2A} receptors: a review of in vivo studies. *Behav. Brain Res.*, *73*,.177-181.

[25] Escobar-Morreale, H.F., del Rey, F.E., Obregon, M.J., & Escobar, G.M. (1996). Only the combined treatment with thyroxine and triiodthyronine ensures euthyroidism in all tissues of the thyroidectomized rat. *Endocrinology*, *137*, 2490-2502.

[26] Fundaro, A. (1989). Behavioral modifications in relation to hypothyroidism and hyperthyroidism in adult rats. *Prog. Neuropsychopharmacol. Biol. Psychiatry*, *13,* 927-940.

[27] Goldey, E.S., & Crofton, K.M. (1998). Thyroxine replacement attenuates hypothyroxinemia, hearing loss, and motor deficits following developmental exposure to Aroclor 1254 in rats. *Toxicol. Sci., 45,* 94-105.

[28] Goldey, E.S., Kehn, L.S., Rehnberg, G.L., & Crofton, K.M. (1995). Effect of developmental hypothyroidism on auditory and motor function in the rat. *Toxicol. Appl. Pharmacol., 135*, 67-76.

[29] Gordon, C.J. (1997). Behavioral and autonomic thermoregulation in the rat following propylthiouracil-induced hypothyroidism. *Pharmacol. Biochem. Behav., 58*, 231-236.

[30] Gordon, J.T., Kominski, D.M., Rozanov, C.B., & Dratman, M.B. (1999). Evidence that 3,3',5-triiodothyronine is concentrated in and delivered from the locus coereleus to its noradrenergic targets via anterograde axonal transport. *Neurosci., 93*, 943-954.

[31] Gorzalka, B.B., & Hanson, L.A. (1998). Sexual behavior and wet dog shakes in the male rat: regulation by corticosterone. *Behav. Brain Res., 97*, 143-151.

[32] Green, A.R, & Heal, D.J. (1985). The effects of drugs on serotonin-mediated behavioural models. In: A.R. Green (Ed.),. *Neuropharmacology of serotonin* (pp. 326-365), Oxford: Oxford University.

[33] Heal, D.J., & Smith, S.L. (1988). The effects of acute and repeated administration of T3 to mice on 5-HT1 and 5-HT2 function in the brain and its influence on the actions of repeated electroconvulsive shock. *Neuropharmacol., 27*, 1239-1248.

[34] Henley, W.N., Bellush, L.L., & Tressler, M. (1998). Bulbospinal serotonergic activity during changes in thyroid status. *Can. J. Physiol. Pharmacol., 76*, 1120-1131.

[35] Henley, W.N., Chen, X., Klettner, C., Bellush, L.L., & Notestine, M.A. (1991). Hypothyroidism increases serotonin turnover and sympathetic activity in the adult rat. *Can. J. Physiol. Pharmacol.., 69*, 205-210.

[36] Hjorth, S., & Sharp, T. (1991). Effect of the 5-HT_{1A} receptor agonist 8-OH-DPAT on the release of 5-HT in dorsal and median raphe-innervated rat brain regions as measured by in vivo microdialysis. *Life Sci., 48*, 1779-1786.

[37] Hood, A., Liu, Y.P., GattoneII, V.H., & Klaassen, C.D. (1999). Sensitivity of thyroid gland growth to thyroid stimulating hormone (TSH) in rats treated with antithyroid drugs. *Toxicol. Sci., 49*, 263–271.

[38] Hrdina, P.D., Demeter, E., Vu, T.B., Sotonyi, P., & Palkovits, M. (1993). 5-HT uptake sites and 5-HT_2 receptors in brain of antidepressant-free suicide victims/ depressives: increase in 5-HT_2 sites in cortex and amygdale. *Brain Res., 614*, 37-44.

[39] Jaya Kumar, B., Khurana, M.L., Ammini, A.C., Karmarkar, M.G., & Ahuja, M.M.S. (1990). Reproductive endocrine functions in men with primary hypothyroidism: effect of thyroxine replacement. *Horm. Res., 34*, 215-218.

[40] Jefferys, D., & Funder, J.W. (1989). Thyroid hormones and the acquisition and retention of behavioural responses. *Endocrinol., 125*, 1103 –1105.

[41] Jiang, J.-Y., Umezu, M., & Sato, E. (2000). Characteristics of infertility and the improvement of fertility by thyroxine treatment in adult male hypothyroid *rdw* rats. *Biol. Reprod., 63*, 1637-1641.

[42] Joffe, R.T., & Levitt A.J. (1993). The thyroid and depression. In: R.T. Joffe, & A.J. Levitt (Eds.), *The thyroid axis and psychiatric illness* (pp. 195-253), Washington: American Psychiatric Press Inc.

[43] Joffe, R.T., Sokolov, T.H., & Singer, W. (1995). Thyroid hormone treatment of depression. *Thyroid, 5*, 235 –239.

[44] Karmanova, I.G. (1982). *Evolution of Sleep. Stages of the Formation of the Wakefulness-Sleep Cycle in Vertebrates*, Basel: Karger.

[45] Kirkegaard, C., & Faber, J. (1998). The role of thyroid hormones in depression. *Eur. J. Endocrinol.*, *138*, 1-9.

[46] Kizer, J.S., Palkovits, M., Kopin, I.J., Saavedra, J.M., & Brownstein, M.J. (1976). Lack of effect of various endocrine manipulations on tryptophan hydroxylase activity of individual nuclei of the hypothalamus, limbic system, and midbrain of the rat. *Endocrinology*, *98*,743-747.

[47] Klemm, W.R. (1989). Drug effects on active immobility responses: What they tell us about neurotransmitter systems and motor functions. *Progress in Neurobiology*, *32*, 403-422.

[48] Klemm, W.R. (1990). Behavioral inhibition. In W.R. Klemm, & R.P. Vertes (eds.), *Brainstem Mechanisms of Behavior* (pp.497-533). N.Y.: John Wiley & Sons.

[49] Kobayashi, K., Tsuji, R., Yoshioka, T., Kushida, M., Yabushita, S., Sasaki, M., Mino, T., & Seki, T. (2005). Effects of hypothyroidism induced by perinatal exposure to PTU on rat behavior and synaptic gene expression. *Toxicology*, *212*, 135-147.

[50] Kolpakov, V., Barykina, N.N., Alekhina, T.A., & Chepkassov, I. (1985). Catalepsy in rats: Its inheritance and relationship to pendulum movements and audiogenic epilepsy. *Behav. Proc.*, *10*, 63-76.

[51] Kolpakov, V.G., Barykina, N.N., & Chepkasov, I. (1981). Genetic predisposition to catatonic behaviour and methylphenidate sensitivity in rats. *Behav. Proc.*, *6*, 269-281.

[52] Kulikov, A.V., Avgustinovich, D.F., Kolpakov, V.G., Maslova, G.B., & Popova, N.K. (1995). 5-HT_{2A} serotonin receptors in the brain of rats and mice hereditarily predisposed to catalepsy. *Pharmacol. Biochem. Behav.*, *50*, 383-387.

[53] Kulikov, A.V., Barykina, N.N., Tikhonova, M.A., Chuguy, V.F., Kolpakov, V.G., & Popova, N.K. (2002). Effect of chronic thyroxine treatment on catalepsy in rats. *Neurosci. Lett.*, *330*, 207-209.

[54] Kulikov, A., & Jeanningro, R. (2001). The effects of hypothyroidism on 5-HT1A and 5-HT2A receptors and the serotonin transporter protein in the rat brain. *Neurosci. Behav. Physiol.*, *31*, 445-449.

[55] Kulikov, A.V., Kolpakov, V.G., & Popova, N.K. (2006). The genetic cataleptic (GC) rat strain as a model of depressive disorders. In A. Kaluev (Ed.), *Animal Models in Biological Psychiatry* (pp. 59-73), N.Y.: Nova Science Publishers.

[56] Kulikov, A., Kozlachkova, E., & Popova, N. (1992). The activity of tryptophan hydroxylase in brain of hereditary predisposed to catalepsy rats. *Pharmacol. Biochem. Behav.*, *43*, 999-1003.

[57] Kulikov, A.V., Maksyutova, A.V., Ivanova, E.A., Khvorostov, I.B., & Popova, N.K. (2002). The effect of thyroidectomy on the expression of the mRNA of 5-HT2A serotonin receptors in the rat frontal cortex. *Dokld. Biochem. Biophys.*, *383*, 116-118.

[58] Kulikov, A.V., Moreau, X., & Jeanningros, R. (1999). Effects of experimental hypothyroidism on 5-HT_{1A}, 5-HT_{2A} receptors, 5-HT uptake sites and tryptophan hydroxylase activity in mature rat brain. *Neuroendocrinol.*, *69*, 453-459.

[59] Kulikov, A.V, Tikhonova, M.A., Barykina, N.N., Alekhina, T.A., Chugui, V.F., Kolpakov, V.G., & Popova, N.K. (2002). Effect of chronic administration of imipramine on 2A-serotonin receptor mRNA in brain cortex of rats predisposed and resistant to catalepsy. *Bull. Exp. Biol. Med.*, *134*, 168-170.

[60] Kulikov, A.V., Tikhonova, M.A., Lebedeva, E.I., Chugui, V.F., & Popova, N.K. (2005). Effect of experimental increases and decreases in thyroxine levels on the extend cataleptic freezing reactions in rats. *Neurosci. Behav. Physiol.*, *35*,763-767.

[61] Kulikov, A., Torresani, J., & Jeanningros, R. (1997). Experimental hypothyroidism increases immobility in rats in the forced swim paradigm. *Neurosci. Lett.*, *234*, 111-114.

[62] Maes, M., & Meltzer, H.Y. (1995). The serotonin hypothesis of major depression. In F.R. Bloom, & D.J. Kupfer (Eds.), *Psychopharmacology: The Fourth Generation of Progress* (pp. 933-944). N.Y.: Raven Press.

[63] Martin, J.V., Williams, D.B., Fitzgerald, R.M., Im, H.K., & Von Voigtlander, P.F. (1996). Thyroid hormonal modulation of the binding and activity of the GABAA receptor complex in brain. *Neurosci.*, *73*, 705-713.

[64] Mason, G.A., Bondy, S.C., Nemeroff, C.B., Walker, C.H., & Prange, A.J. (1987). The effects of thyroid state on beta-adrenergic and serotonergic receptors in rat brain. *Psychoneuroendocrinol.*, *12*, 261-270.

[65] Mason, G.A., Walker, C.H., & Prange, A.J. (1993). L-Triiodothyronine: Is this peripheral hormone a central neurotransmitter? *Neuropsychopharmacol.*, *8*, 253-258.

[66] Muller, Y., & Clos, J. (1997). L'hormone thyroidienne est-elle un donneur de temps au cours du developpement du systeme nerveux central. *Medecine Sciences*, *13,* 116-122. (In French)

[67] Murphy, D.L., Lerner, A., Rudnick, G., & Lesch, K.P. (2004). Serotonin transporter; gene, genetic disorders and pharmacogenetics. *Molecular Interventions*, *4*, 109-123.

[68] Musselman, D.L., Nemeroff, C.B. (1996). Depression and endocrine disorders: focus on the thyroid and adrenal system. *Br. J. Psychiatry* (Suppl.), *168*, 123-128.

[69] Negishi, T., Kawasaki, K., Sekiguchi, S., Ishii, Y., Kyuwa, S., Kuroda, Y., & Yoshikawa, Y.(2005). Attention-deficit and hyperactive neurobehavioural characteristics induced by perinatal hypothyroidism in rats. *Behav Brain Res.*, *159*, 323-31.

[70] Nikulina, E., Popova, N., Kolpakov, V., & Alekhina, T. (1987). Brain dopaminergic system in rats with a genetic predisposition to catalepsy. *Biogenic Amines*, *4*, 399-406.

[71] Pineyro, G., & Blier, P. (1999). Autoregulation of serotonin neurons: Role in antidepressant drug action. *Pharmacol Reviews*, *51*, 533-591.

[72] Popova, N.K., & Amstislavskaya, T.G. (2002). 5-HT_{2A} and 5-HT_{2C} serotonin receptors differentially modulate mouse sexual arousal and the hypothalamo-pituitary-testicular response to the presence of a female. *Neuroendocrinol.*, *76*, 28-34.

[73] Popova, N., Barykina, N., Plyusnina, T., Alekhina, T., & Kolpakov, V. (2000). Expression of the startle reaction in rats genetically predisposed towards different types of defensive behavior. *Neurosci. Behav. Physiol.*, *30*, 321-325.

[74] Popova, N, & Kulikov, A. (1995). On the role of brain serotonin in expression of genetic predisposition to catalepsy in animal models. *Am. J. Med. Genetics (Neuropsychiatric Genetics)*, *60*, 214-220.

[75] Porsolt, R.D., Le Pichon, M., & Jalfre, M. (1977). Depression: a new animal model sensitive to antidepressant treatments. *Nature*, *266*, 730-732.

[76] Prange, A.J. (1996). Novel uses of thyroid hormones in patients with affective disorders. *Thyroid*, *6*, 537-543.

[77] Ramos, A., Berton, O., Mormede, P., & Chaouloff, F. (1997). A multiple-test study of anxiety-related behaviours in six inbred rat strains. *Behav. Brain Res.*, *85*, 57-69.

[78] Rastogi, R.B., & Singhal, R.L. (1976). Influence of neonatal and adult hyperthyroidism on behavior and biosynthetic capacity for norepinephrine, dopamine and 5-hydroxytryptamine in rat brain. *J. Pharmacol. Exp. Ther.*, *198*, 609-618.

[79] Rastogi, R.B., & Singhal, R.L. (1978). The effect of thyroid hormone on serotonergic neurones: depletion of serotonin in discrete brain areas of developing hypothyroid rats. *Naunyn Schmiedebergs Arch. Pharmacol.*, *304*, 9-13.

[80] Redei, E.E., Solberg, L.C., Kluczynski, J.M., & Pare, W.P. (2001). Paradoxical hormonal and behavioral responses to hypothyroid and hyperthyroid states in the Wistar-Kyoto rat. *Neuropsychopharmacol.*, *24*, 632-639.

[81] Ruiz-Marcos, A., Sanchez-Toscano, F., Escobar del Rey, F., & Morreale de Escobar, G. (1980). Reversible morphological alterations of cortical neurons in juvenile and adult hypothyroidism in the rat. *Brain Res.*, *185*, 91-102.

[82] Sala-Roca, J., Marti-Carbonell, M.A., Garau, A., Darbra, S., & Balada, F. (2002). Effects of chronic dysthyroidism on activity and exploration. *Physiol. Behav.*, *77*, 125-133.

[83] Sanberg, P.R., Bunsey, M.D., Giordano, M., & Norman, A.B. (1988). The catalepsy test: Its ups and downs. *Behav. Neurosci.*, *102*,748-759.

[84] Sanders-Bush, E., & Canton, H. (1995). Serotonin receptors. Signal transduction pathways. In F.R. Bloom, & D.J. Kupfer (Eds.), *Psychopharmacology: The Fourth Generation of Progress* (pp. 431-441). N.Y.: Raven Press.

[85] Sandrini, M., Vitale, G., Vergoni, A.V., Ottani, A., & Bertolini, A. (1996). Effect of acute and chronic treatment with triiofothyronine on serotonin levels and serotonergic receptors subtypes in the rat brain. *Life Sci.*, *58*, 1551-1559.

[86] Sarkar, P.K., & Ray, A.K. (1994). Synaptosomal T3 content in cerebral cortex of adult rat in different thyroidal states. *Neuropsychopharmacol.*, *11*, 151-155.

[87] Saudou, F., & Hen, R. (1994). 5-Hydroxytryptamine receptor subtypes: Molecular and functional diversity. *Adv. In Pharmacol.*, *30*, 327-380.

[88] Savard, P., Merand, Y., Di Paolo, T., & Dupont, A. (1983). Effect of thyroid state on serotonin, 5-hydroxyindolacetic acid and substance P in discrete brain nuclei of adult rats. *Neurosci.*, *10*, 1399-1404.

[89] Savard, P., Merand, Y., Di Paolo, T., & Dupont, A. (1984). Effect of neonatal hypothyroidism on serotonin system in the rat brain. *Brain Res.*, *292*, 99-108.

[90] Schneider, B.F., & Golden, W.L. (1987). Acquisition of acoustic startle response in relation to growth and thyroid function in rats. *Int. J. Dev. Neurosci.*, *5*, 99-106.

[91] Sharp, T., & Hjorth, S. (1990). Application of brain microdialysis to study the pharmacology of the 5-HT_{1A} autoreceptor. *J. Neurosci. Methods*, *34*, 83-90.

[92] Silva, J.E., Gordon, M.B., Crantz, F.R., Leonard, J.L., & Larsen, P.R. (1984). Qualitative and quantitative differences in the pathways of extrathyroidal triiodothyronine generation between euthyroid and hypothyroid rats. *J. Clin. Invest.*, *73*, 898-907.

[93] Tamasy, V., Meisami, E., Vallerga, A., & Timiras, P.S. (1986). Rehabilitation from neonatal hypothyroidism: spontaneous motor activity, exploratory behavior, avoidance learning and responses of pituitary--thyroid axis to stress in male rats. *Psychoneuroendocrinology*, *11*, 91-103.

[94] Tejani-Butt, S.M., Yang, J., & Kaviani, A. (1993). Time course of altered thyroid states on 5-HT1A receptors and 5-HT uptake sites in rat brain: An autoradiographic analysis. *Neuroendocrinol.*, *57*, 1011-1018.

[95] Tikhonova, M.A., Kulikov, A.V., Lebedeva, E.I., Barykina, N.N., Amstislavskaya, T.G., & Popova, N.K. (2005). On association between cortical 5-HT2A receptors and behavior in rats with experimental thyroid disturbances. *Pharmacol. Biochem. Behav.*, *82*, 506-514.

[96] Tohei, A., Watanabe, G., & Taya, K. (1998). Effects of thyroidectomy or thiouracil treatment on copulatory behavior in adult male rats. *J. Vet. Med. Sci.*, *60*, 281-285.

[97] Walther, D., & Bader, M. (2003). A unique central tryptophan hydroxylase isoform. *Biochem. Pharmacol.*, *66*, 1673-1680.

[98] Walther, D.J., Peter, J.-U., Bashammakh, S., Hortnagl, H., Voits, M., Fink, H., & Bader, M.(2003). Synthesis of serotonin by a second thryptophan hydroxylase isoform. *Science, 299*, 76.

[99] Whybrow, P.C., & Prange, A.J. (1981). A hypothesis of thyroid-catacholimine-receptor interaction. *Arch. Gen. Psychiatry, 38*, 105-113.

[100] Willner, P. (1990). Animal models of depression: an overview. *Pharmacol. Ther.*, *45*, 425-455.

[101] Wortsman, J., Rosner, W., & Dufau, M.L.. (1987). Abnormal testicular function in men with primary hypothyroidism. *Am. J. Med.*, *82*, 207–212.

[102] Wright, D.E., Seroogy, K.B., Lundgren, K.H., Davis, B.M., & Jennes, L. (1995). Comparative localization of serotoni1A, 1C and 2 receptor subtype mRNAs in rat brain. *J. Comp. Neurol.*, *351*, 357-373.

[103] Zifa, E. & Fillion, G. (1992). 5-Hydroxytryptamine receptors. *Pharmacol. Rev.*, 1992, 44, 401-458.

In: Psychoneuroendocrinology Research Trends
Editor: Martina T. Czerbska, pp. 343-360

ISBN: 978-1-60021-665-7

Chapter XI

INTERLEUKIN-1 DEFICIENCY AND AGGRESSIVENESS IN MALE MICE

Akira Tamagawa[1], Irina Kolosova[2], Yasuo Endo[3], Ludmila Gerlinskaya[2], Yoichiro Iwakura[4] and Mikhail Moshkin[2,]*

[1]Department of Behavioral Medicine, Tohoku University, Sendai, Japan;
[2]Institute of Systematic and Ecology of Animals, Novosibirsk, Russia;
[3]Department of Pharmacology, School of Dentistry, Tohoku University, Sendai, Japan;
[4]Laboratory Animal Research Center, Institute of Medical Science, University of Tokyo, Japan.

ABSTRACT

The influence of interleukin-1 (IL-1) deficiency on aggressive behavior, social investigation, and endocrine status was studied in IL-1α/β deficient (IL-1-KO) male mice derived from the BALB/cA strain. In social investigation tests adult IL-1-KO males spent more time on chasing and sniffing and other form of interactions with juvenile intruders in comparison with wild type males. In contrast to very peaceful wild type males, which did not fight in intra-strain pair-wise test conducted 3-5 days after interactions with juveniles, IL-1-KO males showed fights in each test. Encounter in inter-strain pair-wise tests IL-1 KO and BALB/cA males experienced in contacts with juveniles also demonstrated higher aggressiveness in IL-1 deficient males. The difference in aggressiveness between the males of these strains was correlated with differences in androgen secretion. Concentration of testosterone in feces collected during 5 days of social isolation, including period of the short-term contacts with juvenile males, was higher in IL-1-KO males than in BALB/cA males. Males of both strains showed the similar concentrations and similar decline of fecal corticosterone after contacts with juveniles. Therefore, the reciprocal relationships between immunocompetence and

* Correspondence concerning this article should be addressed to Mikhail Moshkin Fax: +81-3832-170-973; E-mail address: mmp@eco.nsc.ru.

aggressiveness predicted by life history trade-off could be established through the primary changes of the certain immunoregulatory functions such as cytokines.

Keywords: aggressiveness; interleukin-1 α/β; knockout mice; fecal testosterone; fecal corticosterone; life history trade-off

Introduction

Interleukin-1 (IL-1) is a cytokine produced by activated monocytes and macrophages in response to infection or injury. In addition to its immune functions IL-1 posses the wide pleiotropic effects on the other physiological systems, including neuro-endocrine control of the behavior (Blalock, 1994). Role of the pro-inflammatory cytokines in behavioral modifications were demonstrated in several papers dedicated to studies of the psycho-physiological changes under an infection. In particularly, stimulation of cytokine secretion by means bacterial endotoxin (lipopolysaccharide – LPS) or injection with pro-inflammatory cytokines (e.g. IL-1) reduced male social response to juvenile intruders and intermale aggression (Crestani et al., 1991; Dantzer et al., 1992; Cirulli et al., 1998; Dantzer, 2001). The neuro-endocrine effects of cytokine, which considered as main reason for the behavioral changes in the infected animals, usually ignored as plausible reason for the immuno-behavioral correlations that were found in healthy animals and human. Very common proximate explanations of these correlations base on unidirectional influences of the hormones and neurotransmitters on the immune system (Devoino et al., 1993; Petitto et al., 1994, 1999; Granger et al., 2000).

Such position dominates in extensive ummino-behavioral studies that were stimulated by hypothesis of the immunocompetence handicap (Folstad and Karter, 1992). According to Folstad and Karter (1992) the dual effects of androgens on the development of sexual ornaments or behavioral displays (Andersson, 1994; Owens and Short, 1995; Kimball and Ligon, 1999) and on depressing the acquired immunity (Grossman, 1985; Zuk, 1996) shapes immuno-morphological and immuno-behavioral correlations that were found in wild and laboratory mammals and birds (Barnard et al., 1996; Petitto et al., 1999; Weatherhead et al., 1993; Verhulst et al., 1999; Zuk and Johnsen, 2000; Faivre et al., 2001). Several hypotheses were offered for the ultimate explanation of the androgen-dependent immunosupression. There are the resource re-allocation in favor of the breeding efforts (Wedekind and Folstad, 1994), the sperm protection against autoimmune attack (Hillgarth et al., 1997), and the re-distribution immune defense between adaptive and innate immunity (Braude et al., 1999). Most studies dedicated to immunocompetence handicap hypothesis have been done on bird species due to their bright ornamentations. In nocturnal mammal such as mice, males and females are decorated equally modestly, but they show sexual dimorphism in intra-species aggression. Aggressiveness of males determines reproductive success through the influences on social rank, home range size and odor attraction for the females (Gerlinskaya et al., 1995; Qvarstrom and Forsgren, 1998). Like other secondary sexual traits aggressiveness of male mice is an androgen-dependent feature (Brain et al., 1980; Naumenko et al., 1983). Alongside with androgens, many other hormones and neurotransmitters modify inter-male aggression, (Adams, 1983; Brain and Benton, 1983; Leshner, 1983; Cirulli et al., 1998).

Discovering neuroendocrine and behavioral effects of cytokines (Dantzer et al., 1992; Dantzer, 2001; Cirulli et al., 1998; Kelley et al., 2003) let us to suppose bi-directional links between immunocompetence and secondary sexual traits, including sex specific behavior. Following this assumption we can expect modifications of the male sexual displays and intermale aggression due to primary changes of immunocompetence. Verhulst et al. (1999) were one of the first, who undertook this proposal. They were measured comb size as secondary sexual trait and plasma testosterone in males of domestic chickens *Gallus domesticus* selected for high and low humoral immune response. Highly responsive individuals showed smaller comb size and lower plasma testosterone than those of the lowly responsive individuals.

Parallel changes of both comb size and testosterone stay open question about cause and effect of the selection to high and low immunity. Based on these data we can not exclude that selection, first of all, separates chickens according to their the gonad secretion and only then affects immunoresponsiveness. Gene knockout technology now provides a means for the direct manipulation of the certain immuno-physiological function. In the present study, we tested behavioral effects induced by knockout of the IL-1 genes (IL-1α and IL-1β). The homozygous IL-1α/β knockout (IL-1-KO) BALB/cA mice were originally produced by Iwakura and his co-workers (Horai et al., 1998). Deficient of the pro-inflammatory cytokines is resulted in the significantly reduced humoral immune response to sheep red blood cells (Nakae et al., 2001).

Peripheral or central administration of IL-1 simulates sickness behavior that occurs transitorily under infection or stress conditions (Dantzer, 2001). Besides this current view, critically reviewed literature provides the evidences that IL-1 genes are constitutively expressed in brain cells and permanently modulated neuronal activities (Vitkovic et al., 2000). The constitutional production of the pro-inflammatory cytokines proposes the behavioral effects, which could be caused by deficiency of the IL-1. There are several reasons for the individual variations in the IL-1 production in healthy subjects. It may be polymorphism reflected on the gene expression and polymorphism reflected on the reception and functional efficiency of the cytokines (Bailly et al., 1996; Loughrey et al., 1998; Daly et al., 2002; Nicklin et al., 2002; Yucesoy et al., 2002). Recent study on human was demonstrated association of the IL-1 polymorphism and psychic disorder and individual response to antidepressants (Yu et al., 2003). Thus we can expect appreciable behavioral changes due to IL-1 knockout, which abolishes constitutive expression of IL-1 genes.

The effect of IL-1 deficiency on inter-male aggression were studied on adult males of both IL-1 KO and wild type strains that were encountered in the pair-wise tests conducted after previous contacts with juvenile males. Since many physiological and behavioral effects of cytokines can be resulted in modulation of the hypothalamic-pituitary gonad (HPG) and HPA systems we measured the concentrations of testosterone and corticosterone in feces collected from socially isolated male mice. Validation of fecal testosterone and corticosterone as indexes reflected gonad and adrenal secretion have been done in several studies on rodents including mice (Billitti et al., 1998; Gerlinskaya et al., 1993; Harper and Austad, 2000; Muir et al., 2002; Zav'yalov et al., 2003).

In previous experiments we have found that both BALB/c and IL-1 KO males show exclusively low level of inter- male aggressions (Moshkin et al., 2001). They had no any attacks and fights in pair-wise tests established after 1 or 2 weeks of social isolation, which is usually enough to maximize aggressiveness in male mice (Goldsmith et al., 1976). Due to extremely low aggressiveness of these strains we study agonistic behavior in males that were experienced in contacts with juvenile males. Short-term interactions with juvenile intruders remarkably increase inter-male aggression that singly kept adult males showed in pair-wise test (Moshkin et al., 2001).

MATERIALS AND METHODS

Animals

IL-1-KO mice were bred in the Department of Pharmacology (School of Dentistry, Tohoku University). Control BALB/cA mice were also obtained from stock in Tohoku University. Adult males of the both strains aged 8-10 weeks and juvenile males of BALB/cA aged 3-4 weeks were used in the behavioral experiments. Animals were kept in standard plastic cages on 12:12 h light-dark cycle (lights on 0700) at a controlled temperature (23±1°C). Standard food pellets (LabMR Stock; Nihon Nosan, Yokohama, Japan) and water were available *ad libitum*. During 7-10 days before experiments, the animals were kept in groups (4-6 adult males per cage and 10-12 juvenile males per cage).

Procedures

Experiment 1: Effects of Previous Contacts with a Juvenile Males on the Aggressive Behavior of Adult Male Mice in Intra-strain Pair-wise Tests

Six IL-1-KO mice and 6 BALB/cA mice were individually housed in standard plastic cages at 1200 h. Behavioral responses to the introduction of a juvenile male were investigated 12 hours after isolation. Four-minute social investigation tests were conducted at 2100 h and 2400 h on the day of isolation and 0300 h and 2100 h on the following day. Then the adult males were kept individually and were investigated through intra-strain pair-wise tests performed 3 and 5 days after the last introduction of a juvenile. Each male was tested twice using a different contestant.

Experiment 2. Effect of Previous Contacts with Juvenile Males on the Fecal Steroids and Inter-strain Competition of Adult Male Mice

Before experiment adult male mice were caged in intra-strain groups of 6. Thirty-three hours before introduction of juveniles (social investigation tests) adult males of IL-1-KO mice and BALB/cA mice were housed individually in stainless steel cages (30X25X20 cm) with netted bottom at 1200 h. Entire time of individually housing we collected fecal samples twice a day at 0700 and 1900 h. Sampling was started 7 h after social isolation at the beginning of the dark phase of photoperiod (1900). Then we followed the protocol of

Experiment 1, except the pair-wise test, where adult males of different strains were encountered. Fifteen min inter-strain competitions were studies 3 days after the last introduction of a juvenile. After 15 min pair-wise test males we re-grouped in intra-strain groups. Due to extremely intensive fights animals in both cages were humanely killed 5 days later and skin wounds were calculated.

Pair-wise Test

Two adult males of the same strain were placed in a standard mouse cage with fresh sawdust for the 15-min pair-wise test. A small area of fur on the back was sheared in one of the 2 animals for distinction. The tests were performed using the animals in randomly chosen pairs. New pairings were made for each test. Agonistic behavior was recorded by a videotape recorder. The videotape was played back twice for separate observation of the behavioral acts of each individual. Each behavioral act was recorded by press on the selected key in the keyboard and inputted to computer. The program of Ethology 225 was used for calculation of number and duration of acts. We recorded attacks, fights, chases and escapes, tail-rattles, aggressive grooming and submissive freezing in response to allogrooming, defensive uprights, and naso-nasal and naso-genital sniffing. Often, before aggression, males of both strains showed specific gait, which looks like waddling. At the stop of playback we have found that ratio of length to wide in mice body was much higher during the common walk (3.09 ± 0.05; $n = 115$) in comparison with waddling (1.65 ± 0.04; $n = 75$; $p < .001$). This easy recognized behavioral act was included in recorded list. Since number of waddling is correlated positively with aggressive acts such as attacks ($r = 0.61$; $p < .01$), chases ($r = 0.62$; $p < .01$) and tail-rattles ($r = 0.59$; $p < .01$), it can consider as one form of aggressive demonstration. All tests were performed at nighttime between 2100 h and 2400 h.

Social Investigation Test

A juvenile male was introduced into the cage of an adult male as an intruder, and behavioral acts were recorded by videotape for 4 min. The social exploration time (Fishkin and Winslow, 1997), number of the direct aggressive acts (bites) and the demonstrations of aggression by adult males (tail-rattles and waddling) were scored.

Testosterone and Corticosterone in Feces

Fecal samples collected in experiment 2 at the night and day periods were dried at 30-40°C and stored in closed tube at room temperature. The concentrations of corticosterone and testosterone in dry feces were measured by radioimmunoassay using Sigma antibodies (Rabbit Anti-Corticosterone and Rabbit Anti-Testosterone) and Amersham labeled hormones ([1,2,6,7-H^3]-Corticosterone and [1,2,6,7-H^3]-Testosterone). Except for extraction (Gerlinskaya et al., 1993), the analysis was made according to the Sigma's instruction. Dry feces (80 mg) were homogenized in bidistilled water (2 ml) in glass grinder. After centrifugation, supernatants were harvested and stored at -20°C until assayed. Supernatant (0.3 ml for corticosterone or 0.25 ml for testosterone) was extracted with 3 ml ethyl ether; then 2 ml extract was removed, transferred to new tube, vacuum dried at 55°C, and the residue was re-suspended in 100 μl phosphate buffer, pH 7.4.

The extraction yield was checked for every set of assay using [^{3}H]-labeled steroids and varied from 70 to 76% for both hormones. Sensitivity of the assay was determined from the 95% confidence interval of zero standards that were 30 pg/tube for corticosterone and 5 pg/tube for testosterone. The inter- and intra-assay variations were 11.6 and 7.9 %, respectively, for corticosterone and 9.9 and 6.2 % for testosterone. To determine the parallelism, a five-point, two-fold dilution series of fecal samples in phosphate buffer were prepared and compared with the standard curves of each steroid. There were no significant differences between the slopes of standard curves and the slopes of lines generated from fecal samples of assayed mice.

Statistical Analysis

Since in experiment 1 and experiment 2 we followed the same protocol of the social investigation tests data of both experiments were combined for statistical analysis. Mann-Whitney rank test was used for assessment between strain differences in means of the behavioral traits that were detected in both social investigation test and pair-wise test. A principal components calculation was applied to behavioral traits detected in pair-wise tests (Kendall and Stuart, 1966). Three-way ANOVA was calculated for the evaluation of the effects of strain, day of social isolation and time of day on the concentrations of testosterone and corticosterone in feces. LSD test were used for multiple comparison of the mean levels of steroid hormones in feces collected at different days. Probability values of not more than 0.05 were considered to be significant. The differences in the hormonal levels between mouse strains were examined using Student's *t*-test. Spearman rank correlation coefficients were calculated for analysis of the relationships among behavioral traits.

RESULTS

Social Investigation Test

Adult IL-1-KO males spent more time investigating juvenile intruder in the 1st and 3rd trials in comparison with BALB/cA males (Figure 1A). Total duration of social investigation also was significantly longer in IL-1-KO males than in wild BALB/cA males (262.2 ± 22.4 sec and 196.8 ± 20.8 sec, $p = .023$, Mann-Whitney rank test). In all trials IL-1 KO males surpassed BALB/c males in numbers of waddling (Figure 1B). And total amount of waddling also was higher in IL-1 deficient males (55.7 ± 10.9) in comparison with wild type mice (10.7 ± 3.4, $p = .0051$, Mann-Whitney rank test). Rare aggressive acts such as bites and tail-rattles were summarized for robust comparison direct and indirect aggression of adult males toward juvenile intruders. In 3rd trial IL-1 KO males showed more aggressive acts than BALB/c males (Figure 1C). Sum of aggressive acts demonstrated in four trials also was higher in IL-1 KO males (22.5 ± 9.0) in comparison with BALB/c males (4.6 ± 2.4, p = .035, Mann-Whitney rank test).

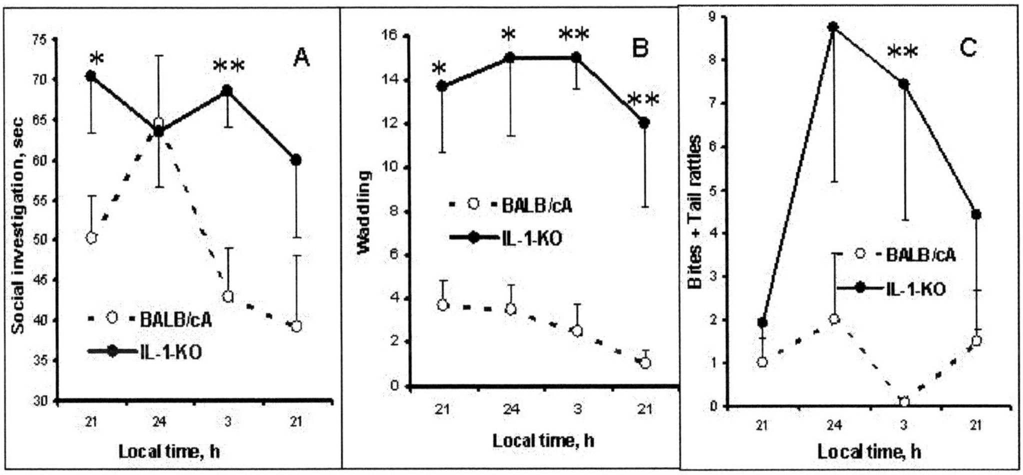

Figure 1. Behavioral response of the adult males BALB/c and IL-1 KO stain on the repeatedly introductions of the juvenile males: social investigation (A), waddling (B), sum of bites and tail rattles (C). Data are expressed as means ± SE. (*) Significant differences in the values were observed between IL-1-KO and BALB/cA males (* - $p < .05$; ** $p < .01$; Mann-Whitney rank test).

Principal Component Analysis of the Agonistic Behavior

Behavioral repertoire demonstrated in pair-wise tests was more reach in comparison with social investigation tests. For reduction of the number analyzed traits ten behavioral variables were treated by PCA. Two principal components extracted from the model with ten variables explained 52.3% of variance in the number of behavioral acts in 15 min pair-wise tests (Table 1). Since the first principal component axis (PC1) correlated positively with both aggressive acts (attacks, chases, tail-rattles, and waddling) and inspection of partner (naso-genital sniffing), we used PC1 as an aggression score. Main load to the second component (PC2) gave numbers of defensive uprights and escapes. For convenient, the inverse values of PC2 were used as a submission score.

Agonistic Behavior in Intra-strain Pair-wise Tests

There were observed statistically significant between strains differences in aggression score that adult males had in pair-wise tests conducted 3-5 days after the contacts with juvenile males (Figure 2A). We found also that IL-1-KO males showed more fights (5.3 ± 1.7) and spent more time for fights (66.5 ± 20.9) than wild BALB/cA males (0 for both numbers and duration of fights; p = .01, Mann-Whitney rank test,).

Table 1. Summary of principal components analysis of ten behavioral variables

Information	PC1	PC2
Eigenvalue	3.35	1.88
Percentage of total variance explained	33.5	18.8
Factor loading		
Attacks	0.85	-0.18
Tail-rattles	0.71	-0.27
Chases	0.85	0.04
Escapes	0.40	-0.69
Aggressive grooming	0.03	0.58
Submissive freezing	-0.32	-0.47
Waddling	0.75	0.09
Defensive uprights	-0.03	-0.80
Naso-nasal sniffing	0.25	0.03
Naso-genital sniffing	0.71	0.30

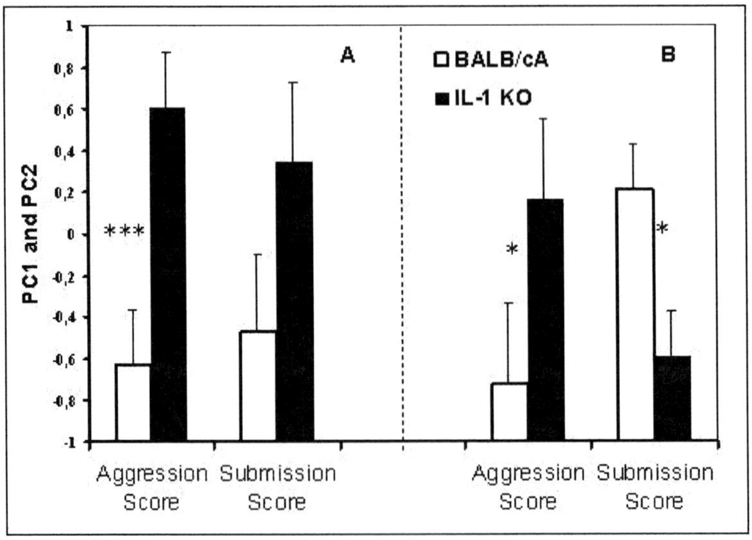

Figure 2. Aggression score (PC1) and submission score (inverted PC2) in intra-strain (A) and inter-strain (B) pair-wise tests. (*) Significant differences in the values were observed between IL-1-KO and BALB/cA males (* - $p < .05$; *** - $p < .001$, Student *t*-test).

As it was shown above IL-1-KO adult males demonstrated more aggressive acts toward juvenile intruders in comparison with BALB/c adult males. Does manifestation of aggression in social investigation tests correlate with aggression in pair-wise tests? The numbers of direct and indirect aggressive acts toward juveniles (bites, tail rattles and waddling) did not correlate with agonistic behavior in pair-wise tests. However, the total time spent adult males in 4 trials for social investigation of juvenile intruders correlated with the aggression score in pair-wise tests (Figure 3). This correlation was negative in wild BALB/cA males, and positive in IL-1-KO males.

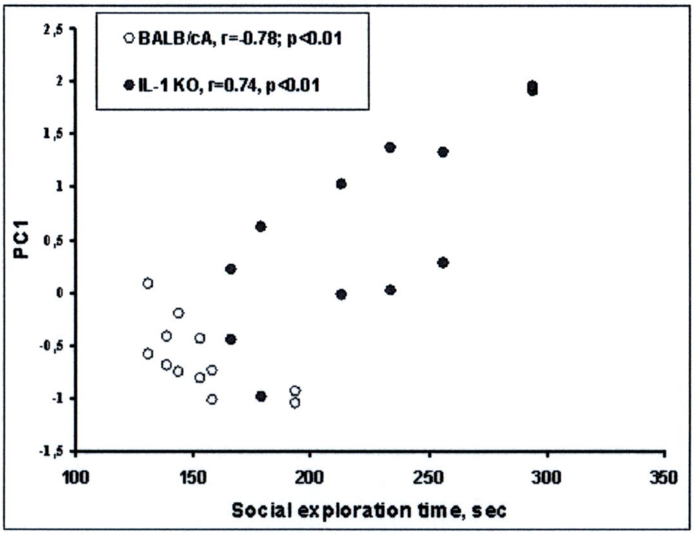

Figure 3. Correlations between the summed-up duration of social investigation time in 4 consequent social investigation tests and the aggression score in intra-strain pair-wise tests (experiment 1).

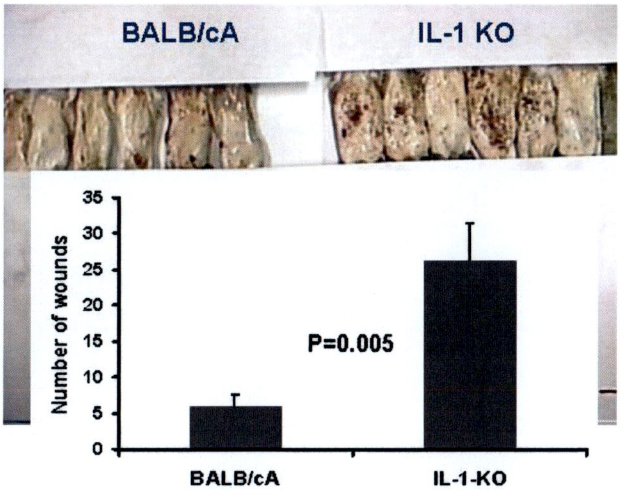

Figure 4. Skin wounds in male mice that were sacrificed 5 days after re-grouping. The inside surfaces of skin with few wounds in BALB/c males and with multiple wounds in IL-KO males are shown above and numbers of wounds (mean ± SEM) are presented below ($p = .005$; Mann-Whitney rank test).

Agonistic Behavior in Inter-strain Pair-wise Tests

IL-1-KO males significantly surpassed wild BALB/cA males in aggression score in inter-strain pair-wise tests (Figure 1B). In contrast, BALB/cA males had higher level of submission score than IL-1-KO males. In inter-strain test numbers (2.00 ± 0.98) and duration (21.7 ± 10.5 sec) of fights did not differ in comparison with intra-strain encounter of IL-1-KO males in experiment 1 ($p = .15$ and $p = .10$, respectively, Mann-Whitney rank test).

Skin Wounds in Male mice Re-grouped after Social Isolation

Inspection of skin in male mice sacrificed 5 days after re-grouping in intra-strain groups revealed high significant between strains difference in numbers of wounds (Figure 4). BALB/c males had 6.0 ± 1.5 (n = 6) wounds per individual, and IL-1 KO had 26.3 ± 5.2 (n = 6; $p = .005$, Mann-Whitney rank test).

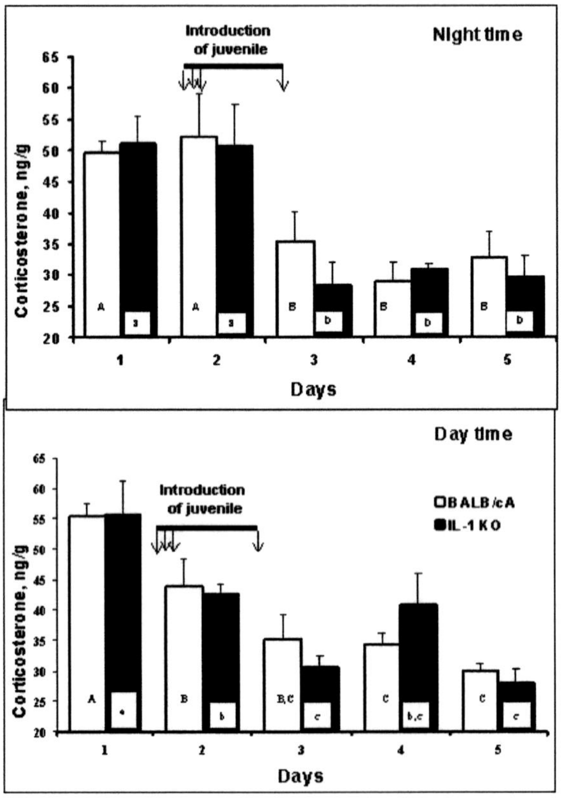

Figure 5. Concentration of corticosterone (ng/g, mean ± SEM) in feces of IL-1-KO and wild BALB/cA males collected during social isolation at the night and daytime. Here and on Fig 6, bars with different letters (A, B, etc. for BALB/c males and a, b, etc. for IL-1 KO males) are significantly different (LSD-test, $p < .05$).

Fecal Corticosterone and Testosterone

Three-way ANOVA with strain, day of isolation and time of day (day/night) as factors revealed statistically significant effects of isolation day ($F_{4,98}$= 25.97, $p < .001$) on the concentration of corticosterone in feces. Males of both strains showed the decline of the fecal corticosterone after introductions of juveniles. Then it stayed on the low level up to the pair-wise tests (Figure 5).

Strain ($F_{1,98}$= 14.43, $p < .001$), day of isolation ($F_{4,98}$= 10.71, $p < .001$) and interactions of entire factors ($F_{4,98}$= 3.43, $p = .011$) influenced a concentration of testosterone in feces.

Average level of fecal testosterone was significantly higher in IL-1 KO males than in wild BALB/c males (7.02 ± 0.40 ng/g and 5.59 ± 0.20 ng/g, p = .0023, Student t-test). Introduction of juveniles resulted in the decline of nocturnal level of fecal testosterone in IL-1 KO males. Concentration of testosterone was returned to initial level in the 5th day of housing individually (Figure 6). In BALB/cA males, concentration of testosterone in nocturnal samples did not change during 4 nights and then it increased in 5th night. Concentration of testosterone in feces sampled in daytime had uniform dynamics in mice of both strains. It was declined at the next day after introductions of juveniles and returned to initial level in day 5. We have to accentuate that just before pair-wise test (5th night and 5th day) concentrations of fecal testosterone were significantly higher in IL-1 KO males than in BALB/cA males (Figure 6).

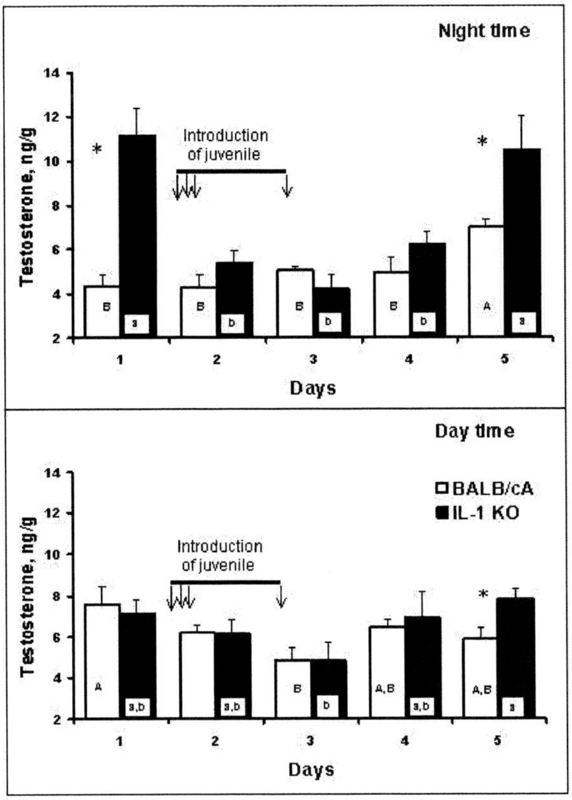

Figure 6. Concentration of testosterone (ng/g, mean ± SEM) in feces of IL-1-KO and wild BALB/cA males collected during social isolation at the night and daytime. (*) Significant differences in the values were observed between IL-1-KO and wild BALB/cA males ($p < .05$, Student t-test).

Discussion

IL-1 plays a key role in the LPS-induced sickness behavior (Fishkin and Winslow, 1997; Swiergiel et al., 1997). Similar to bacterial endotoxin, injection of IL-1 reduces social investigation time (Crestani et al., 1991; Segreti et al., 1997). In turn, how it was found in our study, the constitutional deficiency of IL-1 enhanced the behavioral responses of adult male on the introduction of juvenile males. Adult IL-1-KO males spent more time for inspection of juvenile intruders and demonstrated more aggressive acts in comparison with wild type adults.

As seen in preliminary studies (Moshkin et al., 2001), the previous contact with juvenile males stimulated inter-male aggression in pair-wise tests. But it was right only for IL-1-KO males that showed attack and fights in intra-strain pair-wise tests. Adult males of BALB/cA strain were tamed even after previous contacts with juveniles. These between strain differences were confirmed in inter-strain pair-wise tests, where IL-1 deficient males showed higher aggression score in competition with wild type males. The calculation of skin wounds 5 days after re-grouping of males provided additional proof of higher aggressiveness of IL-1-KO males, which had more skin wounds than wild BALB/cA males.

It is well known that the male mice experienced in competitions with weak partners become more aggressive than the non-experienced males (Scott and Fredericson, 1951). Nevertheless the number of aggressive acts directed towards juveniles in social investigation tests did not correlate with aggressiveness of adult males in pair-wise tests. On the other hand, the total duration of the close contact of IL-1-KO males with intruded juvenile males (social exploration time) correlated positively with their aggressiveness in pair-wise tests.

Usually, but not always, social isolation of adult male mice is accompanied with growth of plasma concentration of corticosterone and testosterone (Frances et al., 2000; Bartolomucci et al., 2003; Misslin et al., 1982; Sayegh et al., 1990). In contrast these data of literature, fecal corticosterone was progressively declined after short-term contacts with juveniles in isolated IL-1 KO and BALB/cA males. Despite to the well-known IL-1 dependent upregulation of HPA activity (Blalock, 1994; del Rey and Besedovsky, 2000; Beishuizen and Thijs, 2003), there were no differences in the fecal corticosterone between IL-1 deficient and wild type male mice. Early the equal basal concentrations of serum corticosterone and similar glucocorticoid increase in response to injection with inflammatory agent (turpentine-oil) were found in IL-1 deficient and wild type mice. Moreover IL-1 deficient mice need more time, than control, for decline till basal level of the turpentine-induced serum corticosterone (Horai et al., 1998).

Concentration of fecal testosterone was transitory reduced in the nights, when juveniles were introduced into cage of adult males, but only in IL-1 deficient mice. Wild type males showed the increase of nocturnal concentration of fecal testosterone on the 5th day of housing singly. Anyway it was less than concentration of testosterone in feces collected from IL-KO males at the same time. Thus, IL-1 KO males significantly surpassed BALB males in androgen concentration, especially just before the pair-wise tests. Increase of the gonad secretion in IL-1-KO males can be explained by participation of this cytokine in down-regulation of the androgen secretion on the testicular level as well as on the hypothalamo-pituitary level (Kmicikiewicz et al., 1999; Jonsson et al., 2001; Ogilvie et al., 1999; Svechnikov et al., 2001). Since aggressive behavior is androgen-dependent trait, the higher level of fecal testosterone in IL-1-KO males in comparison with wild BALB/cA males was in a good concordance with the higher aggressiveness of IL-1-KO males.

Behavioral and endocrine deviations in IL-1 KO mice might be caused either the current deficiency of IL-1 or lack of this cytokine in a certain period of the ontogenesis. Recently A.

Tamagawa and E. Litvinova (personal communication) used anti-IL-1 antibodies for study the behavioral effect of temporary IL-1 deficiency in singly kept adult males of BALB/cA strain (n=8). They observed aggressive behavior in repeated 15-min pair-wise tests, which were executed in three consecutive nights (one test per night with new contestants). Three hour before second test (night two) males were injected intraperitoneally with Anti-Mouse Interleukin -1β (Sigma, developed in goat) in dose of 0.2 μg/g. Antibodies induced block of IL-1 resulted in temporary increase of aggressiveness in BALB/cA males. In second test they spent more time for fighting (26.2±10.1 sec) in comparison with first and third tests (0±0 and 0.5±0.3 respectively, p=0.05, Kruskel-Wallis test). Also 3 h after administration of the IL-1 antibodies males began fight early (487±157 sec) than it was found 24 h before (900±0 sec) and 24 h after injection (871±19 sec, p=0.05, Kruskel-Wallis test). This data demonstrate that current decline of the circulating IL-1 is actual for aggressiveness of male mice.

Alongside with behavioral effects of peripheral or central IL-1 administration (Cirulli et al., 1998; Crestani et al., 1991; Dantzer et al., 1992), both long-term and short-term deficiency of the cytokine is reflected on the social behavior of male mice. Behavioral effects of IL-1 deficiency are associated with the increase of the androgen secretion. Anyway we cannot exclude the direct influence of IL-1 on the brain structure such as serotonergic neurons, which related with control of the aggressiveness. Recently Hassanain et al (2003) demonstrated in study on the cat that microinjections of relatively low doses of IL-1β into the medial hypothalamus potentiated defensive rage behavior elicited from the midbrain periaqueductal gray in dose-related manner. This effect of IL-1 was abolished by pretreatment with a selective 5-HT2 receptor antagonist, LY-53857. Also IL-1 deficiency could modify aggressive behavior via up-regulation of the brain prostaglandins (Rivest, 1999; Repovic et al., 2003). How it was found recently, knockout gene of prostaglandin receptor EP_1 results in higher aggressiveness of knockout male mice in comparison with background strain (Matsuoka et al., 2005).

Thus, the lack of gene expression related with the regulation of immunoresponsiveness leads to growth of aggressive propensity in male mice. In conclusion, our data add one more proof to evidences of the bi-directional links of the of brain and immune functions, which generated immuno-behavioral correlations in healthy persons.

Acknowledgments

We thank Christopher Barnard for the correction of English and for the helpful comments to the first version of manuscript. These studies were supported by: RFBR #05-04-488225; Siberian Branch of RAS #39; Ministry of Education and Science #RNP. 2.1.1.

References

Adams, D.B., 1983. Hormone-brain interactions and their influence on agonistic behavior. In: Svare, B.B., (Ed.), *Hormones and aggressive behavior*, Plenum Press, New York, London, pp.223-245.

Andersson, M., 1994. *Sexual selection.* Princeton Univ. Press, Princeton, New Jersey.

Bailly, S., Israel, N., Fay, M., Gougerot-Pocidalo, M.A., Duff, G.W., 1996. An intronic polymorphic repeat sequence modulates interleukin-1 α gene regulation. *Mol. Immunol. 33(11-12)*, 999-1006.

Barnard, C.J., Behnke, J.M., Sewell, J., 1996. Social status and resistance to disease in house mice (*Mus musculus)*: Status-related modulation of hormonal responses in relation to immunity costs in different social and physical environments. *Ethology 102*, 63-84.

Bartolomucci, A., Palanza, P., Sacerdote, P., Ceresini, G., Chirieleison, A., Panerai, A. E., Parmigiani, S., 2003. Individual housing induces immuno-endocrine respons to psychological stress in male mice. *Psychoneuroendocrinology 28*, 540-548.

Beishuizen, A., Thijs, L.G., 2003. Endotoxin and hypothalamo-pituitary-adrenal (HPA) axis. *J. Endotoxin Res. 9(1)*, 3-24.

Billitti, J.E., Lasley, B.L., Wilson, B.W., 1998. Development and validation of a fecal testosterone biomarker in *Mus musculus* and *Peromyscus maniculatus. Biol. Reprod. 59*, 1023-1028.

Blalock, J.E., 1994. The Immune system. Our Sixth Sense. *The immunologist 2*, 8 -15.

Brain, P.F., Benton, D., 1983. Conditions of housing, hormones and aggressive behavior. In Svare, B.B., (Ed.), *Hormones and aggressive behavior*, Plenum Press, New York, London, pp.351-372.

Brain, P.F., Bowden, N.J., Kelleway, L.G., 1980. Anti-estrogenic effects on fighting maintained by testosterone or estradiol-replacement in castrated mice. *J. Endocrinol. 87*, 4-5.

Braude, S., Tang-Martinez, Z., Taylor, G.T., 1999. Stress, testosterone, and the immunoredistribution hypothesis. *Behav. Ecol. 10*, 345-350.

Cirulli, F., De Aceti, L., Alleva, E., 1998. Behavioral effects of peripheral interleukin-1 administration in adult CD-1 mice: specific inhibition of the offensive components of intermale agonistic behavior. *Brain Res. 791*, 308-312.

Crestani, F., Seguy, F., Dantzer, R., 1991. Behavioral effects of peripheral injected interleukin-1: Role of prostaglandins. *Brain Res. 542*, 330-335.

Daly, A.K., Day, C.P., Donaldson, P.T., 2002. Polymorphism in immunoregulatory genes: towards individualized immunosuppressive therapy? *Am. J. Pharmacogenomics 2(1)*, 13-23.

Dantzer, R., 2001. Cytokine-induced sickness behavior: where do we stand? *Brain Behav. Immun. 15*, 7-24.

Dantzer, R., Bluthe, R.M., Kent, S., Kelley, K.W., 1992. Behavioral effects of cytokines. In: Rothwell, N., Dantzer, R., (Eds.), *Interleukin-1 in the brain*, Pergamon, New York, pp. 135-151.

Devoino, L., Alperina, E., Kudryavtseva, N., Popova, N., 1993. Immune responses im male with aggressive and submissive behavior patterns: strain differences. *Brain. Behav. Immun.*, *7*, 287-289.

del Rey, A., Besedovsky, H.O., 2000. The cytokine-HPA axis circuit contributes to prevent or moderate autoimmune processes. *Z. Rheumatol. 2*, 31-35.

Elkabir, D.R., Wyatt, M.E., Vellucci, S.V., Herbert, J., 1990. The effects of separate or combined infusions of corticotrophin-releasing factor and vasopressin either intraventricularly or in the amygdala on aggressive and investigative behaviour in the rat. *Regul. Pept. 28*, 199-214.

Faivre, B., Preault, M., Salvadori, F., Thery, M., Gaillardo, M., Cezilly, F., 2001. Bill colour and immunocompetence in the European blackbird. *An. Behav. 65*, 112-1131.

Fishkin, R.J., Winslow, J.T., 1997. Endotoxin-induced reduction of social investigation by mice: interaction with amphetamine and anti-inflammatory drugs. *Psychopharmacology 132*, 335-341.

Folstad, I., Karter, A.J., 1992. Parasites, bright males, and the immunocompetence handicap. *Am. Nat. 139*, 603-622.

Frances, H., Graulet, A., Debray, M., Coudereau, J.P., Gueris, J., Bourre, J.M., 2000. Morphine-induced sensitization of locomotor activity in mice: effect of social isolation on plasma corticosterone levels. *Brain Res. 860*, 136-140.

Gerlinskaya, L.A., Moshkin, M.P., Evsikov, V.I., 1993. The methodical approaches to estimation of stress level in the wild animals. *Ecology 1*, 97-99. (In Russian)

Gerlinskaya, L.A., Rogova, O.A., Yakushko, O.F., Evsikov, V.I., 1995. Female olfactory choice and its influence on pregnancy in mice. *Adv. Biosci. 93*, 297-302.

Goldsmith, J.F., Brain, P.F., Benton, D., 1976. Effects of age at differential housing and the duration of individual housing/grouping on intermale fighting behaviour and adrenocortical activity in TO strain mice. *Aggressive Behavior 2*, 207-223.

Granger, D.A., Booth, A., Johnson, D.R., 2000. Human aggression and enumerative measures of immunity. *Psychosomatic Medicine 62*, 583-590.

Grossman, Ch.J., 1985. Interactions between the gonadal steroids and the immune system. *Science* 227, 257-261.

Harper, J.M., Austad, S.N., 2000. Fecal glucorticoids: a nonivasive method of measuring adrenal activity in wild and captive rodents. *Physiol. Biochem. Zoology 73*, 12-22.

Hassanain, M., Zalcman, S., Bhatt, S., Seigel, A., 2003. Interleukin-1β in the hypothalamus potentiates feline defensive rage: role of serotonin-2 receptors. *Neuroscience 120(1)*, 227-233.

Hillgarth, N., Ramenofsky, M., Wingfield, J., 1997. Testosterone and sexual selection. *Behav. Ecol. 8*, 108-109.

Horai, R., Masahide, A., Katsuko, S., Hirotaka, K., Masatoshi, S., Masugi, N., Takahashi, M., Iwakura, Y., 1998. Production of mice deficient in genes for Interleukin (Il)-1α, Il-1β, Il-1α/β, and Il-1 receptor antagonist shows that Il-1β is crucial in turpentine-induced fever development and glucocorticoid secretion. *J. Exp. Med. 187*, 1463-1475.

Jonsson, C.K., Setchell, B.P., Martinelle, N., Svechnikov, K., Soder, O., 2001. Endotoxin-induced interleukin 1 expression in testicular macrophages is accompanied by downregulation of the constitutive expression in Sertoli cells. *Cytokine 14*, 283-288.

Kelley, K.W., Bluthe, R. M., Dantzer, R., Zhou, J.H., Shen, W.H., Johnson, R.W., Broussard, S.R., 2003. Cytokine-induced sickness behavoir. *Brain Behav. Immun. 17*, S112-S118.

Kendall, M.Y., Stuart, A., 1966. The Advanced Theory of Statistics. Oriffin, London.

Kimball, R.T., Ligon, J.D., 1999. Evolution of avian plumage dichromatism from a proximate perspective. *Am. Nat. 154*, 182-193.

Kmicikiewicz, I., Wojtusiak, A., Bilinska, B., 1999. The effect of testicular macrophages, macrophage-conditioned medium and interleukin-1alpha on bank vole Leydig cell steroidogenesis. *Exp. Clin. Endocrinol. Diabetes 107*, 262-271.

Leshner, A.I., 1983. Pituitary-adrenocortical effects on intermale agonistic behavior. In: Svare, B.B., (Ed.), *Hormones and aggressive behavior*, Plenum Press, New York, London, pp.27-38.

Lorenz, K., 1974. *Das sogenannte Bose (Zur Naturgeschichte der Agression)*. Taschenbuch Verlag, Munchen.

Loughrey, B.V., Maxwell, A.P., Fagarty, D.G., Middleton, D., Harron, J.G., Patterson, C.C., Darke, C., Savage, D.A., 1998. An interleukin-1 β allele, which correlates with a high secretor phenotype, is associated with diabetic nephropathy. *Cytikine 10(12)*, 984-988.

Matsuoka, Y., Furuyashiki, T., Yamada, K., Nagaai, T., Bito, H., Tanaka, Y., Kitaoka, S., Ushikubi, F., Nabeshima, T., Narumia, S., 2005. Prostaglandin E receptor EP1 controls impulsive behaviour under stress. *Proc. Natl. Acad. Sci. USA, 102*, 16066-16071.

Mele, A., Cabib, S., Oliverio, A., Melchiorri, P., Puglisi-Allegra, S., 1987. Effects of corticotropin releasing factor and sauvagine on socio behavior of isolated mice. *Peptides 8*, 935-938.

Misslin, R., Herzog, F., Koch, B., Ropartz, P., 1982. Effects of isolation, handling and novelty on the pituitary-adrenal response in the mouse. *Psychoneuroendocrinology 7*, 217-221.

Moshkin, M.P., Frolova, O.F., Evsikov, V.I., 1991. Hypothalamo-pituitary andrenocortical system and aggressive behavior in male mice during establishing of social hierarchy. *Siberian biological Journal 4*, 46-51. (In Russian)

Moshkin, M.P., Potapov, M.A., Frolova, O.F., Evsikov, V.I., 1993. Changes in aggressive behavior, thermoregulation and endocrine responses in BALB/cLac and C57Bl/6j mice under cold exposure. *Physiol. Behav. 53*, 535-538.

Moshkin, M.P., Tamagawa, A., Kolosova, I.E., Gerlinskaya, L.A., Iwakura, Y., Endo, Y., 2001. Behavioral and endocrine effects of endotoxin in wild type and Il-1 deficient mice. Sickness behavior or adaptive reaction? *Dokladi Akademii Nauk* (*Proc. Russian Acad. Sci.) 379*, 564-566. (In Russian)

Muir, C., Spironello-Vella, E., Pisani, N., deCatanzaro, D., 2002. Enzyme immunoassay of 17-beta-estradiol, estrone conjugates and testosterone in urinary and fecal samples from male and female mice. *Horm. Metabol. Res. 33*, 653-658.

Nakae, S., Asano, M., Horai, R., Sakaguchi, N., Iwakura, Y., 2001. IL-1 enhances T cell-dependent antibody production through induction of CD40 ligand and OX40 on T cells. *J. Immunol. 167*, 90-97.

Naumenko, E.V., Osadchuk, A.V., Serova, L.I., Shishkina, G.T., 1983. *Genetic and physiological mechanisms of testicular functions*. Ed. Popova, N.K. Nauka, Novosibirsk. (In Russian).

Nicklin, M.J., Barton, J.L., Nguyen, M., FitzGerald, M.G., Duff, G.W., Kornma, K., 2002. A sequence-based map of the nine genes of the human interleukin-1 cluster. *Genomics 79(5)*, 718-725.

Ogilvie, K.M., Held, H.K, Roberts, M.E., Hales, D.B., Rivier, C., 1999. The inhibitory effect of intracerebroventricularly injected interleukin 1beta on testosterone secretion in the rat: role of steroidogenic acute regulatory protein. *Biol. Reprod. 60*, 527-533.

Owens, I.P.F., Short, R.V., 1995. Hormonal basis of sexual dimorphism in birds: implications for new theories about sexual selection. *Trends Ecol.Evol. 10*, 44-47.

Petitto, J.M., Lysle, D.T., Gariepy, J.L., Clubb, P.H., Cairns, R.B., 1994. Genetic differences in social behaviour and cellular immune responsiveness: effects of social experience. *Brain Behav. Immun.*, *8*, 111-112

Petitto, J.M., Gariepy, J.L., Gendreau, P.L., Rodriguiz, R., Lewis, M.H., Lysle, D.T., 1999. Differences in NK cell function in mice bred for high and low aggression: genetic linkage between complex behavioral and immunological traits? *Brain Behav. Immun. 13*, 175-186.

Qvarnstrom, A., Forsgren, E., 1998. Should females prefer dominant males? *Trends Ecol. Evol. 13*, 498-501.

Repovic P., Mi K., Benveniste E.N., 2003. Oncostatin M enhances the expression of prostaglandin E_2 and cyclooxygenase-2 in astrocites: Synergy with interleukin-1β, tumor necrosis factor-α, and bacterial lipopolysaccharide. *Glia*, *42*, 433-466.

Rivest S., 1999. What is the cellular source of prostaglandins in the brain in response to systemic inflammation? Facts and controversies. *Mol. Psychiatry*, *4*, 501-507.

Sayegh, J.F., Kobor, G., Lajtha, A., Vadasz, C., 1990. Effects of social isolation and the time of day on testosterone levels in plasma of C57BL/6By and BALB/cBy mice. *Steroids 55*, 79-82.

Scott, J.P., Fredericson, E., 1951. The causes of fighting in mice and rats. *Physiol. Zool. 24*, 273-309.

Segreti, J., Gheusi, G., Dantzer, R., Kelley, K.W., Johnson, R.W., 1997. Defect in interleukin-1α secretion prevents sickness behavior in C3H/HeJ mice. *Physiol. Behav. 61*, 873-878.

Svechnikov, K.V., Sultana, T., Soder, O., 2001. Age-dependent stimulation of Leydig cell steroidogenesis by interleukin-1 isoforms. *Mol. Cell. Endocrinol. 182*, 193-201.

Swiergiel, A.H., Smagin, G.N., Johnson, L.J., Dunn, A.J., 1997. The role of cytokines in the behavioral responses to endotoxin and influenza virus infection in mice: effects of acute and chronic administration of the interleukin-1-receptor antagonist (IL-1 ra). *Brain Res. 776*, 96-104.

Verhulst, S., Dieleman, S.J., Parmentier, H.K., 1999. A tradeoff between immunocompetence and sexual ornamentation in domestic fowl. *Proc. Nat. Acad. Sci. USA 96*, 4478-4481.

Vitkovic, L., Bockaert, J., Jacque, C., 2000. "Inflammatory" cytokines: neuromodulators in normal brain? *J. Neurochemistry 74*, 457-471.

Weatherhead, P., Metz, K., Bennetti, G., Irwin, R., 1993. Parasite faunas, testosterone and secondary sexual traits in male red-winged blackbirds. *Behav. Ecol. Sociobiol. 33*, 13-23.

Wedekind, C., Folstad, I., 1994. Adaptive and non-adaptive immunosupression by sex hormones. *Am. Nat. 143*, 936-938.

Yu, Y.W., Chen, T.J., Hong, C.J., Chen, H.M., Tsai, S.J., 2003. Association study of the interleukin-1 β (C-511T) genetic polymorphism with major depressive disorder, associated symptomatology, and antidepressant response. *Neuropsychopharmacology 28(6)*, 1182-1185.

Yucesoy, B., Sleijffers, A., Kashon, M., Garssen, J., de Gruijl, F.R., Boland, G., van Hattum, J., Simenova, P.P., Luster, M.I., van Loveran, H., 20(25-26), 3193-3196.

Zav'yalov, E.L., Gerlinskaya, L.A., Evsikov, V.I., 2003. Estimation of stress level in the bank vole *Clethrionomys glareolus* (RODENTIDAE, RODENTIA) by fecal corticosterone. *Zool. J. 82*, 508-513. (In Russian).

Zuk, M., Johnsen, T.S., 2000. Social environment and immunity in male red jungle fowl. *Behav. Ecol. 11*, 146-153.

Zuk, M., 1996. Disease, endocrine-immune interactions, and sexual selection. *Ecology 77*, 1037-1042.

In: Psychoneuroendocrinology Research Trends
Editor: Martina T. Czerbska, pp. 361-380
ISBN: 978-1-60021-665-7

Chapter XII

GLUCOCORTICOID RECEPTOR SIGNALING AND BIPOLAR DISORDER: UPDATES

*P. Moutsatsou**
Laboratory of Biological Chemistry, Medical School, University of Athens, GR 115 27 Goudi, Athens

ABSTRACT

The glucocorticoid receptor (GR) has for long been considered in the aetiopathogenesis of bipolar disorder (BD). Reference is made on the GR gene and protein structure, its splicing variants, on important aspects of GR-mediated signaling events, such as the mode of action by gene activation through its glucocorticoid response elements GREs (transactivation effect) or via its interference with the cognate response elements of other transcription factors i.e. AP-1 and NF-kB (transrepression effect). The interaction of GR with other signaling molecules such as the mitogen activated protein kinases or the G-protein generated signals are presented. Recent advances on clinical and preclinical data delineating the role of GR signaling in BD as well as the effect of antidepressants on GR function are reviewed. Data regarding the presence of GR in the mitochondria of brain cells as well as the association of mitochondrial dysfunction with BD are briefly discussed.

INTRODUCTION

Glucocorticoids (GCs), with cortisol being the principal glucocorticoid (GC) in humans, regulate many life-sustaining functions, among which the homeostasis of central nervous

* Correspondence concerning this article should be addressed to P.Moutsatsou, Laboratory of Biological Chemistry, Medical School, University of Athens, 75 Mikras Asias str., GR 115 27 Goudi, Athens. Tel.-Fax: 0030-210-7462682; E-mail: pmoutsatsou@med.uoa.gr.

system (CNS) is of particular importance. GCs play a central role in stress-related disorders such as bipolar disorder and major depression, known to be triggered under stressful life events. Bipolar disorder (BD) patients are characterized by a hyperactivity of the hypothalamus-pituitary-adrenal (HPA) axis, by increased cortisol concentration (plasma or urine), increased release of corticotrophin releasing hormone (CRH) in the hypothalamus, exaggerated cortisol response to ACTH and an enlargement of both the pituitary and the adrenal glands [1,2,3,4]. In BD patients there is an impairment in the GC-mediated feedback inhibition, associated by impaired response in the dexamethasone suppression test or the combined dex/CRH test, data which reveal a loss of sensitivity to glucocorticoids at a higher level, likely at the pituitary level or even at the brain level [5,6,7]. Endogenous glucocorticoids increase in response to stress and via binding to the glucocorticoid receptor (GR) in tissues of the HPA axis such as hippocampus, hypothalamus and pituitary, regulate negatively the HPA axis activity (negative feedback inhibition), resulting thus in synthesis and release of CRH and secretion of ACTH [8].

Given the above, and because chronic antidepressant treatment of patients with depression is associated with normalization of HPA axis function, concominantly with an enhanced GR function, the involvement of GR signaling has for long been considered in BD aetiopathogenesis [9,10,11]. Despite the important advances that have been made in elucidating some of the molecular and cellular actions of GC-GR complex that are relevant in BD, the precise mechanism of action of GC-GR induced effects in BD is not yet understood. However, a deeper insight has recently been given, mainly through the application of molecular biology techniques. To this direction, a plethora of studies have been carried out in peripheral blood mononuclear cells (PBMC) or brain cells in BD patients assessing the protein or gene expression levels of GR, the sensitivity to glucocorticoids and the molecular structure of the GR gene. The interaction of GR with other intracellular signaling molecules, such as mitogen activated protein kinases (MAPK) and transcription factors (e.g. AP-1, NF-kB), which are known to regulate the tissue sensitivity to glucocorticoids and the neuroendocrine function, has also been investigated in BD. In addition, animal studies and cell culture studies in vitro have elucidated the role of GR signaling in animal models of depression and the efficacy of antidepressants in GR function.

This article summarizes recent evidence on important aspects of GR signaling and its implications in BD. The clinical and preclinical data demonstrating the abnormalities in signaling events upstream and downstream of the glucocorticoid receptor in BD patients, are reviewed thoroughly. Recent data presenting the mitochondrial dysfunction in BD along with the rapid advances on the localization of GR in brain mitochondria, point out that the scrutiny of GR signaling in BD aetiopathogenesis and treatment remains a scientific challenging issue.

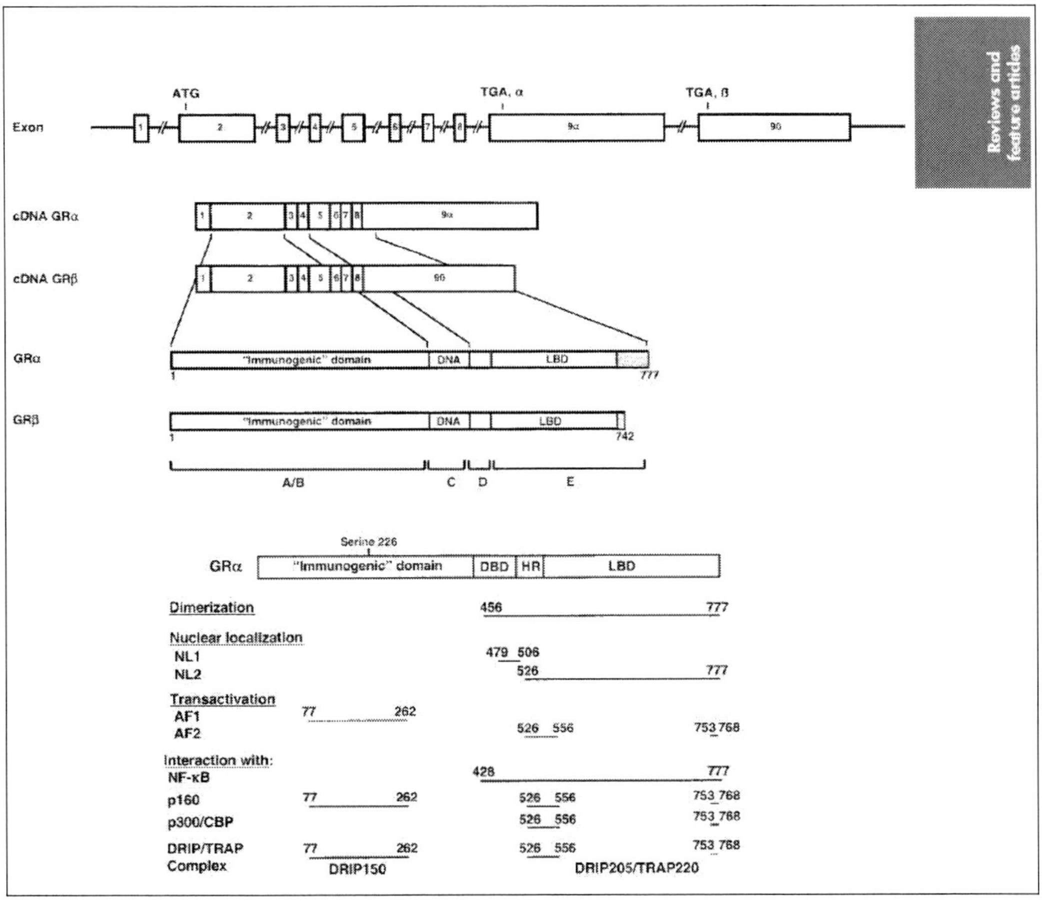

Figure 1. B, Genomic and complementary DNA and protein structure of the human GR. The human GR gene consists of 10 exons. Exon 1 is an untranslated region, exon 2 codes for the immunogenic domain (A and B), exon 3 and 4 code for the DNA-binding domain (C), and exons 5 to 9 code for the hinge region (D) and the ligand-binding domain (E). The gene encoding GR contains two terminal exons 9 (exon 9α and 9β) alternatively spliced to produce the classic GRα and the non-ligand binding GRβ. C-terminal gray-colored domains in GRα and GRβ show their specific portions. C, Functional domains of the GRα. Functional domains and subdomains are indicated. HR, Hinge region; NL1 and NL2, nuclear translocation signals 1 and 2. (From Kino T. and Chrousos GP .(2002). J.Allergy Clin. Immunol. 109,609-613. With permission)

Mode of Action of Glucocorticoids

Glucocorticoids mediate their effects on responsive cells by activating their intracellular receptors, the glucocorticoid receptor alpha (GRα) and beta isoform (GRβ). The classic GR, today known as GRα binds GC, whereas a second GR, now called GRβ does not bind GC [12]. Both, GRα and GRβ are products of the same gene, located on chromosome 5, and result from differential splicing through the alternative use of two distinct terminal exons, 9α and 9β respectively [13]. The classic GRα is α 777-aminoacid protein whereas GRβ isoform contains 742 aminoacids, the first 727 aminoacids from the N-terminus being identical in the two isoforms. GRα possesses an additional 50 aminoacids, whereas GRβ encodes an additional 15 non-homologous aminoacids in the C terminus. The classical GR is characterized by the presence of three main distinct functional domains -the ligand-binding domain (LBD), the DNA-binding domain (DBD), the N-terminal or immunogenic domain-. Furthermore, various subdomains have been identified with functional roles, such as regions responsible for dimerization, for interaction with heat shock proteins, nuclear regulatory proteins and nuclear localization signals (NLS). The classical GR contains two transactivation domains at its N-terminal and LBD, the activation function 1 (AF-1) and AF-2 respectively (Figure 1) [14].

In the absence of hormone, GRα is located primarily in the cytoplasm in the form of a hetero-oligomeric complex with heat shock proteins (HSPs) 90, 70, 50 and 20 and other proteins, such as protein 14-3-3, p53. After binding to glucocorticoids, the GRα undergoes conformational changes, dissociates from the heat shock proteins, homodimerizes and translocates into the nucleus where it interacts directly with its specific DNA sequences, the glucocorticoid-response elements (GREs), in the promoter of target genes [15]. The GRα/GRE complex stimulates or diminishes the GRE-mediated gene transcription (known as transactivation effect). In addition, GC-activated GRα monomers may interact with other transcription factors (TFs), such as nuclear factor-kB (NF-kB), activator protein 1 (AP-1), CRE-binding protein (CREB), signal transducer and activator of transcription 5 (STAT5) and others, via protein-protein interactions, thus influencing indirectly the activity of these transcription factors on their target genes (known as transrepression effect) (Figure 2) [14,16]. Moreover, the GC-GR complex may interact through AF-1 and AF-2 activation functions with various proteins in the cellular compartment, the coactivators, such as the nuclear receptor coactivator complexes p160, p300/CREB-binding protein (CBP), p300/CBP-associated factor (P/CAF), as well as the corepressors [17]. These coactivators are often chromatin remodelling complexes since they contain histone acetyltransferase (HAT) activity, thus loosening the chromatin structure and facilitating the binding of transcriptional machinery to DNA (18). The p160 proteins play a central role in the transactivation by GR since they are first attracted to the DNA-bound GR and via mutual interactions help the p300/CBP and p/CAF to move to the promoter region. AF-2 function is activated upon ligand binding whereas AF-1 function interacts with components of the basic transcriptional machinery by way of the TAF proteins and through these with the TATA-binding protein TBP, thus stabilizing the transcription initiation complex.

Inflammatory mediators, on the other hand, such as tumor necrosis factor-α (TNF-α) and stress signals activate several transcription factors (such as NF-kB and AP-1), thus affecting negatively the GR-mediated transactivation effects and peripheral and CNS tissue sensitivity to glucocorticoids [19]. The N-terminal domain contains a number of phosphorylation sites activated by the extracellular signal-regulated kinase (ERK), the c-Jun-N-terminal kinase (JNK) and p38 mitogen-activated protein kinase (p38 MAPK) at aminoacids (serines) 211 and 226. Cyclin-dependent kinases (CDKs) phosphorylate serines at aminoacids 203 and 211. The aforementioned MAPK -mediated signals affect the GR evoked transcriptional effects, resulting in enhancement or inhibition of GR-induced gene expression, thus influencing tissue sensitivity to glucocorticoids [20]. Of note, the stress induced JNK affects negatively the GR evoked transcriptional enhancement, thus affecting peripheral and CNS tissue sensitivity to glucocorticoids [21,22]. The N-terminal domain contains also a novel site that interacts with the β component of the heterotrimeric guanine nucleotide-binding protein (G protein) complex. In particular, the region of GR containing aminoacids 263 to 419 physically interacts with the β component of the G protein complex resulting in the suppression of GR-induced transactivation of the glucocorticoid responsive genes [23]. In this way, G-protein coupled receptor (GRPC)-signals activated by extracellular hormones and other compounds may also influence GR transcriptional activity and tissue sensitivity to glucocorticoids. Since GR interacts with G protein β subunit at the inner surface of the plasma membrane, this may explain some nongenomic actions of glucocorticoids [24].

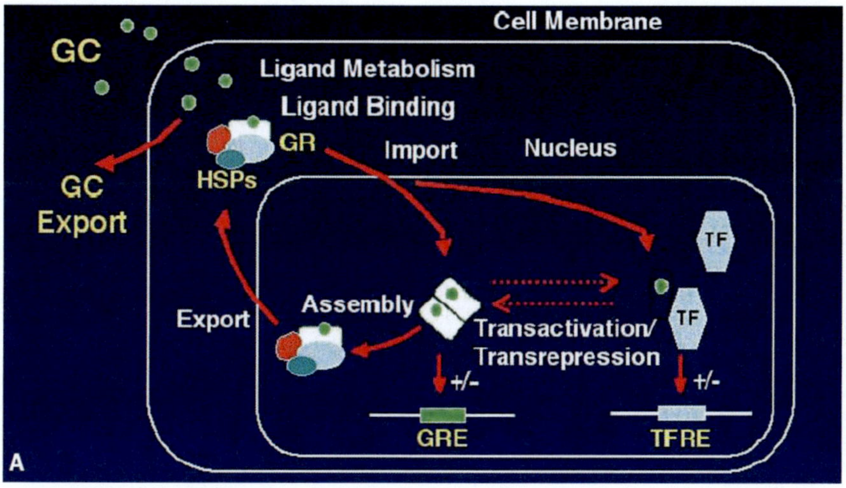

Figure 2. Circulation of the GR between the cytoplasm and the nucleus and its transactivating or transrepressive activities. GRE, glucocorticoid responsive elements; TF, transcription factor; TFRE, transcription factor responsive element; HSP, heat shock protein. (From Kino T. and Chrousos GP (2002) J. Allergy Clin. Immunol. 109, 609-613. With permission).

The isoform GRβ has been shown to be a potential endogenous inhibitor of GC-GRα action in humans, implicating that variations in the relative tissue levels of GRα and GRβ may be an important determinant of tissue sensitivity to glucocorticoids [12]. Recently, Lu and Cidlowski [25] showed that each GRα mRNA and GRβ mRNA is translated from at least eight initiation sites into multiple GRα and possibly GRβ isoforms. The existence of variable amounts of GRα isoforms in target tissues, having varying transcriptional activities and glucocorticoid-responsive gene induction, implicate that the GC/GR signaling cascade is highly complex [24,25].

Today it is believed that the mechanisms which may potentially result in alterations of tissue-sensitivity to glucocorticoids may be at any step of the GR signaling pathway resulting in diminished or enhanced sensitivity to glucocorticoids. Mutations of the GR gene have been described in the primary generalized glucocorticoid resistance syndrome (familial or sporadic) [26]. Several other factors might change the sensitivity of tissue to glucocorticoids: chaperones and cochaperons (e.g. heat shock proteins), transcription factors (e.g. NF-kB, AP-1, CREB, STATs, C/EBP, p53, some of which are induced by various stimuli such as TNF-α, inflammatory cytokines, NO and others), the receptor isoform (GRβ), coregulators (e.g.

coactivators, corepressors), inflammatory cytokines (such as TNF-α), as well as the phosphorylation and nitrosylation status of GR play a key role and may contribute to the induction of glucocorticoid resistance or hypersensitivity [14,24,26]. The glucocorticoid resistance reported in autoimmune-inflammatory diseases, bronchial asthma, septic shock and respiratory distress syndrome, are often attributed to defects in the cross talk between GR and the transcription factors NF-kB, AP-1 and probably several STATs [14]. The GR isoform, GRβ, has also been correlated with glucocorticoid insensitivity in several pathologic conditions. A plethora of data support that the sensitivity to glucocorticoids may be either increased (known as glucocorticoid hypersensitivity) or decreased (known as glucocorticoid resistance) and may be generalized, or tissue-specific [14,26]. Evidence points out that BD is characterized by glucocorticoid resistance, which is rather "localized resistance in CNS". Other tissues in BD patients, such as bone and abdominal tissue, retain their sensitivity to GCs, resulting in increased-intra-abdominal fat deposition and decreased bone mineral density (BMD) [27,28,29].

Glucocorticoid Receptors and Mitochondria

Energy production is co-ordinated by the cross-talk between the nuclear and mitochondrial processes, the enzymes of the respiratory chain playing a key role in cellular energy production. Mitochondria are the main energy suppliers to cells producing ATP mainly by the enzymes of oxidative phosphorylation (OXPHOS), some of their subunits being encoded by nuclear and some by mitochondrial genes. Human mitochondrial DNA (mtDNA) encodes 13 subunits of the respiratory chain, 22 tRNAs and 2 rRNAs [30]. Evidently, factors affecting ATP production and its control in mitochondria would influence high energy consuming cellular functions and organs.

Glucocorticoids play a major role in brain functions, i.e. mediate the stress response, influence synaptic plasticity, they influence psychiatric disorders such as depression, post-traumatic stress disorder and Alzheimer disease while they have detrimental effects in brain development [31,32,33,34]. All the aforementioned processes require high energy demands, the mitochondria being the major providers of the energy that cells need to function. Accumulating evidence has pointed out that glucocorticoid hormones affect mitochondria functions and energy metabolism. Some of the respiratory enzyme subunits are encoded solely by mitochondrial genes and glucocorticoids have been shown to control the expression of a number of these genes [35]. These early findings demonstrating that glucocorticoids regulate the oxidative phosphorylation (OXPHOS) enzyme subunits encoded by the mitochondrial genome raised the question: Is the mitochondrion itself the primary site of action of glucocorticoids or the effects on mitochondrial processes are secondary to a primary nuclear action?

This led to a series of research studies aiming to document the presence of glucocorticoid receptors in mitochondria of various cell lines and the possible presence of glucocorticoid response elements in mitochondrial genome. Nowdays, a great amount of data demonstrate the presence of glucocorticoid receptors in mitochondria of animal and human cells [36]. Nucleotide sequences in the human mitochondrial genome, with strong similarities to glucocorticoid respone elements (GREs), have also been detected (Table 1). The position of the putative GREs in the mitochondrial genome is either in the D-loop or within the structural genes. Given that the D-loop is the only initiation site in the mitochondrial genome, their role could be in transcription initiation as well as in other steps of the transcription process [37]. Gel shift analysis has demonstrated the potential of GR to bind to putative mitochondrial GREs, while in transfection experiments mitochondrial GREs confered dexamethasone inducibility on hybrid reporter constructs [37]. Such data have given support to hypothesis that mitochondrial genes may be sites of primary action of steroid hormones [38].

Table 1. Sequences showing partial similarity to the GRE consensus sequence detected in the human (H) mitochondrial genome

		AGAGGA	NNN	TGTTCT		Consensus sequence
(H)	1195	AGAGGA	NNN	TGTTCT	1209	12 S rRNA
	3228	AACAAA	NNN	TGTTCT	3242	16 S rRNA/tRMA-Leu
	4102	TAACCT	NNN	TGTTCT	4116	URF1
	16,489	CCGACA	NNN	GGTTCC	16,503	D-loop

The numbers refer to the position of the sequences in the genome. Denoted also are the genes in which the various GREs have been detected.

Given the key role of GR in CNS function , Moutsatsou et al [39] investigated the possibility that GR might be localized in brain mitochondria. The presence of GR in brain tissue has been documented by a variety of techniques, such as radioligand binding assays, in situ hybridization and immunohistochemistry, the brain cortex and hippocampus being the regions highly rich in GR content. By using Western blot analysis and immunogold electron microscopy, Moutsatsou et al [39] revealed the presence of GR in mitochondria and synaptosomal mitochondria in rat brain cortex and hippocampus, thus supporting the concept of a potential role of GR in mitochondrial gene transcription and energy regulation. The importance of mitochondrial GR in brain cells was further documented by Koufali et al [40], who showed that treatment of rat C6 glioma cells with dexamethasone resulted in a dramatic decrease in mitochondrial GR levels in parallel with those of the cytosolic receptor and concomitantly with the import of GR into nuclear compartment. A marked stimulation of the expression of the mitochondrialy-encoded cytochrome oxidase-1 (COX-1) gene was also found following GR activation and its export from mitochondria. Taken together, Koufali et al [40] demonstrated that the dex-associated translocation of GR out of the mitochondria was associated with alterations in the transcription of elements of the mitochondrial genome in a glioma cell line, thus emphasizing the potential importance of mitochondrial GR in CNS diseases.

Mitochondrial Function in Bipolar Disorder

Mitochondria have their own genetic material, the mitochondrial DNA (mtDNA), which is a circular DNA comprised of 16,569 base pair (bp) with no introns, no histones. The total mtDNA is about 10% of the nuclear DNA, characterized by a poor DNA repairing system and susceptibility to somatic mutations. Mitochondrial DNA is highly polymorphic while certain polymorphisms have been considered to be risk factors for many diseases such as diabetes mellitus, Alzheimer disease, Parkinson disease as well as some kinds of cancer [41,42].

Increasing amount of data support that mitochondrial dysfunction is also associated with bipolar disorder. Mothers, but not fathers, transmit manic depressive illness in some families, which suggests that mothers in these families may transmit disease-related mutations in non-nuclear, i.e. mitochondrial, DNA [43]. Indeed, a mtDNA 3243A>G mutation has been detected in the postmortem brains of patients with BD and it has been associated with increased expression of LARS2 gene which encodes the enzyme that catalyzes the aminoacylation of $tRNA^{Leu}$ [44]. The rate of 5178C polymorphism in mtDNA has been significantly higher in patients with BD compared to controls, especially in maternally transmitted cases [45,46]. Furthermore, molecular genetic analysis has revealed association of BD with mtDNA 10398A>G polymorphism and 3644C mutation. Both, the mtDNA 5178C and mtDNA 10398A polymorphisms cause aminoacid substitution in subunits (ND2 or ND3) of the mitochondrial complex I (NADH: ubiquinone oxidoreductase). Complex I (NADH) is the largest enzyme in the mitochondrial electron transport chain consisting of at least 43 subunits, 7 of which are mtDNA encoded while the others are nuclear encoded. In postmortem brains, increased levels of mtDNA 4977 bp deletion have been demonstrated while the existence of 3243G mutation and altered expression of mitochondria-related genes have also been reported [46,47].

Recent findings showing that 1) The 8701A and 10398A mtDNA polymorphisms were associated with lower mitochondrial matrix pH and higher responses in intracellular calcium dynamics and 2) that there is comorbidity of affective disorders in certain types of mitochondrial disorders, such as mitochondrial diabetes mellitus with the 3243 mutation, support further the mitochondrial dysfunction hypothesis for BD [48]. Finally, the results by Washizuka et al [49] pointing out that the mtDNA polymorphisms might be related to maintenance lithium treatment response, highlight further the importance of mitochondrial genome in future therapeutic strategies in BD.

Alterations in various biochemical parameters, which reflect mitochondrial function, have also been reported in bipolar disorder. For example, gene array analysis has revealed that the hippocampus in subjects with bipolar disorder were characterized by an extensive decrease in the expression of genes regulating oxidative phosphorylation and the adenosine triphosphate-dependent processes [50]. The decreased expression of NDUFV2 mitochondrial-related gene (mitochondrial complex I subunit gene) in BD patients, associated with decreased expression of other subunit genes (nuclear encoded) in the mitochondrial respiratory chain, present further evidence for the mitochondrial dysfunction in BD [51]. Finally, magnetic resonance spectroscopy (MRS) studies point out alterations in brain metabolism in brain of patients with affective disorders , such as reduced creatine phosphate and ATP levels, reduced phosphomonoesters, accumulation of lactate and decreased intracellular pH [46,47,52]. The above evidence has led to a hypothesis of mitochondrial dysfunction for the pathophysiology of BD that involves impaired oxidative phosphorylation, a shift towards glycolytic energy production, a decrease in total energy production and/or substrate availability, and altered phospholipid metabolism [53].

The documented mitochondrial dysfuntion in BD and the presence of GR in brain mitochondria implicate that a deeper understanding of the molecular mechanism(s) underlying the role of mitochondrial GR in brain energy production may provide important information for the development of a therapeutic drug which would specifically target mitochondrial function and energy supply for the treatment of BD.

GLUCOCORTICOID RECEPTOR SIGNALING IN BIPOLAR DISORDER

a) Clinical Studies

A great number of studies have assessed GR in patients with BD and major depression. Given the limited access to brain GRs in clinical populations, the GR assessments have been made mainly on peripheral cell types, such as mononuclear and polymorphonuclear leucocytes and fibroblasts. Information regarding GR function in CNS is limited, however, given that GR changes induced by a variety of stimuli are similar in the brain and immune cells, the use of peripheral blood mononuclear cells (PBMC) to assess GR signaling is a valid method and remains a viable option [54,55].

Various types of assays have been used to assess GRs in depression, such as the radioligand binding assays (cytosolic or whole cell assays), glucocorticoid receptor functional assays and western blot analysis. Due to differences in the techniques used to assess GR as well as to heterogenicity in the population studied (i.e. medication status, disease state) variable results have been obtained. Briefly, the radioligand binding assay-based studies support either that GR numbers in depressed patients are not different compared to controls or that depressed patients had lower GR numbers. Of note, none of the reports showed alterations in the affinity of GR for ligand (Kd) in depressed patients [10]. In support to findings from studies using PBMC, two CNS studies also showed that there were either no differences in GR mRNA levels in the hippocampus of postmortem brain tissue obtained from depressed patients relative to controls, or that GR mRNA levels were decreased in the hipppocampus and frontal cortex of depressives [56,57]. The GR functional assays , on the other hand, evaluate the ability of glucocorticoids to inhibit the mitogen stimuli-induced proliferation of PBMC. These studies, using mainly dexamethasone, have given consistent results exhibiting that depressed patients had reduced responses to dexamethasone. Such data support that there is resistance to in vitro GR-mediated responses and a loss of sensitivity of PBMC to dexamethasone. Accumulating evidence supports that the glucocorticoid resistance in depression is "localized GR resistance in CNS". Indeed, depressed patients exhibit increased intra-abdominal fat deposition and decreased bone mineral density, data which point out that some tissues retain their sensitivity to GCs and that in depressed patients resistant and sensitive cells may coexist [27,28,29]. It is important to state that clinical recovery has been associated with normalization of PBMCs' sensitivity to dexamethasone and increase in GR numbers [10]. However, changes in GR number do not occur always together with impaired or normalised GR function. Calfa et al [58] showed a decrease in GR density of PBMC in free-medication patients with major depression but GR function comparable to healthy controls, suggesting a lack of relationship between GR density and glucocorticoid sensitivity as reflected by the cortisol-induced inhibition of lymphocyte proliferation. In their study , the intake of antidepressants resulted in increased GR numbers, the increase being higher in patients than in controls, thus highlighting that GR levels might be used as markers of a successful treatment.

Spiliotaki et al [59] assessed whole cell and nuclear GR levels in lymphocytes of BD patients. Using Western blot analysis, which reflects alterations in upregulation or downregulation of the total cellular GRs, they showed that depressed patients (being under multiple antidepressants) had significantly higher whole cell as well as nuclear GR levels than controls. However, euthymic patients (being under lithium monotherapy) had not altered whole cell GR levels but they did show increased nuclear GR levels, indicating that various types of antidepressants may induce effective nuclear GR translocation.

Apart from the possible role of GR numbers in mediating the GR resistance in BD patients, another major possibility that has been considered to explain the GR resistance seen in depression is the primary defect(s) in the genetic structure of the GR. The reduced HPA axis suppression by dexamethasone in first-degree relatives of depressed patients further supports that alterations in feedback inhibition may be related to a genetic vulnerability to depressive disorders [60]. To this direction, Moutsatsou et al [61] demonstrated that the molecular structure of GRα and GRβ gene is intact in BD patients, thus excluding the possibility that mutations in GRα and GRβ may be responsible for the diminished sensitivity to glucocorticoids observed during the "manic" or "depressed" state of the disease.

Abnormalities in signaling events downstream of the GR molecule such as abundance of coactivators/corepressors or the GR-DNA binding ability, could also be a causative factor for the GR-resistance in BD patients [11,14,26]. To this effect, Spiliotaki et al [59] addressed the question: are depressed patients characterized by a reduced GR-DNA binding activity? The novel findings emerging from their study were a) the impaired GR-DNA binding concurrent with multiple drug therapy and the depressed state of the disease, b) the normalization of GR-DNA binding concurrent with lithium monotherapy and behavioural improvement in euthymia. Taken together, their data indicated that BD patients are characterized by a defective GR signaling at the level of GR-DNA binding activity and implicate that GR-DNA binding may be an index of the course of illness and the effect of therapy. Henning et al [62] have also demonstrated an increase in GR numbers in depressed patients being under antidepressant therapy. Moreover, they showed a greater GR downregulation with hydrocortisone (GC autoregulation) in B-lymphoblastoid cells of the depressed patients as compared to controls. Given that hydrocortisone is a glucocorticoid receptor modulator of variable transactivation/ transrepression potencies [63,64], its greater reducing effect on patients' GR numbers than in controls, may be related to the potential of hydrocortisone to modulate GR to restore an impaired GR signaling in patients rather than in controls.

It is now appreciated that the interplay (cross-talk) between the GR and other signaling molecules , such as MAPK kinases (JNK), ubiquitously distributed cytoplasmic transcription factors (AP-1, NF-kB) and components of the G-protein complex, play a crucial role in determining the tissue sensitivity to glucocorticoids and the neuroendocrine function [24,65,66,67]. In this regard, Spiliotaki et al [59] hypothesized that dysregulation of JNK, AP-1 and NF-kB signaling could be causally related to bipolar disorder. Their investigation, in lymphocytes of BD patients, revealed that the depressed state of the disease (under multiple drug therapy) was associated with an impaired AP-1-DNA binding activity whereas euthymia (under lithium monotherapy) was accompanied by normalization of the AP-1-DNA binding activity. Concomitant with the observed reduction in AP-1-DNA binding, the c-fos protein was significantly lower in the depressed state. Their study showed also that JNK was significantly altered in the depressed state, characterized by reduced total cellular and nuclear JNK levels as compared to euthymic and controls. Such data implicate a possible role of JNK/AP-1 signaling in BD disease and that lithium favors JNK/AP-1 signaling compared to multiple antipsychotic regimes. However, no changes were observed in NF-kB-DNA binding activity between depressed patients, euthymics and controls, while total cellular NF-kB levels were higher in the depressed state. In agreement, Sun et al [68] have also reported that the levels of RNA transcripts, encoding components of the NF-kB transcription factor complex, were significantly higher in the frontal cortex in BD patients compared to controls.

Another important target in aetiopathogenesis and therapeutics of BD is considered the family of G-proteins. The heterotrimeric guanine nucleotide-binding proteins (G proteins, Gα, Gβ, Gγ) are downstream signal transducers for the G protein-coupled receptors (GPCRs), which form a large family [69]. The family of G proteins include stimulatory (Gs) and inhibitory (Gi) types. Different molecular weight variants exist in different tissues and Gsα arise as a result of alternative splicing of a single gene. G proteins, present in all cell types, are the "coupling" agents between cell membrane receptors and internal "second messenger" systems. Extracellular molecules, such as hormones, neurotransmitters, cytokines, growth factors, transduce their effects into the intracellular compartment via specific cell membrane receptors .These receptors in turn activate signal-transducing molecules located on the cytoplasmic side of the plasma membrane, which then activate downstream effector molecules. The G proteins are downstream signal transducers for the G protein coupled receptors (GPCRs). Many studies have examined G proteins in patients with mood disorders [70]. The Gα subunit has been mostly studied in bipolar disorder. Young et al [71] were the first to report that there are elevated levels of Gαs protein (one of Gα-subtypes) in postmortem brain tissue from patients with BD. Similar findings have been reported by other independent laboratories i.e. enhanced Gs/ adenylate cyclase (AC) signaling in postmortem brains in BD or elevated Gαs levels in frontal cortical membrane preparations from BD patients [72,73]. Studies examining G protein in PBMC exhibited also elevated Gαs protein and mRNA levels [74,75]. Because Gβ proteins play key roles in many signal transduction cascades, the recently shown GR-associated Gβ might play a role in the development of target tissue resistance to glucocorticoids in states in which G proteins are affected such as in bipolar disorder [23].

b) Animal Studies and Cell Culture Studies in vitro

Most of the accumulated data based on animal studies concern 1) the impact of GR gene defects on stress-related hormonal and behavioral responses and 2) the impact of antidepressants on GR-mediated signaling and behavior.

The development of mouse mutants in which the GR knockout was restricted to neural tissue have revealed that these mice showed signs of reduced anxiety and impaired stress response , suggesting that loss of GR-mediated signaling in brain results in alterations in emotional behavior [76]. Another mouse model expressing GR antisense mainly in neural tissue, which was expected to be a well suited animal model of depression associated with impaired GR function, showed that the mice needed higher dexamethasone dosages than control mice in order to suppress basal corticosterone [76]. Taken together, alteration of GR gene expression (knock out GR gene or GR antisense) has resulted in reduced anxiety-like behavior in mice and decreased CRH expression. However, none of these generated GR mutants can be viewed as animal model of psychiatric disease. These mouse mutants have been considered of great value to study symptoms such as anxiety, stress response, behavior and sleep which are associated with HPA disturbances [11]. A recent work by Boyle et al [77] showed that introduction of an acquired disruption of GR signaling into the mouse forebrain, leads to development of a number of physiological and behavioral abnormalities that mimic major depression in humans, such as hyperactivity of HPA axis and impaired negative glucocorticoid feedback.

Regarding the impact of antidepressants on GR signaling, a vast amount of data have shown that most antidepressants, but not all, possess the ability to restore HPA axis sensitivity to glucocorticoids, thus facilitating the GC-mediated feedback inhibition and normalization of HPA axis activity. In addition, long-time treatment with antidepressants have resulted in upregulation of GR protein and mRNA levels in vital brain regions, including hippocampus and hypothalamus [78,79,80,81,82,83]. However, two negative reports showed that certain antidepressants lowered the GR number whereas other had no effect on GR expression [10]. Of note, in an animal model of depression, the GR-DNA binding activity has been shown to increase in brain tissues during successful antidepressant therapy of depressive illness [84,85], being thus in agreement with the findings reported by the clinical study of Spiliotaki et al [59].

In vitro studies have employed mouse fibroblast cell lines or a mouse hippocampal cell line to study the effect of antidepressants on the transcriptional activity of GR in vitro. Herr et al [86], using a mouse hippocampal cell line HT22, showed that a wide range of antidepressants of different substance classes differentially affect steroid-induced GR-mediated gene expression. The majority of antidepressants enhance the steroid-induced GR-mediated gene transcription at concentrations which are clinically relevant [86,87,88]. However, high concentrations of antidepressants, which exceed therapeutic concentration in the brain, have been associated with inhibition of corticosterone-induced gene transcription [89]. Of note, in vitro studies have demonstrated, being in agreement with in vivo studies, that there may be an increased GR-mediated gene transcription without necessarily of an accompanying upregulation of GR protein [86,87,88].

Animal and in vitro studies have further elucidated the impact of antidepressants on signaling molecules/pathways (such as AP-1, NF-kB, JNK), which are known to interact with GR signaling.The AP-1-DNA binding activity has been shown to be either induced or inhibited under the effect of antidepressants, possibly due to different experimental conditions and models used [90,91,92,93,94,95]. Lithium has been shown to enhance c-fos expression indicating that this is probably a mechanism for its therapeutic effect [90,96]. Moreover, lithium alters stress-induced JNK levels and affects JNK-induced phosphorylation, thus indicating that phoshorylation in BD warrants further investigation [91,97]. Lithium as well as other antidepressants have shown differential effects on NF-kB activity and neuronal cell death [98,99].

Conclusion

The role of glucocorticoid receptor (GR) signaling in regulating HPA axis function and associated abnormalities in BD is increasingly being understood, due to rapid advances and amassing of data concerning the molecular biology of GR, its mode of action at the molecular level, as well as its interaction with other key regulatory molecules and the mitochondrial genome. These interactions are currently of high scientific interest and further research is expected to give a deeper insight on the role of GC-GR signaling cascade in BD aetiopathogenesis and therapeutic strategies.

References

[1] Raadsheer, F.C., Hoogendijk, W.J., Stam, F.C., Tilders, F.J. & Swaab, D.F. (1994). Increased numbers of corticotropin-releasing hormone expressing neurons in the hypothalamic paraventricular nucleus of depressed patients. *Neuroendocrinology 60*, 436-444.

[2] Rubin, R.T., Phillips, J.J., Sadow, T.F. & McCracken, J.T. (1995). Adrenal gland volume in major depression. Increase during the depressive episode and decrease with successful treatment. *Arch. Gen. Psychiatry 52*, 213-218.

[3] Young, A.H. (2004). Cortisol in mood disorders. *Stress 7*, 205-208.

[4] Manji, H.K., Drevets, W.C. & Charney, D.S. (2001a). The cellular neurobiology of depression. *Nat. Med. 7*, 541-547.

[5] Lowy, M.T., Reder, A.T., Antel, J.P. & Meltzer, H.Y. (1984). Glucocorticoid resistance in depression: the dexamethasone suppression test and lymphocyte sensitivity to dexamethasone. *Am. J. Psychiatry 141*, 1365-1370.

[6] Heuser, I., Yassouridis, A. & Holsboer, F. (1994). The combined dexamethasone/CRH test: A refined laboratory test for psychiatric disorders. *J. Psychiatr. Res. 28*, 341-356.

[7] Bauer, M.E., Papadopoulos, A., Poon, L., Perks, P., Lightman, S.L., Checkley, S. & Shanks, N. (2003). Altered glucocorticoid immunoregulation in treatment resistant depression. *Psychoneuroendocrinology 28*, 49-65.

[8] Reul, J.M., van den Bosch, F.R. & de Kloet, E.R. (1987). Relative occupation of type-I and type-II corticosteroid receptors in rat brain following stress and dexamethasone treatment: functional implications. *J. Endocrinol. 115*, 459-467.

[9] Barden, N., Reul, J.M.H. & Holsboer, F. (1995). Do antidepressants stabilize mood through actions on the hypothalamic-pituitary-adrenocortical sytems? *Trends Neurosci. 18*, 6-11.

[10] Pariante, C.M. & Miller, A.H. (2001). Glucocorticoid receptors in major depression: Relevance to pathophysiology and treatment. *Biol. Psychiatry 49*, 391-404.

[11] Holsboer, F. (2000). The corticosteroid receptor hypothesis of depression. *Neuropsychopharmacology 23*, 477-501

[12] Bamberger, C.M., Bamberger, A., DeCastro, M. & Chroussos, G.P. (1995). Glucocorticoid receptor β, a potential endogenous inhibitor of glucocorticoid action in humans. J. Clin. Invest. *95*, 2435-2441.

[13] Hollenberg, S.M., Weinberger, C., Ong, E.S., Cerelli, G., Oro, A., Lebo, R., Thompson, E.B., Rosenfeld, M.G. & Evans, R.M. (1985). Primary structure and expression of a functional human glucocorticoid receptor cDNA. *Nature 318*, 635-641.

[14] Kino T. & Chrousos G.P. (2002). Tissue-specific glucocorticoid resistance-hypersensitivity syndromes: Multifactorial states of clinical importance. J. *Allergy Clin. Immunol. 109*, 609-613.

[15] Beato, M. & Sanchez-Pacheco, A. (1996). Interaction of steroid hormone receptors with the transcription initiation complex. *Endocr. Rev. 17*, 587-609.

[16] De Bosscher, K., Van den Berghe, W. & Haegeman, G. (2003). The interplay between the glucocorticoid receptor and nuclear factor-κB or activator protein-1: Molecular mechanisms for gene repression. *Endocr. Rev. 24*, 488-522.

[17] McKenna, N.J., Lanz, R.B. & O'Malley, B.W. (1999). Nuclear receptor coregulators: cellular and molecular biology. *Endocr. Rev. 20*, 321-344.

[18] Blanco, J.C., Minucci, S., Lu, J., Yang, X.J., Walker, K.K., Chen, H., Evans, R.M., Nakatani, Y. & Ozato, K. (1998). The histone acetylase PCAF is a nuclear receptor coactivator. *Genes Dev. 12*, 1638-1651.

[19] Caldenhoven, E., Liden, J., Wissink, S., Van de Stolpe, A., Raaijmakers, J., Koenderman, L., Okreet, S., Gustafsson, J.A. & Van der Saag, P.T., 1995. Negative cross-talk between RelA and the glucocorticoid receptor: a possible mechanism for the antiinflammatory action of glucocorticoids. *Mol. Endocrinol. 9*, 401-412.

[20] Krstic, M.D., Rogatsky, I., Yamamoto, K.R. & Garabedian, M.J. (1997). Mitogen-activated and cyclin-dependent protein kinases selectively and differentially modulate transcriptional enhancement by the glucocorticoid receptor. *Mol. Cell. Biol. 17*, 3947-3954.

[21] Rogatsky, I., Logan, S.K. & Garabedian, M.J. (1998). Antagonism of glucocorticoid receptor transcriptional activation by the c-Jun N-terminal kinase. *Proc. Natl Acad. Sci. USA 95*, 2050-2055.

[22] Wang, X., Wu, H., Lakdawala, V.S., Hu, F., Hanson, N.D. & Miller, A.H. (2004). Inhibition of Jun N-terminal kinase (JNK) enhances glucocorticoid receptor-mediated function in mouse hippocampal HT22 cells. *Neuropsychopharmacology 30(2)*:242-249.

[23] Kino, T., Tiulpakov, A., Ichijo, T., Chheng, L., Kozasa, T. & Chrousos, G.P. (2005). G protein beta interacts with the glucocorticoid receptor and suppresses its transcriptional activity in the nucleus. *J. Cell. Biol. 169*, 885-896.

[24] Chrousos, G.P. & Kino, T. (2005). Intracellular glucocorticoid signaling: a formerly simple system turns stochastic. *Science's stke*. www.stke.org/cgi/content/full/ sigtrans;2005/304/pe48.

[25] Lu, N.Z. & Cidlowski, J.A. (2005). Translational regulatory mechanisms generate N-terminal glucocorticoid receptor isoforms with unique transcriptional target genes. *Mol. Cell 18*, 331-342.

[26] Kino, T., De Martino, M.U., Charmandari, E., Mirani, M. & Chrousos, G.P. (2003). Tissue glucocorticoid resistance/ hypersensitivity syndromes. *J. Steroid Biochem. Mol. Biol. 85*, 457-467.

[27] Thakore, J.H., Richards, P.J., Reznek, R.H., Martin, A. & Dinan, T.G. (1997). Increased intra-abdomnal fat deposition in patients with major depressive illness as measured by computed tomography. *Biol. Psychiatry 41*, 1140-1142.

[28] Horrobin, D.F. & Bennett, C.N. (1999). Depression and bipolar disorder: relationships to impaired fatty acid and phospholipid metabolism and to diabetes, cardiovascular disease, immunological abnormalities, cancer, ageing and osteoporosis. Possible candidate genes. *Prostaglandins Leukot. Essent. Fatty Acids 60*, 217-234.

[29] Misra, M., Papakostas, G.I. & Klibanski, A. (2004). Effects of psychiatric disorders and psychotropic medications on prolactin and bone metabolism. *J. Clin. Psychiatry 65*, 1607-1618.

[30] Anderson, S., Bankier, A.T., Barrell, B.G., de Bruijn, M.H. Coulson., A.R., Drouin, J. et al (1981). Sequence and organization of the human mitochondrial genome. *Nature 290*, 457-465.

[31] Kim, J.J., Lee, H.J., Han, J.S. & Packard, M.G. (2001). Amygdala is critical for stress-induced modulation of hippocampal long-term potentiation and learning. *J. Neurosci. 21*, 5222-5228.

[32] Sapolsky, R.M. (2000). Glucocorticoids and hippocampal atrophy in neuropsychiatric disorders. *Arch. Gen. Psychiatry 57*, 925-935.

[33] Welberg, L.A. & Seckl, J.R. (2001). Prenatal stress, glucocorticoids and the programming of the brain. *J. Neuroendocrinol. 13*, 113-128.

[34] Sapolsky, R.M. (1992). Stress, the ageing brain and the mechanisms of neuron death. Bradford Books, Cambridge, MA.

[35] Scheller, K. & Sekeris, C.E. (2003). The effects of steroid hormones on the transcription of genes encoding enzymes of oxidative phosphorylation. *Experim. Physiol. 1*, 129-140

[36] Scheller, K., Seibel, P. & Sekeris, C.E. (2003). Glucocorticoid and thyroid hormone receptors in mitochondria of animal cells. *Int. Rev. Cytol. 222*, 1-61

[37] Demonacos, C.V., Karayanni, N., Hatzoglou, E., Tsiriyiotis, Chr., Spandidos, D.A. & Sekeris, C.E. (1996). Mitochondrial genes as sites of primary action of steroid hormones. *Steroids 61*, 226-232.

[38] Sekeris, C.E. (1990). The mitochondrial genome: a possible primary site of action of steroid hormones. *In vivo 4*, 317-320.

[39] Moutsatsou, P., Psarra, A.G., Tsiapara, A., Paraskevakou, H., Davaris, P. & Sekeris, C.E. (2001). Localization of the glucocorticoid receptor in rat brain mitochondria. *Arch. Biochem. Biophys. 386*, 69-78.

[40] Koufali, M.-M., Moutsatsou, P., Sekeris, C.E. & Breen, K.C. (2003). The dynamic localization of the glucocorticoid receptor in rat C6 glioma cell mitochondria. *Mol. Cell. Endocrinol. 209*, 51-60.

[41] Tanaka, M., Takeyasu, T., Fuku, N., Li-Jun, G. & Kurata, M. (2004). Mitochondrial genome single nucleotide polymorphisms and their phenotypes in the Japanese. *Ann. N.Y. Acad. Sci. 1011*, 7-20.

[42] Penta, J.S., Johnson, F.M., Wachsman, J.T. & Copeland, W.C. (2001). Mitochondrial DNA in human malignancy. *Mutation Res. 488*, 119-133.

[43] MacKinnon, F., Jamison, K.R. & DePaulo, J.R. (1997). Genetics of manic depressive illness. *Annu. Rev. Neurosci. 20*, 355-373.

[44] Munkata K., Iwamoto, K., Bundo, M. & Kato, T. (2005). Mitochondrial DNA 3243A>G mutation and increased expression of LARS2 gene in the brains of patients with bipolar disorder and schizophrenia. *Biol. Psychiatry 57*, 525-532.

[45] Kato, T., Kunigi, H., Nanko, S. & Kato, N. (2000). Association of bipolar disorder with the 5178 polymorphism in mitochondrial DNA. *Am. J. Med. Genet. 96*, 182-186.

[46] Kato, T., & Kato, N. (2000). Mitochondrial dysfunction in bipolar disorder. *Bipolar Disord. 2*, 180-190.

[47] Kato, T. (2005). Mitochondrial dysfunction in bipolar disorder: from ^{31}p-magnetic resonance spectroscopic findings to their molecular mechanisms. *Int. Rev. Neurobiol. 63*, 21-40.

[48] Kazuno, A.A., Munakata, K., Nagai, T., Shimozono, S., Tanaka, M., Yoneda, M., Kato, N., Miyawaki, A. & Kato, T. (2006). Identification of mitochondrial DNA polymorphisms that alter mitochondrial matrix pH and intracellular calcium dynamics. *PLoS Genet 2*, e128.

[49] Washizuka, S., Ikeda, A., Kato, N., & Kato, T. (2003). Possible relationship between mitochondrial DNA polymorphisms and lithium response in bipolar disorder. *Int. J. Neuropsychopharmacol. 6*,421-424.

[50] Konradi, C., Eaton, M., MacDonald, M.L., Walsh, J., Benes, F.M. & Heckers, S. (2004). Molecular evidence for mitochondrial dysfunction in bipolar disorder. *Arch. Gen. Psychiatry 61*, 300-308.

[51] Washizuka, S., Kakiuchi, C., Mori, K., Tajima, O., Akiyama, T. & Kato, T. (2005). Expression of mitochondria-related genes in lymphoblastoid cells from patients with bipolar disorder. *Bipolar Disord. 7*, 146-152.

[52] Moretti, A., Gorini, A. & Villa, R.F. (2003). Affective disorders, antidepressant drugs and brain metabolism. *Mol. Psychiatry 8*, 773-785.

[53] Storck, C. & Renshaw, P.F. (2005). Mitochondrial dysfunction in bipolar disorder: evidence from magnetic resonance spectroscopy research. *Mol. Psychiatry 10*, 900-919.

[54] Lowy, M.T. (1990). Reserpine-induced decrease in type I and II corticosteroid receptors in neuronal and lymphoid tissues of adrenalectomized rats. *Neuroendocrinology 51*, 190-196.

[55] Spencer, R.L., Miller, A.H., Stein, M. & McEwen, B.S. (1991). Corticosterone regulation of type I and type II adrenal steroid receptors in brain, pituitary, and immune tissue. *Brain Res. 549*, 236-246.

[56] Lopez, J.F., Chalmers, D.T., Little, K.I. & Watson, S.J. (1998). Regulation of serotonin 1A, glucocorticoid and mineralocorticoid receptor in rat and human hipocampus: Implications for the neurobiology of depression. *Biol. Psychiatry 43*, 547-573.

[57] Webster, M.J., Knable, M.B., O'Grady, J., Orthmann, J. & Weickert, C.S. (2002). Regional specificity of brain glucocorticoid receptor mRNA alterations in subjects with schizophrenia and mood disorders. *Mol. Psychiatry 7*, 985-994.

[58] Calfa, G., Kademina, S., Ceschin, D., Vega, G., Rabinovich, G.A. & Volosin, M. (2003). Characterization and functional significance of glucocorticoid receptors in patients with major depression: modulation by antidepressant treatment. *Psychoneuroendocrinology 28*, 697-701.

[59] Spiliotaki, M., Salpeas, V., Malitas, P., Alevizos, V. & Moutsatsou, P. (2006). Altered glucocorticoid receptor signaling cascade in lymphocytes of bipolar disorder patients. *Psychoneuroendocrinology 31*, 748-760.

[60] Modell, S., Lauer, C.J., Schreiber, W., Huber, J., Krieg, J.C., Holsboer, F. (1998). Hormonal response pattern in the combined DEX-CRH test is stable over time in subjects at high familiar risk for affective disorder. *Neuropsychopharmacology 18*, 253-262.

[61] Moutsatsou, P., Tsolakidou, A., Trikkas, G., Troungos, C. & Sekeris, C.E. (2000). Glucocorticoid receptor alpha and beta isoforms are not mutated in bipolar affective disorder. *Mol. Psychiatry 5*, 196-202.

[62] Henning, U., Krieger, K., Loeffler, S., Rivas, F., Orozco, G., de Castro, M.G., Rietschel, M., Noethen, M.M. & Klimke, A. (2005). Increased levels of glucocorticoid receptors and enhanced glucocorticoid receptor auto-regulation after hydrocortisone challenge in B-lymphoblastoids from patients with affective disorders. *Psychoneuroendocrinology 30*, 325-332.

[63] Fruchter, O., Kino, T., Zoumakis, F., Alesci, S., De Martino, M., Chrousos, G. & Hochberg, Z. (2005). The human glucocorticoid receptor (GR) isoform (beta) differentially suppresses GR{alpha}-induced transactivation stimulated by synthetic glucocorticoids. *J. Clin. Endocrinol. Metab. 90*, 3505-3509.

[64] Smit, P., Russcher, H., de Jong, F.H., Brinkmann, A.O., Lamberts, S.W. & Koper, J.W. (2005). Differential regulation of synthetic glucocorticoids on gene expression levels of glucocorticoid-induced leucine zipper and interleukin-2. *J. Clin. Endocrinol. Metab. 90*, 2994-3000.

[65] Wissink, S., Meijer, O., Pearce, D., van Der Burg, B. & van Der Saag, P.T. (2000). Regulation of the rat serotonin-1A receptor gene by corticosteroids. *J. Biol. Chem. 275*, 1321-1326.

[66] Herdegen, T. & Waetzig, V. (2001). AP-1 proteins in the adult brain: facts and fiction about effectors of neuroprotection and neurodegeneration. *Oncogene 20*, 2424-2432.

[67] Marini, A.M., Jiang, X., Wu, X., Tian, F., Zhu, D., Okagaki, P. & Lipsky, R.H. (2004). Role of brain-derived neurotrophic factor and NF-kappa B in neuronal plasticity and survival: From genes to phenotype. *Restor. Neurol. Neurosci. 22*, 121-130.

[68] Sun, Y., Zhang, L., Johnston, N.L., Torrey, E.F. & Yolken, R.H. (2001). Serial analysis of gene expression in the frontal cortex of patients with bipolar disorder. *Br. J. Psychiatry 41*, 137-141.

[69] Cabrera-Vera, T.M., Vanhauwe, J., Thomas T.O., Medkova, M., Preininger, A., Mazzoni, M. & Hamm H.E. (2003). Insights into G protein structure, function and regulation. *Endocr. Rev. 24*, 765-781.

[70] Manji, H.K. & Lenox, R.H. (2000). Signaling: Cellular insights into the pathophysiology of bipolar disorder. *Biol. Psychiatry 48*, 518-530.

[71] Young, L.T., Li, P.P., Kish, S.J., Siu, K.P., Kamble, A., Hornykiewiez, O. & Warsh, J.J. (1993). Cerebral cortex Gs α protein levels and forskolin-stimulated cyclic AMP formation are increased in bipolar affective disorder. *J. Neurochem. 61*, 890-898.

[72] Friedman, E. & Wang, H.Y. (1996). Receptor-mediated activation of G proteins is increased in postmortem brains of bipolar affective disorder subjects. *J. Neurochem. 67*, 1145-1152.

[73] Warsh, J.J., Young, L.T. & Li, P.P. (2000). Guanine nucleotide binding (G) protein disturbances in bipolar affective disorder. In: Manji HK, Bowden CL, Belmaker RH, editors. *Bipolar Medications: Mechanisms of Action.* Washington DC: American Psychiatric Press, 299-329.

[74] Spleiss, O., van Calker, D., Scharer L., Adamovic, K., Berger, M. & Gebicke-Haerter, P.J. (1998). Abnormal G protein alpha(s) and alpha(i2)-subunit mRNA expression in bipolar affective disorder. *Mol. Psychiatry 3*, 512-520.

[75] Chen, G., Hasanat, K.A., Bebehuk, J.M., Moore, G.J., Glitz, D. & Manji H.K. (1999). Regulation of signal transduction pathways and gene expression by mood stabilizers and antidepressants. *Psychosom. Med. 61*, 599-617.

[76] Tronche, R., Kellendonk, C., Kretz, O., Gass, P., Anlag, K., Orban, P.C., Bock, R., Klein, R. & Schütz, G. (1999). Disruption of the glucocorticoid receptor gene in the nervous system results in reduced anxiety. *Nature Genet 23*, 99-103.

[77] Boyle, M.P., Brewer, J.A., Funatsu, M., Wozniak, D.F., Tsien, J.Z., Izumi, L. & Muglia, J. (2005). Acquired deficit of forebrain glucocorticoid receptor produces depression-like changes in adrenal axis regulation and behavior. *Proc. Natl. Acad. Sci. USA 102*, 473-478.

[78] Peiffer, A., Veilleux, S. & Barden, N. (1991). Antidepressant and other centrally acting drugs regulated glucocorticoid receptor messenger RNA levels in rat brain. *Psychoneuroendocrinology 16*, 505-515.

[79] Przegalinski, E. & Budziszewska, B. (1993). The effect of long-term treatment with antidepressant drugs on the hippocampal mineralocorticoid and glucocorticoid receptors in rats. *Neurosci. Lett. 161*, 215-218.

[80] Seckl, J.R. & Fink, G. (1992). Antidepressants increase glucocorticoid and mineralocorticoid receptor mRNA expression in rat hippocampus in vivo. *Neuroendocrinology 55*, 621-626.

[81] Montkowski, A., Barden, N., Wotjak, C., Stec, I., Ganster, J., Meaney, M., Engelmann, M., Reul, J.M., Landgraf, R. & Holsboer, F. (1995). Long-term antidepressant treatment reduces behavioral deficits in transgenic mice and impaired glucocorticoid receptor function. *J. Neuroendocrinology 7*, 841-845.

[82] Greden, J.F., Gardner, R., King, D., Grunhaus, L., Carroll, B.J. & Kronfol, Z. (1983). Dexamethasone suppression tests in antidepressant treatment of melancholia. The process of normalization and test-retest reproducibility. *Arch. Gen. Psychiatry 40*, 493-500.

[83] Holsboer, F., Liebl, R. & Hofschuster, E. (1982). Repeated dexamethasone suppression test during depressive illness. Normalization of test result compared with clinical improvement. *J. Affect. Disord. 4*, 93-101.

[84] Barden, N. (1996). Modulation of glucocorticoid receptor gene expression by antidepressant drugs. *Pharmacopsychiatry 29*, 12-22.

[85] Frechilla, D., Otano, A. & Del Rio, J. (1998). Effect of chronic antidepressant treatment on transcription factor binding activity in rat hippocampus and frontal cortex. *Prog. Neuropsychopharmacol. Biol. Psychiatry 22*, 787-802.

[86] Herr, A.S., Tsolakidou, A.F., Yassouridis, A., Holsborer, F. & Reinm T. (2003). Antidepressants differentially influence the transcriptional activity of the glucocorticoid receptor in vitro. *Neuroendocrinology 78*, 12-22.

[87] Pepin, M.C., Govindan, M.V. & Barden, N. (1992). Increased glucocorticoid receptor gene promoter activity after antidepressant treatment. *Mol. Pharmacol. 41*, 1016-1022.

[88] Pariante, C.M., Pearce, B.D., Pisell, T.L., Owens, M.J. & Miller, A.H. (1997). Steroid-independent translocation of the glucocorticoid receptor by the anti-depressant desipramine. *Mol. Pharmacol. 52*, 571-581.

[89] Budziszewska, B., Jaworska-Feil, L., Kajta, M. & Lason, W. (2000). Antidepressant drugs inhibit glucocorticoid receptor-mediated gene transcription - a possible mechanism. *Br. J. Pharmacol. 130*, 1385-1393.

[90] Brunello, N. & Tascedda, F., 2003. Cellular mechanisms and second messengers: relevance to the psychopharmacology of bipolar disorders. Intern. *J. Neuropsychopharmacol. 6*, 181-189.

[91] Chuang, D.-M., Chen, R.-W., Chalecka-Franaszek, E., Ren, M., Hashimoto, R., Senatorov, V., Kanai, H., Hough, C., Hiroi, T. & Leeds, P. (2002). Neuroprotective effects of lithium in cultured cells and animal models of diseases. *Bipolar Disord. 4*, 129-136.

[92] Tamura, T., Morinobu, S., Okamoto, Y., Kagaya, A. & Yamawaki, S. (2002). The effects of antidepressant drug treatments on activator protein-1 binding activity in the rat brain. *Prog. Neuropsychopharmacol. Biol. Psychiatry 26*, 375-381.

[93] Ozaki, N. & Chuang, D.M. (1997). Lithium increases transcription factor binding to AP-1 and cyclic AMP-responsive element in cultured neurons and rat brain. *J. Neurochem. 69*, 2336-2344.

[94] Okamoto, H., Shino, Y., Hashimoto, K., Kumakiri, C., Shimizu, E., Shirasawa, H. & Iyo, M. (2003). Dynamic changes in AP-1 transcription factor DNA binding activity in rat brain following administration of antidepressant amitriptyline and brain-derived neurotrophic factor. *Neuropharmacology 45*, 251-259.

[95] Manji, H.K., Moore, G.J. & Chen, G. (2001b). Bipolar disorder: leads from the molecular and cellular mechanisms of action of mood stabilizers. *Br. J. Psychiatry 41(suppl 1)*, 107-119.

[96] Chen, G., Hasanat, K.A., Bebchuk, J.M., Moore, G.J., Glitz, D. & Manji, H.K. (1999). Regulation of signal transduction pathways and gene expression by mood stabilizers and antidepressants. *Psychosom. Med. 61*, 599-617.

[97] Chen, R.W., Qin, Z.H., Ren, M., Kanai, H., Chalecka-Franaszek, E., Leeds, P. & Chuang, D.M. (2003). Regulation of c-Jun-N-terminal kinase, p38 kinase and AP-1 DNA binding in cultured brain neurons : roles in glutamate excitotoxicity and lithium neuroprotection. *J. Neurochem. 84*, 566-575.

[98] Bartholoma, P., Erlandsson, N., Kaufmann, K., Rossler, O.G., Baumann, B., Wirth, T., Giehl, K.M. & Thiel, G. (2002). Neuronal cell death induced by antidepressants: lack of correlation with Egr-1, NF-kappa B and extracellular signal-regulated protein kinase activation. *Biochem. Pharmacol. 63*, 1507-1516.

[99] Post, A., Crochemore, C., Uhr, M., Holsboer, F. & Behl, C. (2000). Differential induction of NF-kappa B activity and neural cell death by antidepressants in vitro. *Eur. J. Neurosci. 12*, 4331-4337.

In: Psychoneuroendocrinology Research Trends
Editor: Martina T. Czerbska, pp. 381-410
ISBN: 978-1-60021-665-7

Chapter XIII

INDIVIDUAL DIFFERENCES IN COPING STRATEGIES FOR SOCIAL STRESS, PRIOR EMOTIONAL REACTIVITY AND CORTICOSTERONE LEVELS IN SUBORDINATE MICE

L. Garmendia, A. Azpiroz, Z. De Miguel, E. Gómez and A. Arregi
Department of Basic Psychological Processes and Their development, Basque Country University, San Sebastian 20018, Spain

ABSTRACT

The aim of this research project was to study the relationships between different coping styles adopted by subordinate mice with experience of chronic defeat and the neuroendocrine response to situations of social stress. Also, this study analyses whether different coping strategies for chronic and acute social stress are related to differences in emotional reactivity levels prior to stress. OF1 mice were subjected to different behavioral tests (Open field test, Plus-maze test and Boissier's test) in order to establish emotional reactivity profiles. Subsequently, they were socially stressed by repeated experiences of defeat for 23 days in a sensorial contact model. Our results indicate a relationship between the coping style adopted by subordinate subjects in response to chronic social stress and the physiological response to stress. Subjects with a passive coping strategy for chronic social stress showed higher levels of corticosterone than those which adopted an active coping strategy. The data also indicate that a passive strategy for coping with chronic or acute social stress is related to greater emotional reactivity, and that the strategies adopted in response to acute social stress do not determine the behavior the subject will adopt in response to chronic social stress. The results in general may be of interest in the study of individual differences in susceptibility to illnesses related to social stress in humans, such as depression or anxiety.

Keywords: Social stress, subordinate, defeat, coping strategies, open field test, Boissier's test, plus-maze test, corticosterone

1. Introduction

The importance of the social environment as a stress generator has been repeatedly shown in different investigations focusing on the study of the effects of stress on health (Henry, 1997; Kessler, 1997; Repetti, Taylor, & Seeman, 2002; Sapolsky, 2005; Valencia-Alfonso, Feria-Velasco, Luquin, Diaz-Burke, & Garcia-Estrada, 2004). In human beings, it is now generally accepted that the greatest source of aversive and chronic stress is social in nature, thus contributing to a large extent to the development and expression of diverse systemic illnesses and mental disorders. Studies with animals, in which subjects are constantly exposed to social relationships of dominance and submission, maternal separation and fights with other members of their species, etc. have shown that social events acquire a major biological significance and constitute a source of stress similar to that found in the case of humans. Thus, biological measures such as analyses of corticosteroid and catecholamine concentrations highlight loss of social control as one of the most powerfully stressful social stimuli (Koolhaas, De Boer, De Rutter, Meerlo, & Sgoifo, 1997). This social sphere, which generates possible stress factors, is particularly important due to the fact that each subject is exposed to it throughout the whole of their life. Their response to it is vital to both individual survival and the evolution of the species, since it controls a wide variety of adaptive processes such as the development of personality and behavior, and may have an adverse effect on physiological functions, such as growth, reproduction or the immune response (Chrousos & Gold, 1992; Habib, Gold, & Chrousos, 2001). In general, the response manifested by organisms to stress is functionally oriented towards mobilizing energy in order to cope with the threatening situation, and is, in many ways, adaptive. In basic terms, there are two main systems which mediate the majority of stress response mechanisms. The first is the hypothalamic-pituitary-adrenal axis (HPA axis), which stimulates the adrenal cortex to secrete glucocorticoids such as cortisol and corticosterone into the blood; and the second is the sympathetic-adreno-medullary axis (SAM), which is responsible for the release of catecholamines (Mason, 1968a, 1968b; Sapolsky, 1993).

The paradox of stress lies in the combination of its adaptive nature and its possible maladaptive consequences. The adaptive process which takes place in order to actively maintain stability through change (Sterling & Edelmann, 1988) has been termed "*allostasis*" (McEwen, 2000; McEwen, 2003; McEwen & Stellar, 1993). The brain plays a central role in the maintenance of *allostasis*, executing orders and incorporating influential factors such as experience, memory, anticipation and reassessment of the need for physiological and behavioral requirements (Koob & Le Moal, 2001). One response to change may be beneficial, but may also exact a price from the organism, particularly when demands occur again and again over a long period of time, or when they are handled ineffectively (McEwen, 1998). The price exacted from the organism is termed the "allostatic load", and can be described as the accumulated deterioration or wear generated by this adaptive process. Thus, the price exacted from the organism increases if the allostatic mediators, such as adrenal hormones (glucocorticoids), neurotransmitters or immune cytokines, etc. are secreted too often or are used inefficiently (McEwen, 1998; McEwen, 2003). Recent studies support the idea that a state of allostatic load is more likely to develop in response to social stressors that are unpredictable and chronic, generate physiological and psychological adjustments and may damage or wear underlying physiological functions (Blanchard, McKittrick, Hardy, & Blanchard, 2002; Koolhaas et al., 1997). Different data indicate that repeated challenges, failure to habituate with repeated challenges, failure to finalize the stress response after the challenge is over and failure to produce an adequate response are important factors in the generation of allostatic load (McEwen, 2003; McEwen & Wingfield, 2003).

Currently, the study of individual differences in the effects of social stress reveal that, even when a certain number of subjects are exposed to the same situation of stress, not all develop a state of allostatic load to the same degree. Different data indicate that psychological and physiological responses vary a great deal from one subject to the next following a process of social stress (Carere, Welink, Drent, Koolhaas, & Groothuis, 2001; Grootendorst, De Kloet, Vossen, Dalm, & Oitzl, 2001; Nielsen, Arnt, & Sanchez, 2000; Ruis et al., 2001; Touyarot, Venero, & Sandi, 2004). For example, it has been found that subordinate subjects exposed to the same social stress treatment may manifest differences in motor activity and aggressiveness levels, as well as in the cardiovascular and immune functions (Avitsur, Stark, Dhabhar, Kramer, & Sheridan, 2003; Bartolomucci et al., 2005). Understanding the causes of these individual differences and their consequences in terms of wellbeing, adaptive capacity and individual susceptibility to illness is certainly one of the main objectives of biomedical research. Indeed, research in this field may reveal mechanisms of vital importance to gaining a better understanding of the factors which underlie susceptibility to stressful events.

Furthermore, it is now generally accepted that inter-individual variability with regard to stress response depends more on coping styles than on the physical characteristics of the stressors. In other words, the impact of the stressors is determined to a large extent by the organism's ability to cope with the situation (Ursin, 1998; Ursin & Olff, 1995). The term "coping" has been defined as the physiological and behavioral effort aimed at handling a situation (Lazarus, 1966; Wechsler, 1995). Wide-ranging literature from both medicine and psychology, as well as various animal models, have shown that individuals may indeed differ as regards their coping capacity, and that the factors which affect these differences include genotype, development, early experiences and social support, among others. Thus, it is important to understand the mechanisms and factors which underlie an individual's capacity to cope with environmental challenges.

Many studies have shown that in humans also, coping styles are important to health and illness, and have attempted to determine individual vulnerability to stress-related illnesses by assessing different coping capacities (Eriksen, Olff, Murison, & Ursin, 1999; Olff, 1991, 1999; Ursin & Olff, 1995). A coping style has been defined as a coherent set of behavioral and physiological stress responses which is consistent over time and which is characteristic of a certain group of individuals (Koolhaas et al., 1999). In humans, it has been observed that the reactivity shown to stress is a stable characteristic in individuals. Results reveal that many individual indicators of the physiological stress response (cardiovascular response, SAM and HPA axis hormones) are also fairly stable over time and in response to different situations, thus suggesting the possible existence of underlying dispositional characteristics (Cohen & Hamrick, 2003).

Basically, two different stress coping response patterns have been identified. The first is the active response, originally described by Cannon (1915) as the "fight-flee" response, which from a behavioral point of view is characterized by territorial control and aggression. The second type of response, the "conservation-withdrawal" response, was originally described by Engel and Schmale (1972) and is characterized by immobility and low levels of aggression. Different works have revealed that individual levels of aggressive behavior, such as, for example, the tendency of males to defend their territory, is related to the way in which said individuals react in general to a wide variety of environmental challenges. Thus, some authors have suggested the existence of at least two coping styles: active and passive, more recently termed proactive and reactive coping styles (Benus, Bohus, Koolhaas, & Van Oortmerssen, 1991b; Van Oortmerssen & Bakker, 1981; Veenema, Meijer, De Kloet, & Koolhaas, 2003). The proactive or active style is mainly characterized by a short attack latency and high aggression, routine behavior, low immobility and low behavioral flexibility. The reactive or passive style is characterized by high attack latency and low aggression, as well as by the adoption of immobility behaviors. Non-aggressive, more reactive males seem to be more adaptive and flexible, responding only when absolutely necessary. The difference in the flexibility of the response may explain why the proactive style is more successful in situations in which the environment does not change, while the reactive style is more successful in situations involving unpredictable changes, such as migration, for example (Koolhaas et al., 1999). Both types of behavioral pattern are considered coping styles in the sense that they are aimed at successfully controlling the individual's immediate environment (Koolhaas & Bohus, 1989; Koolhaas et al., 1999).

2. Emotional Reactivity and Coping Style

Individual differences in animals' emotional reactivity to stress situations have been related to different coping styles. Emotional reactivity has been defined as the way in which an individual perceives and reacts to a potentially stressful situation (Boissy, 1995), and some authors have considered the possibility that it may constitute a predictive factor for the coping style employed in response to other situations, such as chronic stress. Thus, rats genetically selected for their high and low degree of avoidance behavior, Roman high avoidance (RHA) and Roman low avoidance (RLA), employ different coping styles in response to stress. In comparison with the RHA line, the RLA line show a greater response to stress: freezing behavior and high secretions of adrenocorticotropic hormone (ACTH), corticosterone and prolactin. They also adopt a more passive coping style when faced with a new environment (Steimer & Driscoll, 2003; Steimer, Driscoll, & Schulz, 1997).

Other studies using rats genetically selected for their high (HAB) and low (LAB) anxiety levels in the plus-maze test, have suggested a close link between the emotional assessment of a novel and stressful situation and an individual's coping strategy. Thus, in the open field test (Liebsch, Montkowski, Holsboer, & Landgraf, 1998) and the modified hole-board test (Landgraf & Wigger, 2002; Ohl, Toschi, Wigger, Henniger, & Landgraf, 2001), the ratio of time spent and distance traveled in the center, compared to the border zone, was lower in the HAB line than the LAB line. Furthermore, studies have revealed that in the forced swim test, HAB rats remain immobile earlier and spend more time in this posture than LAB rats (Liebsch et al., 1998), and show signs of hyperactivity of the HPA axis similar to those found in psychiatric patients. In general, the data indicate that in addition to showing a major difference in the anxiety trait in different behavioral tests, HAB and LAB rats also differ as regards their strategies for coping with stress, with the former being more vulnerable to exposure to stressors and adopting a more passive strategie (Landgraf & Wigger, 2002).

Data also exist which indicate that the behavioral coping style adopted in response to a sub-chronic mild stress is related to the initial emotional reactivity shown in the elevated plus-maze test. Thus, BALB/c ByJ mice selected for their high emotionability, are characterized by the predominance of inhibited behaviors (consumption of a palatable food, physical state index and grooming behavior) following the application of the stress, in comparison with their less emotional counterparts (C57BL/6) (Ducottet & Belzung, 2004; Ducottet, Griebel, & Belzung, 2003). Similarly, it has also been suggested that mice with high emotional reactivity may assess negative stimuli as more threatening and assess repeated stressors as more demanding than those with a lower level of emotional reactivity (Ducottet & Belzung, 2004).

The different levels of emotional reactivity manifested by mice genetically selected for their short (SAL) or long attack latency (LAL) has been related to different coping styles used in response to stress. Attack latency is a characteristic of the coping strategy adopted by these animals in response to an environmental challenge. LAL mice adopt a passive coping style while SAL mice adopt an active one in response to different stressful situations (Benus, Bohus, Koolhaas, & van Oortmerssen, 1989, 1991a; Benus et al., 1991b; Sluyter, Korte, Bohus, & Van Oortmerssen, 1996; Veenema, Koolhaas, & De Kloet, 2004). In this sense, some data show that LAL mice are less active in the plus-maze test, tend to engage more in freezing behaviors in the sudden silence test, and are more immobile in the open field and forced swimming tests (Veenema et al., 2003). The coping styles adopted by these two mouse lines are in turn associated with observable physiological differences in the regulation of the HPA axis, serotoninergic neurotransmission and the degree of hippocampal proliferation (Veenema et al., 2004).

More recently, researchers have shown an interest in studying individual differences in non genetically-selected populations encompassing the whole range of behavioral patterns. Similarly in said works, it is important to assess the consistency of behavioral strategies adopted in response to various types of stressors applied through different tests (sawdust-digging escape task, open field test, novel item emergent test, forced swimming test, tail-clip test, lipopolysaccharide test, fox urine test), and to determine the degree of generalization of said strategies. There is also interest in determining which behavioral variables are important for discriminating between different coping strategies for stress. Data exist which show that the initiative to explore a new environment is one of the most discriminative tasks for distinguishing between coping strategies (Campbell, Lin, DeVries, & Lambert, 2003; Koolhaas et al., 1999), and reveal a high level of correlation between different behavioral tests which require exploration or movement. Thus, the initiative shown by the animal as regards exploring a new environment seems to be an important aspect of coping (Campbell et al., 2003). It has been found that individual differences, manifested through a high or low exploration profile for mice in relation to a new environment, are related to a characteristic behavioral profile. Mice with a high exploration profile were found to be less anxious in the light/dark test and the elevated plus-maze test, demonstrated a greater degree of locomotion in the open field test, engaged in more avoidance behavior in aversive conditioning tasks (electric footshock) and showed a more appetitive approach to reward (food). These mice were also more aggressive in the intruder test, while low exploration profile mice were either non-aggressive or submissive (Kazlauckas et al., 2005). The pattern of exploratory behavior is then, according to Kazlauckas, et al., (2005), a combination of harm avoidance (fear) and novelty-seeking (curiosity) behavior. Thus, animals can be classified as either shy (tending towards avoidance) or bold (showing interest when exposed to the challenge of a new situation). These behavioral styles are often consistent both over time and in response to a variety of different experimental situations (Engel & Schmale, 1972; Sih, Bell, & Johnson, 2004). In accordance with this finding, physiological data exist to suggest that the response of the HPA axis to stressors is also a consistent individual trait. Thus, studies in non-genetically-selected rats reveal a good intra-subject reliability of ACTH and corticosterone responses to stressors involving exposure to a new environment (elevated plus-maze, hole-board and circular corridor) (Marquez, Nadal, & Armario, 2005, 2006).

3. Coping Styles and Susceptibility to Illness

Coping styles in response to stress may be important when studying health consequences. In this way, if the organism is unable to cope with the stressor or if the efforts required in order to cope are very high, the consequences will be negative. In accordance with this, the coping strategy adopted (either active or passive) is of enormous importance and seems to be associated with the development of different pathologies (Koolhaas et al., 1999; Korte, Koolhaas, Wingfield, & McEwen, 2005).

In humans, in addition to the coping style adopted, the subject's perception and interpretation of the stressors in question may determine a pathological reaction to stress. While an active, problem-centered coping style is often associated with a good mood, a passive, emotion-centered style often correlates with depressive symptoms (Haghighatgou & Peterson, 1995; Turner, King, & Remblay, 1992). The depressive reaction pattern has been described as a less adaptive coping style and has been found to correlate negatively with self-reported health and different immune parameters (Olff, 1991, 1999). Subjects who adopt an efficient coping style, i.e. those with positive expectations of results, tend to show a shorter catecholamine response. Individuals who adopt a less efficient coping style, i.e. those who experience feelings of defenselessness and despair in response to the situation, tend to show signs of prolonged activation that may involve many of the organism's systems and eventually give rise to illness (Eriksen et al., 1999).

In animals, when studying the factors that determine the different impact of stressors on health, researchers have used mice strains with different emotional reactivity in response to stress. The greater emotional reactivity of the BALB/cByJ strain in comparison with the C57BL/6ByJ strain is manifested through behavioral and biochemical changes which have been related to disorders such as anxiety and depression. Thus, in the BALB/cByJ strain, acute stress caused by physical restriction results in an increase in anxiety in a series of behavioral tests (Anisman, Hayley, Kelly, Borowski, & Merali, 2001; Belzung & Griebel, 2001; Crawley et al., 1997; Griebel, Belzung, Misslin, & Vogel, 1993; Griebel, Belzung, Perrault, & Sanger, 2000), increases in ACTH and corticosterone levels (Shanks, Griffiths, Zalcman, Zacharko, & Anisman, 1990; Zaharia, Kulczycki, Shanks, Meaney, & Anisman, 1996) and an increase in monoamine turnover in various regions of the brain (Herve et al., 1979; Shanks, Griffiths, & Anisman, 1994a, 1994b; Shanks, Zalcman, Zacharko, & Anisman, 1991). Furthermore, the neurochemical and behavioral changes associated with acute stress are more persistent and pronounced in the BALB/cByJ strain following exposure to chronic stress, a finding which may reflect this strain's greater susceptibility to the development of behavioral pathologies (Tannenbaum & Anisman, 2003). Similarly, rats selected for their low avoidance behavior (RLA) also show greater anxiety behavior in tests such as the open-field test, the plus-maze test, the black/white box test and the new light/dark open field test. It has been suggested that the anxiety shown by these animals may be the result of their specific psycho-physiological profile, i.e. high emotionability and a passive coping style (Steimer & Driscoll, 2003; Steimer et al., 1997).

Individual differences in anxiety levels have also been related to different coping styles in response to stress and a greater or lesser susceptibility to the effects of said stress. Thus, rats selected for their high anxiety levels (HAB) in the elevated plus-maze test, remain immobile for longer in the forced swimming test than their low anxiety counterparts (LAB), thereby denoting a passive coping style similar to that observed in depressed patients. Furthermore, in HAB rats, treatment with antidepressants such as paroxetine or repetitive transcranial magnetic stimulation results in a more active coping style, with a level of activity similar to that shown by LAB rats in the forced swimming test and a normalization of the HPA axis activity (Keck et al., 2003).

The selection of mice for their attack latency (SAL and LAL lines) also serves to highlight the relationship between coping styles and susceptibility to the effects of chronic, inescapable psychosocial stress. Following the application of stress, LAL mice show lower activity levels in the elevated plus-maze and sudden silence tests, as well as a greater reactivity of the HPA axis and a lower hippocampal ratio of mineralocorticoid and glucocorticoid receptors (MR:GR ratio) (Veenema et al., 2004; Veenema et al., 2003). In relation to this, it has been found that suicide victims with a history of depression presented a significantly low MR:GR ratio (Lopez, Chalmers, Little, & Watson, 1998). It has been hypothesized that a reduced MR capacity, especially in the hippocampus, is related to the deregulation of the HPA system characteristic of human depression (Reul et al., 2000; Reul, Stec, Soder, & Holsboer, 1993).

Bearing in mind the fact that different levels of emotional reactivity are associated with different neuroendocrine and neurobiological reactivity, it is possible that specific pathologies may develop in accordance with different emotional reactivity profiles. In this sense, it has been suggested that the behavioral profile of high exploration mice (high novelty-seeking/ low harm avoidance) resembles the hyperthymic temperament (Maremmani et al., 2005) or uninhibited phenotype that seems to be related to a predisposition to bipolar disorder (Hirshfeld-Becker et al., 2003).On the other hand, the behavioral characteristics of low exploration mice (low novelty-seeking/ high harm avoidance) are similar to those found in depressive temperaments (Maremmani et al., 2005) or inhibited phenotype, that has been related to unipolar depression (Hirshfeld-Becker et al., 2003; Kazlauckas et al., 2005). Furthermore, data exist which indicate that, in rats, low locomotor reactivity in the open-field test predicts the degree of behavioral sensitization induced by a single stressful experience of foot-shock in tests such as the noise test and the shock prod test, thus suggesting that an anxiety-prone personality or a passive coping style may increase the risk of developing stress-related disorders such as post-traumatic stress disorder (Geerse, Van Gurp, Wiegant, & Stam, 2006).

4. Characteristics of Coping Styles in Response to Stress in Subordinate Subjects and Susceptibility to Illness

In all group-organized animal species, relationships of dominance are established and maintained through agonistic behavior, thus resulting in differences in social status. Dominant, more aggressive animals normally adopt a proactive coping style, while subordinate, less aggressive animals are characterized by their more reactive style (Koolhaas et al., 1999). Social status influences the way in which an individual copes with physical and social stressors, thus affecting both quality of life and health, particularly in relation to stress-related illnesses (Sapolsky, 2005).

Dominant and subordinate subjects are different as regards their behavioral and physiological response to stressors, and, in general, the physiological cost of being subordinate is evident and is reflected in a greater susceptibility to different disorders. Thus, it has been observed that, in subordinate subjects, defeat generates a stress response that makes them more susceptible to psychopathologies such as depression, anxiety and drug abuse (Fuchs & Flugge, 1995; Fuchs, Kramer, Hermes, Netter, & Hiemke, 1996; Kudryavtseva & Bakshtanovskaia, 1991; Kudryavtseva, Madorskaya, & Bakshtanovskaya, 1991; Miczek, Covington, Nikulina, & Hammer, 2004a). For their part, although they are also subject to the negative effects of chronic social stress (Bartolomucci et al., 2001; Bartolomucci et al., 2004; Blanchard et al., 1995; Sapolsky, 1993), dominant individuals seem more able to become habituated to the situation, while their subordinate counterparts tend to experience the more severe effects of exposure to chronic stress (Sapolsky, 1993). Dominant status has been less widely studied in relation to psychopathologies, although data do exist which relate it to anxiety disorders and mania, as well as to violent and antisocial behavior (Arregui, Azpiroz, Fano, & Garmendia, 2006).

As stated above, following repeated situations of defeat, subordinate subjects experience drastic physiological and behavioral changes which may even affect the integrity of the organism itself (Blanchard et al., 1995). Of the behavioral changes observed which are considered signs of depression, we could highlight hypophagia (Berton, Aguerre, Sarrieau, Mormede, & Chaouloff, 1998) and the alteration of circadian rhythms and the sleep-wake cycle (Harper, Tornatzky, & Miczek, 1996; Meerlo, De Boer, Koolhaas, Daan, & Van den

Hoofdakker, 1996; Meerlo, Pragt, & Daan, 1997). A reduction in exploratory activity has also been observed, alongside an increase in immobility in the Porsolt test (Avgustinovich, 2003; Bartolomucci et al., 2003; Berton et al., 1998; Kudryavtseva & Bakshtanovskaia, 1991; Kudryavtseva et al., 1991) and a decrease in sensitivity to reward (Von Fritag et al., 2000; Von Fritag, Van den Bos, & Spruijt, 2002). One behavioral deficit in mice with repeated experiences of defeat is the predominance of passive defensive postures and immobility, as opposed to active defense and withdrawal behaviors which characterize the first confrontation (Kudryavtseva & Bakshtanovskaia, 1991; Kudryavtseva et al., 1991). Equally, an increase in anxiety has been observed in subordinate subjects exposed to defeat, not only through a drop in social interactions but also through other behavioral tests which measure anxiety in rodents (Avgustinovich, 2003; Avgustinovich, Gorbach, & Kudryavtseva, 1997; Berton et al., 1998; Haller & Halasz, 2000; Heinrichs et al., 1994; Heinrichs, Pich, Miczek, Britton, & Koob, 1992; Menzaghi et al., 1994; Rodgers & Cole, 1993; Ruis et al., 1999).

Subordinate subjects are not only characterized by their behavior, but also present characteristic physiological and neuroendocrine profiles. Subordinate animals subjected to repeated experiences of defeat show a chronic activation of the HPA axis, with high levels of plasmatic corticosterone (Blanchard, Sakai, McEwen, Weiss, & Blanchard, 1993; Cacho et al., 2003; Fano et al., 2001; Fuchs et al., 1996; Kudryavtseva & Avgustinovich, 1998). These physiological changes are associated with an increase in emotional reactivity, which includes an increase in both submissive behavior and anxiety-related behaviors (File, Zangrossi, & Andrews, 1993; Heinrichs et al., 1992). Similarly, it has been observed that subordinate animals show increased levels of corticotrophin-releasing factor (CRF) and alterations in the balance of central glucocorticoid receptors (MR and GR), thus rendering them more susceptible to the development of psychopathologies such as depression (Buwalda, Van Kalkeren, De Boer, & Koolhaas, 1998; Groenink et al., 2002; Korte, 2001). In parallel, a significant reduction has been found in the secretion of testosterone in subordinate subjects (Korte et al., 2005). Similar decreases in testosterone levels have also been found in patients suffering from depression (Schweiger et al., 1999).

In situations of social stress, subordinate subjects experience dynamic changes in their monoaminergic systems which lead to alterations in the number and functionality of the central receptors. Of these changes which occur after exposure to chronic stress, we can highlight the up-regulation of alpha-2 adrenergic receptors in the prefrontal cortex and the down-regulation of beta adrenergic receptors in the hippocampus and parietal cortex (Flugge, 2000; Flugge, Ahrens, & Fuchs, 1997a, 1997b), as well as a reduction in the number of 5HT1A serotoninergic receptors in the hypothalamus, hippocampus and cortex and an increase in 5HT2A receptors in the cortex (Berton et al., 1998; Fuchs & Flugge, 1995; Korte, Buwalda, Meijer, De Kloet, & Bohus, 1995; Korte, Smit, Bouws, Koolhaas, & Bohus, 1990; Kudryavtseva & Avgustinovich, 1998; McKittrick, Blanchard, Blanchard, McEwen, & Sakai, 1995), similar to those found in patients suffering from depression (Arango et al., 1990; Arias, Gasto et al., 2001; Arias, Gutierrez, Pintor, Gasto, & Fananas, 2001; Deparmentier, Crompton, Katona, & Horton, 1993; Garcia-Sevilla et al., 1999; Lesch, 1991; Lesch et al., 1990; Little, Clark, Ranc, & Duncan, 1993; Meana, Barturen, & Garcia-Sevilla, 1992; Meyer et al., 2003; Ordway, Widdowson, Smith, & Halaris, 1994). Furthermore, it has also been suggested that the glucocorticoids found in high concentrations in subordinate subjects may mediate the changes observed in the monoaminergic systems in situations of social stress (Chaouloff, 1995; Jhanwar-Uniyal & Leibowitz, 1986; McKittrick et al., 1995; Meijer & De Kloet, 1994, 1995).

In addition to monoaminergic changes, subordinate subjects also undergo neurogenesis-related alterations which have been linked to affective disorders. It has been suggested that the decrease in hippocampal 5HT1A receptors may be responsible for the suppression of neurogenesis observed in tree shrews with experience of chronic defeat (Czeh et al., 2001). Data reveal that aggressive mice with a greater expression and linkage of 5HT1A hippocampal receptors show greater resistance to the suppression of cellular proliferation than their non-aggressive counterparts (Veenema et al., 2003). In humans, it has been hypothesized that the loss of neuronal plasticity and the drop in neuron survival rates are key factors in the pathophysiology of depression (Duman, Heninger, & Nestler, 1997; Manji, Moore, & Chen, 2000).

The social status of dominance-subordination seems to be an important variable that has been related to the initiation of drug consumption. Rats repeatedly exposed to the social stress of defeat through aggressive attacks by an opponent showed a decreased latency in the acquisition of the behavior of self-administration of cocaine and a high rate of self-administration (Haney, Maccari, Le Moal, Simon, & Piazza, 1995; Kabbaj et al., 2001; Miczek & Mutschler, 1996). Furthermore, social stress accelerates the transition from regulated self-administration to deregulated or uncontrolled consumption patterns (Miczek et al., 2004a; K. A. Miczek, H. E. Covington, E. A. Nikulina, & R. P. Hammer, 2004b). It has been suggested that this effect is probably mediated by the increase in the dopaminergic activity in the mesolimbic dopaminergic system observed in subordinate subjects (Haney, Noda, Kream, & Miczek, 1990; Tidey & Miczek, 1996), and is considered crucial to the acquisition of self-administration behavior in relation to psychostimulants (Tidey & Miczek, 1997).

In addition to data which indicate that prolonged subordination may result in a tendency to adopt more passive or immobile defense strategies, which may in turn give rise to pathological states, it has also been found that characteristic individual differences may interact with the experience of subordination stress to determine more drastic changes in subjects with a characteristic subordination profile. Specifically, studies with rats and mice have found that within the group of subordinates, some individuals show a flat corticosterone response to an acute non-social stressor such as physical restriction, thereby indicating a deficit in the HPA axis response. These non-responsive subordinates also differed from their responsive counterparts in that they demonstrated fewer active defense behaviors and a greater degree of immobility in response to mid-level threatening stimuli, such as handling, and a greater suppression of activity in the open field (Blanchard, Yudko, Dulloog, & Blanchard, 2001). The interaction between individual differences and social factors has also been observed in other studies. In our laboratory we have found that different coping styles are related to a greater or lesser degree of tumor development. Not only did all subjects exposed to social stress through confrontation with a dominant subject develop a greater number of tumor foci than their control counterparts, but furthermore, more passive and subordinate subjects which adopted a strategy characterized by an absence of attack and low levels of non social exploration, and which engaged in more defense, submission and avoidance behavior, developed a greater number of metastases (Vegas, Fano, Brain, Alonso, & Azpiroz, 2006).

5. Study of the Individual Differences in Coping Strategies for Social Stress in Subordinate Subjects, Prior Emotional Reactivity and Corticosterone Levels

As stated above, chronic stress is a threat to both individuals' physical and psychological health. Therefore, a understanding of how the different behavioral responses adopted by subjects to stress situations affect their condition of health or illness may be of enormous value. Similarly, it is important to explore the influence of all the variables relevant at the moment at which the complex response to stress is manifested.

The proven importance of social relations as a source of chronic stress has prompted the use of animal social stress models, which are extremely useful in studying the causes and mechanisms involved in the development of psychopathological disorders (Blanchard et al., 1995; Blanchard, Spencer, Weiss, Blanchard, McEwen, & Sakai, 1995; Van Kampen, Kramer, Hiemke, Flugge, & Fuchs, 2002; Van Kampen, 2002), as well as individual differences in relation to the effects of stress.

Our research project focuses on the behavioral study of different coping styles adopted by subordinate subjects with experience of chronic social defeat and their possible relationship with a characteristic reactivity profile of the HPA axis. We are also interested in analyzing whether the different strategies adopted by subordinate subjects are associated with individual differences observed in another context, i.e. whether they correspond to a coping style characteristic of said subjects. For this reason, we will analyze whether the coping style adopted by subordinate subjects in response to acute and chronic social stress is related to the emotional reactivity measured prior to the application of stress. Bearing in mind that acute stress does not constitute the same challenge as chronic stress, the strategy adopted in response to each situation may be different. Thus, we will also study the evolution of the coping strategies adopted by subordinate subjects in response to acute and chronic stress.

5.1. Procedure

For our study we used 6-week-old OF1 mice (30-32gr) (n= 140) which, after a 5-day period of adaptation, were subjected to different behavioral tests to establish their emotional reactivity profiles (Ramos, Berton, Mormede, & Chaouloff, 1997). The Open Field test was used to record behaviors such as frequency, latency and percentage of the total time they remained in the center or on the periphery, the distance traveled and immobility on the periphery, the number of times they adopted an upright posture on the periphery or in the center and the amount of feces. The Elevated Plus Maze test recorded latency and frequency of the number of times the subject entered the closed arm or the open arm, and the percentage of time it remained in them, as well as the number of times it sniffed the closed arm or the open arm. Finally, Boissier's test was used to record the number of times the mouse inserted its nose into a hole. Two days after the emotional reactivity tests, the animals were subjected to social confrontations using the resident-intruder model (Miczek, 1979), in order to obtain subordinate subjects. Furthermore, subordinate subjects remained in the cage of a dominant resident without opportunity for physical interaction but with permanent sensorial contact (Kudryavtseva, Bakshtanovskaia, & Koryakina, 1991). Subordinate animals were subjected to 23 days of defeat experience in order to establish different coping strategy profiles (active or passive). To this end, the behaviors shown in response to acute stress (2 days of confrontation), sub-chronic stress (10 days of confrontation) and chronic stress (23 days of confrontation) were recorded. The behavioral assessment was carried out using an ethogram (Brain, McAlliste, & Walmsley, 1989) in which 48 behaviors are described, divided into the following behavioral categories: immobility, social exploration, non social exploration, exploration from a distance, defense-submission, flee-avoidance, aggression (attack, threat) and other behaviors (body care and digging). Forty minutes after the last social stress session, the animals were put down in order for a blood sample to be taken and corticosterone levels determined using an ELISA.

A cluster analysis (n= 63) using the mean percentage of time allocated to each assessed behavioral element was carried out on all subordinate subjects in terms of the behavioral characteristics they showed in the social stress situation. In order to analyze the extent to which each of the behavioral variables enabled a significant differentiation between subjects from different clusters, a one-way ANOVA was performed. A multivariate discriminant analysis was performed (Wilk's Lambda method with step entry) to investigate the integrity of the groups derived from the cluster analysis and to determine which behavioral variables enabled the clusters to be distinguished most efficiently. We also analyzed whether different coping strategies of subordinated subjects were related to either corticosterone levels or the emotional reactivity using one-way ANOVAs. When appropriate, specific comparisons were made using post hoc Tukey tests. Finally, in order to study the evolution of the coping strategies adopted by subordinate subjects, a two-factor repeated-measures ANOVA with coping strategy (between-subjects variable) x time of stress (within-subjects) was performed. All statistics involved SPSS (14.0) for Windows (SPSS Inc. Chicago, Illinois, USA) with the level of significance set at $p \leq 0.05$.

5.2. Results

The results revealed that the cluster analysis distinguished three groups of subordinate subjects, in accordance with their behavioral characteristics when coping with acute and chronic social stress.

5.2.1. Acute Stress

5.2.1.1. Acute Stress and Coping Strategies Adopted by Subordinate Subjects

The cluster analysis carried out discriminated between subordinate subjects in accordance with the behaviors manifested in the situation of acute social stress. As a cut-off criterion, the inflection point was established at a distance equal to 15, resulting in three clusters characterized by the following behaviors:

Group 1 (n=15). Encompasses active subjects with low aggression which reacted to acute social stress by widely exploring the environment (non social exploration) and exploring their opponent both directly and from a distance (exploration from a distance and social exploration). These subjects did not dedicate much time to immobility behaviors and engaged to a greater degree in other behaviors such as digging or body care. It is, however, a group with a low level of aggression (aggression) which engaged in few flee-avoidance behaviors.

Group 2 (n=34). This is the most passive group with low levels of aggression. It is the group which spent the least percentage of time exploring both their environment (non social exploration) and their opponent (social exploration and exploration from a distance), engaged less in other behaviors and spent the greatest percentage of time engaging in immobility behaviors. The group also spent more time avoiding and fleeing from the opponent.

Group 3 (n=12). This is the intermediate group with regard to activity, since it showed the same levels of social and non social exploration as group 2 (more passive) but also the same levels of immobility as group 1 (more active). Nevertheless, it is mainly characterized by engaging in more aggressive behaviors than the other two groups.

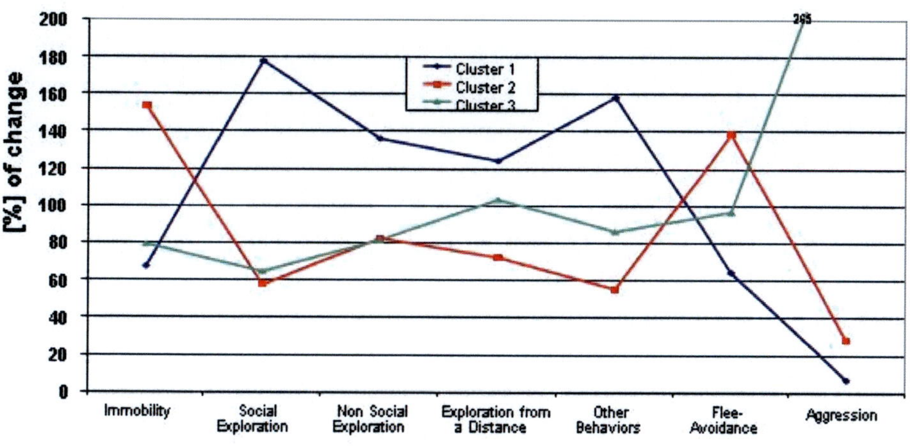

Figure 1. Behavioral coping profiles in response to acute social stress, for subordinate subjects in cluster 1 (active strategy), cluster 2 (passive strategy) and cluster 3 (intermediate strategy). Percentage of change with regard to the mean percentage of time dedicated by the 3 groups to each of the behaviors manifested during a 5 min exposure to stress. *Immobility*: clusters $1<2>3<1$ ($p \leq 0.000$). *Social exploration*: cluster $1>2$ and $3<1$ ($p \leq 0.000$). *Non social exploration*: cluster $1>2$ and $3<1$ ($p \leq 0.000$). *Exploration from a distance*: cluster $1>2$ ($p \leq 0.000$). *Other behaviors*: cluster $1>2$ ($p=0.019$). *Avoidance-Flee:* cluster $1<2$ ($p=0.037$). *Aggression*: cluster $2<3$ and $3>1$ ($p \leq 0.000$).

The analysis of variance performed in order to analyze the differences between behavioral variables in each of the subordinate groups obtained (Figure 1) revealed significant differences with regard to non social exploration ($F[2,58]=18.951$; $p<0.0001$), social exploration ($F[2,58]= 8.771$; $p<0.0001$), exploration from a distance ($F[2,58]= 5.610$; $p<0.006$), immobility ($F[2,58]= 37.486$; $p<0.0001$), flee-avoidance ($F[2,58]= 3.389$; $p<0.041$), other behaviors ($F[2,58]= 3.921$; $p<0.025$) and aggression ($F[2,58]= 63.913$; $p<0.0001$). The defense-submission behavior was the only one for which no differences were revealed between the three established groups. The post hoc test revealed significant differences between the three groups in immobility: clusters 1<2>3<1 ($p\leq0.000$); social exploration: cluster 1>2 and 3<1 ($p\leq0.000$); non social exploration: cluster 1>2 and 3<1 ($p\leq0.000$); exploration from a distance: cluster 1>2 ($p\leq0.000$); other behaviors: cluster 1>2 ($p=0.019$); avoidance-flee: cluster 1<2 ($p=0.037$); aggression: cluster 2<3 and 3>1 ($p\leq0.000$).

The discriminant model applied accounted for 100% of cases obtained by the cluster solution, thus confirming the statistical validity of the three groups of subordinates, as well as the behavioral description. Aggression was the variable that best discriminated between the three clusters, followed by Immobility and non social exploration. The remaining elements made no significant contribution to discriminating between the three clusters and were eliminated from the explanatory model. Although aggression was the variable that best discriminated between the 3 groups, in the distribution map of the subjects offered by the discriminant analysis in Figure 2, it can be seen that discriminant function 2 (ordinate axis) was the one which mainly contributed to distinguishing between group 1 (animals which explored the environment and showed low levels of immobility and aggression) and group 2 (more passive animals which showed greater levels of immobility and lower levels of exploration of the environment and aggression).

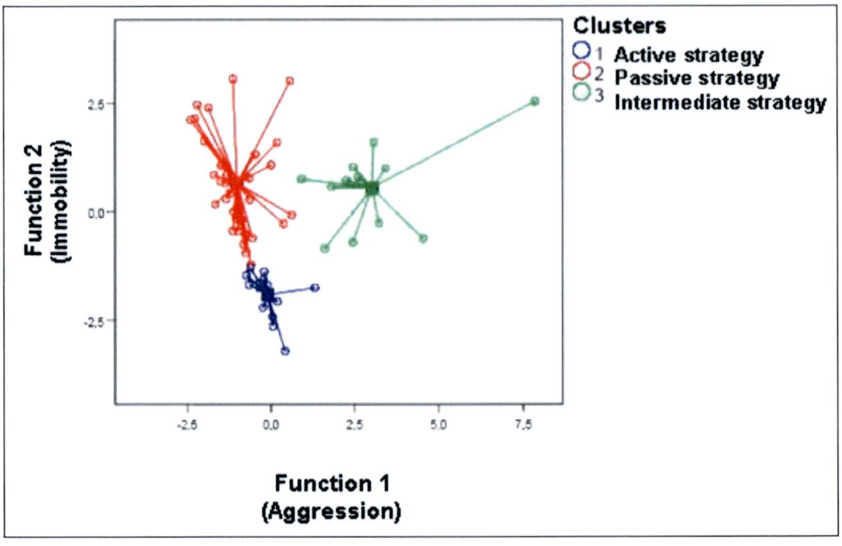

Figure 2. Spatial representation of all subordinate subjects in acute social stress grouped into clusters 1, 2 and 3 in accordance with two discriminant functions. Function 1: represented by positive aggression values. Function 2: represented by positive immobility values.

5.2.1.2. Emotional Reactivity prior to Social Stress and coping Styles

The results obtained revealed that the coping strategies adopted by subordinate subjects in response to acute stress differed in accordance with the reactivity levels measured prior to stress using the open field test. In specific terms, these differences lay in the Number of times they entered the center (F[2,57]= 3.290; p<0.044), the distance traveled in the center (F[2,57]= 3.155; p<0.05) and latency in engaging in the first immobility behavior (F[2,57]= 7.245; p<0.002). The exact nature of these differences is shown in Figure 3, which reveals levels of emotional reactivity in those subjects which adopted a more passive coping style, and a lower level of emotional reactivity in more active subjects. None of the other tests carried out revealed significant differences for any of the behaviors assessed. The post hoc test revealed significant differences between the three groups in frequency in the center: clusters 1>2 (p=0.035), frequency of freezing in the center: cluster 1>2 and 3 (p=0.024, p=0.007), distance traveled in the center: cluster 1>2 (p=0.040), latency of immobility in the center: cluster 1>2 and 3 (p=0.035, p=0.044).

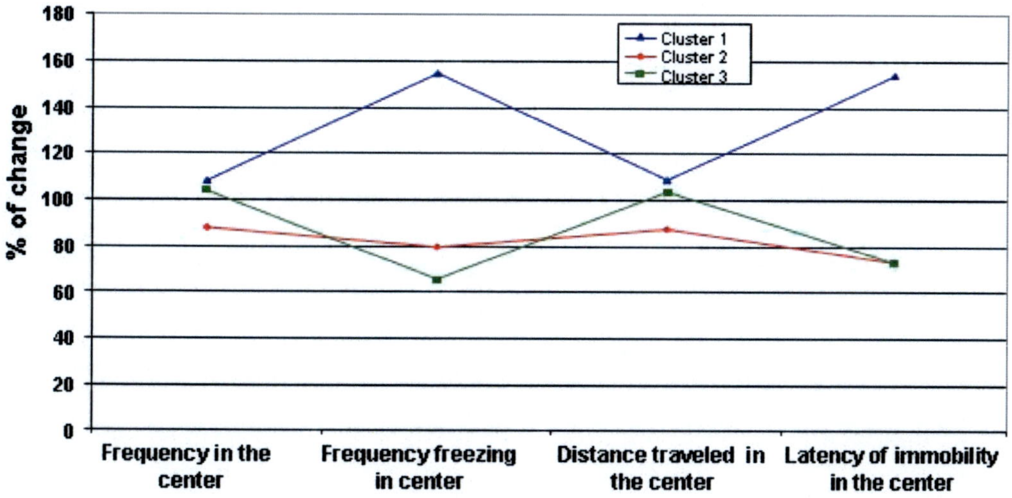

Figura 3. Behavioral emotional reactivity of subordinate subjects grouped into clusters, in accordance with the behaviors manifested in response to acute social stress. Cluster 1 (active strategy), cluster 2 (passive strategy) and cluster 3 (intermediate strategy). Percentage of change with regard to the mean percentage of time dedicated by the 3 groups to each of the behaviors manifested during a 5 min period in the open field test. *Frequency in the center*: clusters 1>2 (p=0.035). *Frequency of freezing in the center*: cluster 1>2 and 3 (p=0.024, p=0.007). *Distance traveled in the center:* cluster 1>2 (p=0.040). *Latency of immobility in the center*: cluster 1>2 and 3 (p=0.035, p=0.044).

5.2.2. Chronic Stress

5.2.2.1. Chronic Stress and Coping Styles Adopted by Subordinate Subjects

The cluster analysis performed discriminated between subordinate subjects in accordance with the behaviors shown in the situation of chronic social stress. As a cut-off criterion, the inflection point was established at a distance equal to 7, resulting in three clusters characterized by the following behaviors:

Group 1 (n=17). This group encompasses subordinate subjects which responded to chronic exposure to an aggressive opponent in a basically active manner, engaging in social exploration, non social exploration, exploration at a distance, avoidance, and engagement in other behaviors such as digging and body care, and dedicating very little time to immobility.

Group 2 (n=24). This group encompasses more passive subordinate subjects which responded to chronic social stress mainly by engaging in immobility behaviors and which spent very little time engaging in non social exploration, social exploration and exploration from a distance, and almost no time at all engaging in defense-submission behavior.

Group 3 (n=22). This is the intermediate group between the two previous ones, although it is more similar in nature to group 2. Its response differs from group 2 mainly in that its members defended themselves against their opponent by engaging in defense-submission behaviors and spent more time on social exploration and less time on immobility behaviors.

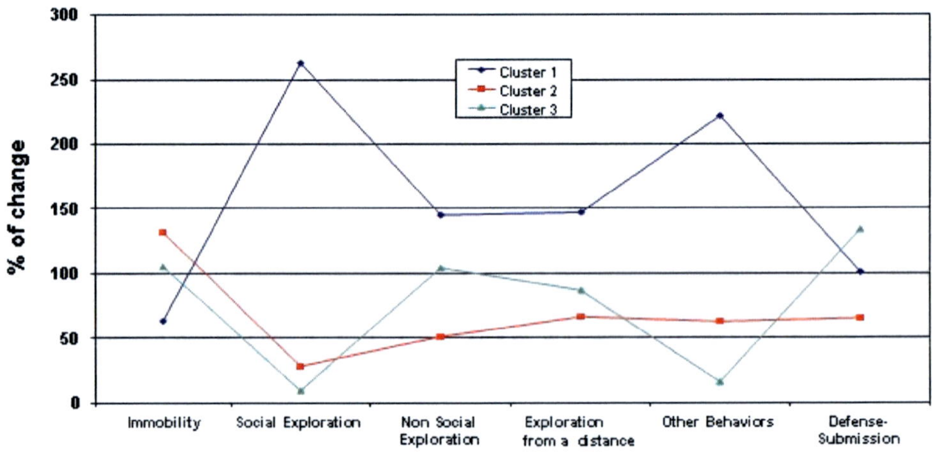

Figure 4. Behavioral coping profiles in response to chronic social stress, for subordinate subjects in cluster 1 (active strategy), cluster 2 (passive strategy) and cluster 3 (intermediate strategy). Percentage of change with regard to the mean percentage of time dedicated by the 3 groups to each of the behaviors manifested during a 5 min exposure to stress. *Immobility*: clusters 1<2>3<1 (p≤0.000). *Social exploration*: cluster 1>2 and 3<1 (p≤0.000). *Non social exploration*: cluster 1>2<3<1 (p≤0.000, p≤0.000, p=0,01). *Exploration from a distance*: cluster 1>2 and 3<1 (p≤0.001, p=0.023). *Other behaviors*: cluster 3<1 (p=0.016). *Defense-submission*: cluster 2<3 (p=0.008).

The analysis of variance performed in order to analyze the differences in behavioral variables observed in each of the subordinate groups obtained (Figure 4) revealed significant differences in non social exploration (F[2,60]= 26.320; p<0.0001), social exploration (F[2,60]= 39.197; p<0.0001), exploration from a distance (F[2,60]= 7.085; p<0.002), immobility (F[2,60]= 94.360; p<0.0001), defense-submission (F[2,60]= 4.827; p<0.011) and other behaviors (F[2,60]= 4.388; p<0.017). The flee-avoidance behavior was the only one for which no differences were observed between the three established groups. Behaviors such as attack and threat were not manifested by subordinate subjects after exposure to chronic stress. The post hoc test revealed significant differences between the three groups in immobility: clusters 1<2>3<1 (p≤0.000); social exploration: cluster 1>2 and 3<1 (p≤0.000); non social exploration: cluster 1>2<3<1 (p≤0.000, p≤0.000, p=0,01); exploration from a distance:

cluster 1>2 and 3<1 ($p \leq 0.001$, $p=0.023$); other behaviors: cluster 3<1 ($p=0.016$); defense-submission: cluster 2<3 ($p=0.008$).

The discriminant model applied accounted for 100% of cases obtained by the cluster solution, thus confirming the statistical validity of the three groups of subordinates, as well as the behavioral description. Immobility was the variable that best discriminated between the three clusters, followed by social exploration. The remaining elements made no significant contribution to discriminating between the three clusters and were eliminated from the explanatory model. In the distribution map of the subjects offered by the discriminant analysis in Figure 5, it can be seen that discriminant function 1 (abscissa axis) was the one which mainly contributed to distinguishing between the three groups, especially (and conclusively) between group 1 (animals which showed low levels of immobility) and group 2 (animals which showed greater levels of immobility and lower levels of exploration of the environment), with group 3 being situated in an intermediate position.

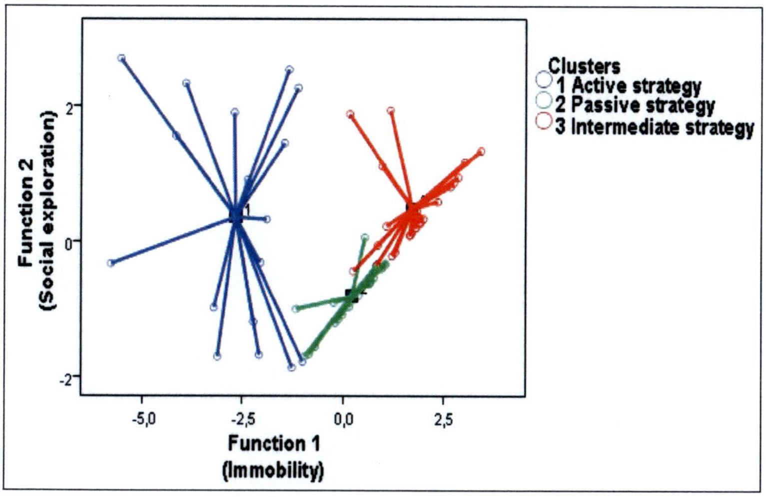

Figure 5. Spatial representation of all subordinate subjects in chronic social stress grouped into clusters 1, 2 and 3 in accordance with two discriminant functions. Function 1: represented by positive immobility values. Function 2: represented by positive social exploration values.

5.2.2.2. Coping Styles and Corticosterone Levels

The results revealed significant differences in corticosterone levels (F [2,41]= 3.643; $p<0.035$) between the subordinate groups which adopted different strategies for coping with social stress. The post hoc analysis indicated that corticosterone levels were lower in group 1 (active strategy) than in group 2 (passive strategy) ($p<0.033$) (Figure 6).

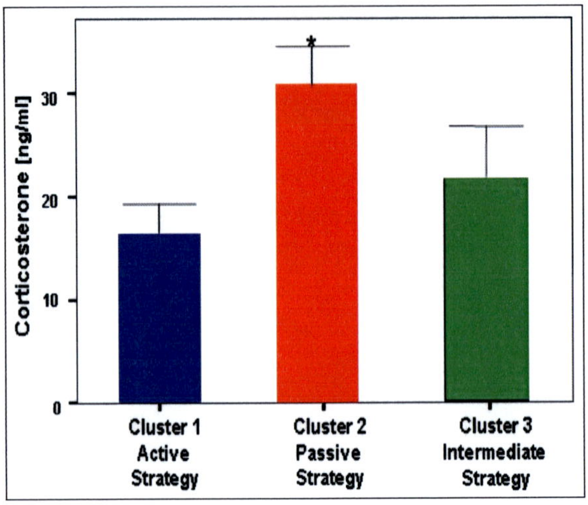

Figure 6. Mean levels (± E.E.M.) of corticosterone in serum (ng/ml) of the 3 clusters of subordinate subjects selected when exposed to chronic social stress. Clusters 2>1 (p=0.033)

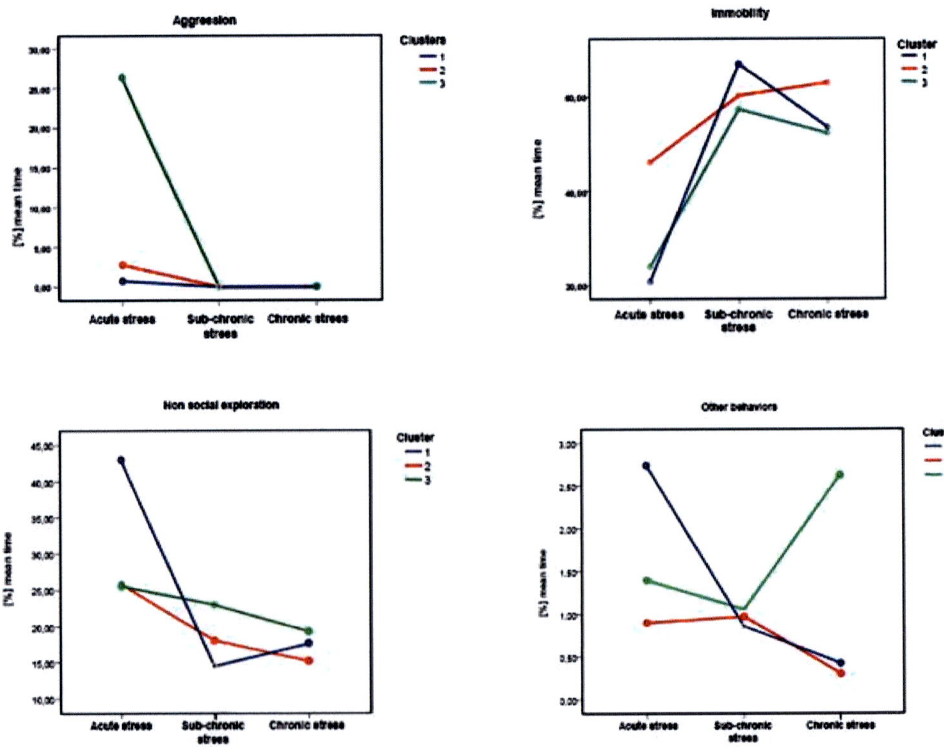

Figure 7. Evolution of *aggression, immobility, non social exploration* and *other behaviors* in the 3 clusters of subordinate subjects selected in acute stress. Percentage of time dedicated, out of a total of 5 minutes, by each of the 3 groups when exposed to acute, sub-chronic and chronic stress.

5.2.2.3. Emotional Reactivity prior to Social Stress and Coping Styles

The results revealed that the coping strategies adopted by subordinate subjects in response to chronic stress differed in accordance with their emotional reactivity levels measured prior to said stress, specifically in the distance traveled on the periphery in the open field test ($F[2,59]= 3.755$; $p<0.029$). The post hoc analysis revealed that subordinate subjects which adopted a passive coping strategy (group 2) also had the highest levels of reactivity in the open field test (less motor activity) ($p<0.035$). None of the other tests carried out revealed significant differences in any of the assessed behaviors.

5.2.3. Evolution of Coping Strategies for Acute Social Stress in Subordinate Subjects

The results of the study of the evolution of the coping strategies adopted by subordinate subjects in response to acute social stress indicate a main effect of group x time of stress in immobility ($F[2,57]= 3.851$; $p<0.027$), aggression (attack and threat) ($F[2,57]= 62,295$; $p<0.000$), non social exploration ($F[2,57]= 10.580$; $p<0.000$) and other behaviors (digging and body care) ($F[2,57]= 5.649$; $p<0.006$).

The ANOVA performed on the significant behaviors at each of the stress times revealed that the differences observed in the acute stress between the three subordinate groups as regards both immobility and aggression (the variables which best discriminated between the groups), as well as non social exploration (see results section 5.2.1.1.), disappeared after exposure to sub-chronic and chronic stress. The differences found between the three groups as regards other behaviors also disappeared after sub-chronic stress, although they reappeared once again following chronic stress, with the time spent engaged in this behavior being greater in the intermediate group than in the other two groups ($F[2,57]= 7,823$; $p< 0.001$) (Figure 7).

6. CONCLUSION

The results of our research reveal that subordinate subjects adopt different strategies for coping with acute and chronic social stress, and that these strategies are clearly distinguishable as regards the behaviors manifested during social confrontations. The group characterized by its passive coping style, encompasses subjects which show little interest in their environment, their opponent and their own body care, remain practically immobile and do not attempt to flee, escape or defend themselves from the situation. The other group, which adopts an active coping style, is characterized by high levels of exploration of both the environment and the opponent, high levels of body care and low levels of immobility. These groups are not only different from a behavioral point of view. Following exposure to chronic social stress, subjects with a passive coping style present a characteristic reactivity of the HPA axis, manifested through higher corticosterone levels. Thus, it is reasonable to assume that animals with a passive coping style perceive the same stressful situation as more intense than those with a more active coping style, and may therefore be more susceptible to the effects of said stress.

Subordinate subjects which adopt different coping strategies in response to acute and chronic social stress also show different levels of emotional reactivity prior to said stress. Subjects which adopt a passive coping style in response to acute stress show less locomotor activity in the central part of the open field and remain immobile quicker in said area, indicating a more inhibited behavior than the group which adopts a more active strategy. The subordinate group which adopts a passive coping strategy in response to chronic stress also shows a high level of prior emotional reactivity, manifested through less locomotor activity on the periphery of the open field. Thus, the passive coping style adopted by subordinate subjects in response to both acute and chronic stress is related to a higher level of emotional reactivity, manifested through changes in locomotor behavior and perhaps indicative of higher anxiety levels.

Furthermore, the study of the evolution of the coping strategies adopted by subordinate subjects in response to social stress reveals that subjects which adopt a passive strategy in response to acute stress do not necessary maintain the same strategy in response to inescapable chronic stress. Thus, subjects which adopt a passive strategy in response to acute stress, may react differently in response to the requirements of a more prolonged stressful situation. Nevertheless, even through the strategy adopted in response to acute stress does not seem to determine that adopted in subsequent situations of stress, our data show that a subject's emotional reactivity is associated with a passive coping strategy for both acute and chronic stress. Thus, behavioral traits existing in the subject prior to exposure to stress may influence the type of behavior manifested in response to different stressful situations.

ACKNOWLEDGEMENTS

This work was supported by grants from Spanish Ministry of Science and Technology SEJ2005-03981.

REFERENCES

Anisman, H., Hayley, S., Kelly, O., Borowski, T., & Merali, Z. (2001). Psychogenic, neurogenic, and systemic stressor effects on plasma corticosterone and behavior: mouse strain-dependent outcomes. *Behav Neurosci, 115*(2), 443-454.

Arango, V., Ernsberger, P., Marzuk, P. M., Chen, J. S., Tierney, H., Stanley, M., et al. (1990). Autoradiographic demonstration of increased serotonin 5-HT2 and beta-adrenergic receptor binding sites in the brain of suicide victims. *Arch Gen Psychiatry, 47*(11), 1038-1047.

Arias, B., Gasto, C., Catalan, R., Gutierrez, B., Pintor, L., & Fananas, L. (2001). The 5-HT(2A) receptor gene 102T/C polymorphism is associated with suicidal behavior in depressed patients. *Am J Med Genet, 105*(8), 801-804.

Arias, B., Gutierrez, B., Pintor, L., Gasto, C., & Fananas, L. (2001). Variability in the 5-HT(2A) receptor gene is associated with seasonal pattern in major depression. *Mol Psychiatry, 6*(2), 239-242.

Arregui, A., Azpiroz, A., Fano, E., & Garmendia, L. (2006). Aggressive behavior: implications of dominance and subordination for the study of mental disorders. *Aggress Violent Behav, 11*, 394-413.

Avgustinovich, D. F. (2003). [Anxiety in females caused by long-term psycho-emotional stress]. *Ross Fiziol Zh Im I M Sechenova, 89*(7), 858-867.

Avgustinovich, D. F., Gorbach, O. V., & Kudryavtseva, N. N. (1997). Comparative analysis of anxiety-like behavior in partition and plus-maze tests after agonistic interactions in mice. *Physiol Behav, 61*(1), 37-43.

Avitsur, R., Stark, J. L., Dhabhar, F. S., Kramer, K. A., & Sheridan, J. F. (2003). Social experience alters the response to social stress in mice. *Brain Behav Immun, 17*(6), 426-437.

Bartolomucci, A., Palanza, P., Costoli, T., Savani, E., Laviola, G., Parmigiani, S., et al. (2003). Chronic psychosocial stress persistently alters autonomic function and physical activity in mice. *Physiol Behav, 80*(1), 57-67.

Bartolomucci, A., Palanza, P., Gaspani, L., Limiroli, E., Panerai, A. E., Ceresini, et al. (2001). Social status in mice: behavioral, endocrine and immune changes are context dependent. *Physiol Behav, 73*(3), 401-410.

Bartolomucci, A., Palanza, P., Sacerdote, P., Panerai, A. E., Sgoifo, A., Dantzer, R., et al. (2005). Social factors and individual vulnerability to chronic stress exposure. *Neurosci Biobehav Rev, 29*(1), 67-81.

Bartolomucci, A., Pederzani, T., Sacerdote, P., Panerai, A. E., Parmigiani, S., & Palanza, P. (2004). Behavioral and physiological characterization of male mice under chronic psychosocial stress. *Psychoneuroendocrinology, 29*(7), 899-910.

Belzung, C., & Griebel, G. (2001). Measuring normal and pathological anxiety-like behaviour in mice: a review. *Behav Brain Res, 125*(1-2), 141-149.

Benus, R. F., Bohus, B., Koolhaas, J. M., & Van Oortmerssen, G. A. (1989). Behavioural strategies of aggressive and non-aggressive male mice in active shock avoidance. *Behavl Processes, 20*(1-3), 1-12.

Benus, R. F., Bohus, B., Koolhaas, J. M., & Van Oortmerssen, G. A. (1991a). Behavioural differences between artificially selected aggressive and non-aggressive mice: response to apomorphine. *Behav Brain Res, 43*(2), 203-208.

Benus, R. F., Bohus, B., Koolhaas, J. M., & Van Oortmerssen, G. A. (1991b). Heritable variation for aggression as a reflection of individual coping strategies. *Experientia, 47*(10), 1008-1019.

Berton, O., Aguerre, S., Sarrieau, A., Mormede, P., & Chaouloff, F. (1998). Differential effects of social stress on central serotonergic activity and emotional reactivity in Lewis and spontaneously hypertensive rats. *Neuroscience, 82*(1), 147-159.

Blanchard, D. C., McKittrick, C. R., Hardy, M. P., & Blanchard, R. J. (2002). Effects of social stress on hormones, brain, and behavior. In D. W. Pfaff, A. P. Arnold, A. M. Etgen, S. E. Fahrbach & R. T. Rubin (Eds.), *Hormones, Brain and Behavior* (pp. 735-772). San Diego: Academic Press.

Blanchard, D. C., Sakai, R. R., McEwen, B., Weiss, S. M., & Blanchard, R. J. (1993). Subordination stress: behavioral, brain, and neuroendocrine correlates. *Behav Brain Res, 58*(1-2), 113-121.

Blanchard, D. C., Spencer, R. L., Weiss, S. M., Blanchard, R. J., McEwen, B., & Sakai, R. R. (1995). Visible burrow system as a model of chronic social stress: behavioral and neuroendocrine correlates. *Psychoneuroendocrinology, 20*(2), 117-134.

Blanchard, D. C., Spencer, R. L., Weiss, S. M., Blanchard, R. J., McEwen, B. S., & Sakai, R. R. (1995). Visible borrow system as a model of chronic social stress: behavioral and neuroendocrine correlates. *Psychoneuroendocrinology, 20*, 117-134.

Blanchard, R. j., Yudko, E., Dulloog, L., & Blanchard, D. C. (2001). Defense changes in stress nonresponsive subordinate males in a visible burrow system. *Physiol Behav, 72*, 635-642.

Boissy, A. (1995). Fear and fearfulness in animals. *Q Rev Biol, 70*(2), 165-191.

Brain, P., McAlliste, K. H., & Walmsley, S. V. (1989). Drugs effects on social behavior: methods in ethopharmacology. In A. A. Boulton, G. B. Baker & A. J. Grennshaw (Eds.), *Neuromethods. Psychopharmacology.* (pp. 687-739). Clifton, NJ: The Humana Press Inc.

Buwalda, B., Van Kalkeren, A. A., de Boer, S. F., & Koolhaas, J. M. (1998). Behavioral and physiological consequences of repeated daily intracerebroventricular injection of corticotropin-releasing factor in the rat. *Psychoneuroendocrinology, 23*(3), 205-218.

Cacho, R., Fano, E., Areso, P., Garmendia, L., Vegas, O., Brain, P. F., et al. (2003). Endocrine and lymphoproliferative response changes produced by social stress in mice. *Physiol Behav, 78*(3), 505-512.

Campbell, T., Lin, S., DeVries, C., & Lambert, K. (2003). Coping strategies in male and female rats exposed to multiple stressors. *Physiol Behav, 78*(3), 495-504.

Cannon, W. B. (1915). *Bodily changes in pain, hunger fear and rage.* New York: Appleton.

Carere, C., Welink, D., Drent, P. J., Koolhaas, J. M., & Groothuis, T. G. (2001). Effect of social defeat in a territorial bird (Parus major) selected for different coping styles. *Physiol Behav, 73*(3), 427-433.

Chaouloff, F. (1995). Regulation of 5-HT receptors by corticosteroids: where do we stand? *Fundam Clin Pharmacol, 9*(3), 219-233.

Chrousos, G. P., & Gold, P. W. (1992). The concepts of stress and stress system disorders. Overview of physical and behavioral homeostasis. *Jama, 267*(9), 1244-1252.

Cohen, S., & Hamrick, N. (2003). Stable individual differences in physiological response to stressors: implications for stress-elicited changes in immune related health. *Brain Behav Immun, 17*(6), 407-414.

Crawley, J. N., Belknap, J. K., Collins, A., Crabbe, J. C., Frankel, W., Henderson, N., et al. (1997). Behavioral phenotypes of inbred mouse strains: implications and recommendations for molecular studies. *Psychopharmacology (Berl), 132*(2), 107-124.

Czeh, B., Michaelis, T., Watanabe, T., Frahm, J., De Biurrun, G., Van Kampen, M., et al. (2001). Stress-induced changes in cerebral metabolites, hippocampal volume, and cell proliferation are prevented by antidepressant treatment with tianeptine. *Proc Natl Acad Sci U S A, 98*(22), 12796-12801.

Deparmentier, F., Crompton, M. R., Katona, C. L. E., & Horton, R. W. (1993). Beta-adrenoreceptors in brain and pineal from depressed suicide victims. *Pharmacol Toxicol, 71*(86-95).

Ducottet, C., & Belzung, C. (2004). Behaviour in the elevated plus-maze predicts coping after subchronic mild stress in mice. *Physiol Behav, 81*(3), 417-426.

Ducottet, C., Griebel, G., & Belzung, C. (2003). Effects of the selective nonpeptide corticotropin-releasing factor receptor 1 antagonist antalarmin in the chronic mild stress model of depression in mice. *Prog Neuropsychopharmacol Biol Psychiatry, 27*(4), 625-631.

Duman, R. S., Heninger, G. R., & Nestler, E. J. (1997). A molecular and cellular theory of depression. *Arch Gen Psychiatry, 54*(7), 597-606.

Engel, G. L., & Schmale, A. H. (1972). Conservation-withdrawal: a primary regulatory process for organismic homeostasis. *Ciba Found Symp, 8*, 57-75.

Eriksen, H. R., Olff, M., Murison, R., & Ursin, H. (1999). The time dimension in stress responses: relevance for survival and health. *Psychiatry Res, 85*(1), 39-50.

Fano, E., Sanchez-Martin, J. R., Arregi, A., Castro, B., Alonso, A., Brain, P., et al. (2001). Social stress paradigms in male mice: Variations in behavior, stress and immunology. *Physiol Behav, 73*(1-2), 165-173.

File, S. E., Zangrossi, H., Jr., & Andrews, N. (1993). Novel environment and cat odor change GABA and 5-HT release and uptake in the rat. *Pharmacol Biochem Behav, 45*(4), 931-934.

Flugge, G. (2000). Regulation of monoamine receptors in the brain: dynamic changes during stress. *Int Rev Cytol, 195*, 145-213.

Flugge, G., Ahrens, O., & Fuchs, E. (1997a). Beta-adrenoceptors in the tree shrew brain. II. Time-dependent effects of chronic psychosocial stress on [125I]iodocyanopindolol bindings sites. *Cell Mol Neurobiol, 17*(4), 417-432.

Flugge, G., Ahrens, O., & Fuchs, E. (1997b). Monoamine receptors in the prefrontal cortex of Tupaia belangeri during chronic psychosocial stress. *Cell Tissue Res, 288*(1), 1-10.

Fuchs, E., & Flugge, G. (1995). Modulation of binding sites for corticotropin-releasing hormone by chronic psychosocial stress. *Psychoneuroendocrinology, 20*(1), 33-51.

Fuchs, E., Kramer, M., Hermes, B., Netter, P., & Hiemke, C. (1996). Psychosocial stress in tree shrews: clomipramine counteracts behavioral and endocrine changes. *Pharmacol Biochem Behav, 54*(1), 219-228.

Garcia-Sevilla, J. A., Escriba, P. V., Ozaita, A., La Harpe, R., Walzer, C., Eytan, A., et al. (1999). Up-regulation of immunolabeled alpha2A-adrenoceptors, Gi coupling proteins, and regulatory receptor kinases in the prefrontal cortex of depressed suicides. *J Neurochem, 72*(1), 282-291.

Geerse, G. J., van Gurp, L. C., Wiegant, V. M., & Stam, R. (2006). Individual reactivity to the open-field predicts the expression of stress-induced behavioural and somatic pain sensitisation. *Behav Brain Res, 174*(1), 112-118.

Griebel, G., Belzung, C., Misslin, R., & Vogel, E. (1993). The free-exploratory paradigm: an effective method for measuring neophobic behaviour in mice and testing potential neophobia-reducing drugs. *Behav Pharmacol, 4*(6), 637-644.

Griebel, G., Belzung, C., Perrault, G., & Sanger, D. J. (2000). Differences in anxiety-related behaviours and in sensitivity to diazepam in inbred and outbred strains of mice. *Psychopharmacology (Berl), 148*, 164-170.

Groenink, L., Dirks, A., Verdouw, P. M., Schipholt, M., Veening, J. G., Van der Gugten, J., et al. (2002). HPA axis dysregulation in mice overexpressing corticotropin releasing hormone. *Biol Psychiatry, 51*(11), 875-881.

Grootendorst, J., De Kloet, E. R., Vossen, C., Dalm, S., & Oitzl, M. S. (2001). Repeated exposure to rats has persistent genotype-dependent effects on learning and locomotor activity of apolipoprotein E knockout and C57Bl/6 mice. *Behav Brain Res, 125*(1-2), 249-259.

Habib, K. E., Gold, P. W., & Chrousos, G. P. (2001). Neuroendocrinology of stress. *Endocrinol Metab Clin North Am, 30*(3), 695-728; vii-viii.

Haghighatgou, H., & Peterson, C. (1995). Coping and depressive symptoms among Iranian students. *J Soc Psychol, 135*(2), 175-180.

Haller, J., & Halasz, J. (2000). Effects of two acute stressors on the anxiolytic efficacy of chlordiazepoxide. *Psychopharmacology (Berl), 151*(1), 1-6.

Haney, M., Maccari, S., Le Moal, M., Simon, H., & Piazza, P. V. (1995). Social stress increases the acquisition of cocaine self-administration in male and female rats. *Brain Res, 698*(1-2), 46-52.

Haney, M., Noda, L., Kream, R., & Miczek, K. A. (1990). Regional serotonin and dopamine activity: Sensitivity to amphetamine and aggressive behavior in mice. *Aggressive Behavior, 16*(259-270).

Harper, D. G., Tornatzky, W., & Miczek, K. A. (1996). Stress induced disorganization of circadian and ultradian rhythms: comparisons of effects of surgery and social stress. *Physiol Behav, 59*(3), 409-419.

Heinrichs, S. C., Menzaghi, F., Pich, E. M., Baldwin, H. A., Rassnick, S., Britton, K. T., et al. (1994). Anti-stress action of a corticotropin-releasing factor antagonist on behavioral reactivity to stressors of varying type and intensity. *Neuropsychopharmacology, 11*(3), 179-186.

Heinrichs, S. C., Pich, E. M., Miczek, K. A., Britton, K. T., & Koob, G. F. (1992). Corticotropin-releasing factor antagonist reduces emotionality in socially defeated rats via direct neurotropic action. *Brain Res, 581*(2), 190-197.

Henry, J. P. (1997). Stress, health and the social environment. James P. Henry's ethologic approach to medicine, reflected by recent research in animals and man. In memory of a great 20th century physiologist. *Acta Physiol Scand Suppl, 640*, 1-179.

Herve, D., Tassin, J. P., Barthelemy, C., Blanc, G., Lavielle, S., & Glowinski, J. (1979). Difference in the reactivity of the mesocortical dopaminergic neurons to stress in the BALB/C and C57 BL/6 mice. *Life Sci, 25*(19), 1659-1664.

Hirshfeld-Becker, D. R., Biederman, J., Calltharp, S., Rosenbaum, E. D., Faraone, S. V., & Rosenbaum, J. F. (2003). Behavioral inhibition and disinhibition as hypothesized precursors to psychopathology: implications for pediatric bipolar disorder. *Biol Psychiatry, 53*(11), 985-999.

Jhanwar-Uniyal, M., & Leibowitz, S. F. (1986). Impact of circulating corticosterone on alpha 1- and alpha 2-noradrenergic receptors in discrete brain areas. *Brain Res, 368*(2), 404-408.

Kabbaj, M., Norton, C. S., Kollack-Walker, S., Watson, S. J., Robinson, T. E., & Akil, H. (2001). Social defeat alters the acquisition of cocaine self-administration in rats: role of individual differences in cocaine-taking behavior. *Psychopharmacology (Berl), 158*(4), 382-387.

Kazlauckas, V., Schuh, J., Dall'Igna, O. P., Pereira, G. S., Bonan, C. D., & Lara, D. R. (2005). Behavioral and cognitive profile of mice with high and low exploratory phenotypes. *Behav Brain Res, 162*(2), 272-278.

Keck, M. E., Welt, T., Muller, M. B., Uhr, M., Ohl, F., Wigger, A., et al. (2003). Reduction of hypothalamic vasopressinergic hyperdrive contributes to clinically relevant behavioral and neuroendocrine effects of chronic paroxetine treatment in a psychopathological rat model. *Neuropsychopharmacology, 28*(2), 235-243.

Kessler, R. C. (1997). The effects of stressful life events on depression. *Annu Rev Psychol, 48*, 191-214.

Koob, G. F., & Le Moal, M. (2001). Drug addiction, dysregulation of reward, and allostasis. *Neuropsychopharmacology, 24*(2), 97-129.

Koolhaas, J. M., & Bohus, B. (1989). Social control in relation to neuroendocrine and immunological responses. In S. A & A. A (Eds.), *Stress, personal control and health* (pp. 295-304). Brussels: Wiley.

Koolhaas, J. M., De Boer, S. F., De Rutter, A. J., Meerlo, P., & Sgoifo, A. (1997). Social stress in rats and mice. *Acta Physiol Scand Suppl, 161*, 69-72.

Koolhaas, J. M., Korte, S. M., De Boer, S. F., Van Der Vegt, B. J., Van Reenen, C. G., Hopster, H., et al. (1999). Coping styles in animals: current status in behavior and stress-physiology. *Neurosci Biobehav Rev, 23*(7), 925-935.

Korte, S. M. (2001). Corticosteroids in relation to fear, anxiety and psychopathology. *Neurosci Biobehav Rev, 25*(2), 117-142.

Korte, S. M., Buwalda, B., Meijer, O., De Kloet, E. R., & Bohus, B. (1995). Socially defeated male rats display a blunted adrenocortical response to a low dose of 8-OH-DPAT. *Eur J Pharmacol, 272*(1), 45-50.

Korte, S. M., Koolhaas, J. M., Wingfield, J. C., & McEwen, B. S. (2005). The Darwinian concept of stress: benefits of allostasis and costs of allostatic load and the trade-offs in health and disease. *Neurosci Biobehav Rev, 29*(1), 3-38.

Korte, S. M., Smit, J., Bouws, G. A., Koolhaas, J. M., & Bohus, B. (1990). Behavioral and neuroendocrine response to psychosocial stress in male rats: the effects of the 5-HT 1A agonist ipsapirone. *Horm Behav, 24*(4), 554-567.

Kudryavtseva, N. N., & Avgustinovich, D. F. (1998). Behavioral and psysiological markers of experimental depression induced by social conflicts (DISC). *Aggressive Behav, 24*(271-286).

Kudryavtseva, N. N., & Bakshtanovskaia, I. V. (1991). [The neurochemical control of aggression and submission]. *Zh Vyssh Nerv Deiat Im I P Pavlova, 41*(3), 459-466.

Kudryavtseva, N. N., Bakshtanovskaia, I. V., & Koryakina, L. A. (1991). Social model of depression in mice of C57BL/6J Strain. *Pharmacol Biochem Behav, 133*, 315-320.

Kudryavtseva, N. N., Madorskaya, I. A., & Bakshtanovskaya, I. V. (1991). Social success and voluntary ethanol consumption in mice of C57BL/6J and CBA/Lac strains. *Physiol Behav, 50*(1), 143-146.

Landgraf, R., & Wigger, A. (2002). High vs low anxiety-related behavior rats: an animal model of extremes in trait anxiety. *Behav Genet, 32*(5), 301-314.

Lazarus, R. (1966). *Psychosocial stress and the coping process.* New York.

Lesch, K. P. (1991). 5-HT1A receptor responsivity in anxiety disorders and depression. *Prog Neuropsychopharmacol Biol Psychiatry, 15*(6), 723-733.

Lesch, K. P., Mayer, S., Disselkamp-Tietze, J., Hoh, A., Wiesmann, M., Osterheider, M., et al. (1990). 5-HT1A receptor responsivity in unipolar depression. Evaluation of ipsapirone-induced ACTH and cortisol secretion in patients and controls. *Biol Psychiatry, 28*(7), 620-628.

Liebsch, G., Montkowski, A., Holsboer, F., & Landgraf, R. (1998). Behavioural profiles of two Wistar rat lines selectively bred for high or low anxiety-related behaviour. *Behav Brain Res, 94*(2), 301-310.

Little, K. Y., Clark, T. B., Ranc, J., & Duncan, G. E. (1993). Beta-adrenergic receptor binding in frontal cortex from suicide victims. *Biol Psychiatry, 34*(9), 596-605.

Lopez, J. F., Chalmers, D. T., Little, K. Y., & Watson, S. J. (1998). A.E. Bennett Research Award. Regulation of serotonin1A, glucocorticoid, and mineralocorticoid receptor in rat and human hippocampus: implications for the neurobiology of depression. *Biol Psychiatry, 43*(8), 547-573.

Manji, H. K., Moore, G. J., & Chen, G. (2000). Clinical and preclinical evidence for the neurotrophic effects of mood stabilizers: implications for the pathophysiology and treatment of manic-depressive illness. *Biol Psychiatry, 48*(8), 740-754.

Maremmani, I., Akiskal, H. S., Signoretta, S., Liguori, A., Perugi, G., & Cloninger, R. (2005). The relationship of Kraepelian affective temperaments (as measured by TEMPS-I) to the tridimensional personality questionnaire (TPQ). *J Affect Disord, 85*(1-2), 17-27.

Marquez, C., Nadal, R., & Armario, A. (2005). Responsiveness of the hypothalamic-pituitary-adrenal axis to different novel environments is a consistent individual trait in adult male outbred rats. *Psychoneuroendocrinology, 30*(2), 179-187.

Marquez, C., Nadal, R., & Armario, A. (2006). Influence of reactivity to novelty and anxiety on hypothalamic-pituitary-adrenal and prolactin responses to two different novel environments in adult male rats. *Behav Brain Res, 168*(1), 13-22.

Mason, J. W. (1968a). A review of psychoendocrine research on the pituitary-adrenal cortical system. *Psychosom Med, 30*(5), Suppl:576-607.

Mason, J. W. (1968b). A review of psychoendocrine research on the sympathetic-adrenal medullary system. *Psychosom Med, 30*(5), Suppl:631-653.

McEwen, B. S. (1998). Protective and damaging effects of stress mediators. *N Engl J Med, 338*(3), 171-179.

McEwen, B. S. (2000). Effects of adverse experiences for brain structure and function. *Biol Psychiatry, 48*(8), 721-731.

McEwen, B. S. (2003). Interacting mediators of allostasis and allostatic load: towards an understanding of resilience in aging. *Metab Clin Exp, 52*(Supplement 2), 10-16.

McEwen, B. S., & Stellar, E. (1993). Stress and the individual. Mechanisms leading to disease. *Arch Intern Med, 153*(18), 2093-2101.

McEwen, B. S., & Wingfield, J. C. (2003). The concept of allostasis in biology and biomedicine. *Horm Behav, 43*(1), 2-15.

McKittrick, C. R., Blanchard, D. C., Blanchard, R. J., McEwen, B. S., & Sakai, R. R. (1995). Serotonin receptor binding in a colony model of chronic social stress. *Biol Psychiatry, 37*(6), 383-393.

Meana, J. J., Barturen, F., & Garcia-Sevilla, J. A. (1992). Alpha 2-adrenoceptors in the brain of suicide victims: increased receptor density associated with major depression. *Biol Psychiatry, 31*(5), 471-490.

Meerlo, P., De Boer, S. F., Koolhaas, J. M., Daan, S., & Van den Hoofdakker, R. H. (1996). Changes in daily rhythms of body temperature and activity after a single social defeat in rats. *Physiol Behav, 59*(4-5), 735-739.

Meerlo, P., Pragt, B. J., & Daan, S. (1997). Social stress induces high intensity sleep in rats. *Neurosci Lett, 225*(1), 41-44.

Meijer, O. C., & de Kloet, E. R. (1994). Corticosterone suppresses the expression of 5-HT1A receptor mRNA in rat dentate gyrus. *Eur J Pharmacol, 266*(3), 255-261.

Meijer, O. C., & de Kloet, E. R. (1995). A role for the mineralocorticoid receptor in a rapid and transient suppression of hippocampal 5-HT1A receptor mRNA by corticosterone. *J Neuroendocrinol, 7*(8), 653-657.

Menzaghi, F., Howard, R. L., Heinrichs, S. C., Vale, W., Rivier, J., & Koob, G. F. (1994). Characterizacion of a novel and potent corticotropin-releasing factor antagonist in rats. *J Pharmacol Ex Ther, 269*, 564-572.

Meyer, J. H., McMain, S., Kennedy, S. H., Korman, L., Brown, G. M., DaSilva, J. N., et al. (2003). Dysfunctional attitudes and 5-HT2 receptors during depression and self-harm. *Am J Psychiatry, 160*(1), 90-99.

Miczek, K. A. (1979). New test for aggression in rats without aversive-stimulation differential effects of D-amphetamine and cocaine. *Psychopharmacology, 60*, 253-259.

Miczek, K. A., Covington, H. E., 3rd, Nikulina, E. M., Jr., & Hammer, R. P. (2004a). Aggression and defeat: persistent effects on cocaine self-administration and gene expression in peptidergic and aminergic mesocorticolimbic circuits. *Neurosci Biobehav Rev, 27*(8), 787-802.

Miczek, K. A., Covington, H. E., Nikulina, E. A., & Hammer, R. P. (2004b). Aggression and defeat: persistent effects on cocaine self-administration and gene expression in peptidergic and aminergic mesocorticolimbic circuits. *Neurosci Biobehav Rev, 27*(8), 787-802.

Miczek, K. A., & Mutschler, N. H. (1996). Activational effects of social stress on IV cocaine self-administration in rats. *Psychopharmacology (Berl), 128*(3), 256-264.

Nielsen, C. K., Arnt, J., & Sanchez, C. (2000). Intracranial self-stimulation and sucrose intake differ as hedonic measures following chronic mild stress: interstrain and interindividual differences. *Behav Brain Res, 107*(1-2), 21-33.

Ohl, F., Toschi, N., Wigger, A., Henniger, M. S., & Landgraf, R. (2001). Dimensions of emotionality in a rat model of innate anxiety. *Behav Neurosci, 115*(2), 429-436.

Olff, M. (1991). Defence and Coping: Self-reported Health and Psychological Correlates. Utrecht. *ISOR.*

Olff, M. (1999). Stress, depression and immunity: the role of defense and coping styles. *Psychiatry Res, 85*(1), 7-15.

Ordway, G. A., Widdowson, P. S., Smith, K. S., & Halaris, A. (1994). Agonist binding to alpha 2-adrenoceptors is elevated in the locus coeruleus from victims of suicide. *J Neurochem, 63*(2), 617-624.

Ramos, A., Berton, O., Mormede, P., & Chaouloff, F. (1997). A multiple test study of anxiety-related behaviors in six inbred rat strains. *Behav Brain Res, 85*, 57-69.

Repetti, R. L., Taylor, S. E., & Seeman, T. E. (2002). Risky families: family social environments and the mental and physical health of offspring. *Psychol Bull, 128*(2), 330-366.

Reul, J. M., Gesing, A., Droste, S., Stec, I. S., Weber, A., Bachmann, C., et al. (2000). The brain mineralocorticoid receptor: greedy for ligand, mysterious in function. *Eur J Pharmacol, 405*(1-3), 235-249.

Reul, J. M., Stec, I., Soder, M., & Holsboer, F. (1993). Chronic treatment of rats with the antidepressant amitriptyline attenuates the activity of the hypothalamic-pituitary-adrenocortical system. *Endocrinology, 133*(1), 312-320.

Rodgers, R. J., & Cole, J. C. (1993). Anxiety enhancement in the murine elevated plus maze by immediate prior exposure to social stressors. *Physiol Behav, 53*(2), 383-388.

Ruis, M. A., de Groot, J., te Brake, J. H., Dinand Ekkel, E., van de Burgwal, J. A., Erkens, J. H., et al. (2001). Behavioural and physiological consequences of acute social defeat in growing gilts: effects of the social environment. *Appl Anim Behav Sci, 70*(3), 201-225.

Ruis, M. A., te Brake, J. H., Buwalda, B., De Boer, S. F., Meerlo, P., Korte, S. M., et al. (1999). Housing familiar male wildtype rats together reduces the long-term adverse behavioural and physiological effects of social defeat. *Psychoneuroendocrinology, 24*(3), 285-300.

Sapolsky, R. M. (1993). Endocrinology alfresco: psychoendocrine studies of wild baboons. *Recent Prog Horm Res, 48*, 437-468.

Sapolsky, R. M. (2005). The influence of social hierarchy on primate health. *Science, 308*(5722), 648-652.

Schweiger, U., Deuschle, M., Weber, B., Korner, A., Lammers, C. H., Schmider, J., et al. (1999). Testosterone, gonadotropin, and cortisol secretion in male patients with major depression. *Psychosom Med, 61*(3), 292-296.

Shanks, N., Griffiths, J., & Anisman, H. (1994a). Central catecholamine alterations induced by stressor exposure: analyses in recombinant inbred strains of mice. *Behav Brain Res, 63*(1), 25-33.

Shanks, N., Griffiths, J., & Anisman, H. (1994b). Norepinephrine and serotonin alterations following chronic stressor exposure: mouse strain differences. *Pharmacol Biochem Behav, 49*(1), 57-65.

Shanks, N., Griffiths, J., Zalcman, S., Zacharko, R. M., & Anisman, H. (1990). Mouse strain differences in plasma corticosterone following uncontrollable footshock. *Pharmacol Biochem Behav, 36*(3), 515-519.

Shanks, N., Zalcman, S., Zacharko, R. M., & Anisman, H. (1991). Alterations of central norepinephrine, dopamine and serotonin in several strains of mice following acute stressor exposure. *Pharmacol Biochem Behav, 38*(1), 69-75.

Sih, A., Bell, A., & Johnson, J. C. (2004). Behavioral syndromes: an ecological and evolutionary overview. *Trends Ecol Evol, 19*(7), 372-378.

Sluyter, F., Korte, S. M., Bohus, B., & Van Oortmerssen, G. A. (1996). Behavioral stress response of genetically selected aggressive and nonaggressive wild house mice in the shock-probe/defensive burying test. *Pharmacol Biochem Behav, 54*(1), 113-116.

Steimer, T., & Driscoll, P. (2003). Divergent stress responses and coping styles in psychogenetically selected Roman high-(RHA) and low-(RLA) avoidance rats: behavioural, neuroendocrine and developmental aspects. *Stress, 6*(2), 87-100.

Steimer, T., Driscoll, P., & Schulz, P. E. (1997). Brain metabolism of progesterone, coping behaviour and emotional reactivity in male rats from two psychogenetically selected lines. *J Neuroendocrinol, 9*(3), 169-175.

Sterling, S., & Edelmann, R. J. (1988). Reactions to anger and anxiety-provoking events: psychopathic and nonpsychopathic groups compared. *J Clin Psychol, 44*(2), 96-100.

Tannenbaum, B., & Anisman, H. (2003). Impact of chronic intermittent challenges in stressor-susceptible and resilient strains of mice. *Biol Psychiatry, 53*(4), 292-303.

Tidey, J. W., & Miczek, K. A. (1996). Social defeat stress selectively alters mesocorticolimbic dopamine release: an in vivo microdialysis study. *Brain Res, 721*(1-2), 140-149.

Tidey, J. W., & Miczek, K. A. (1997). Acquisition of cocaine self-administration after social stress: role of accumbens dopamine. *Psychopharmacology (Berl), 130*(3), 203-212.

Touyarot, K., Venero, C., & Sandi, C. (2004). Spatial learning impairment induced by chronic stress is related to individual differences in novelty reactivity: search for neurobiological correlates. *Psychoneuroendocrinology, 29*(2), 290-305.

Turner, R. A., King, P. R., & Remblay, P. F. (1992). Coping styles and depression among psychiatric outpatients. *Pers Individ Dif, 13*, 1145-1147.

Ursin, H. (1998). The psychology in psychoneuroendocrinology. *Psychoneuroendocrinology, 23*(6), 555-570.

Ursin, H., & Olff, M. (1995). Aggression, defense and coping in humans. *Aggressive Behav, 21*, 13-19.

Valencia-Alfonso, C. E., Feria-Velasco, A., Luquin, S., Diaz-Burke, Y., & Garcia-Estrada, J. (2004). [The effects of the social environment on the brain]. *Rev Neurol, 38*(9), 869-878.

Van Kampen, M., Kramer, K. A., Hiemke, C., Flugge, G., & Fuchs, E. (2002). The chronic psychosocial stress paradigm in male tree shrews: Evaluation of a novel animal model for depressive disorders. *Stress, 5*, 37-46.

Van Oortmerssen, G. A., & Bakker, T. C. (1981). Artificial selection for short and long attack latencies in wild Mus musculus domesticus. *Behav Genet, 11*(2), 115-126.

Veenema, A. H., Koolhaas, J. M., & De Kloet, E. R. (2004). Basal and stress-induced differences in HPA axis, 5-HT responsiveness, and hippocampal cell proliferation in two mouse lines. *Ann N Y Acad Sci, 1018*, 255-265.

Veenema, A. H., Meijer, O. C., De Kloet, E. R., & Koolhaas, J. M. (2003). Genetic selection for coping style predicts stressor susceptibility. *J Neuroendocrinol, 15*(3), 256-267.

Vegas, O., Fano, E., Brain, P., Alonso, A., & Azpiroz, A. (2006). Social stress, coping strategies and tumor development in male mice: Behavioral, neuroendocrine and immunological implications. *Psychoneuroendocrinology, 31*, 69-79.

Von Fritag, J. C., Reijmers, L. G., Van der Harst, J. E., Leus, I. E., Van den, B. R., & Spruijt, B. M. (2000). Defeat followed by individual housing results in long-term impaired reward- and cognition-related behaviours in rats. *Behavi Brain Res, 117*, 137-146.

Von Fritag, J. C., Van den Bos, R., & Spruijt, B. M. (2002). Imipramine restores the long-term impairment of appetitive behavior in socially stressed rats. *Psychopharmacology, 162*, 232-238.

Wechsler, B. (1995). Coping and coping strategies: a behavioural view. *Appl Anim Behav Sci, 43*(2), 123-134.

Zaharia, M. D., Kulczycki, J., Shanks, N., Meaney, M. J., & Anisman, H. (1996). The effects of early postnatal stimulation on Morris water-maze acquisition in adult mice: genetic and maternal factors. *Psychopharmacology (Berl), 128*(3), 227-239.

In: Psychoneuroendocrinology Research Trends
Editor: Martina T. Czerbska, pp. 411-432
ISBN: 978-1-60021-665-7

Chapter XIV

ANDROGENS, COGNITION AND SOCIAL BEHAVIOR IN CHILDREN

Aitziber Azurmendi, Aizpea Sorozabal and J.R. Sánchez-Martín*
Area of Psychobiology, Faculty of Psychology, University of the Basque Country, San Sebastian, Spain.

INTRODUCTION

We currently know that androgens act in the brain during its development, affecting both its structure and its neural function. In other words, in addition to the activating effects of adolescence, these hormones also have diverse organizational effects that, since they mold the central nervous system (CNS), influence both cognitive processes and the subject's behavior. Thus, a wide variety of data gathered over recent decades show that gonadal hormones affect the cognitive abilities of both men and women. Furthermore, recent studies carried out with normal populations instead of selected groups with hormonal abnormalities, have provided evidence that associates endogen hormone levels with behavior.

This chapter reviews existing research into both the relationship between androgen levels and diverse cognitive abilities, and the relationship between androgen levels and social behavior during childhood. First of all, it presents an overview of the principal androgens and their ontogenetic development, their main action mechanisms and their influence in the sexual differentiation of the brain and behavior. Next, it reviews the existing literature on the relationship between androgen levels and diverse cognitive abilities, paying special attention to social cognition (theory of mind, etc.) in children, before reviewing those research projects focusing on the relationship between androgens and social behavior in children, paying

* Correspondence concerning this article should be addressed to José Ramón Sánchez-Martín. Area de Psicobiología, Facultad de Psicología, Universidad del País Vasco. Av/ Tolosa, 70. 20018 San Sebastián (Spain). Phone. 0034 943 01 5729; Fax: 0034 943 015670; E-mail: joseramon.samchez@ehu.es.

special attention to behaviors of dominance and aggression. Finally, the principal conclusions are discussed from an evolutionist or ultimate causes perspective.

ANDROGENS

Androgens have diverse effects at both a physiological and behavioral level. These steroid hormones are present right from the first weeks of gestation and, in addition to playing a key role in the development of the male sex organs (both internal and external), they are also involved in the development of the CNS in both males and females, and therefore influence subjects' subsequent behavior.

The principal androgens found in the human body are testosterone, androstenedione, dehydroepiandrosterone (DHEA) and its sulfate (DHEA-S). Testosterone is the principal product of the testicles, while DHEA and DHEA-S are predominantly of adrenal origin and androstenedione is produced by both the adrenal glands and the gonads [1]. The following is a brief overview of the most important characteristics of the main androgens.

Testosterone

Testosterone has androgenic effects in the organism, being responsible for virilizing its internal structures, while one of its most active metabolites, dihydrotestosterone, is responsible for virilizing the external genitals. Another important metabolite of testosterone is estradiol. The conversion of testosterone to estradiol (the principal female hormone), a process which takes place in the CNS thanks to the enzyme aromatase, is considered important for the early development of the 'masculine brain'. Aromatase is mainly concentrated in cerebral areas related to the sexual differentiation of the CNS, such as the hypothalamus. Many sexual differences in behavior, especially in aggression, are, in all probability, the result of the different extent to which the CNS is exposed to testosterone and estradiol in the development of males and females [2].

There are three periods in the life of human males in which the plasmatic concentration of testosterone is relatively high: during embryonic development (the phase during which sexual differentiation is established), during the neonatal period and throughout the whole of the adult sexual life. During fetal development, by around 10-12 weeks gestation, testosterone levels are higher in males than in females, reaching their peak concentration between 12 and 18 weeks. This sexual difference is the result of the fact that circulating testosterone is produced by the testicles, since no evidence exists to suggest that human fetal ovaries synthesize testosterone, and fetal adrenal production in both sexes is very low. From weeks 22-28 onwards, sexual differences are not so apparent. A few days after birth, a rapid increase occurs in the testosterone levels of boys, reaching maximum levels (comparable to those observed in adolescents and approximately half of that found in adult males) between 20 and 60 days after delivery. From hereon, the levels start to drop and no sexual differences are observed again until puberty. In girls, testosterone levels at birth are similar to those found in adult women, although they begin to decrease after two weeks of life and then remain unchanged throughout childhood. During puberty, between 12 and 18 years of age, testosterone levels in boys increase by a factor of 20 to 30. In adulthood, testosterone concentrations in men are 15 times higher than in women. Finally, a gradual decrease has been observed in the levels of this hormone in men from the age of 60 onwards. In postmenopausal women, testosterone levels are also significantly lower than during their reproductive life [1].

Total testosterone levels are subject to diurnal and seasonal variations, reaching their daily peak between 06:00-08:00 h, and then dropping slowly down to 35% over the course of the day, before starting to rise again at midnight. This daily peak is tempered in old age [3]. As regards seasonal variations, it is in the months of November and December that adult men experience their highest annual peaks [4]. Furthermore, we should not forget that diverse environmental factors may also influence day-time testosterone concentrations; for example, levels may fluctuate from minutes up to hours depending on the subject's sexual activity and experiences, as well as his experiences of competitive success or failure [5].

Androstenedione

The main biological actions of androstenedione (which works together with DHEA and DHEA-S) are anabolic. This hormone stimulates erythropoiesis (the production of red corpuscles in the hemopoietic - blood producing - organs) and the development and maintenance of underarm and pubic hair. Under normal conditions, its virilizing action in the human male is weak, due to the high concentrations of testosterone present in the male body. Nevertheless, it can virilize human females if secreted in excess. Furthermore, androstenedione plays other key roles in both sexes, regulating lipids, reducing susceptibility to atherosclerosis and osteoporosis and fostering longevity [6].

During the fetal period, androstenedione levels are relatively low and there are no sexual differences. Although there is evidence that this hormone is produced by the testicles, the majority of circulating fetal androstenedione is of adrenal origin and, as in the case of testosterone, there is no evidence to suggest that fetal ovaries synthesize androstenedione. Postnatal changes in androstenedione levels are more modest than in the case of testosterone. In both sexes, levels decrease rapidly during the first week of life, but in boys they stop dropping after the first month, whereas in girls, they continue to decrease until the third month. This is the period in which sexual differences can be observed in the concentrations of this hormone. During childhood, the levels remain very low until they begin to increase with adrenarche (around the age of 8), and then continue rising gradually until they reach double or even triple their initial level during puberty. In young adults, levels stabilize and then remain the same throughout the rest of the subject's life (not including natural variations caused by the menstrual cycle) [1].

Androstenedione concentrations follow a circadian rhythm, with the highest levels being reached between 05:30 and 09:30 h and the lowest concentrations between 19:00 and 02:30 h. Furthermore, there is a concentration peak that coincides with each meal (lunch and dinner) [7].

Dehydroepiandrosterone (DHEA)

In addition to the aforementioned biological actions that it enables alongside androstenedione and DHEA-S, DHEA also influences the working of diverse systems, such as the immune system, the cardiovascular system, the endocrine system and the central nervous system [8]. We know that as well as the gonads and the adrenal glands, the brain also synthesizes DHEA and DHEA-S, causing these hormones to act as neurosteroids [9]. For example, in the CNS, DHEA stimulates axonal growth, while DHEA-S stimulates dendritic growth [2].

Human and non-human primates are the only creatures whose adrenal glands secrete large quantities of DHEA and, above all, DHEA-S, which can be converted into powerful androgens and/or estrogens in the peripheral tissues. Indeed, plasmatic levels of DHEA-S in adult men and women are between 100 and 500 times as high as testosterone levels and between 1000 and 10000 times as high as estradiol levels; this provides a large reserve pool for the conversion of androgens and/or estrogens in peripheral tissues which contain the enzymatic machinery required to transform DHEA into active sex steroids [10].

In the fetal stage, DHEA levels increase significantly during the final trimester of gestation, and do not show any sexual differences. In the neonatal stage, levels are also high (the same ranges as those found in adults) and again, no sexual differences are observed [1]. Later on, from the first month of life to the age of 5, DHEA concentrations decline. From the age of 7 in girls and 9 in boys, the levels start to increase rapidly, reaching their maximum concentration at the age of 20 in girls and between 20 and 30 in boys. In adulthood, women experience another concentration peak at the age of 40, after which time DHEA levels begin to drop in both sexes [11].

Normally, plasmatic levels of DHEA and DHEA-S follow a modest circadian rhythm. In humans, the highest concentrations are reached during the day, regardless of whether the hormones are produced by the adrenal gland or the brain [9].

ANDROGEN-SECRETING GLANDS

Suprarenal Glands

Adrenal or suprarenal glands owe their name to their anatomical location at the upper pole of each kidney [6]. In many mammalian species, including humans, each suprarenal gland consists of two embryologically and biochemically different organs: the external layer of the gland, the adrenal cortex, consisting of three different zones (glomerulosa, fasciculata and reticularis); and the inside of the gland, the adrenal medulla, which is actually a part of the autonomic nervous system [12,13].

In order to enable them to carry out their function, the suprarenal glands are well supplied with arterial blood from diverse branches of the aorta, the inferior phrenic artery and the renal arteries, thus obtaining the highest rate of blood per gram of tissue in the whole body [14].

The outermost layer of the adrenal cortex, the zona glomerulosa, is the thinnest and its cells are arranged in rounded groups. It is in this zone that mineralocorticoids (such as, for example, aldosterone) are produced. The middle layer, known as zona fasciculata, is the largest and its cells are arranged radially in parallel cords. It is in this zone that glucocorticoids (such as, for example, cortisol) are produced. The innermost later, the zona reticularis, is only slightly thinner than the fasciculata. Its most important function is the secretion of adrenal androgens (DHEA, DHEA-S and androstenedione), and, in much lesser quantities, of estrogens [6]. The zona reticularis is smaller in infants and only begins to grow between the ages of 6 and 7, its development being parallel to the increase in adrenal androgens in plasma during the prepubertal period [15].

As regards the regulation of the quantity of circulating cortical adrenal hormones, it is the adrenocorticotropic hormone (ACTH) of the pituitary gland that promotes its synthesis. Adrenal steroids have a negative feedback effect on the release of ACTH. Thus, as the level of hormones secreted by the adrenal cortex rises, the secretion of ACTH drops, thus reducing the rate of hormonal release in the adrenal cortex. When adrenal steroid levels drop, the secretory cells of the pituitary gland are freed from their inhibition and the concentration of ACTH in the blood rises. This in turn results in an increase in the release of hormones from the adrenal cortex [13].

Gonads

Testicles

The testicle is an endocrine gland that fulfils two different functions: the production and secretion of sex steroids (especially testosterone) by the Leydig cells, and spermatogenesis in the seminiferous tubules, which provides the fertilizing capacity. The testicle also secretes small quantities of estradiol, although the majority (approximately 75% to be exact) of the estradiol circulating around the male body comes from the peripheral aromatization of testosterone in diverse tissues, particularly the adipose [16].

The production and release of testosterone is regulated by a hormone of the anterior pituitary gland known as Luteinizing hormone (LH) [13]. During infancy, the pituitary gland secretes LH sporadically, causing the testicle to remain in a state of 'functional repose' practically until the onset of puberty. During this stage, due to mechanisms we still do not really understand, the biological clock is activated and as a result, the testicle begins gradually to mature, both functionally and morphologically [16].

Ovaries

The ovaries also produce mature gametes (ovules) and hormones. Given the correct conditions, i.e. the presence of the required enzymes, the progesterone produced in the ovary is transformed into androgens and estrogens [17].Note that the hormonal activities of the ovaries are more complex than those of the testicles and their production, in humans, is carried out in cycles that last around four weeks [13].

The ovary has three functional subunits: the follicles, each of which contains a developing egg or ovule; the corpora lutea, structures that evolve from the follicles when the ovule is released; and the support tissue or stroma. Around the follicular cells, although separated from them by a membrane, are the cells of the theca interna and the theca externa [12]. The stroma of the ovaries secretes androstenedione, usually in small quantities, although it can lead to virilization in diseases in which the ovarian stroma is overly developed (polycystic ovary syndrome) [17].

It is mainly the theca cells that produce the highest concentrations of androgens, particularly DHEA and androstenedione [18], which are then used for synthesizing estrogens, the principal sex steroids in females.

As with the testicles, the production of hormones in the ovaries is controlled by LH [13].

Biosynthesis of Steroid Hormones

The adrenal glands, testicles and ovaries share the same steroidogenic capacity, although each one specializes in the production of one specific group of steroids. The fact that these three pairs of glands have the same hormone synthesis mechanism is the result of their sharing the same embryonic origin (all three derive from areas near to the neural crest).

The precursor of all steroid hormones in vertebrates is cholesterol, which in turn is produced in large quantities from acetate in our livers [12]. Through a complex enzymatic process of hydroxylation, which takes place in the mitochondria, cholesterol is converted to pregnenolone. It is during this phase that the aforementioned specific regulation occurs, in the adrenals through ACTH and in the gonads through LH [17]. From hereon, both in the gonads and the suprarenal glands, specific enzymes convert pregnenolone into androgens [12].

Demolase transforms 17-hydroxypregnenolone into DHEA, which can be esterified reversibly into DHEA-S. Alternatively, the 3β-hydroxyl group of DHEA can be oxidized, thereby producing androstenedione. This product can also be obtained directly from 17-hydroxiprogesterone. The reduction of androstenedione produces testosterone [17]. Finally, the reduction of testosterone by the 5α-reductase enzyme results in dihydrotestosterone.

Action Mechanisms of Androgens

Since steroid hormones are fat soluble, they move easily through cellular membranes. However, at the same time, they also have the disadvantage of not being very water soluble, which makes their movement around the circulatory system somewhat complicated. Therefore, steroid hormones tend, in general, to bind with water soluble carrier proteins which increase their solubility and carry them through the bloodstream to the target tissues (tissues whose cells contain specific receptors for a specific hormone) [12].

Thus, steroid hormones are transported in the bloodstream in two fractions, one bound to plasmatic proteins and the other, much smaller, dissolved and free [19]. DHEA and androstenedione bind mainly with albumin, unlike testosterone and estrogens which combine with great affinity with one specific globulin, sex hormone binding globulin (SHBG). However, only free hormones are able to act and penetrate cells in order to bind to specific cytoplasmic receptors [19]. Although hormones can influence target cells in many different ways, they hardly ever change their function, altering rather the rhythm of the normal cellular function, stimulating growth and cellular [20,21].

Once inside the target tissues, the steroid hormone dissociates itself from the carrier molecule (if it was bound to it) and spreads through the cellular membrane, where it binds to the cytoplasmic receptors [12]. Next, the steroid-receptor complex is transported to the nucleus of the cell where it regulates the synthesis of specific proteins, mainly altering the transcription rate of specific genes [20].

The brain contains receptors for all kinds of steroid hormones, including androgens, and these receptors are similar to those found in any other part of the body [19]. Thus, steroids can also affect cerebral functions, both during the fetal-perinatal period and during adulthood, since the central nervous system contains many places to which steroids can bind. Of these, the most extensively studied are the classic intracellular receptors mentioned above (steroid-receptor complexes); however, more recently, a number of specific places have been described in the membrane also, where some families of steroid hormones can bind. This may be very important for explaining the extremely quick and fleeting effects induced by steroid hormones, effects which cannot be accounted for by slower genomic actions. Finally, steroid hormones and their metabolites may interact with the receptors for some neurotransmitters [21].

The cellular action mechanism of androgens frequently includes a local metabolism in the target cells. For example, in specific organs (prostate, seminal vesicles, skin, etc.) testosterone is reduced by the enzyme 5α-reductase, located in the endoplasmic reticulum, thus turning into dihydrotestosterone (DHT) and binding to the receptors. As a result, in these organs testosterone is a prohormone, and DHT is the true active component. In other organs (kidneys, muscles), the amount of DHT formed is very low, and testosterone binds directly to the receptor. In these organs, testosterone is the active hormone. In the hypothalamus, on the other hand, part of the testosterone is aromatized into estradiol by the aromatase enzyme. Thus, some authors are inclined to believe that estradiol or one of its derivatives is the active intermediary of the specific effects of testosterone in diverse cerebral structures [17].

Sexual Differentiation of the Brain and Behavior

Sexual Differentiation

The process which takes place during development that turns an individual into either a male or a female is known as sexual differentiation [22]. The chain of events which occur during differentiation is well known in placental mammals such as humans. Sexual differentiation begins with the establishment of chromosomal sex in fertilization, which is followed by the development of gonadal sex and culminates in the development of secondary sexual characteristics, commonly called masculine and feminine phenotypes [23].

Genetic Sex

During early development, both XX and XY individuals have gonads that have still not turned into testicles or ovaries, and which are simply called "non-differentiated gonads". One crucial task carried out by at least one gene of the Y chromosome is the transformation of these non-differentiated gonads into testicles. This gene of the Y chromosome is called the sex determining region of the Y gene, or SRY [22]. SRY possesses the testicular determination factor (TDF) which induces the testicular development of the non-differentiated gonad and consequently, the development of the Leydig cells, which are responsible for secreting testosterone and androstenedione around the eighth week of development [24]. If the cells of the non-differentiated gonad contain a Y chromosome with the SRY gene, testicles will develop. In the absence of the SRY gene, the gonad will develop as an ovary. Thus, genetic sex (i.e. whether an individual is XX or XY) determines gonadal sex (whether the individual will have ovaries or testicles). After this, the rest of the sexual differentiation process is not determined directly by genetic sex, but rather by the hormones secreted by the gonads [22].

Gonadal Sex

As mentioned above, the differentiation of the gonads into testicles begins in humans around the eighth week after conception; if the gonads are to become ovaries, this process begins later, around the twelfth week. Testicles are differentiated earlier than ovaries because the androgens produced by the testicles are necessary for the subsequent phases of male development. In this phase (7/8-12 weeks after conception) each embryo has two primitive duct systems: the Müllerian duct system, which may evolve into feminine reproductive structures (oviduct, uterus and upper part of the vagina) and the Wolffian duct system, which may evolve into masculine structures (epididymis, vas deferens and seminal vesicles). It is the presence or absence of testicles which determines which of these duct systems evolves. Subsequently, the testicles promote the differentiation of the external genitals in the masculine form, through the secretion of the androgen DHT. If the androgens do not act on these tissues, the external genitals will be differentiated as female [13].

Hormonal Sex

Once the testicles are formed, they begin to produce sex hormones, particularly testosterone. As for the ovaries, they mainly secrete estrogens and progesterone. The important thing here is to note that the difference between males and females does not lie in the type of sex hormones they produce, but rather in the relative proportion of the set of sex hormones produced by each sex. Women do produce testosterone (in this case in the suprarenal glands, ovaries and through peripheral conversion), but in much lower quantities than those produced by the testicles in human males. Similarly, men also produce estrogens and progesterone, but only in very small amounts [24].

Therefore, hormonal sex is determined by the quantity or level of hormones produced in the gonads. If there is enough testosterone, then masculine differentiation will occur, otherwise, the embryo will follow the female course regardless of chromosomal or gonadal sex [24]. Even if a fetus is female, and therefore does not possess testicles, its genitals may assume a masculine form under the influence of androgens in certain clinical conditions such as congenital adrenal hyperplasia: the adrenal cortex normally produces small amounts of androgens, but a hyperactive adrenal or maternal gland may produce sufficient androgens during the critical period of fetal development to cause the genitals to adopt a masculine form in genetically female subjects [13].

Effects of Organization and Activation

The sexual differentiation of the nervous system and, consequently, of behavior, is guided by the same steroids that cause the sexual differentiation of the body [22]. Thus, in the development and differentiation of the brain, a series of extremely long and complex events occur, beginning during gestation and continuing, at least in some species, during the early postnatal period [25].

The action of steroid hormones in the central nervous system can be classified as organizational and activating. Organizational effects refer to the steroids' ability to sculpt the structure of the central nervous system during development. Structural organization is permanent, remaining long after the period of development during which it is exposed to the hormone [26]. It is believed that this occurs because the hormones direct basic neural development processes, thus determining which nervous cells live and which die, with which other cells they should connect automatically and which neurotransmitters they should use [27]. Activating effects, on the other hand, refer to the steroids' ability to modify the activity of target cells so as to facilitate behavior in a specific social context. Activating effects are temporary, they come and go according to the presence or absence of the hormones in question and are typically associated with the action of steroids during adulthood [26].

We know that androgens act on the developing brain, producing sexual differences in both the neural structure and function. All the cells of the sexually dimorphic regions of the brain are rich in androgen receptors, and their development is mainly affected by testosterone, both during early fetal life and later on [28].

Many sexual differences become more evident after puberty as the result of the activation effects of steroid hormones [26]. However, as mentioned above, the organizational influence is already present during early fetal development, organizing the nervous system in such a way that the processing of information for specific aspects with adaptive implications is different for each sex. Furthermore, due to the early organizational effects of hormones, major sexual differences may also occur in an individual's sensitivity to sex steroids from a very early age. In fact, evidence exists to suggest that individuals differ not only with regard to their levels of circulating hormones, but also with regard to their sensitivity to said levels [29]. According to Baron-Cohen et al. [28],there seems to be a very close relationship between sexual differences in the brain and androgen levels, since the sexually dimorphic regions of the brain (amygdala, corpus callosum, etc.) contain numerous androgen receptors and their development seems to be influenced by androgens during the fetal stage and subsequent development phases.

For example, studies carried out with clinical populations in humans have found that prenatal exposure to atypically high levels of androgens, as in the case of women suffering from congenital adrenal hyperplasia (CAH), masculinizes both behavior and cognitive skills

[30]. Moreover, studies carried out with healthy populations have found that inter-individual hormonal variations in prenatal levels of androgens are associated with subsequent behavior typical of the specific sex [28].

As we will see below, the hormones responsible for the sexual differentiation of the organism are also involved in the sexual dimorphism of behavior and diverse cognitive abilities, since because they have an organizational effect on the CNS, they help mold both cognition and adult behavior.

RELATION OF ANDROGENS TO PSYCHOLOGICAL PROCESSES

Although there is no difference between human males and females as regards general intelligence, sexual differences have been observed in relation to specific cognitive tasks. For example, men perform better in mental rotation and spatial navigation tests, while women score higher in emotional recognition, social sensitivity and verbal fluency tests. Moreover, girls begin to speak before boys [28]. A number of different biological and socio-cultural factors have been used to explain these differences; one such biological factor proposed is related to the role played by hormones in early development [31].

Based on the studies carried out in other species, it is reasonable to assume that the presence of high levels of masculinizing hormones (androgens) during critical periods of cerebral development cause masculinization or defeminization in cognitive development, while low levels of the same hormones result in demasculinization or feminization. Thus, it has been hypothesized that hormone levels atypical of the sex to which the individual belongs are associated with cognitive patterns also atypical of that sex. In this model of sexual differences in cognition, high levels of androgens (as observed in typical males and females with an atypical exposure to androgens) would be associated with high scores in spatial ability, and relatively low scores in verbal fluidity, perceptual rapidity and memory; on the other hand, low levels of androgens (such as those found in typical women and men with reduced levels of androgens) would be associated with relatively poor spatial abilities and high scores in verbal fluidity, perceptual rapidity and memory.

Studies have also been carried out on the possible activating influences of hormones on human cognition. Many of these works suggest a non-linear (inverted U-shaped) relationship between circulating testosterone levels and various cognitive abilities, mainly visuospatial abilities in adult men. These studies suggest the existence of an optimum level of testosterone for the correct performance of cognitive tests. Thus, an increase in testosterone would not necessarily mean greater spatial skills, in fact, just the opposite [32]. In a study carried out by Gouchie and Kimura [33], in which women were included in the study sample, the authors found that men with low levels of T performed better in spatial tests and mathematics than those with high levels of T. However, women with high levels of T performed better at said tests than men with low levels. It is also worth noting that a number of other studies failed to find any association between androgens and the cognitive function [34]. Nevertheless, the idea that hormones have an activating influence on the human cognitive function is widely accepted [35].; although the precise nature of this influence is not yet clear [36].

In studies carried out with elderly individuals who were administered exogenous hormones for different reasons, replacement testosterone was observed to have diverse beneficial effects on a number of cognitive abilities, including spatial cognition [37] and the visual working memory [38]. Studies taking into account endogenous testosterone levels found a positive relationship between verbal memory and testosterone in older women, and a negative one between verbal fluidity and testosterone in older men [39]. Some studies have also analyzed the effects of administering DHEA on cognition, although results are inconsistent [40,41]. Hirshman et al. [42] suggest that the influence of DHEA on diverse cognitive abilities may be mediated by its estrogenic and androgenic metabolites. A study carried out with postmenopausal women found that those taking replacement DHEA had varying levels of circulating estrogens and androgens (thus confirming previous studies regarding the rapid metabolism of DHEA); it was therefore impossible for the authors to observe the direct effect of DHEA on diverse cognitive tests, since its estrogenic and androgenic metabolites also influenced subjects' performance.

The idea that emerges from the data provided by diverse sources is that prenatal hormone levels affect the configuration of cognition differently depending on the sex of the subject (for example, better visuospatial abilities in boys and better verbal skills in girls), and that the normal daily and monthly fluctuations in the levels of these hormones in healthy human adults correlate with subjects' performance in diverse cognitive tests [43].

Nevertheless, could androgens influence those cognitive abilities for which no sexual differences have been observed? It seems that the answer is yes, at least in the case of fluid intelligence. This kind of intelligence is related to a person's adaptability and flexibility in coping with unexpected situations during problem solving, and is typically assessed through tasks containing matrixes, series and classifications, etc. Tan and Tan [44] found that very low and very high levels of testosterone may be disadvantageous for fluid intelligence in women, while in men, very low or medium levels of testosterone have a positive relationship with fluid intelligence, and very high levels of testosterone have a negative relationship with this type of cognition.

The relationship between androgens and cognitive abilities in prepubertal children has not been researched in any great depth, given that, as we have seen, the majority of studies have focused on young people and adults. We can only highlight three studies in this field. The first was carried out by Jacklin et al. [45], who analyzed the relationship between perinatal testosterone levels (obtained from the umbilical cord) and performance in spatial tasks at the age of 6. The authors found that girls with higher levels of testosterone scored lower in said tests, while no relationship was observed for boys. In this sense, it is worth highlighting the fact that hormone levels obtained from the umbilical cord do not necessarily reflect the levels to which the developing fetus has been exposed. The second study, carried out by Finegan et al. [46], found a negative non-linear relationship between prenatal testosterone levels (obtained from the amniotic fluid) and performance in a spatial visualization test in 4-year-old girls, while no relationship was observed for boys. Here also we should point out that, as in the previous case, hormone levels taken from the amniotic fluid may not reflect the levels that have influenced the fetus. Finally, Grimshaw et al. [47] observed a positive correlation between testosterone levels (obtained from the amniotic fluid) and the mental rotation rate of 7-year-old girls, while these correlations were not significant for boys, although they did suggest a positive relationship.

Another very interesting aspect of cognition in children, that is also relevant from the point of view of subjects' social adaptation, is that related to theory of mind skills. Between the ages of 3 and 5, children acquire theory of mind, i.e. the ability to understand others' mental states, including beliefs, emotions, desires, etc. [48,49]. Over recent years, a wide variety of tests have been developed to assess different traits of theory of mind. Thus, various authors have found sexual differences in relation to some of these abilities, such as the understanding of emotions, for example, in which girls tend to score better than boys [50,51]. Geary [52] suggests that endocrine correlates may underlie these sexual differences, although this has not received any experimental attention to date. Nevertheless, it is worth noting that other researchers have failed to discover any sexual differences in the performance of diverse theory of mind tests [53,54].

Our research team recently analyzed the relationship between circulating levels of androgens (testosterone, DHEA and androstenedione) and performance in diverse cognitive tests in 5-year-old children [55].

As seen earlier, in prepubertal children, adrenal androgens such as DHEA, DHEA-S and androstenedione play a very important role (as much as or more than testosterone). Given the technological progress of recent years, which enables us to analyze androgens in saliva samples gathered in a non-invasive manner, we decided to measure also testosterone levels, DHEA levels and androstenedione levels (as far as we know, no kits are available which measure DHEA-S in saliva, since this molecule is too large to enter the oral cavity and is therefore not found in saliva, unless, of course, there is blood contamination).

In our opinion, another very important aspect of our study is that we selected a sample of 129 healthy normal children, 60 boys and 69 girls. As we know, many studies in this field have focused on subjects who have developed in atypical hormonal environments. Without wishing to detract from the importance of all these studies, which have provided much useful information, it is nevertheless necessary to carry out studies with healthy subjects in order to see how normal inter-individual variations in hormone levels are related to diverse cognitive abilities.

The data for this study were gathered over the course of two months, during March and April 2004. Two saliva samples were obtained (one each month) for subsequent analysis of hormone levels. The levels of salivary androgens were measured in duplicate using an enzyme immunoassay kit. The two values for each hormone were averaged, as they presented a positive correlation, with the result serving as a base line for each androgen in each subject. This base line was then used in the subsequent statistical analyses.

Cognitive abilities were assessed over a series of sessions deliberately kept short in order to avoid tiring the subjects, thus maintaining their motivation to perform. Cognitive tests were administered to all subjects between 09:00 and 10:00 h in a room adjacent to the classroom in each of the schools. The Kaufman intelligence test or K-BIT was used to measure intelligence. This test provides three IQ measures, assessing fluid intelligence through the Matrices subtest and crystallized intelligence through the Vocabulary subtest, as well as giving an IQ composite or global intelligence measure, which is the combination of the two previous subtests. Four further tests were used to assess theory of mind or subjects' understanding of others' mental states (beliefs, emotions, intentions, etc.): a) False belief, which measures whether or not the child is capable of attributing to others beliefs that are different from their own; b) Emotional labeling, which assesses the child's ability to recognize emotions; c) Appearance-reality distinction, which assesses whether or not the child understands the critical distinction between the appearance of an emotion and the reality of that same emotion; and finally, d) Rules of expression, which assesses whether or not the child comprehends the coexistence of contradictory feelings towards another person in the same situation.

As regards the results, despite the fact that no sexual differences were found for any of the cognitive abilities assessed or in the levels of testosterone and androstenedione (in the case of DHEA, girls' levels were significantly higher than boys'), the correlation analysis revealed a positive relationship between fluid intelligence and testosterone levels in boys, a negative relationship between crystallized intelligence and androstenedione levels in girls and a negative relationship between emotional labeling and androstenedione levels in boys. Furthermore, a multiple regression analysis indicated that androstenedione was one of the best predictors for some of the cognitive abilities assessed; in specific terms, it predicted negatively both emotional labeling in boys and global intelligence for the whole sample group (both sexes combined).

It is worth highlighting that almost all the relationships observed in our study between hormone measures and cognitive performance were negative. In other words, the higher the androgen levels, the poorer the cognitive performance (except in the case of fluid intelligence in boys).

Relationship between Androgens and Social Behavior

Over recent decades, much progress has been made in the study of the relationship between hormones and behavior. Recent studies focusing on normal populations instead of selected groups with hormonal abnormalities have established a relationship between the circulating levels of various androgens (mainly testosterone) and social behavior in humans.

Although the specific mechanisms through which hormones affect behavior are still unknown, Susman and Ponirakis [56] offer a detailed description of the four main models that have been developed in order to try and explain hormone-behavior interactions. The first model proposes that hormones have a causal influence on behavior, and is based on research carried out with animal models. These studies support the organizational and activating effects of hormones on behavior. The second model postulates that it is behavior which has a causal influence on hormone concentrations. According to this model, the environment in which an organism develops affects both its structures and biological functions. Many studies have observed hormone changes resulting from diverse behaviors and experiences. The third model focuses on hormone-behavior interactions and assumes that any changes in one dominion will influence changes in the other. Raine et al. [57] call this sequential biosocial effect. Finally, the fourth model takes context into account also. According to this model, the role of contextual factors, such as psychological and social factors, influence both behavior and hormone levels. In other words, such factors may moderate or mediate hormone-behavior interactions.

In addition to sexual behavior, there are also three main types of social behavior for which relationships have been established with androgen levels: aggression [58], dominance [59] and prosocial behavior [60]. The following is a summary of the different models or hypotheses which have been developed to explain the relationship between androgens and the display of aggression or dominance behaviors in humans. The reader should have no trouble identifying similarities with the models described above. There are three main models or hypotheses:

1. *Mazur and Booth's* [58] *biosocial hypothesis of status*. This model postulates a bidirectional relationship between testosterone levels and dominance. Thus, in the same way that high testosterone levels can encourage behavior aimed at dominating others, an increase in the experience of dominance can also increase testosterone levels.
2. *The challenge hypothesis*. This hypothesis suggests that aggression and testosterone correlate in moments of social instability or when an individual is challenged by a conspecific [61]. In the case of humans, evidence obtained to date indicates a low (but inconsistent) correlation between aggression and testosterone levels, and a higher and more consistent association between dominance and testosterone levels [62].
3. *The multivariate association model between dominance and testosterone levels*. Nyborg [63,64] proposed a multivariate theoretical model of association between testosterone and dominance (the general trait covariance model), which integrates testosterone, dominance and intelligence. According to this author, individuals with a high IQ and low levels of testosterone may have high levels of formal dominance and enjoy a high status in areas in which their analytical abilities combine favorably with sensitivity. For their part, individuals with a high IQ and high levels of testosterone may also possess formal dominance and enjoy a high status, although in areas in which a combination of high intelligence and a certain degree of aggression and insensitivity are valued.

An association has also been found between androgen levels and affiliation and prosocial behavior, although the existing literature is a lot more scarce than in the case of aggression and dominance. In specific terms, a negative correlation has been found between testosterone and prosocial personality in adult men and women [60]. Furthermore, a negative correlation has also been found between testosterone levels and the emission of smiles and kindness/pleasantness [65,66].

The majority of studies on the relationship between social behavior and androgen levels have focused on pubescent or post-pubescent subjects (especially males), and the few that were carried out with prepubertal individuals tend to focus on the relationship between hormones and aggression and between hormones and dominance (for a review on hormones and aggression in childhood and adolescence, see [67]).

Not enough studies have focused on the question of whether or not prenatal androgen levels influence human aggression, i.e. whether they have organizational effects [68]. Therefore, there is no reliable evidence regarding the androgen-aggression relationship in children in this respect.

As regards the possible relationship between circulating androgens and aggressive behavior, only a very few studies have focused on this question in preschool children. Of these, the majority have focused on testosterone. Constantino et al. [69] failed to find any relationship between testosterone levels and aggression in children aged between 4 and 10. Van Goozen et al. [70] also failed to find any significant relationships between testosterone levels and aggression and delinquency scores in the CBCL test. However, Sánchez-Martín et al. [71] found a significant positive correlation between testosterone levels and the aggression observed in social interactions in healthy normal four-year-old boys. The results obtained in the study carried out by Chance et al. [72] do nothing to clarify things: the authors found a positive association between testosterone and aggression in children aged between 9 and 11, but not in children aged between 5 and 8.

Even more scarce are studies which focus on the relationship between adrenal androgens and aggressive behavior in prepubertal children. Van Goozen et al. [70] found that DHEA-S levels in boys with conduct disorder were significantly higher than in the control group and were closely related to aggression and delinquency scores. In another study carried out by the same research group (Van Goozen et al., 2000b) the authors also found that boys with oppositional-defiant disorder had significantly higher DHEA-S levels than normal boys.

In light of these data, it is not unreasonable to assume that adrenal androgens, which are characteristic of childhood, contribute to initiating and maintaining human aggression [71,74,75]. It is important to remember that androgens are at their lowest levels during the preschool period [1], and the majority of circulating androgens in prepubertal children are produced by the adrenal glands. We therefore believe that it is very important to take adrenal androgens into account when studying the development of aggression in children.

Recently, our research team explored the potential relationship between social behavior (aggression, dominance and affiliation) and androgen levels (testosterone, DHEA and androstenedione). We also aimed to analyze whether or not intelligence played a moderating or mediating role in said potential relationship [76].

Given that there is evidence to suggest that the data obtained from behavioral observation enable researchers to establish a more consistent relationship between androgen levels and aggressive and dominance behavior than data obtained through self-reports [77], we decided to study children's social behavior through the systematic observation of their behavior during interaction with their peers in contexts of free play.

Subjects' social interactions with their peers were recorded on video daily throughout an entire school year, in the context of free play in the school playground. We obtained 30 minutes of recorded behavior per child, of which 15 minutes were assessed using the behavioral analysis software Observer 4.1 (Noldus).

The behavioral patterns considered were grouped into 3 main categories: aggression, governance and affiliation, with each category containing behaviors with the same functional area. Within the categories selected, we opted for more compact variables. To this end, we calculated the principal components of the behavioral traits, and the solutions obtained were as follows: victimization, offensiveness and provocation (for the Aggression behavioral category); subordination and dominance (for the Governance category); and linking, prosociality and affectivity (for the Affiliation category). Given that in the section on the relationship between androgens and cognitive abilities we present part of our work, we will not repeat here how the hormonal and cognitive data were collected.

Despite the fact that there were no sexual differences in the social behavior displayed by the subjects of our sample, we did find androgen-behavioral relationships specific to each sex. The correlation analysis showed that, in boys, androstenedione levels are associated negatively with victimization and positively with provocation; while in girls, testosterone levels are related positively with affectivity.

Next, we carried out various regression analyses in order to see whether or not intelligence acted as a moderator in those hormone-behavior relationships which were found to have significant effects in the correlation analyses. One regression analysis was carried out for each social behavior category (Provocation, Victimization and Affectivity), with one behavioral factor being introduced into each regression as a dependent variable, and one hormone, IQ and the interaction between the two as predictors. We found that, in the case of boys, the androstenedione-IQ interaction negatively predicted provocation; while for girls, the testosterone-IQ interaction positively predicted affectivity. Subsequently, in those cases in which the interaction of the variables assessed was significant, the association between hormones and behavioral factors was analyzed through a regression analysis for both levels of intelligence, high and low. Thus, we found that it was in boys with a high IQ that androstenedione predicted provocative behavior, and it was in girls with a low IQ that testosterone predicted affective behavior.

CONCLUSION

The existence of sexual differences in diverse spheres of human behavior is a fact supported by a large body of evidence [78]. From a phylogenetic or ultimate causes perspective, the existence of differentiated reproductive strategies in males and females in our species is associated, among other things, with sexual dimorphism in the behavior favored by sexual selection [79]. Evolutionist Psychology has also proposed that diverse evolutionary-based information processing mechanisms (such as, for example, those related to theory of mind and spatial abilities, etc.), have been molded by sexual selection [80].

From the perspective of proximate causes, the hormones known as sex hormones (a group which includes androgens) are among the principal elements responsible, at a biological level, for these different cognitive and behavioral manifestations [31]. These hormonal influences are even more evident both in those stages in which their organizational effects are more pronounced, and in those in which circulating levels in males and females are markedly different.

The results obtained from our research support the idea that, as suggested by other authors, even in stages in which hormone levels (in this case androgen levels) are low and present hardly any sexual differences, a relationship can be observed between hormones and cognition and hormones and behavior, precisely in those aspects of cognition and behavior in which, later on in the life cycle, sexual differences can be observed between individuals.

Firstly, it is likely that early organizational effects result in each sex being particularly sensitive to specific hormones (or at least to certain levels of such hormones). And secondly, as Archer [67] points out, it is probable that hormones (in this case androgens) are associated with the lifetime trajectories of specific subjects (for example, certain levels of androgens will be associated with specific levels of cognitive abilities and/or aggressive behaviors).

Finally, the biopsychosocial approach [82,83] suggests that the hormone-behavior relationship is moderated by other factors also, both those that are strictly psychological in nature (cognitive, emotional, etc.) and social ones (family ecology, peers, etc.). Our work supports the moderating effect of intelligence on the relationship between hormones and behavior (both aggressive and affective) even during very early stages of development.

REFERENCES

[1] Forest, M. G., (1989). Physiological changes in circulating androgens. *Pediatr Adolesc Endocrinol, 19*, 104-129.

[2] Pfaff, D. W., Phillips, M. I. y Rubin, R. T., (2004). *Principles of hormone/behavior relations*. Elsevier Academic Press, London.

[3] Mooradian, A. D., Morley, J. E., y Korenman, S. G., (1987). Biological actions of androgens. *Endocrine Reviews, 8(1)*, 1-28.

[4] Dabbs, J. M., (1990). Age and seasonal variation in serum testosterone concentration among men. *Chronobiol Int, 7(3)*, 245-249.

[5] Dabbs, J. M., (1990). Salivary testosterone measurements: Reliability across hours, days, and weeks. *Physiology and Behavior, 48*, 83-86.

[6] Jara, A. y González, G., (2001). *Fisiopatología de la corteza suprarrenal.* In: A. Jara (Ed.), Endocrinología. Editorial Médica Panamericana, Madrid.

[7] Goldman, J., Wajchenberg, B. L., Liberman, B., Nery, M., Achando, S. y Germek, O. A., (1985). Contrast analysis for the evaluation of the circadian rhythms of plasma cortisol, androstenedione, and testosterone in normal men and the possible influence of meals. *Journal of Clinical Endocrinology and Metabolism, 60,* 164-167.

[8] Granger, D. A., Schwartz, E. B., Booth, A., Curran, M. y Zakaria, D., (1999). Assesing dehydroepiandrosterone in saliva: A simple radioimmunoassay for use in studies of children, adolescents and adults. *Psychoneuroendocrinology, 24*, 567-579.

[9] Majewska, M. D., (1995). Neuronal actions of dehydroepiandrosterone. Possible roles in brain development, aging, memory, and affect. *Annals New York Academy of Sciences, 774*, 111-120.

[10] Labrie, F., Luu-The, V., Bélanger, A., Lin, S. X., Simard, J., Pelletier, G. y Labrie, C., (2005). Is dehydroepiandrosterone a hormone? *Journal of Endocrinology, 187*, 169-196.

[11] Kroboth, P. D., Salek, F. S., Pittenger, A. L., Fabian, T. J. y Frye, R. F., (1999). DHEA and DHEA-S: A Review. *Journal of Clinical Pharmacology, 39*, 327-348.

[12] Nelson, R. J., (1996). *Psicoendocrinología*. Ariel Psicología, Barcelona.

[13] Rosenweig, M. R. y Arnold, L., (1994). *Psicología fisiológica*. McGraw-Hill, Madrid.

[14] Kaplan, N. M., (1992). The adrenal glands. In: J. E. Griffin y S. R. Ojeda (Eds.), *Textbook of endocrine physiology*. Oxford University Press, New York.
[15] Bertrand, J. y Saez, M., (1992). Adrenal steroids: Biochemistry and physiology. In: J. Bertrand, R. Rappaport, and P. C. Sizonenko (Eds.), *Pediatric Endocrinology*. Williams and Wilkins, Baltimore.
[16] Miralles, J. M., (2001). Testículo. Exploración. Fisiopatología. In: A. Jara (Ed.), *Endocrinología*. Editorial Médica Panamericana, Madrid.
[17] Milgrom, E., (1990). Steroid Hormones. In: E. Baulieu and P. A. Kelly (Eds.), *Hormones*. Chapman and Hall, London.
[18] Chiva, L. y Chiva de Agustín, L., (2001). *Ovario. Fisiopatología. Exploración*. In: A. Jara (Ed.), Endocrinología. Editorial Médica Panamericana, Madrid.
[19] Donovan, B. T., (1985). *Hormones and human behavior*. Cambridge University Press, New York.
[20] Mendelson, C. R., (1992). Mechanisms of hormone action. In: J. E. Griffin y S. R. Ojeda (Eds.), *Textbook of endocrine physiology*. Oxford University Press, New York.
[21] Negri-Cesi, P., Poletti, A., Martini, L. y Piva, F., (1999). Steroid metabolism in the brain: Role in sexual differentiation. In: A. Matsumoto (Ed.), *Sexual differentiation of the brain*. CRC Press, Boca Raton, Florida.
[22] Breedlove, S. M. y Hampson, E., (2002). Sexual differentiation of the brain and behavior. In: J. B. Becker, S. M. Breedlove, D. Crews, and M. M. McCarthy (Eds.), *Behavioral Endocrinology*. MIT press, London.
[23] George, F. W., (1992). Sexual differentiation. In: *Textbook of Endocrine Physiology* (Ed.), Griffin, J.E.
[24] Colom, R. y Zaro, M. J., (2004). *La psicología de las diferencias de sexo*. Biblioteca Nueva, Madrid.
[25] Cooke, B., Hegstrom, D., Villeneuve, L. S. y Breedlove, S. M., (1998). Sexual differentiation of the vertebrate brain: Principles and mechanisms. *Frontiers in Neuroendocrinology, 19*, 323-362.
[26] Sisk, C. L. y Zehr, J. L., (2005). Pubertal hormones organize the adolescent brain and behavior. *Frontiers Neuroendocrinology, 26*, 163-174.
[27] Hines, M., (1999). Gonadal hormones and sexual differentiation of human behavior: Effects on psychosexual and cognitive development. In: A. Matsumoto (Ed.), *Sexual differentiation of the brain*. CRC Press, Boca Raton, Florida.
[28] Baron-Cohen, S., Knickmeyer, R. C. y Belmonte, M. K., (2005). Sex differences in the brain: Implications for explaining autism. *Science, 310(4)*, 819-823.
[29] Cohen-Bendahan, C.C., van den Beek, C. y Berenbaum, S. A., (2005). Prenatal sex hormona effects on child and adult sex-typed behavior: Methods and findings. *Neuroscience and Biobehavioral Reviews, 29*, 353-384.
[30] Collaer, M. L. y Hines, M., (1995). Human behavioral sex differences: A role for gonadal hormones during early development. *Psychological Bulletin, 118(1)*, 55-107.
[31] Berenbaun, S. A., Korman, K. y Leveroni, C., (1995). Early hormones and sex differences in cognitive abilities. *Learning and Individual Differences, 7(4)*, 303-321.
[32] Christensen, K. y Knussman, R., (1987). Sex hormones and cognitive functioning in men. *Neuropsychobiology,* 18, 27-36.

[33] Gouchie, C. y Kimura, D., (1991). The relationship between testosterone levels and cognitive ability patterns. *Psychoneuroendocrinology, 16(4)*, 323-334.
[34] McKeever, W. F., Rich, D. A., Deyo, R. A. y Conner, R. L., (1987). Androgens and spatial ability: Failure to find a relationship between testosterone and ability measures. *Bulletin of the Psychonomic Society, 25*, 438-440.
[35] Kimura, D. y Hampson, E., (1994). Cognitive pattern in men and women is influenced by fluctuations in sex hormones. *Current Directions in Psychological Science, 3*, 57-61.
[36] Hines, M., (2004). *Brain gender*. Oxford University Press, New York.
[37] Janowsky, J. S., Oviatt, S. K. y Orwoll, E. S., (1994). Testosterone influences spatial cognition in older men. *Behavioral Neuroscience, 108(2)*, 325-332.
[38] Janowsky, J. S., Chavez, B., y Orwoll, E., (2000). Sex steroids modify working memory. *Journal of Cognitive Neuroscience, 12*, 407-414.
[39] Wolf, O. T. y Kirschbaum, C., (2002). Endogenous estradiol and testosterone levels are associated with cognitive performance in older women and men. *Hormones and Behavior, 41*, 259-266.
[40] Wolf, O. T., Naumann, O., Hellhammer, D., Geiben, A. C., Strasburger, C. J., Dressendörfer, R. A., Pirke, K. M. y Kirschbaum, C., (1997). Effects of a two-week physiological dehydroepiandrosterone substitution on cognitive performance and well-being in healthy elderly women and men. *Journal of Clinical Endocrinology and Metabolism, 82(7)*, 2363-2367.
[41] Wolf, O. T., Naumann, O., Hellhammer, D. y Kirschbaum, C., (1998). Effects of dehydroepiandrosterone replacement in elderly men on event-related potential, memory and well-being. *Journal of Gerontology: Medical Sciences, 53*, M385-M390.
[42] Hirshman, E., Merrit, P., Wang, C., Wierman, M., Budescu, D., Kohrt, W., Templin, J. y Bhasin, S., (2004). Evidence than androgenic and estrogenic metabolites contribute to the effects of dehydroepiandrosterone on cognition in postmenopausal women. *Hormones and Behavior, 45*, 144-155.
[43] Halpern D. F. y LaMay, M. L., (2000). The smarter sex: A critical review of sex differences in intelligence. *Educational Psychology Review, 12(2)*, 229-246.
[44] Tan, Ü. y Tan, M., 1998. Curvilinear correlation between total testosterone levels and fluid intelligence in men and women. *International Journal of Neuroscience, 95*, 77-83.
[45] Jacklin, C. N., Wilcox, K. T. y Maccoby, E. E., (1988). Neonatal sex-steroid hormones and cognitive abilities at six years. *Developmental Psychology, 21*, 567-574.
[46] Finegan, J. K., Niccols, G. A. y Sitarenios, G., (1992). Relations between prenatal testosterone levels and cognitive abilities at 4 years. *Developmental Psychology, 28*, 1075-1089.
[47] Grimshaw, G. M., Sitarenios, G. y Finegan, J. K., (1995). Mental rotation at 7 years: Relations with prenatal testosterone levels and spatial play experiences. *Brain and Cognition, 29*, 85-100.
[48] Astington, J., (1993). *The child's discovery of the mind.* Harvard University Press, Cambridge, MA.
[49] Wellman, H., (1990). *The child's theory of mind.* MIT Press, Cambridge, MA.
[50] Brown, J. y Dunn, J., (1996). Continuities in emotion understanding from three to six years. *Child Development, 67*, 789-802.

[51] Dunn, Y., Brown, J., Slomkowski, C., Tesla, C. y Youngblade, L., (1991). Young children's understanding of other people's feelings and beliefs: Individual differences and their antecedents. *Child Development, 62*, 1352-1366.

[52] Geary, D., (1999). *Male, female. The evolution of human sex differences*. American Psychological Association, Washington, DC.

[53] Roazzi, A. y Santana, S., (1999). Theory of mind: the effect of age, sex, and the use of animate and inanimate "actors" in inferring mental states. *Psicologia: Reflexao e Critica, 12(2)*, 307-330.

[54] Simpson, D. L., (2003). The role of culture and gender in children's development of display rules and theory of mind. *Dissertation Abstracts International: Section B: The Sciences and Engineering, 63(7-B)*, 3498.

[55] Azurmendi, A., Braza, F., Sorozabal, A., García, A., Braza, P., Carreras, M. R., Muñoz, J. M., Cardas, J. y Sánchez-Martín, J. R., (2005). Cognitive abilities, androgen levels, and body mass index in 5-year-old children. *Hormones and Behavior, 48*, 187-195.

[56] Susman, E. J. and Ponirakis, E., (1997). *Hormones-context interactions and antisocial behavior in youth*. In: Raine, A., Brennan, P.A., Farrington, D.P., and Mednick, S.A.

[57] Raine, A., Brennan, P., and Farrington, D.P, (1997). *Biosocial bases of violence: Conceptual and theoretical issues*. In: Raine, A., Brennan, P.A., Farrington, D.P., and Mednick, S.A.

[58] Archer, J., (1991). The influence of testosterone on human aggression. *British Journal of Psychology, 82*, 1-28.

[59] Mazur, A. y Booth, A., (1998). Testosterone and dominance in men. *Behavioral and Brain Sciences,* 21, 355-397.

[60] Harris, J. A., Rushton, J. P., Hampson, E. y Jackson, D. N., (1996). Salivary testosterone and self-report aggressive and pro-social personality characteristics in men and women. *Aggressive Behavior, 22*, 321-331.

[61] Wingfield, J. C., (2005). A continuing saga: The role of testosterone in aggression. *Hormones and Behavior, 48*, 253-255.

[62] Archer, J., (2006). Testosterone and human aggression: an evaluation of the challenge hypothesis. *Neuroscience and Biobehavioral Reviews, 30*, 319-345.

[63] Nyborg, H., (1994). *Hormones, sex, and society: The science of physicology*. Praeger, Westport, CT.

[64] Nyborg, H., (2004). Multivariate modelling of testosterone-dominance associations. *Behavioral and Brain Sciences, 27(1)*, 155-159.

[65] Dabbs, J. M., Hargrove, M. F. y Heusel, C., (1996). Testosterone differences among college fraternities. Well-behaved vs. rambunctious. *Personality and Individual Differences, 20(2)*, 157-161.

[66] Dabbs, J. M., (1997). Testosterone, smiling, and facial appearance. *Journal of Nonverbal, Behavior 21*, 45-55.

[67] Ramirez, J., (2003). Hormones and aggression in childhood and adolescence. *Aggression and Violent Behavior*, Nov-Dec 8(6), 621-644.

[68] Van Goozen, S., (2005). Hormones and the developmental origins of aggression. In Tremblay, R.E., Hartup, W.W., and Archer, J. (Eds.) *Developmental origins of aggression*. Guilford Publications, New York.

[69] Constantino, J. N., Grosz, D., Saenger, P., Chandler, D. W., Nandi, R. y Earls, F. J., (1993). Testosterone and aggression in children. *Journal of The American Academy of Child and Adolescent Psychiatry, 32*, 1217-1222.

[70] Van Goozen, S. H. M., Matthys, W., Cohen-Kettenis, P., Thijssen, J. H. H. y van Engeland, H., (1998). Adrenal androgens and aggression in conduct disorder prepuberal boys and normal controls. *Biological Psychiatry, 43(2)*, 156-158.

[71] Sánchez-Martín, J. R., Fano, E., Ahedo, L., Cardas, J., Brain, P. F. y Azpíroz, A., (2000). Relating testosterone levels and free play social behavior in male and female preschool children. *Psychoneuroendocrinology, 25*, 773-783.

[72] Chance, S. E., Brown, R. T., Dabbs, J. M. y Casey, R., (2000). Testosterone, intelligence and behavior disorders in young boys. *Personality and Individual Differences, 28,* 437-445.

[73] Van Goozen, S., Van den Ban, E., Matthys, W., Cohen-Kettenis, P.T., Thijssen, J. and Van Engeland, H., (2000). Increased adrenal androgen functioning in children with oppositional defiant disorder: A comparison with psychiatric and normal controls. *Journal of the American Academy of Child and Adolescent Psychiatry, 39(11)*, 1446-1451.

[74] Chance, S.E., Brown, R.T., Dabbs, J.M., and Casey, R., (2000). Testosterone, intelligence and behavior disorders in young boys. *Personality and Individual Differences*, *28*, 437-445.

[75] Scerbo, A. y Kolko, D. J., (1994). Salivary testosterone and cortisol in disruptive children: Relationship to aggressive, hyperactive, and internalizing behaviors. *Journal of The American Academy of Child and Adolescent Psychiatry, 33(8)*, 1174-1184.

[76] Azurmendi, A., Braza, F., García, A., Braza, P., Muñoz, J. M. y Sánchez-Martín, J. R., (2006). Aggression, dominance, and affiliation: their relationships with androgen levels and intelligence in 5-year-old children. *Hormones and Behavior, 50(1)*, 132-140.

[77] Archer, J., Graham-Kevan, N., y Davies, M., (2005). Testosterone and aggression: A reanalysis of Book, Starzyk and Quinsey´s (2001) study. *Aggression and Violent Behavior, 10*, 241-261.

[78] Rose, A. J. y Rudolph, K. D., (2006). A review of sex differences in peer relationships processes: Potential trade-offs for the emotional development of girls and boys. *Psychological Bulletin, 132(1)*, 98-131.

[79] Geary, D., (2006). Sex differences in social behavior and cognition: Utility of sexual selection for hypothesis generation. *Hormones and Behavior*, *49*, 273-275.

[80] Buss, D. M., 1995. Psychological sex differences. Origins through sexual selection. *American Psychologist, 50*, 164-168.

[81] Archer, J., (1996). Sex differences in social behaviour: Are the social role and evolutionary explanations compatible? *American Psychologist, 51*, 909-917.

[82] Booth, A., Carver, K. y Granger, D. A., (2000). Biosocial perspectives on the family. *Journal of Marriage and Family*, *62*, 1018-1034.

[83] Granger, D. A. y Kivlighan, K. T., (2003). Integrating biological, behavioral, and social levels of analysis in early child development: Progress, problems, and prospects. *Child Development 74(4)*, 1058-1063.

In: Psychoneuroendocrinology Research Trends
Editor: Martina T. Czerbska, pp. 433-449

ISBN: 978-1-60021-665-7

Chapter XV

STRESS AND THE AWAKENING CORTISOL RESPONSE (ACR) IN MENTAL DISORDERS

T.J. Huber[1], *O.T. Wolf*[2] *and K. Issa*[1]
[1]Medical School Hannover, Carl-Neuberg-Strasse 1, 30625 Hannover, Germany;
[2]Department of Psychology, University of Bielefeld, Germany.

ABSTRACT

Stress is a universally known and well studied phenomenon that can be regarded as both necessary and potentially detrimental for a person's mental and physical well-being. Most mental disorders can be both triggered and worsened by stressful experiences. An important role of stress and its consequences has been assumed for a variety of mental afflictions ranging from depression to adjustment disorders. Accordingly, stress and the physiological stress response including changes of the hypothalamus-pituitary-adrenal (HPA) axis have been investigated extensively in mental health patients. In this course, stress hormone levels have been assessed in urine and blood and suppression and stimulation tests have been employed with important but often conflicting results. The assessment of salivary cortisol and the awakening cortisol response (ACR) has provided a new method to detect even subtle changes of the HPA axis by a reliable and non-invasive method, but the complexity of the system and the contradictory data imply that much about the role of stress and the HPA axis in mental disorders is so far not sufficiently understood. The present article gives an overview of the current literature concerning the HPA axis in mental disorders with a focus on depression and on the ACR. New results are presented on the ACR in psychotherapy inpatients. These preliminary results suggest that the ACR is a promising marker for evaluating the influence of therapeutic interventions on the HPA axis in psychiatric patients.

INTRODUCTION

The development of subclinical depressive-like symptoms following significant or repeated stress is a universally known phenomenon that can be regarded as part of most human's life experience. Depressive disorders also can be triggered by stressful life events and can be viewed as a chronic, out of balance stress reaction.

It is generally accepted to regard the etiology of depressive disorders as multicausal with a greatly varying impact of factors like genetic predisposition, social influence, learning experiences and personality as well as stress experiences. Akiskal and McKinney (1973) have developed a model integrating these influences to a bio-psycho-social model of depression. It assumes a genetic predisposition being able to lower the threshold for a depressive disorder without necessarily causing an actual episode to manifest itself. Early childhood incidents like deprivation, abuse or viral infections can cause biological alterations such as brain receptor or HPA axis function changes and thus further lower this threshold. These biological alterations can in themselves influence personality and stress responsiveness in the long run. Biological or social stressors in combination with the factors mentioned could lead to a lasting stress response and ultimately to a depressive episode as a final common pathway if these challenges cannot be coped with adequately.

Stress and the Stress Response

Stress can be defined as a situation resulting from a perceived physical or mental threat (the "stressor") that is followed by a multitude of physical and emotional consequences called the stress response. If the stressor is temporary, the acute stress response will pass. However, long lasting stressors may lead to a chronic stress response. A stress response can follow not only actual threats like an accident or a physical disorder, but can also result from an imagined or anticipated stress (Nitsch 1981). Therefore, to develop stress an individual's subjective perception and assessment is much more essential than the quality of the stressor itself (Lazarus 1981). Similarly varying is the nature of the stress response. The term "eustress" has been phrased to describe a stressor leading to a positive emotional response training the organism and leading to enhanced performance (Selye). In contrast, if a stressor is perceived as potentially dangerous and answered by a negative emotional response, the term distress has been suggested. In this chapter, "stress" refers to the latter situation.

An acute stress response consists of an activation of a neuroendocrinological (the hypothalamus-pituitary-adrenal (HPA) axis) and a neurochemical system being represented by norepinephrine, dopamine and possibly epinephrine. Both systems interact with each other and converge in a final common pathway in the adrenal gland. Physiologically, the stress response results in cold, clammy skin, raised blood pressure and pulse, platelet activation as well as changes in the immune system, enhanced lipolysis and glucogenolysis, reduced hunger, fatigue and sexual desire. These changes are necessary for an organism to answer the perceived stress by the classical "fight or flight" reaction.

Focus of this chapter is the HPA axis. In the paraventricular nucleus of the hypothalamus corticotrophin releasing factor (CRF) is released stimulating the release of adrenocorticotropic hormone (ACTH) in the pituitary. This leads to the release of glucocorticoids in the adrenal cortex and of norepinephrine and epinephrine in the medulla. The most important stress hormone is cortisol, which activates glucogenolysis and lipolysis, reduces protein production, stimulates proteolysis and has anti-inflammatory effects. Cortisol receptors are found throughout the human organism, which produces about 27.3 umol/l cortisol daily from cholesterine (Esteban et al. 1991). In blood plasma roughly 90 % of cortisol are bound to corticosteroid binding globuline, albumin and erythrocytes, whereas between 2 and 15 % remain free (Kirschbaum & Hellhammer 2000). Its half life in the blood is 60 to 110 minutes (Kugler 1991). An acute stress response is terminated by cortisol, which inhibits the release of CRF and ACTH via a feedback mechanism at the pituitary, hypothalamus but also "higher brain centres" like the hippocampus and the anterior cingulate (Kirschbaum & Hellhammer 1999).

In healthy subjects, an enhanced release of ACTH can be found in 7 to 10 daily periods happening primarily in the early morning. These explain the high cortisol levels around the time of awakening. In the context of a normal sleeping pattern cortisol levels fall thereafter to a minimum around midnight. Routine testing for Cushing's disease therefore includes assessment of blood cortisol levels at 8 o'clock a.m. (normal values 5-25 ug/dl) and 0 o'clock p.m. (normal values 0-5 ug/dl). Superimposed on this circadian rhythm is a pulsatile or episodic secretion pattern with multiple cortisol peaks (Gallagher et al. 1973, Lenz 1987). Both are dependent on the individual sleep pattern, so that any disturbance like an intercontinental flight will cause the rhythm to shift. Cortisol levels are also influenced by light and darkness, meals or fasting, physical exercise, hypoglycaemia, heat or cold (Olehansky 1990). Many drugs as well as pregnancy, oestrogens and gestagens also have an impact on cortisol levels.

Chronic stress in contrast to acute stress is hallmarked in most subjects by a persisting activation of the HPA axis (Imaki et al 1991), reduced expression of cortisol receptors and lasting physiological changes. This form of a chronic stress response has been demonstrated in chronic mental disorders, in persisting mental conflicts as well as following the loss of a spouse (Spiegel 1991). In 20 to 25 % of subjects, a relative hypocortisolism is found under chronic stress conditions, which can result from chronic mental as well as physical strain or trauma. This condition is assumed to be able to lead to chronic fatigue, stress intolerance and increased pain sensitivity (Hellhammer & Schommer 2003).

Salivary Cortisol and the Awakening Cortisol Response (ACR)

Free cortisol is a liphophilic glucocorticosteroid with a low molecular weight that can easily cross the lipid membrane of cells and is found in blood, liquor, urine, sudor and saliva. While blood contains free and bound cortisol, only free cortisol can be detected in the saliva and can serve as an index for the biologically active fraction since assays for its assessment have become available. Its concentration is unimpaired by salivary flow rate (Umeda et al. 1981, Kirschbaum & Hellhammer 2000). Absolute salivary cortisol values are lower than free cortisol levels in the blood because of the high concentration of 11-Beta-hydroxysteroid dehydrogenase in the saliva, which is important for cortisol metabolisation (Jahn et al. 2003).

Thus, salivary cortisol levels amount to about two to five percent of the total plasma levels of cortisol (Wuest et al. 2000a) with a correlation of free blood and salivary cortisol levels of more than 80 % (Kirschbaum & Hellhammer 2000). This high correlation results from the mainly passive diffusion of cortisol.

Assessing salivary cortisol as a marker for the HPA axis is painless and thus avoids stimulation of the HPA axis by venous puncture as is necessary for obtaining blood cortisol levels. It also allows repeated measurements in a short time period without significant impact on wellbeing.

Single salivary cortisol levels have a low intraindividual stability (Schulz & Knabe 1994) and high interindividual variation with a significant overlap between healthy subjects and Cushing or Addison patients (Laudat et al. 1988). While more reliable, assessment of the diurnal profile of salivary cortisol would be costly and time consuming (Wuest et al. 2000b). Stimulation tests have widely been used to detect changes of the HPA axis, most notably with the dexamethasone suppression test (DST). 1-2 mg dexamethasone are given at night and suppress morning cortisol levels by the described feedback mechanism in healthy subjects. In depressed patients a nonsuppression in the DST has repeatedly but not consistently been demonstrated (Henry 1987). The DST is hampered by a low sensitivity compared to salivary cortisol or urine sampling (Viardot 2005). Because dexamethasone constitutes a very strong feedback signal to the HPA axis it has also been suggested that the DST is unsuitable to reflect the magnitude of endogenous HPA responses (Burke et al. 2005a). A low dose dexamethasone test (0.5 mg) has been suggested to be more sensitive in conditions with suspected hypocortisolism like post traumatic stress disorder for the detection of an enhanced feedback sensitivity (Yehuda 2002).

A more subtle, but easily detectable challenge to the HPA system is the awakening in the morning which is answered by a steep rise of cortisol levels in healthy subjects (Spaeth-Schwalbe et al. 1992, Van Cauter et al. 1994). Within the first 30 minutes after awakening free cortisol levels rise by 50 to 100 % above baseline and remain thus for about 60 minutes followed by a slow continuous decline (Pruessner et al. 1997). This awakening cortisol response (ACR) has a higher intraindividual stability than single or fixed time measurements (Pruessner et al. 2003, Wuest et al. 2000b). It also offers a high consistency between studies (Clow et al. 2004, Wuest et al. 2000b) and is widely accepted as a reliable method to detect even subtle changes of the HPA axis. While the exact causes for the ACR are not sufficiently understood as yet it is believed to be distinct from the diurnal cycle of cortisol secretion (Clow et al. 2004).

An enhanced ACR has been demonstrated in healthy subjects under chronic stress like high work load or social stress (Wuest et al. 2000b). In 219 workers high work load and persisting worries were associated with an enhanced ACR on working days but not on the weekend (Schlotz et al. 2004).

The ACR was found to be blunted in healthy college students having lost a near relative by death or separation before they were 14 years old (Meinlschmidt & Heim 2005). Mobbing victims also showed a lower ACR than colleagues without mobbing experiences (Hansen et al. 2006). Another study found a significant correlation of the financial situation with the ACR with the lowest ACR in the most financially challenged subjects (Ranjit et al. 2005).

The ACR in Affective Distress and Depression

The onset and perpetuation of depressive disorders is universally assumed to be linked to stress and to HPA axis alterations, particularly hypercortisolism (Kendler et al. 1992, Ryan et al. 2004). The neurobiological changes following stressful early life events are similar to those found in depressed subjects. In chronically stressed healthy subjects as well as depressed patients hypertrophic adrenal glands, an impaired negative feedback of cortisol, smaller hypocampi and cognitive impairment are all found (Sheline et al. 1996, Thase et al. 1996, Heim et al. 2004). In an experimental stress study by Burke et al. (2005b), induced stress lead to similarly elevated cortisol concentrations in healthy controls and depressed patients. However, in a recreational phase 25 minutes after the end of the stressor, depressed subjects had significantly higher cortisol levels. In contrast, other studies found normal or lowered cortisol levels in response to experimental stress in depressed patients (Gotthardt et al. 1995, Young et al. 2000, Kudoh et al. 2000).

Studies on the HPA axis in depressive disorders are numerous and conflicting. The most consistent findings have been elevated blood cortisol levels and non-suppression in the dexamethasone test and an exaggerated response in the combined dexamethasone/CRF test in many, but not all depressed patients (Holsboer 2003, Ising et al., 2005). In contrast observations on the salivary cortisol response to awakening in depression are rare. Teachers suffering from burnout have been found to exhibit a blunted morning cortisol rise (Pruessner et al. 1999). Similarly, in a community study comparing 103 healthy volunteers and 74 subjects with chronic health problems, the seven subjects suffering from "psychiatric chronic health problems" showed a trend for a blunted ACR (Kudielka et al. 2003). In contrast more severe levels of self-reported depressive symptoms were associated with a higher ACR in healthy college students (Pruessner et al. 2003a) and higher neuroticism scores were accompanied by an exaggerated ACR (Portella et al. 2005). Acutely depressed subjects in the community also demonstrated an enhanced ACR in comparison to controls (Bhagwagar et al. 2005). Other studies found unchanged or blunted cortisol responses to a stressor in association with depressive symptoms (Strickland et al. 2002, Burke et al. 2005a). Data on the influence of psychosocial variables and mental disorders on the ACR are therefore still scarce and inconsistent (Clow et al. 2004).

HPA Axis and Psychotherapy

Even fewer data are available on the HPA axis in psychotherapy patients. One study examined 29 depressed patients taking part in a three week outpatient cognitive behavioural therapy program. Hypercortisolism was found in these patients using the dexamethasone suppression test and urinary cortisol levels. Success of the psychotherapy program was inversely correlated with baseline urinary cortisol concentrations, but there was no association with the results of the suppression test (Thase et al. 1996). It was concluded that hypercortisolic patients responded less favourably to psychotherapy and would profit more from somatic interventions.

In summary, the evidence for HPA axis function changes in depressive disorders in the form of a chronic stress response is strong, but there are conflicting results suggesting distinct patterns of HPA function in different populations or subtypes of depressive disorders. The investigations presented in this chapter therefore studied only inpatients who participated in a psychotherapy program and did not suffer from a melancholic subtype of depressive disorders. On the basis outlined above, a chronic stress response was assumed and the following hypotheses were to be tested:

1. The ACR in patients taking part in an inpatient psychotherapy program is principally characterized by a high level of non-responders and a low rise of salivary cortisol levels from baseline.
2. Subjects suffering from an affective disorder exhibit a lower ACR in comparison to a group of mixed diagnoses other than depression with a higher rate of non-responders in the former.
3. During inpatient psychotherapy, the ACR increases significantly.
4. There is an inverse correlation of ACR and symptom severity.

METHODS

72 participants of an integrated inpatient psychotherapy program were included into the study after approval of the local ethics committee and written consent of all subjects. Free salivary cortisol levels were assessed at waking by a member of the hospital staff between 7.00 and 7.15 a.m. and at 15 minute intervals for the first half hour afterwards using a commercially available immunoassay (IBL, Hamburg, Germany) with an inter-assay and intra-assay variation of less than 15 %. Subjects were asked to remain in bed for the first 15-minutes. Sampling took place on the day following admission and at three week intervals until discharge. Samples were collected under supervision and subjects were asked to fill in a protocol stating subjective sleep quality on a ten point scale, time of going to bed, waking up and collecting the samples to ensure adherence. In the case of non-adherence (i.e. awakening before the waking, rising before the second sample was collected), the procedure was repeated on the following day, which was necessary in less than 10 % of patients.

On the basis of the three cortisol levels assessed for each patient and sample day, mean free cortisol levels (mean) and change from baseline to 30-minute sample (delta) were computed. Following the suggestions of Pruessner (Pruessner et al. 2003b) and Clow (Clow et al. 2004) the area under the curve relative to zero (AUC_G) and the area under the curve with respect to increase (AUC_i) were also assessed. Since correlations between AUC_G and mean free cortisol and between AUC_i and delta were high ($r > 0.9$), only the mean and delta are given with emphasis on the delta. The rationale is the delta (or AUC_i, respectively) being linked to the sensitivity of the system and the relation to perceived stress – which is the focus of the present chapter – whereas mean cortical levels (AUC_G respectively) seem to be more an index for physical complaints (Pruessner et al. 2003a).

Psychometric testing included the Hamilton Depression Rating Scale (HAMD) administered by trained interviewers, the Beck Depression Inventory (BDI) and the Symptom

Check List in its 90 item revised version, the primary variable of interest being the general symptomatic index (GSI). Correlation between interviewers was assessed regularly and was > 0.9.

The psychotherapy program investigated consists of a slow open psychodynamically oriented group with new patients being admitted at three-week intervals. At this point of time subjects already in therapy for three weeks move into the next treatment phase. The program lasts for six to nine weeks with the option of an extension to 12 weeks in special circumstances when this seems feasible. Additionally to the psychodynamic group, patients took part in groups with social skills training, role play, occupational therapy and a sports program as well as weekly individualized sessions with a trained psychotherapist.

The statistical analysis for between group comparisons used non parametrical tests (mann-whitney-u-test) because of the scewnesss of data. Similarly correlations were computed with Spearman rank order correlations. The level of probability was set at $p < 0.05$.

RESULTS

The results presented in this chapter have partly been previously published (Huber et al. 2006). They are supplemented by new data and results on the time course of the ACR and below are given by order of the hypothesis stated above:

1. The ACR in Patients taking Part in an inpatient Psychotherapy Program is Principally Characterized by a high Level of Non-responders and a low rise of Salivary Cortisol Levels from Baseline

The mean baseline total free cortisol level directly after waking was 16.93 nmol/l. In the entire group it increased to a mean of 20.87 nmol/l 30 minutes after waking. The delta (AUC_i, respectively) therefore was 4.28 (4.69) nmol/l (table 1). In the literature a delta in healthy subjects of roughly 9 nmol/l is suggested (Clow et al. 2004).

With 36 % a significant number of subjects (n = 26) did not show the typical awakening cortisol response seen in healthy volunteers, but according to Wuest and colleagues (2000b) can be defined as non-responders (increase of free cortisol after waking less than 2.5 nmol/l).

2. Subjects Suffering from a Depressive Episode exhibit a lower ACR in Comparison to a Group of Mixed Diagnoses other than Depression

57 of the subjects (79.2 %) suffered from for a major depressive episode according to DSM-IV and ICD-10 criteria. There were no patients classified as “with melancholic features” or “somatic syndrome”, which is of importance as these have previously been suggested to be linked with HPA overdrive in contrast to a hypoactive HPA axis in atypical depression (Antonijevic 2006). These patients were grouped together and compared with a mixed group of 15 patients devoid of depressive symptoms allowing the diagnosis of a

depressive disorder. Of these, nine suffered from a personality disorder, eight from an anxiety disorder and two from a somatoform disorder with some overlap because of comorbidity. In both groups, 33 subjects were without medication (24 in the depressed group and 9 in the mixed diagnostic group), whereas 39 patients received different antidepressants on admission (24 selective serotonine reuptake inhibitors, nine mirtazepine, six trimipramine or amitryptiline) for a mean duration of 7.44 weeks. The age of all subject ranged from 20 to 69 years and 43 (59.7 %) were female. Between the depressed and the non-depressed subjects, there were no significant differences in age or gender (Chi2 = 1.459) or subjective sleep quality ($p = 0.786$). However, depressed subjects reported a longer duration of symptoms in concordance with the often chronic course of this disorders in the sample presented. 49.1 % (28 subjects) reported recurrent depressive episodes. In table 1 an overview of the relevant data on both subgroups and the entire study population is given.

Table 1. Age, gender distribution, HAMD scores and duration of symptoms (months) for the entire study population and the depressed and non-depressed subgroup; increase of free cortisol between wake up and 30 minutes (delta; nmol/l), mean free cortisol of the three collected samples (mean; nmol/l)

	entire study population	depressed subgroup	non-depressed subgroup	p
age (years)	40.54 (11.87)	41.40 (12.4)	37.27 (9.3)	0.232
percent female	59.7	56.1	73.3	0.227
HAMD score	20.26 (5.73)	20.33 (5.93)	20.0 (4.99)	0.835
duration of symptoms (months)	52.41 (76.16)	44.51 (73.05)	84.62 (83.0)	**0.005**
cortisol increase (delta)	4.28 (9.0)	3.47 (9.6)	7.66 (4.7)	**0.049**
mean cortisol	19.01 (8.4)	19.42 (8.5)	17.32 (8.2)	0.194

The standard deviation is presented in brackets. p = level of probability for a difference between the depressed and non-depressed group (taken from Huber et al. 2006).

The rise in free cortisol levels in depressed patients was significantly lower than that in non-depressed subjects ($p = 0.049$) with a delta of 3.47 nmol/l in the depressed and of 7.66 nmol/l in the non-depressed subgroup. There was also a trend for higher baseline levels in depressed patients ($p = 0.072$) which appeared in correlation (Spearman rank $r = -0.46$, $p < .001$) with this finding. For the significant differences in the ACR delta as well as for the trend in differences in the wake up value Cohen's d for pooled variances were calculated (Cohen 1988). The effect size for the delta was -0.47 indicating a medium sized effect. The effect size for the differences in wake up levels was similar (0.47). Both groups did not differ in their cortisol levels at 15 minutes ($p = 0.49$) or 30 minutes ($p = 0.85$) or in their mean cortisol levels ($p = 0.39$; figure 1).

25 out of the 26 non-responders (92 %) were found in the depressed subjects, although these accounted for only 79.2 % of all patients. The rate of non-responders was significantly higher in depressed as compared to non-depressed subjects (Pearson $Chi^2 = 7.329$; $p = 0.007$).

In the non-responders higher basal cortisol levels of 20.52 (SD 10.32) nmol/l as compared to the responders with 13.56 (SD 7.61) nmol/l ($p = 0.003$) were demonstrated, whereas levels at 15 minutes ($p = 0.959$) were comparable and 30 minute values with 17.05 (SD 8.93) nmol/l were lower than in responders with 24.33 (SD 8.34) nmol/l ($p = 0.003$). Between non-responders and responders no differences were detectable concerning Hamilton Depression Rating Scale scores ($p = 0.245$) or Beck Depression Inventory scores ($p = 0.707$), but the non-responders were significantly older at 43.85 (SD 12.72) years than the responders at 36.06 (SD 10.02) years. On calculation of the observed significantly different ACR between depressed and non-depressed subjects only in responders, this significance was reduced to a trend ($p = 0.083$). The ACR did not differ significantly between patients on and off medication in the entire sample as well as in depressed subjects only ($p > 0.1$).

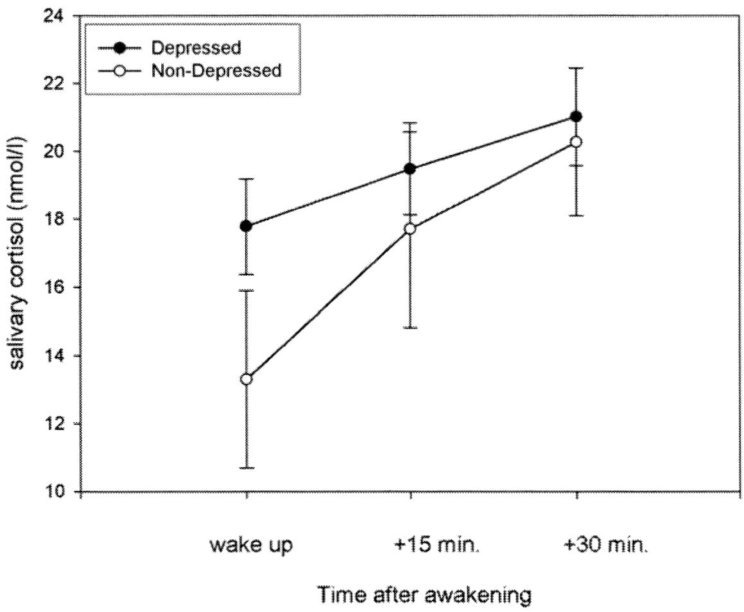

Figure 1. Cortisol awakening response (ACR) in depressed and non-depressed subjects. Depressed patients tended to have a higher wake up level ($p < .10$) and had a significantly lower ACR delta increase ($p < .05$). Error bars represent standard error of the mean (SE).

3. During inpatient Psychotherapy, the ACR increases Significantly

While there was no change in mean cortisol levels during the therapy program, a rise-decline-rise pattern was observed concerning the delta (figure 2). It increased significantly during the first three weeks, declined slightly to week 6 and reached its highest point in week 9. There was a significant difference between the delta on admission and just before discharge ($p = 0.028$).

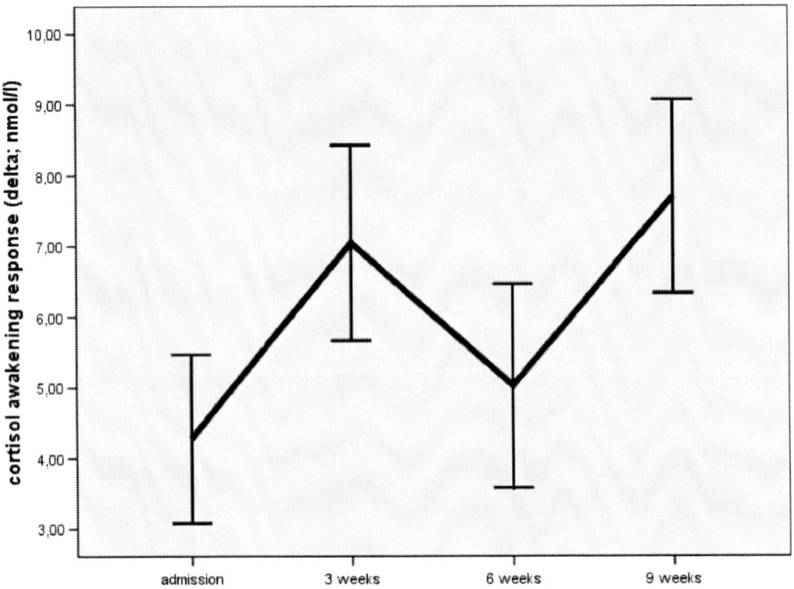

Figure 2. ACR (delta) over the course of the therapy in the entire study population (n = 72). Error bars indicate standard error of the mean. Significant differences from admission to three weeks and to discharge at nine weeks.

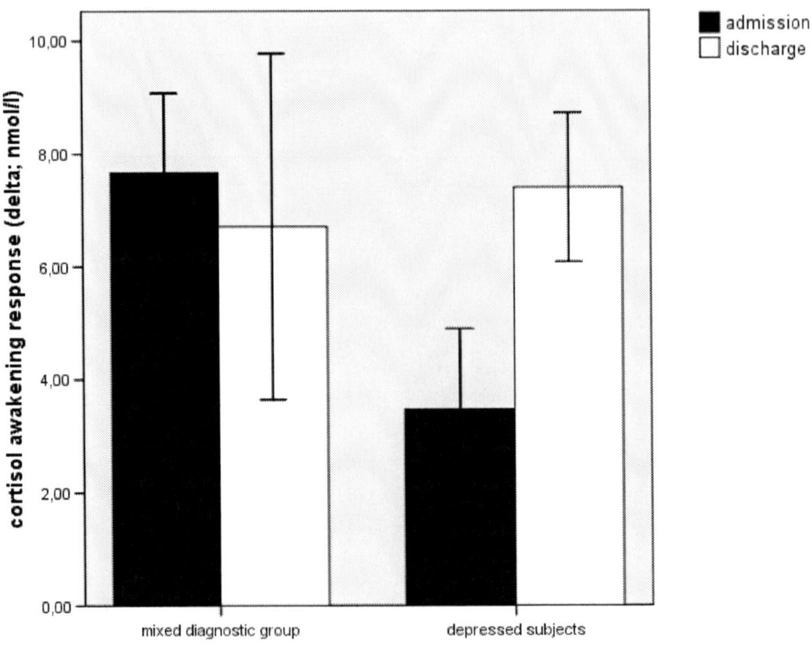

Figure 3. Cortisol awakening response (delta) on admission and after at discharge in the control group of mixed diagnoses and in depressed patients. Significant rise from admission to discharge in the depressed subjects ($p = 0.013$), but not in the mixed diagnostic group.

The hypothesised significant rise of the ACR from admission to discharge was found in the entire group as well as in the depressed group only ($p = 0.013$) as shown in figure 3.

However, in the control group, the ACR did not change significantly during therapy ($p = 0.753$).

4. There is a Correlation of ACR and Symptom Severity

In the course of the treatment symptom severity reflected by GSI scores inversely correlated significantly but weakly (correlation coefficient - 0.222) with the delta ($p = 0.034$). Higher GSI scores were therefore associated with a lower ACR. At no point of time a difference could be observed between patients dropping out of therapy and those concluding it.

CONCLUSION

A variety of studies using different methods has been undertaken in order to investigate the doubtlessly important role of the hypothalamo-pituitary-adrenal (HPA) axis in depressive disorders. The results are conflicting and suggest different patterns of HPA axis function changes depending on the population and subtype of depression studied. Since the assessment of salivary cortisol has been recognized as an easy and reliable method, the awakening cortisol response (ACR) is increasingly regarded as a useful measure to detect subtle changes in the HPA axis. However, data on the ACR in depressed inpatients are still scarce and conflicting and there are virtually no studies regarding inpatient psychotherapy.

The present study was performed on 72 patients admitted to an inpatient psychotherapy program in order to control for influences of daily routine and to investigate an as homogenous as possible depressed subgroup. The entire study population including a control group of mixed diagnoses exhibited a lower response to awakening in comparison to literature data for healthy subjects (Clow et al 2004); however, due to different methods and study populations, this comparison must be made with extreme caution. The ACR was significantly lower in depressed subjects as compared to non-depressed patients, which appeared not to be due to an enhanced ACR in the latter. In the current study there was a trend for higher wake-up cortisol levels in the depressed subjects and both groups reached comparable 30-minute levels. This is in accordance with previous reports of an attenuated ACR in association with higher wake-up cortisol levels (Williams et al. 2005) and hints to a stronger chronic stress response in the depressed subgroup. From admission to discharge, the ACR rose significantly in the depressed subjects but not in the group of mixed diagnoses leading to similar discharge deltas in both groups. Even though somewhat preliminary this is the first evidence that psychotherapy can lead to a normalized ACR in depressed patients showing a blunted ACR at the time of admission.

The study presented has some limitations which have to be taken into account on interpretating its results: subjects were not medication free, which could explain the differences between the depressed and non-depressed subgroups, particularly since the former received antidepressants more often. Antidepressants have been suggested to influence the HPA axis activity (Schule et al. 2003) and the ACR (Laakmann et al. 2004), most notably in the case of mirtazepine. The latter was taken by 9 of the depressed patients. Medication-free subjects are too few in the present investigation to allow separate statistical analysis; nonetheless no significant differences in the ACR of the entire group was detected between patients on and off medication. Another point of concern is the possibly impaired sleep quality in the depressed patients. These might have woken up due to depressive sleep disturbances before the waking so that the actual ACR was missed by the investigation. This would explain the higher baseline cortisol levels, the lower ACR and the high percentage of non-responders in the depressed group. However, subjective sleep quality ratings did not differ between depressed and non-depressed subjects and analysis of only the responders still showed a trend for an attenuated ACR in the depressed group.

A particular strength of the study is the design allowing to control for the stress of a hospital admission and a variety of factors like food intake, daily routine and physical exercise that are all known to have an impact on the ACR. Thus, the observed blunted ACR in the depressed subgroup cannot be explained by these factors and appears to be more likely dependent on the depressive symptomatology. However, an alternative variable possibly accounting for the blunted ACR apart from the medication mentioned could be the longer symptom duration in the depressed group. Another advantage of the present investigation appears to be the use of an everyday stimulating influence (waking) on the HPA axis in contrast to a challenge with dexamethasone. This is designed to test the negative feedback of the HPA axis resulting from a very strong glucocorticoid signal and may thus not reflect an endogenous HPA response like that resulting from suprahypothalamic influences like the limbic system (Burke et al. 2005b).

Concerning the time course of the ACR, the significant increase in the first three weeks could be interpreted as a normalisation reflecting the familiarisation on the ward with the initially more supportive therapy. The slight decline in the following three weeks could be explained by the more confronting and revealing therapy phase following, while stabilisation and symptom reduction in the last therapy phase could account for the afresh rise of the ACR in this last therapy segment. This is supported by the observed although weak inverse correlation of subjective symptom severity and cortisol rise after waking, but must be regarded as highly speculative. The ACR did not allow to predict premature treatment termination.

The data presented allow some interesting conclusions: when the current literature is taken into account, psychotherapy inpatients appear to be distinct from acutely depressed outpatients and healthy subjects with some depressive symptoms as well as from the depressed subjects of some studies (Pruessner et al. 2003a) as these exhibit an increased as opposed to the blunted ACR seen in the study presented. In contrast the latter appear to demonstrate a pattern more similar to that of teachers suffering from burnout (Pruessner et al. 1999), chronically ill patients (Kuldielka and Kirschbaum 2003) and college students heaving lost a near relative in their childhood (Meinlschmidt & Heim 2005) in spite of the fact that all patients in the depressed group fulfilled criteria for a major depressive episode. A recent meta analysis has fittingly demonstrated more severely depressed patients to be more likely to have a blunted reactivity pattern than less severely depressed subjects (Burke et al. 2005b). Because of the limitations mentioned, the data have to be interpreted with caution. Possibly distinct subtypes of depression are associated with different alterations of the HPA axis, as has been suggested by Antonijevic (2006). The patients investigated might therefore reflect the hypoactive state seen in atypical depression rather than the melancholic subtype. This would be in accordance with the fact that psychotherapy patients are less likely to suffer from a melancholic depression and that the present sample did not include any subjects fulfilling criteria for this. The assumption of distressing early life events being of significance in more neurotic forms of depression are nicely supported by the similarity of the ACR in the patients investigated and the college students with early life losses studied by Meinlschmidt and Heim (2005).

On the causes and pathophysiological processes involved in this hyporeactivity in some depressed inpatients only speculation is possible. One candidate mechanism would be a stress-related hippocampal atrophy in the patients investigated as demonstrated in a variety of imaging studies (Videbech and Ravnkilde 2004). In patients with selective hippocampal damage (Buchanan et al. 2004) and amnesia resulting from brain injury (Wolf et al. 2005) a similar pattern with a blunted or missing ACR has been shown. Of course other brain regions are critically involved in the pathogenesis of depression as well as in HPA regulation. Among them are the amygdala and several prefrontal regions (Gold et al. 2002).

Of interest is the time course of the ACR during therapy. Firstly, successful psychotherapy seems to be associated with an increase of the blunted ACR suggesting psychotherapy to be accompanied by a normalisation of HPA axis activity. This seems to happen not by a linear course, but rather by variations possibly reflecting treatment phases. Future studies using salivary cortisol assessment might possibly be able to detect patients unable to move into the next treatment phase or in danger of not benefiting from it.

The role of the HPA axis in depressed inpatients and most particularly in patients undergoing non-pharmacological treatment is still not sufficiently understood. Further research on the salivary cortisol in depression possibly including structural imaging techniques seems justified to verify the findings presented and to control for the possible effect of medication and sleep quality as well as symptom duration. These future studies could try to relate alterations of the ACR to other markers of HPA activity. In addition the combination of these neuroendocrine markers with structural or functional imaging techniques, but also with genetic risk markers (Caspi et al. 2003) appears to be promising.

References

Akiskal, H.S., & McKinney, W.T. (1973). Depressive disorders: toward a unified hypothesis. *Science 182*, 20-29.

Antonijevic, I.A. (2006). Depressive disorders – is it time to endorse different pathophysiologies? *Psychoneuroendocrinology 31*, 1-15.

Bhagwagar, Z., Hafizi, S., & Cowen P.J. (2005). Increased salivary cortisol after waking in depression. *Psychopharmacology 182*, 54-57.

Buchanan, T.W., Kern, S., Allen, J.S., Tranel, D., & Kirschbaum, C. (2004). Circadian regulation of cortsiol following hippocampal damage in humans. *Biol. Psychiatry 56*, 651-656.

Burke, H.M., Fernald, L.C., Gertler, P.J., & Adler, N.E. (2005a). Depressive symptoms are associated with blunted cortisol stress responses in very low-income women. *Psychosom. Med. 67*, 211-216.

Burke, H.M., Davis, M.C., Otte, C., & Mohr, D.C. (2005b). Depression and cortisol responses to psychological stress: a meta-analysis. *Psychoneuroendocrinology 30*, 846-856.

Caspi, A., Sugden, K., Moffitt, T. E., Taylor, A., Craig, I. W., Harrington, H. et al. (2003). Influence of life stress on depression: moderation by a polymorphism in the 5-HTT gene. *Science 301,* 386-389.

Clow, A., Thorn, L., Evans, P., & Hucklebridge, F. (2004). The awakening cortisol response: methodological issues and significance. *Stress 7*, 29-37.

Cohen, J. (1988). Statistical power analysis for the social sciences. Hilsdale, NJ: Erlbaum.

Esteban, N.V., Loughlin, T., Yergey, A.L., Zawadzki, J.K., Booth, J.D., Winterer, J.C., & Loriaux, D.L. (1991). Daily cortisol production rate in man determined by stable isotope dilution/mass spectrometry. *J Clin Endocrinol Metab 72 (1)*, 39-45

Gallagher, T.F., Yoshida, K., Roffwarg, H.D., Fukushima, D.K., Weitzman, E.D., & Hellman, L. (1973). ACTH and cortisol secretory patterns in man. *J Clin Endocrinol Metab 36*, 1058-1068.

Gold, P.W., Drevets, W., Charney, D., & Drevets, W. (2002). New insights into the role of cortisol and the glucocorticoid receptor in severe depression. *Biol Psychiatry 52,* 381-385.

Gotthardt, U., Schweiger, U., Fahrenberg, J., Lauer, C.J., Holsboer, F., & Heuser, I. (1995). Cortisol, ACTH, and cardiovascular response to a cognitive challenge paradigm in aging and depression. *Am J Physiol 268 (4Pt2)*, R865-873.

Hansen, A.M., Hogh, A., Persson, R., Karlson, B., Garde, A.H., & Orbaek, P. (2006). Bullying at work, health outcomes, and physiological stress response. *J Psychosom Res 60 (1),* 63-72.

Heim, C., Plotsky, P.M., & Nemeroff, C.B. (2004). Importance of studying the contributions of early adverse experience to neurobiological findings in depression. *Neuropsychopharmacology 29*, 641-648.

Henry, E., Aimard, G., Claustrat, B., Guillot, M., Rousselle, C., & Devic, M. (1987). Dexamethasone suppression test in psychiatric pathology. *Presse Med 16 (14),* 669-671.

Holsboer, F. (2003). Corticotropin-releasing hormone modulators and depression. *Curr. Opin. Investig. Drugs 4*, 46-50.

Huber, T.J., Issa, K., Schik, G., & Wolf, O.T. (2006). The cortisol awakening response is blunted in psychotherapy inpatients suffering from depression. *Psychoneuroendocrinology 31,* 900-904.

Jahn, H., Kiefer, F., Schick, M., Yassouridis, A., Steiger, A., Kellner, M., & Wiedemann, K. (2003). Sleep endocrine effects of the 11-beta-hydroxysteroiddehydrogenase inhibitor metyrapone. *Sleep 26 (7)*, 823-829.

Kendler, K.S., Neale, M.C., Kessler, R.C., Heath, A.C., & Eaves, L.J. (1992). Childhood parental loss and adult psychopathology in women. A twin perspective. *Arch Gen Psychiatry 49,* 109-116.

Kirschbaum, C. & Hellhammer, D.C. (1999). Hypothalamus-Hypophysen-Nebennierenrindenachse. In C Kirschbaum, & D. Hellhammer (Eds.), *Enzyklopädie der Psychologie: Psychoendokrinologie und Psychoimmunologie.* (pp. 80-140) Göttingen, Bern, Toronto, Seattle: Hogrefe-Verlag.

Kirschbaum, C. & Hellhammer, D.H. (2000). Salivary Cortisol. *Encyclopedia of Stress 3*, 379-383.

Kudielka, B.M., & Kirschbaum, C. (2003). Awakening cortisol responses are influenced by health status and awakening time but not by menstrual cycle phase. *Psychoneuroendocrinology 28*, 35-47.

Kudoh, A., Ishihara, H., & Matsuki, A. (2000). Inhibition of the cortisol response to surgical stress in chronically depressed patients. *J Clin Anesth 12*, 383-387.

Kugler, J. (1991). Radioimmunologische Bestimmung von Cortisol im Speichel: Messgüte und psychoendokrinologische Einsatzfelder. *Psychologische Beiträge 33*, 132-144.

Laakmann, G., Hennig, J., Baghai, T., & Schule, C. (2004). Mirtazepine acutely inhibits salivary cortisol concentrations in depressed patients. *Ann. N. Y. Acad. Sci. 1032*, 279-292.

Laudat, H.M., Cerdas, S., Fournier, C., Gouiban, D., Guilhaume, B., & Luton, J.P. (1988). Salivary cortisol measurement: a practical approach to assess pituitary adrenal function. *J Clin Endocrin Metab 66*, 343-348

Lazarus, R.S. (1981). Streß und Streßbewältigung – Ein Paradigma. In: Filipp H-S (Ed.), *Kritische Lebensereignisse*. (pp. 198-232). München: Urban und Schwarzenberg.

Lenz, H.J., Raedler, A., & Greten, H. (1987). Corticotropin-Releasing-Faktor (CRF). *Dtsch. med. Wochenschr. 112*, 1588-1591.

Meinlschmidt, G., & Heim, C. (2005). Decreased cortisol awakening response after early loss experience. *Psychoneuroendocrinology 30 (6),* 568-576

Nitsch, J.R. (1981). Streßtheoretische Modellvorstellungen. In: Nitsch, J.R. (Ed.), *Streß: Theorien, Untersuchungen, Maßnahmen* (pp. 52-160). Bern: Huber-Verlag.

Olehansky, M.A., Zoltick, J.M., Herman, R.H., Mougey, E.H., & Meyerhoff, J.L. (1990). The influence of fitness on neuroendocrine responses to exhaustive treadmill exercise. *Eur J Appl Physiol 59 (6)*, 405-410.

Portella, M.J., Harmer, C.J., Flint, J., Cowen, P., & Goodwin, G.M. (2005). Enhanced early morning salivary cortisol in neuroticism. *Am.J.Psychiatry 162,* 807-809.

Pruessner, J.C., Wolf, O.T., Hellhammer, D.H., Buske-Kirschbaum, A., von Auer, K., Jobst, S., Kaspers, F., & Kirschbaum, C. (1997). Free cortisol levels after awakening: a reliable biological marker for the assessment of adrenocortical activity. *Life Sci. 61*, 2639-2549.

Pruessner, J.C., Hellhammer, D.H., & Kirschbaum, C. (1999). Burnout, perceived stress, and cortisol responses to awakening. *Psychosom. Med. 61*, 197-204.

Pruessner, M., Hellhammer, D.H., Pruessner, J.C., & Lupien, S.J. (2003a). Self-reported depressive symptoms and stress levels in healthy young men: associations with the cortisol response to awakening. *Psychosom. Med. 65*, 92-99.

Pruessner, J.C., Kirschbaum, C., Meinlschmid, G., & Hellhammer, D.H., 2003b. Two formulas for computation of the area under the curve represent measures of total hormone concentration versus time-dependent change. *Psychoneuroendocrinology 28*, 916-931.

Ranjit, N., Young, E.A., & Kaplan, G.A. (2005). Material hardship alters the diurnal rhythm of salivary cortisol. *Int J Epidemiol, 34 (5),* 1138-1143.

Ryan, M.C., Shafari, N., Condren, R., Thakore, J.H. (2004). Evidence of basal pituitary-adrenal overactivity in first episode, drug naïve patients with schizophrenia. *Psychoneuroendocrinology 29 (8)*, 1065-1070

Schlotz, W., Hellhammer, J., Schulz, P., & Stone, A.A. (2004). Perceived work overload and chronic worrying predict weekend-weekday differences in the cortisol awakening response. *Psychosom Med 66 (2)*, 207-214.

Schule, C., Baghai, T., Rackwitz, C., & Laakmann, G. (2003). Influence of mirtazepine on urinary free cortisol excretion in depressed patients. *Psychiatry Res. 120*, 257-264.

Schulz, P., & Knabe, R. (1994). Biological uniqueness and the definition of normality. Part 2 – The endocrine 'fingerprint' of healthy adults. *Med Hypotheses 42*, 63-68.

Sheline, Y.I., Wang, P.W., Gado, M.H., Csernansky, J.G., & Vannier, M.W. (1996). Hippocampal atrophy in recurrent major depression. *Proc Natl Acad Sci USA 93 (9),* 3908-3913

Späth-Schwalbe, E., Scholler, T., Kern, W., Fehm, H.L., & Born, J. (1992). Nocturnal adrenocorticotropin and cortisol secretion depends on sleep duration and decreases in association with spontaneous awakening in the morning. *J Clin Endocrin Metab 75*, 1431-1435.

Strickland, P.L., Deakin, J.F., Percival, C., Dixon, J., Gater, R.A., & Goldberg, D.P. (2002). Bio-social origins of depression in the community: interactions between social adversity, cortisol and serotonin neurotransmission. *Br. J. Psychiatry 180*, 168-173.

Thase, M.E., Dubé, S., Bowler, K., Howland, R.H., Myers, J.E., Friedman, E., & Jarrett, D.B. (1996). Hypothalamic-pituitary-adrenocortical activity and response to cognitive behaviour therapy in unmedicated, hospitalised depressed patients. *Am J Psychiatry 153 (7),* 886-891.

Umeda, T., Hiramatsu, R., Iwaoka, T., Shimada, T., Miura, F., & Salo, T. (1981). Use of saliva for monitoring unbound free cortisol levels in serum. *Clinica chimica Acta 110*, 245-253.

Van Cauter, E.V., Polonsky, K.S., Blackman, J.D., Roland, D., Sturis, J., Byrne, M.M., & Sheen, A.J. (1994). Abnormal temporal patterns of glucose tolerance in obesity: relationship to sleep-related growth-hormone secretion and circadian cortisol rhythmicity. *J Clin Endocrin Metab 79*, 1797-1805.

Viardot, A., Huber, P., Puder, J.J., Zulewski, H., Keller, U., & Muller, B. (2005). Reproducibility of nighttime salivary cortisol and ist use in the diagnosis of hypercortisolimus compared with urinary free cortisol and overnight dexamethasone suppression test. *J Clin Endocrinol Metab 90 (10)*, 5730-5736.

Videbech, P., & Ravnkilde, B. (2004). Hippocampal volume and depression: a meta-analysis of MRI studies. *Am. J. Psychiatry 161*, 1957-1966.

Williams, E., Magid, K., & Steptoe, A. (2005). The impact of time of waking and concurrent subjective stress on the cortisol response to awakening. *Psychoneuroendocrinology 30*, 139-148.

Wolf, O.T., Fujiwara, E., Luwinski, G., Kirschbaum, C., & Markowitsch, H.J. (2005). No morning cortisol response in patients with severe global amnesia. *Psychoneuroendocrinology 30*, 101-105.

Wuest, S., Wolf. J., Hellhammer, D.H., Federenko, I., Schommer, N., & Kirschbaum, C. (2000a). The cortisol awakening response – normal values and confounds. *Noise & Health 7,* 77-85.

Wuest, S., Federenko, I., Hellhammer, D.H., & Kirschbaum, C. (2000b). Genetic factors, perceived chronic stress, and the free cortisol response to awakening. *Psychoneuroendocrinology 25*: 707-720.

Yehuda, R. (2002). Post-traumatic stress disorder. *N.Engl.J.Med. 346,* 108-114.

Young, E.A., Lopez, F.J., Murphy-Weinberg, V., Watson, S.J., & Akil, H. (2000). Hormonal evidence for altered responsiveness to social stress in major depression. *Neuropsychopharmacology 23*, 411-418.

In: Psychoneuroendocrinology Research Trends
Editor: Martina T. Czerbska, pp. 451-462
ISBN: 978-1-60021-665-7

Chapter XVI

THE DEXAMETHASONE SUPPRESSION TEST IN BORDERLINE PERSONALITY DISORDER: THE IMPACT OF COMORBID DEPRESSIVE AND PTSD SYMPTOMS

Katja Wingenfeld and Martin Driessen
Department of Psychiatry and Psychotherapy Bethel, Ev. Hospital Bielefeld, Germany

ABSTRACT

Patients with borderline personality disorder (BPD) frequently report early, multiple, and chronic adverse or even traumatic experiences like repeated sexual or physical abuse or emotional neglect. Early life stress has been suggested to be an important risk factor in the development of BPD, although this may not be the case in all patients. However, traumatization is not a specific risk factor for BPD. It has been shown that women with a history of childhood sexual or physical abuse are also more likely to exhibit anxiety or mood disorders and of cause posttraumatic stress disorder (PTSD). Noteworthy, PTSD and MDD are also common in BPD.

In patients suffering from PTSD and MDD alterations of the hypothalamic-pituitary-adrenal (HPA) axis were observed repeatedly. While several studies found decreased cortisol suppression after ingestion of 1mg dexamethasone (DEX) in MDD, PTSD was characterized by an enhanced suppression after a low dose (0.5mg) of DEX. In borderline personality disorder some authors reported low rates of cortisol non-suppressors after 1mg dexamethasone while others found high rates of non-suppressors in such patients. Up to now only few studies used the low dose DST in BPD. Findings indicate an impact of comorbid PTSD symptoms and in part of depressive symptoms on HPA axis feedback regulation.

In sum, the current literature does not suggest a clear pattern of HPA axis feedback dysregulation in BPD patients. Trauma-related and depressive symptoms seem to interact with regard to their effects on HPA axis regulation or might even cancel out each other by opposite effects. Further studies should continue to evaluate such interactions in the relationships of early life stress, BPD, PTSD, and MDD when investigating HPA axis function.

BORDERLINE PERSONALITY DISORDER: PSYCHOPATHOLOGY AND CLINICAL FEATURES

Borderline personality disorder (BPD) is a complex and serious mental disorder which is characterized by intense and rapid changing mood states, as well as impulsivity, suicidal and self-injurious behaviors, fear of abandonment, as well as unstable relationships and unstable self-image (Saß, Wittchen, Zaudig, & Houben, 2003). Suicidal behavior is also highly prevalent in BPD patients and seems to be related to affective instability and depressive mood states (Soloff, Lynch, Kelly, Malone, & Mann, 2000; Yen et al., 2004). The occurrence of BPD is estimated to 2% in the general population, 10% in psychiatric outpatients and up to 20% in psychiatric inpatients. BPD is predominantly found in women (75%). In most cases maladapted behavior already exists in childhood and youth (Saß et al., 2003; Skodol et al., 2002; Soloff et al., 2000; Stone, 1993).

Patients with BPD often suffer from comorbid axis I disorders, with mood disorders (96.3%) and anxiety disorders (88.4%) being the most prominent. But also substance use disorders (64.1%) and eating disorders (53.0%) are of high prevalence (Zanarini et al., 1998). Interestingly, they found this "complex comorbidity" to have strong predictive power for the borderline diagnosis and it seems that a pattern of complex axis I comorbidity itself might be a useful marker for the borderline diagnosis (Zanarini et al., 1998). Furthermore, absence of lifetime axis I disorder, especially mood disorder, anxiety disorder, or posttraumatic stress disorder (PTSD) was found to be a predictor of earlier remission of the BPD symptoms (Zanarini, Frankenburg, Hennen, Reich, & Silk, 2006).

Patients with BPD frequently report early, multiple, and chronic adverse or even traumatic experiences like repeated sexual or physical abuse or emotional or physical neglect (Golier et al., 2003; Herman, Perry, & van der Kolk, 1989; McLean & Gallop, 2003; Ogata et al., 1990; Zanarini et al., 1997). It has been suggested, that early life stress might be an important risk factor in the development of BPD (Driessen et al., 2002; Johnson, Cohen, Brown, Smailes, & Bernstein, 1999; McLean & Gallop, 2003; Ogata et al., 1990), although this may not be the case in all BPD patients (Grossman et al., 2003). However, early life stress/traumatization is not a specific risk factor for BPD. Traumatization is discussed to be a risk factor for many psychiatric, psychosomatic, and physical complaints (Glod, 1993; Goldberg, Pachas, & Keith, 1999; Goodwin & Stein, 2004; Heim & Nemeroff, 2001; Heim, Wagner et al., 2006). It has been shown that women with a history of childhood sexual or physical abuse are more likely to exhibit symptoms of anxiety and of depression or even full DSM IV axis I and II mental disorders, e.g. major depression (MDD) or posttraumatic stress disorder (PTSD), than woman without such adverse early life experiences (Heim & Nemeroff, 1999; Johnson et al., 1999; Kendler, Bulik et al., 2000; Kendler, Thornton, & Gardner, 2000; McCauley et al., 1997; Owens & Nemeroff, 1991). Noteworthy, PTSD and major depression disorder (MDD) are also common in BPD and the absence of these disorders, as well as the absence of childhood sexual abuse and adult rape history enhances the chance of early remission in BPD (Hidalgo & Davidson, 2000; McGlashan et al., 2000; Zanarini et al., 1998; Zanarini et al., 2006).

The reported results suggest a strong association between trauma history and BPD, which leads to the discussion, if BPD might be conceptualized as a complex posttraumatic stress disorder (Driessen et al., 2002; Herman et al., 1989; McLean & Gallop, 2003). On the other hand, in a great sample of 379 patients with BPD the co-occurrence of PTSD was about 55.9% (Zanarini et al., 1998). That's why it has been emphasized that BPD might only be in parts conceptualized as a mental disorder related to traumatic experiences or stress. However, traumatization has been proposed to be an important etiological factor for a relatively wide range of psychiatric diseases, including depression, substance abuse, eating disorders, somatization disorders and anxiety disorders, summarized as trauma spectrum disorders (Heim, Bremner, & Nemeroff, 2006).

In sum, borderline personality disorder is a complex and heterogeneous mental disorder with a wide spectrum of symptoms and comorbid disorders. Especially mood and anxiety disorders, including PTSD, seem to play an important role in characterizing clinical features and treatment outcome (Zanarini et al., 2006).

Hypothalamic-Pituitary-Adrenal (HPA) Axis, Negative Feedback Regulation, and Mental Disorders

Due to the fact that early life stress and traumatization are major risk factors for the development and persistence of mental disorders, many studies investigated the functioning of the hypothalamic-pituitary-adrenal (HPA) axis, which is a major coordinator of the regulation of the stress response. Upon stress exposure, corticotropin-releasing factor (CRF) is released from the hypothalamus and is transported to the anterior pituitary where it stimulates the release of adrenocorticotropin (ACTH) which in turn stimulates the synthesis and secretion of glucocorticoids from the adrenal cortex. The neuroendocrine stress response is counter-regulated by circulating glucocorticoids via negative feedback mechanisms e.g. at the pituitary, hypothalamus, and hippocampus. This negative feedback loop is essential for the regulation of the HPA axis and therefore for the regulation of the stress response (Carrasco & Van de Kar, 2003; Plotsky, Owens, & Nemeroff, 1998).

To investigate the functioning of HPA axis feedback sensitivity the dexamethasone suppression test (DST) is widely used (Carroll, 1982, 1984). When administering 1 to 2mg dexamethasone, a synthetic glucocorticoid, at the evening when cortisol levels are low, a nearly complete suppression of cortisol release is observed at the following morning. Normally, in the morning cortisol levels are high due to the circadian rhythm of the HPA axis hormone release. It must be mentioned that dexamethasone is not able to pass the blood brain barrier and therefore the DST measures HPA axis feedback sensitivity only at pituitary level (Pariante et al., 2002).

In the standard version of the DST 1mg of dexamethasone (DEX) is used (Carroll, 1982). Especially in major depression disorder (MDD) the standard DST has been used in order to identify non-suppression or disinhibition of the HPA axis. In many but not all patients with MDD a decreased cortisol suppression after ingestion of 1mg DEX has been found repeatedly (Aguilar et al., 1984; Carroll, 1984; Musselmann et al., 1998; Plotsky et al., 1998). This reduced negative feedback sensitivity has been attributed to increased levels of circulating corticosteroids and in turn reduced glucocorticoids and mineralocorticoid receptors (Holsboer, 2000; Pariante & Miller, 2001; Pariante, Thomas, Lovestone, Makoff, & Kerwin, 2004). Most studies investigating HPA axis functioning in MDD are cross-sectional, but there is evidence for a normalization of abnormal DST results within clinical remission (Holsboer, Liebl, & Hofschuster, 1982; Holsboer, Wiedemann, Gerken, & Boll, 1986).

In posttraumatic stress disorder (PTSD), where decreased basal adrenal activity has been found in several studies (Mason, Giller, Kosten, Ostroff, & Podd, 1986; Yehuda, Kahana et al., 1995; Yehuda et al., 1990), Yehuda and coworkers suggested an association between hypocortisolism and enhanced feedback sensitivity of the HPA axis. Accordingly, they found a greater number of glucocorticoid receptors in these patients (Yehuda, Lowy, Southwick, Shaffer, & Giller, 1991). Consequently, they proposed to use a lower dose of DEX (0.5mg) to detect enhanced feedback sensitivity or "super-suppression" of cortisol release (Yehuda et al., 1993). Up to now enhanced cortisol suppression after 0.5mg DST has been replicated in several studies in PTSD and persons with history of childhood trauma (Newport, Heim, Bonsall, Miller, & Nemeroff, 2004; Stein, Yehuda, Koverola, & Hanna, 1997; Yehuda, Boisoneau, Lowy, & Giller, 1995; Yehuda, Golier, Halligan, Meaney, & Bierer, 2004; Yehuda, Halligan, Grossman, Golier, & Wong, 2002). Interestingly, super-suppression in the DST has been also found in several stress-related psychosomatic complaints as chronic pelvic pain, fibromyalgia, chronic fatigue syndrome (Gaab et al., 2002; Heim, Ehlert, Hanker, & Hellhammer, 1998; Wingenfeld et al., 2007). In these disorders high rates of trauma and PTSD have been reported (Heim, Ehlert, & Hellhammer, 2000; Heim, Wagner et al., 2006; Lampe et al., 2000; Walker et al., 1997). Up to now little is known about the longitudinal course of feedback sensitivity in PTSD patients. Two case studies suggest that symptom improvement is accompanied by a normalization of HPA axis functioning in PTSD, namely the basal cortisol levels as well as cortisol suppression in the DST (Heber, Kellner, & Yehuda, 2002; Kellner, Yehuda, Arlt, & Wiedemann, 2002).

In sum, MDD and PTSD are characterized by different patterns of HPA axis feedback regulation. While in MDD reduced feedback sensitivity has been reported, several studies found super-suppression or enhanced feedback sensitivity in PTSD. Both, depressive mood states and PTSD are prominent in borderline personality disorder, which leads to the question if and how HPA axis feedback regulation is altered in these patients.

The HPA Axis Feedback Regulation in BPD

In BPD the standard DST (1mg DEX) has been performed repeatedly yielding varying results. Many of these studies found high rates of non-suppressors: e.g. 73% (Baxter, Edell, Gerner, Fairbanks, & Gwirtsman, 1984), 62% (Carroll et al., 1981), 54% (Kontaxakis et al., 1987), 62% (Beeber, Kline, Pies, & Manring, 1984), 54% (Sternbach, Fleming, Extein, Pottash, & Gold, 1983). However, most of these results suggested an association of reduced feedback inhibition with affective dysregulation in these patients or even comorbid MDD. It has been emphasized that in many of these studies no sufficient diagnostic procedure was applied so that the data are difficult to interpret (Lahmeyer, Reynolds, Kupfer, & King, 1989). Interestingly, in a sample of BPD patients which were well diagnosed concerning axis I disorders it has been shown that BPD patients with comorbid MDD had highest cortisol levels after dexamethasone (1.5mg) compared to those without comorbid MDD and BPD patients with PTSD or both (Rinne et al., 2002). On the other hand, BPD patients with comorbid PTSD had the lowest cortisol release after dexamethasone, but the sample size of this subgroup was rather small. Accordingly, in BPD patients without comorbid MDD, low rates of DST non-suppressor have been reported (Lahmeyer et al., 1988). De la Fuente & Mendlewicz (1996) for example found only 25% in patients with BPD group compared to 65% in patients with MDD. Another study reported 26% non-suppressors in BPD patients with major depression disorder compared to 11% in the BPD group without comorbid MDD (Korzekwa, Steiner, Links, & Eppel, 1991). However, normal DST results have been also found in BPD despite high levels of depressive symptoms (Soloff, George, & Nathan, 1982).

In sum, the 1mg DST in BPD resulted in varying results and it remains still unclear whether the found alterations are specific to BPD or are related to comorbid axis I symptoms. Thus the relevance of the DST in borderline personality disorders has been discussed controversially. Korzekwa et al. (1991) for example criticized the low sensitivity of the DST in these patients but emphasized the possibility of an affective borderline subtype, which - if validated - might give important indications for treatment. Up to now and despite of the study of Rinne et al. (2002), only comorbid depression was taken into account when discussing DST results in BPD. As described before traumatization and PTSD are also common features in these patients. In PTSD - in contrast to MDD - enhanced feedback sensitivity has been found when using a lower dose of DEX. Therefore it might be inadequate using only the standard dose when investigating HPA axis feedback sensitivity in BPD.

Grossman, Yehuda, & Siever (1997) used the low dose (0.5mg) DST as proposed by Yeduda and co-workers (Yehuda et al., 1993) and found enhanced cortisol suppression in four patients with BPD. PTSD was not found to have an influence on cortisol suppression, but the sample was very small. In a more recent study with patients suffering from different personality disorders (from whom 42% had BPD), the same working group showed that higher suppression of cortisol after 0.5mg DEX was associated with PTSD, while depression had no significant effect on the cortisol suppression (Grossman et al., 2003). Unfortunately, in this study no control group of healthy participants was examined and it remains unclear how exactly cortisol suppression was altered. Recently, we did not find differences of the HPA axis feedback regulation between BPD patients and controls (Lange et al., 2005). However, when subdividing the BPD group in subgroups with and without comorbid PTSD, we observed reduced cortisol suppression after 0.5mg DEX only in those without PTSD. On the other hand, comorbid major depression was not associated with altered cortisol suppression in this sample. Another study also found reduced rather than enhanced cortisol suppression after 0.5mg of dexamethasone in BPD patients compared to healthy controls (Lieb et al., 2004). In this study none of the patients had comorbid major depression disorder. Unfortunately, PTSD was not assessed, which makes the data difficult to interpret.

Up to now only few studies used the low dose DST in patients with BPD and these few results are also diverging. However, the low dose DST seems to be sufficient to detect both, enhanced and reduced suppression and therefore might be more appropriate when investigating such a heterogeneous population like patients with BPD.

As mentioned before traumatization is very frequent in BPD, which therefore may in part be conceptualized as "complex post traumatic stress disorder" or at least complex trauma-associated disorder (Driessen et al., 2002; Sabo, 1997). If so, it may be rather formal or even artificial to categorize samples of BPD in patients with and without PTSD. In our own study for example (Lange et al., 2005) BPD patients without PTSD showed remarkable higher PTSD symptom scores than the control group as measured by the Impact of Event Scale (IES-R) suggesting a dimensional distribution of the number of PTSD symptoms in BPD patients. However, apart from the study of Grossmann et al. (2006) up to now PTSD symptoms were not specifically taken into account when investigating HPA axis feedback regulation in BPD.

We previously performed the 0.5mg DST in a sample of BPD patients and healthy controls (Wingenfeld et al., subm.). Following the hypothesis of a dimensional distribution of PTSD (and depressive) symptoms in BPD we performed partial correlations between the percentage of cortisol suppression in the DST and number of PTSD symptoms and depression scores, respectively. When statistically controlling for depressive symptoms, we found a positive association between percentage of cortisol suppression and PTSD symptoms. On the other hand a negative association was found between percentage of cortisol suppression and depressive symptoms when controlling for PTSD symptoms. A similar pattern of HPA axis dysregulation was found when measuring free urinary cortisol: cortisol release was positively associated with depression but had a negative correlation with PTSD symptoms (Wingenfeld, Driessen, Adam, & Hill, 2006). It may be that trauma-related and depressive symptoms interact with regard to their effects on HPA axis regulation.

As mentioned before only few studies evaluated the longitudinal course of HPA axis feedback regulation in psychiatric disorders, while a relative intra-individual stability of plasma cortisol and feedback sensitivity was shown in non-psychiatric samples (Huizenga et al., 1998; Yerevanian, Anderson, & Milanese, 1985). Noteworthy, impaired cortisol suppression in major depression disorder is known to normalize after symptom improvement (Holsboer et al., 1982; Holsboer et al., 1986; Ising et al., 2005) but not within the depressive episode (Charles, Wilmotte, Quenon, & Mendlewicz, 1982). To our knowledge, in neither PTSD nor BPD such longitudinal investigations do exist. However, in two case studies it has been shown that within clinical improvement of PTSD symptoms, HPA axis functioning tend to normalize (Heber et al., 2002; Kellner et al., 2002). These first longitudinal reports suggest that it might be useful to include endocrine measures when evaluating treatment outcome. Overall, these studies as well as the results in major depression disorder suggest a relative stability of HPA axis function during illness but also its capability to adjust. In one of our own studies, we evaluated the stability of the dexamethasone suppression test in a sample of patients with BDP in a one year follow-up study (Wingenfeld et al., subm.). We found that cortisol concentrations before and after dexamethasone remain stable over time, when there was no clinical change of PTSD symptoms and general psychopathology in the patients. As we have previously shown (Lange et al., 2005), BPD patients with and without comorbid

PTSD differ in cortisol suppression with PTSD patients also showing more pronounced cortisol suppression after DEX one year after baseline examination. The here reported results support the use of the low-dose DST as a reliable and valid marker of alterations of HPA axis dysregulation, especially because of its capacity to discriminate between patients with and without PTSD.

CONCLUSION

In sum, the current literature does not suggest a clear pattern of HPA feedback dysregulation in BPD patients. There is evidence for both, hypo- and hyper-suppression in the DST. However, it seems that these alterations are strongly related to comorbid or dominant symptoms, like depressive mood states and PTSD symptoms, which both are very prominent in BPD patients. In these patients trauma-related and depressive symptoms seem to interact with regard to their effects on HPA axis regulation or might even cancel out each other by opposite effects. It might be suggested that within the borderline population different subtypes exist, which might be characterized by predominant co-occurring syndromes, e.g. affective or trauma-associated symptoms. Following this hypothesis, diverging pattern of HPA axis dysregulation would be not surprising, possibly reflecting these subtypes. Further studies should continue to evaluate the relationship between early life stress, BPD, PTSD and MDD when investigating HPA axis function and further endocrine reaction patterns to physiological and psychological stimuli.

REFERENCES

Aguilar, M. T., Lemaire, M., Castro, P., Libotte, M., Reynders, J., & Herchuelz, A. (1984). Study of the diagnostic value of the dexamethasone suppression test in endogenous depression. *J Affect Disord, 6*(1), 33-42.

Baxter, L., Edell, W., Gerner, R., Fairbanks, L., & Gwirtsman, H. (1984). Dexamethasone suppression test and Axis I diagnoses of inpatients with DSM-III borderline personality disorder. *J Clin Psychiatry, 45*(4), 150-153.

Beeber, A. R., Kline, M. D., Pies, R. W., & Manring, J. M., Jr. (1984). Dexamethasone suppression test in hospitalized depressed patients with borderline personality disorder. *J Nerv Ment Dis, 172*(5), 301-303.

Carrasco, G. A., & Van de Kar, L. D. (2003). Neuroendocrine pharmacology of stress. *Eur J Pharmacol, 463*(1-3), 235-272.

Carroll, B. J. (1982). The dexamethasone suppression test for melancholia. *Br J Psychiatry, 140*, 292-304.

Carroll, B. J. (1984). Dexamethasone suppression test for depression. *Adv Biochem Psychopharmacol, 39*, 179-188.

Carroll, B. J., Greden, J. F., Feinberg, M., Lohr, N., James, N. M., Steiner, M., et al. (1981). Neuroendocrine evaluation of depression in borderline patients. *Psychiatr Clin North Am, 4*(1), 89-99.

Charles, G., Wilmotte, J., Quenon, M., & Mendlewicz, J. (1982). Reproducibility of the dexamethasone suppression test in depression. *Biol Psychiatry, 17*(7), 845-848.

De la Fuente, J. M., & Mendlewicz, J. (1996). TRH stimulation and dexamethasone suppression in borderline personality disorder. *Biol Psychiatry, 40*(5), 412-418.

Driessen, M., Beblo, T., Reddemann, L., Rau, H., Lange, W., Silva, A., et al. (2002). Ist die Borderline-Persönlichkeitsstörung eine komplexe posttraumatischen Belastungsstörung? *Nervenarzt, 73*, 820-829.

Gaab, J., Huster, D., Peisen, R., Engert, V., Schad, T., Schurmeyer, T. H., et al. (2002). Low-dose dexamethasone suppression test in chronic fatigue syndrome and health. *Psychosom Med, 64*(2), 311-318.

Glod, C. A. (1993). Long-term consequences of childhood physical and sexual abuse. *Arch Psychiatr Nurs, 7*(3), 163-173.

Goldberg, R. T., Pachas, W. N., & Keith, D. (1999). Relationship between traumatic events in childhood and chronic pain. *Disabil Rehabil, 21*(1), 23-30.

Golier, J. A., Yehuda, R., Bierer, L. M., Mitropoulou, V., New, A. S., Schmeidler, J., et al. (2003). The relationship of borderline personality disorder to posttraumatic stress disorder and traumatic events. *Am J Psychiatry, 160*(11), 2018-2024.

Goodwin, R. D., & Stein, M. B. (2004). Association between childhood trauma and physical disorders among adults in the United States. *Psychol Med, 34*(3), 509-520.

Grossman, R., Yehuda, R., New, A., Schmeidler, J., Silverman, J., Mitropoulou, V., et al. (2003). Dexamethasone suppression test findings in subjects with personality disorders: associations with posttraumatic stress disorder and major depression. *Am J Psychiatry, 160*(7), 1291-1298.

Grossman, R., Yehuda, R., & Siever, L. (1997). The dexamethasone suppression test and glucocorticoid receptors in borderline personality disorder. *Ann N Y Acad Sci, 821*, 459-464.

Heber, R., Kellner, M., & Yehuda, R. (2002). Salivary cortisol levels and the cortisol response to dexamethasone before and after EMDR: a case report. *J Clin Psychol, 58*(12), 1521-1530.

Heim, C., Bremner, J. D., & Nemeroff, C. B. (2006). Trauma spectrum disorders. In M. S. R. C. Patterson (Ed.), *Principles of Molecular Medicine* (pp. 1203-1210): Totowa, NJ: Humana Press.

Heim, C., Ehlert, U., Hanker, J. P., & Hellhammer, D. H. (1998). Abuse-related posttraumatic stress disorder and alterations of the hypothalamic-pituitary-adrenal axis in women with chronic pelvic pain. *Psychosom Med, 60*(3), 309-318.

Heim, C., Ehlert, U., & Hellhammer, D. H. (2000). The potential role of hypocortisolism in the pathophysiology of stress-related bodily disorders. *Psychoneuroendocrinology, 25*(1), 1-35.

Heim, C., & Nemeroff, C. B. (1999). The impact of early adverse experiences on brain systems involved in the pathophysiology of anxiety and affective disorders. *Biol Psychiatry, 46*(11), 1509-1522.

Heim, C., & Nemeroff, C. B. (2001). The role of childhood trauma in the neurobiology of mood and anxiety disorders: preclinical and clinical studies. *Biol Psychiatry, 49*(12), 1023-1039.

Heim, C., Wagner, D., Maloney, E., Papanicolaou, D. A., Solomon, L., Jones, J. F., et al. (2006). Early Adverse Experience and Risk for Chronic Fatigue Syndrome: Results From a Population-Based Study. *Arch Gen Psychiatry, 63*(11), 1258-1266.

Herman, J. L., Perry, J. C., & van der Kolk, B. A. (1989). Childhood trauma in borderline personality disorder. *Am J Psychiatry, 146*(4), 490-495.

Hidalgo, R. B., & Davidson, J. R. (2000). Posttraumatic stress disorder: epidemiology and health-related considerations. *J Clin Psychiatry, 61 Suppl 7*, 5-13.

Holsboer, F. (2000). The corticosteroid receptor hypothesis of depression. *Neuropsychopharmacology, 23*(5), 477-501.

Holsboer, F., Liebl, R., & Hofschuster, E. (1982). Repeated dexamethasone suppression test during depressive illness. Normalisation of test result compared with clinical improvement. *J Affect Disord, 4*(2), 93-101.

Holsboer, F., Wiedemann, K., Gerken, A., & Boll, E. (1986). The plasma dexamethasone variable in depression: test-retest studies and early biophase kinetics. *Psychiatry Res, 17*(2), 97-103.

Huizenga, N. A., Koper, J. W., de Lange, P., Pols, H. A., Stolk, R. P., Grobbee, D. E., et al. (1998). Interperson variability but intraperson stability of baseline plasma cortisol concentrations, and its relation to feedback sensitivity of the hypothalamo-pituitary-adrenal axis to a low dose of dexamethasone in elderly individuals. *J Clin Endocrinol Metab, 83*(1), 47-54.

Ising, M., Kunzel, H. E., Binder, E. B., Nickel, T., Modell, S., & Holsboer, F. (2005). The combined dexamethasone/CRH test as a potential surrogate marker in depression. *Prog Neuropsychopharmacol Biol Psychiatry, 29*(6), 1085-1093.

Johnson, J. G., Cohen, P., Brown, J., Smailes, E. M., & Bernstein, D. P. (1999). Childhood maltreatment increases risk for personality disorders during early adulthood. *Arch Gen Psychiatry, 56*(7), 600-606.

Kellner, M., Yehuda, R., Arlt, J., & Wiedemann, K. (2002). Longitudinal course of salivary cortisol in post-traumatic stress disorder. *Acta Psychiatr Scand, 105*(2), 153-155; discussion 155-156.

Kendler, K. S., Bulik, C. M., Silberg, J., Hettema, J. M., Myers, J., & Prescott, C. A. (2000). Childhood sexual abuse and adult psychiatric and substance use disorders in women: an epidemiological and cotwin control analysis. *Arch Gen Psychiatry, 57*(10), 953-959.

Kendler, K. S., Thornton, L. M., & Gardner, C. O. (2000). Stressful life events and previous episodes in the etiology of major depression in women: an evaluation of the "kindling" hypothesis. *Am J Psychiatry, 157*(8), 1243-1251.

Kontaxakis, V., Markianos, M., Vaslamatzis, G., Markidis, M., Kanellos, P., & Stefanis, C. (1987). Multiple neuroendocrinological responses in borderline personality disorder patients. *Acta Psychiatr Scand, 76*(5), 593-597.

Korzekwa, M., Steiner, M., Links, P., & Eppel, A. (1991). The dexamethasone suppression test in borderlines: is it useful? *Can J Psychiatry, 36*(1), 26-28.

Lahmeyer, H. W., Reynolds, C. F., 3rd, Kupfer, D. J., & King, R. (1989). Biologic markers in borderline personality disorder: a review. *J Clin Psychiatry, 50*(6), 217-225.

Lahmeyer, H. W., Val, E., Gaviria, F. M., Prasad, R. B., Pandey, G. N., Rodgers, P., et al. (1988). EEG sleep, lithium transport, dexamethasone suppression, and monoamine oxidase activity in borderline personality disorder. *Psychiatry Res, 25*(1), 19-30.

Lampe, A., Solder, E., Ennemoser, A., Schubert, C., Rumpold, G., & Sollner, W. (2000). Chronic pelvic pain and previous sexual abuse. *Obstet Gynecol, 96*(6), 929-933.

Lange, W., Wulff, H., Berea, C., Beblo, T., Saavedra, A. S., Mensebach, C., et al. (2005). Dexamethasone suppression test in borderline personality disorder--effects of posttraumatic stress disorder. *Psychoneuroendocrinology, 30*(9), 919-923.

Lieb, K., Rexhausen, J. E., Kahl, K. G., Schweiger, U., Philipsen, A., Hellhammer, D. H., et al. (2004). Increased diurnal salivary cortisol in women with borderline personality disorder. *J Psychiatr Res, 38*(6), 559-565.

Mason, J. W., Giller, E. L., Kosten, T. R., Ostroff, R. B., & Podd, L. (1986). Urinary free-cortisol levels in posttraumatic stress disorder patients. *J Nerv Ment Dis, 174*(3), 145-149.

McCauley, J., Kern, D. E., Kolodner, K., Dill, L., Schroeder, A. F., DeChant, H. K., et al. (1997). Clinical characteristics of women with a history of childhood abuse: unhealed wounds. *Jama, 277*(17), 1362-1368.

McGlashan, T. H., Grilo, C. M., Skodol, A. E., Gunderson, J. G., Shea, M. T., Morey, L. C., et al. (2000). The Collaborative Longitudinal Personality Disorders Study: baseline Axis I/II and II/II diagnostic co-occurrence. *Acta Psychiatr Scand, 102*(4), 256-264.

McLean, L. M., & Gallop, R. (2003). Implications of childhood sexual abuse for adult borderline personality disorder and complex posttraumatic stress disorder. *Am J Psychiatry, 160*(2), 369-371.

Musselmann, D. L., DeBattista, C., Nathan, K. I., Kilts, C. D., Schatzberg, A. F., & Nemeroff, C. B. (1998). Biology of Mood disorders. In A. F. Schatzberg, Nemeroff, C.B. (Ed.), *Textbook of Psychopharmacology. 2nd Edition.*: American Psychiatry Press.

Newport, D. J., Heim, C., Bonsall, R., Miller, A. H., & Nemeroff, C. B. (2004). Pituitary-adrenal responses to standard and low-dose dexamethasone suppression tests in adult survivors of child abuse. *Biol Psychiatry, 55*(1), 10-20.

Ogata, S. N., Silk, K. R., Goodrich, S., Lohr, N. E., Westen, D., & Hill, E. M. (1990). Childhood sexual and physical abuse in adult patients with borderline personality disorder. *Am J Psychiatry, 147*(8), 1008-1013.

Owens, M. J., & Nemeroff, C. B. (1991). Physiology and pharmacology of corticotropin-releasing factor. *Pharmacol Rev, 43*(4), 425-473.

Pariante, C. M., & Miller, A. H. (2001). Glucocorticoid receptors in major depression: relevance to pathophysiology and treatment. *Biol Psychiatry, 49*(5), 391-404.

Pariante, C. M., Papadopoulos, A. S., Poon, L., Checkley, S. A., English, J., Kerwin, R. W., et al. (2002). A novel prednisolone suppression test for the hypothalamic-pituitary-adrenal axis. *Biol Psychiatry, 51*(11), 922-930.

Pariante, C. M., Thomas, S. A., Lovestone, S., Makoff, A., & Kerwin, R. W. (2004). Do antidepressants regulate how cortisol affects the brain? *Psychoneuroendocrinology, 29*(4), 423-447.

Plotsky, P. M., Owens, M. J., & Nemeroff, C. B. (1998). Psychoneuroendocrinology of depression. Hypothalamic-pituitary-adrenal axis. *Psychiatr Clin North Am, 21*(2), 293-307.

Rinne, T., de Kloet, E. R., Wouters, L., Goekoop, J. G., DeRijk, R. H., & van den Brink, W. (2002). Hyperresponsiveness of hypothalamic-pituitary-adrenal axis to combined dexamethasone/corticotropin-releasing hormone challenge in female borderline personality disorder subjects with a history of sustained childhood abuse. *Biol Psychiatry, 52*(11), 1102-1112.

Sabo, A. N. (1997). Etiological significance of associations between childhood trauma and borderline personality disorder: conceptual and clinical implications. *J Personal Disord, 11*(1), 50-70.

Saß, H., Wittchen, H. U., Zaudig, M., & Houben, I. (2003). *Diagnostisches und Statistisches Manual Psychischer Störungen - Textrevision - (DSM-IV-TR)*. Göttingen. Bern. Toronto. Seattle.

Skodol, A. E., Gunderson, J. G., Pfohl, B., Widiger, T. A., Livesley, W. J., & Siever, L. J. (2002). The borderline diagnosis I: psychopathology, comorbidity, and personality structure. *Biol Psychiatry, 51*(12), 936-950.

Soloff, P. H., George, A., & Nathan, R. S. (1982). The dexamethasone suppression test in patients with borderline personality disorders. *Am J Psychiatry, 139*(12), 1621-1623.

Soloff, P. H., Lynch, K. G., Kelly, T. M., Malone, K. M., & Mann, J. J. (2000). Characteristics of suicide attempts of patients with major depressive episode and borderline personality disorder: a comparative study. *Am J Psychiatry, 157*(4), 601-608.

Stein, M. B., Yehuda, R., Koverola, C., & Hanna, C. (1997). Enhanced dexamethasone suppression of plasma cortisol in adult women traumatized by childhood sexual abuse. *Biol Psychiatry, 42*(8), 680-686.

Sternbach, H. A., Fleming, J., Extein, I., Pottash, A. L., & Gold, M. S. (1983). The dexamethasone suppression and thyrotropin-releasing hormone tests in depressed borderline patients. *Psychoneuroendocrinology, 8*(4), 459-462.

Stone, M. H. (1993). Long-term outcome in personality disorders. *Br J Psychiatry, 162*, 299-313.

Walker, E. A., Keegan, D., Gardner, G., Sullivan, M., Bernstein, D., & Katon, W. J. (1997). Psychosocial factors in fibromyalgia compared with rheumatoid arthritis: II. Sexual, physical, and emotional abuse and neglect. *Psychosom Med, 59*(6), 572-577.

Wingenfeld, K., Driessen, M., Adam, B., & Hill, A. (2006). Overnight urinary cortisol release in women with borderline personality disorder depends on comorbid PTSD and depressive psychopathology. *Eur Psychiatry*.

Wingenfeld, K., Wagner, D., Schmidt, I., Meinlschmidt, G., Hellhammer, D. H., & Heim, C. (2007). The low-dose dexamethasone suppression test in fibromyalgia. *J Psychosom Res, 62*(1), 85-91.

Yehuda, R., Boisoneau, D., Lowy, M. T., & Giller, E. L., Jr. (1995). Dose-response changes in plasma cortisol and lymphocyte glucocorticoid receptors following dexamethasone administration in combat veterans with and without posttraumatic stress disorder. *Arch Gen Psychiatry, 52*(7), 583-593.

Yehuda, R., Golier, J. A., Halligan, S. L., Meaney, M., & Bierer, L. M. (2004). The ACTH response to dexamethasone in PTSD. *Am J Psychiatry, 161*(8), 1397-1403.

Yehuda, R., Halligan, S. L., Grossman, R., Golier, J. A., & Wong, C. (2002). The cortisol and glucocorticoid receptor response to low dose dexamethasone administration in aging combat veterans and holocaust survivors with and without posttraumatic stress disorder. *Biol Psychiatry, 52*(5), 393-403.

Yehuda, R., Kahana, B., Binder-Brynes, K., Southwick, S. M., Mason, J. W., & Giller, E. L. (1995). Low urinary cortisol excretion in Holocaust survivors with posttraumatic stress disorder. *Am J Psychiatry, 152*(7), 982-986.

Yehuda, R., Lowy, M. T., Southwick, S. M., Shaffer, D., & Giller, E. L., Jr. (1991). Lymphocyte glucocorticoid receptor number in posttraumatic stress disorder. *Am J Psychiatry, 148*(4), 499-504.

Yehuda, R., Southwick, S. M., Krystal, J. H., Bremner, D., Charney, D. S., & Mason, J. W. (1993). Enhanced suppression of cortisol following dexamethasone administration in posttraumatic stress disorder. *Am J Psychiatry, 150*(1), 83-86.

Yehuda, R., Southwick, S. M., Nussbaum, G., Wahby, V., Giller, E. L., Jr., & Mason, J. W. (1990). Low urinary cortisol excretion in patients with posttraumatic stress disorder. *J Nerv Ment Dis, 178*(6), 366-369.

Yen, S., Shea, M. T., Sanislow, C. A., Grilo, C. M., Skodol, A. E., Gunderson, J. G., et al. (2004). Borderline personality disorder criteria associated with prospectively observed suicidal behavior. *Am J Psychiatry, 161*(7), 1296-1298.

Yerevanian, B. I., Anderson, J. L., & Milanese, E. (1985). Effect of weekly testing on dexamethasone suppression test results in normal subjects. *Psychiatry Res, 14*(3), 209-213.

Zanarini, M. C., Frankenburg, F. R., Dubo, E. D., Sickel, A. E., Trikha, A., Levin, A., et al. (1998). Axis I comorbidity of borderline personality disorder. *Am J Psychiatry, 155*(12), 1733-1739.

Zanarini, M. C., Frankenburg, F. R., Hennen, J., Reich, D. B., & Silk, K. R. (2006). Prediction of the 10-year course of borderline personality disorder. *Am J Psychiatry, 163*(5), 827-832.

Zanarini, M. C., Williams, A. A., Lewis, R. E., Reich, R. B., Vera, S. C., Marino, M. F., et al. (1997). Reported pathological childhood experiences associated with the development of borderline personality disorder. *Am J Psychiatry, 154*(8), 1101-1106.

In: Psychoneuroendocrinology Research Trends
Editor: Martina T. Czerbska, pp. 463-495

ISBN: 978-1-60021-665-7

Chapter XVII

Psychoneuroendocrinology of Functional Somatic Disorders

Lineke M. Tak and Judith G.M. Rosmalen
Department of Psychiatry, University Medical Center Groningen, University of Groningen, Groningen, The Netherlands.

Abstract

Functional somatic disorders are syndromes for which no clear or consistent organic pathology can be found. Since functional somatic disorders are associated with early life and chronic stress, the question raises how stress is linked to experiencing complaints. Psychoneuroendocrinological approaches combine psychological and biomedical views and provide a complex but promising new strategy to unravel somatization.

In this chapter, we will first introduce "the big three" functional somatic disorders and discuss the role of stress in these disorders. Next, we will give an overview of current theories and studies about psychoneuroendocrinological alterations in functional somatic disorders, with an emphasis on the two most important stress axes in the human body: the autonomic nervous system (ANS) and the hypothalamic-pituitary-adrenal-axis (HPA-axis).

The majority of alterations in ANS function found in functional somatic disorders consists of a decreased parasympathetic and/or an increased sympathetic activation, nonetheless, there are some dissonant findings. A critical approach will be taken to evaluate these findings.

Assessment of the HPA-axis usually demonstrates a lower cortisol output in patients with functional somatic disorders compared to healthy controls. However, a lot of studies examining the HPA-axis were not able to detect any differences, and few studies detect even an elevated cortisol output. We will make an attempt to show the most consistent evidence out of the heterogeneous methods and results reported.

Because of the heterogeneity in both ANS and HPA-axis findings in functional somatic disorders, we will provide some factors that may explain the discrepancies.

Finally, we discuss lacunas in our current knowledge about the role of the ANS and HPA-axis in somatization. Recommendations for future research include use of standardized research methods and performance of longitudinal cohort studies.

1. INTRODUCTION

Medically unexplained symptoms, symptoms for which no clear or consistent organic pathology can be found, form a substantial part of complaints presented at the general practice and outpatients clinic [1]. These symptoms are difficult to name, conceptualize and classify [2]. Functional somatic disorders are syndromes of related complaints with no known organic pathological basis. The “big three” are chronic fatigue syndrome (CFS), fibromyalgia (FM) and irritable bowel syndrome (IBS). Other examples of functional somatic disorders include whiplash syndrome, functional dyspepsia and temporomandibular dysfunction. Functional somatic disorders can be regarded as the result of accumulation of persistent medically unexplained symptoms.

Different functional somatic disorders share a lot of similarities, for example in case definition, reported symptoms, and in non-symptom specific associations such as sex, prognosis and response to treatment [3]. Because of these similarities, the so-called “lumpers” hypothesize that there is in fact only one reservoir of functional somatic disorders. In contrast, the so-called “splitters”, argue that (a) there are distinctive physical abnormalities in some syndromes, (b) that even if the similarities outnumber the differences, that is no reason to ignore them, and (c) that it is an overstatement that the core features of the syndromes overlap.

Nevertheless, it is clear that functional somatic disorders have common features, including the presence of multiple complaints without any known organic pathology. Although no clear organic cause can be found for functional somatic disorders, the approach in which complaints are considered either somatic or psychiatric in origin can not be preserved in present times. Psychobiological explanations of somatization may mediate the psychological influences on subjective physical health experience. In this chapter, we aim to explore psychoneuroendocrinological pathways as mediators between psychological influences and functional somatic disorders. We will focus on the “big three”, but also include results obtained in other functional somatic disorders characterized by predominantly fatigue, musculoskeletal or gastrointestinal complaints. We will start this chapter by summarizing the evidence for a role of stress in onset and maintenance of functional somatic disorders. Next, we will explore the role of the main stress-responsive systems in the body: the autonomic nervous system and the hypothalamus-pituitary-adrenal axis. For each axis, we will briefly describe the physiology and assessment methods, followed by a discussion of alterations of stress axes function in functional somatic disorders. We will end this chapter with recommendations for further research into the psychoneuroendocrinology of functional somatic disorders.

2. Life Stress in Functional Somatic Disorders

Stress is the consequence of exposure to stressors. Stressors are defined as characteristics, events or conditions that give rise to stress, with the stressfulness of these stressors being highly subjective [4]. When stress exceeds a certain threshold, it evokes a generalized stress system response. The aim of this response is to restore homeostasis. When the load of stressors in an individual is too large or the capacity to adjust for them is diminished, stress system dysfunction may develop [5]. Somatization is sometimes referred to as an expression of psychological stress in physical complaints. This view is probably too one dimensional; a complex model including several other intrinsic and extrinsic factors appears more likely.

Numerous studies on the psychoneuroendocrinology of somatization are based on the premise that chronic stress is involved in functional somatic disorders. In this paragraph, we will shortly summarize the evidence that experienced life stress is associated with development of CFS, FM and IBS.

Concerning CFS, one review suggests that the stress sensitive personality traits neuroticism and introversion are risk factors for CFS. Acute psychological stress, such as the loss of a loved one or a job, may trigger the onset of CFS. Furthermore, evidence for a relatively high prevalence of psychiatric disorders in CFS is abundant [6]. Also, two recent studies suggest a role for stress in the etiology of CFS. One study has supported the hypothesis that psychological factors are implicated in the development of CFS, through demonstrating increased levels of multiple types of childhood trauma in a population-based CFS sample [7]. The other, longitudinal, study shows that the amount of premorbid stress, also if happened several decades ago, predicts CFS later in life [8].

Indications for chronic stress in FM have been reviewed by several authors [9-12]. Population-based studies show that psychological distress, particularly early-life trauma such as parental loss and physical or sexual abuse, is associated with the future development of FM. Not only abuse, but also household disharmony, frequent physical violence between the parents, addictive behaviour in the mother and financial problems are associated with FM later in life. Furthermore, FM patients show high rates of psychiatric comorbidity and often deal with emotion and stress in maladaptive ways. Self-sacrificing lifestyle and the inability to cope with high daily workload are also associated with FM. Although the history of FM patients is frequently characterized by multiple stressors, knowledge about the precise role of stress in FM remains fragmentary.

Concerning IBS, studies have shown that the majority of IBS patients exhibit psychiatric comorbidity. Personality traits such as neuroticism, psychological difficulties and a history of trauma and abuse and other negative life events are also associated with the onset of IBS. The role of these stressors seems increased in the more severe forms of IBS [13-17]. Chronic stress may also be involved in other functional somatic disorders with predominantly gastrointestinal symptoms. In both clinical and population samples, patients with functional dyspepsia report more life events and chronic difficulties than both duodenal ulcer patients and healthy controls. Furthermore, observed higher levels of introversion and suspicion may hamper adequate coping with stress [18]. In children with functional abdominal pain, an association has been found between stress and somatic complaints. Children with functional abdominal pain show elevated rates of anxiety and depression compared to healthy controls. Early loss and separation are also associated with the development of functional abdominal pain. Furthermore, the majority of the parents of children with functional abdominal pain attach the onset of the gastrointestinal complaints to a stressful event [17].

Several problems characterize research towards life stress. Firstly, observed associations between functional somatic disorders and psychological factors may be the result of increased health care seeking by distressed patients. Another problem is the use of self-report measures to assess life stress. The presence of disease may increase the amount of reported life stress through a negative recall bias [19]. Furthermore, it remains unclear whether stress causes functional somatic disorders, or whether patients experience stress because they feel sick.

To summarize, stressful life events, psychiatric comorbidity and stress sensitive personality traits are associated with functional somatic disorders. Stress axes alterations may form a psychoneuroendocrinological bridge between psychosocial difficulties and subjective somatic symptoms. We will continue with discussing the evidence for abnormalities in stress axes function in functional somatic disorders. For clarity, we will start with a brief overview of physiology and assessment methods of the stress-axis.

3. Autonomic Nervous System

3.1. Physiology and Assessment of the ANS

One of the major stress-responsive systems in the body is the ANS. The ANS is responsible for rapid stress responses, since it reacts within seconds after stimulation. The physiology of the ANS is described in detail elsewhere [20]. For the understanding of the potential role of the ANS in somatization, we summarize the most important features of this system. The ANS controls bodily functions such as thermoregulation, breathing and circulation. It helps maintain homeostasis and coordinates responses to external stimuli. The ANS can be divided in two divisions: the sympathetic and the parasympathetic nervous system. The sympathetic nervous system is frequently referred to as the “stress” or “fight or flight" system, as it has a stimulatory effect on bodily systems and organs which are responsible for quick sensory activity and movement. The sympathetic influence is mediated through the postganglionic release of noradrenalin. Roughly said, the parasympathetic nervous system antagonizes the sympathetic nervous system, since it has "rest and digest" activity. Stimulation of the parasympathetic nervous system results in bradycardia and relaxes many bodily systems and organs through the postganglionic release of acetylcholine. The ANS does not function on its own, but is both anatomically and functionally linked to the rest of the nervous system.

There are several methods to assess autonomic function: cardiovascular response tests and analysis of heart rate variability (HRV) or baroreceptor sensitivity (BRS). Descriptions of the tests are outlined in a review about ANS testing by Ravits [21], and shortly reproduced below:

Cardiovascular response to standing. Studying blood pressure (BP) alterations to standing is indicated in testing sympathetic function. Measuring heart rate (HR) alterations to standing (30:15 ratio; the ratio of the slowest HR occurring around 30 beats after standing and the fastest HR about 15 beats after standing) gives an indication of parasympathetic function.

Cardiovascular response to head-up tilt table test. Upon changing from a recumbent to an upright position on a tilt table, a redistribution of venous blood occurs from the central to the peripheral compartment that results in a decrease in BP, after which the baroreflex restores the BP. The BP is regulated by the sympathetic nervous system; the HR by the parasympathetic nervous system.

Cardiovascular response to breathing. This test is based on the variation in HR during the respiratory cycle. The mean difference between the fastest HR during inspiration and the slowest HR during expiration (E/I difference) is expressed in beats per minute. Furthermore, the E/I ratio can also be calculated: the quotient of the HR during inspiration and the HR during expiration. The variation in HR during breathing is regulated by the parasympathetic nervous system.

Cardiovascular response to Valsalva maneuver. The Valsalva maneuver consists of respiratory strain which increases intrathoracic and intra-abdominal pressure and has four different phases. Variation in BP during the Valsalva maneuver is generally considered as a sympathetic measure. The Valsalva ratio is the ratio of the maximal HR in phase two to the minimal HR in phase four and is considered as a parasympathetic measure.

Cardiovascular response to isometric exercise. Sustained muscle contraction withdraws parasympathetic activity and increases sympathetic activity through the exercise reflex. The BP changes are regulated by sympathetic function and the HR changes by parasympathetic function.

Cardiovascular response to mental stress. Mental stress, such as speech tasks or calculating tasks, noise and emotional pressure, increases BP and HR. It has been used as a measure of sympathetic function.

Sympathetic skin response (SSR). This test is indicated in assessing the integrity of peripheral sympathetic function. The SSR is defined as the momentary change of the electrical potential of the skin, spontaneous or reflexively evoked by internal stimuli or external electrical stimulation. The latency is the interval between the onset of the stimulus and the peak of each potential. The amplitude is the potential difference between the maximal positive and the maximal negative deflection. Decreased amplitude indicates diminished sympathetic response. Latency measures are of little value, although central delay in sympathetic neurons may cause relevant alterations [22].

Heart rate variability (HRV). Even if the HR per minute remains stable, the interbeat interval varies from beat to beat. HRV is a measure of these variations in HR, and can be calculated in the time domain or frequency domain. The HRV in the frequency domain is determined by spectral analysis. The power of several components contributing to the HRV can be derived from short as well as 24-hour measurements. In experimental research of the ANS function in somatization, the following bands are important: the Low Frequency (LF)-HRV band usually defined at 0.04 – 0.15 Hz, and the High Frequency (HF)-HRV band defined at 0.15 – 0.40 Hz [23]. The variability in the HF-HRV band is mainly caused by parasympathetic activity, and is driven by respiration. More controversial is the interpretation of the LF-HRV band: some experts consider LF-HRV power as a measure of sympathetic modulation, others interpret it as a combination of sympathetic and parasympathetic activity. The consensus is that LF-HRV reflects a mixture, but is generally a marker of sympathetic activity [24,25]. Some researchers use the LF-HRV/HF-HRV ratio as a measure of sympathovagal balance.

Baroreflex sensitivity (BRS). The ANS is capable of rapidly influencing the BP. These short term BP alterations are regulated by the baroreceptor reflex. Like spectral analysis of HRV, one can also conduct spectral analysis of variation in BP. BRS is the relation between the systolic increase in BP and the increase in interbeat interval. A low BRS means a decreased functioning of the ANS, because it reflects a limited adaptation to BP fluctuations. There are also alternative methods of BRS determination such as measuring BP and HR alterations in response to a stimulus, such as standing or the Valsalva maneuver.

3.2. The ANS in Functional Somatic Disorders

In general, stress decreases activity of the parasympathetic "resting" nervous system, while increasing activity of the sympathetic "stress" nervous system. Since functional somatic disorders are associated with enhanced levels of life stress, ANS alterations would most likely consist of a decreased activity of the parasympathetic nervous system and/or an increased activity of the sympathetic nervous system. In this perspective, ANS dysfunction may either be a mediator between stress and experiencing symptoms, or the organic pathology which causes symptoms. For example, increased sympathetic activity may influence central symptom perception [26]. Another possible mechanism is that dysfunction of the ANS, as a result of chronic stress, could elicit complaints itself, since the ANS innervates multiple organs.

Some studies investigate only one parameter; others conduct a battery of autonomic function tests. The most widely used method is HRV measurement in rest or during autonomic function tests, such as response to standing, tilt table test, Valsalva maneuver or other physical stimuli. Some studies use BRS, others examine the expiration/inspiration (E/I) ratio, BP and HR reactivity or the SSR. Despite the variety of methods, studies reveal a reasonably consistent picture.

3.2.1. Chronic Fatigue Syndrome

Detailed hypotheses explaining how ANS dysfunction may be etiologically related to CFS are scant. Frequently, the motive for studying the ANS is the appearance of autonomic symptoms in CFS.

There are eight CFS studies examining ANS function with HRV. All studies measure HRV under baseline conditions. The majority of studies (including the largest study in 39 CFS patients and 31 controls) reveals no difference between patients and controls in LF-HRV and HF-HRV in rest [27-30]. Two studies only measure HF-HRV without LF-HRV, and report lower [31] or normal HF-HRV [32] resting values in CFS patients compared to healthy controls. One study reports lower HF-HRV but no difference in LF-HRV in CFS patients compared to healthy controls [33]. It is important to note that in the latter researchers select a priori subjects with postural orthostatic tachycardia syndrome (POTS). They stratify a POTS group with CFS and a POTS group without CFS, so this CFS group is not comparable with unselected CFS patients. The only study determining both increased LF-HRV and decreased HF-HRV, applies less stringent fatigue criteria compared to the other CFS studies [34].

Seven studies assess HRV during challenge tests. In three studies, subjects underwent tilt table testing. These tilt table studies reveal no differences in LF-HRV [27,30,33] while two studies observe lower HF-HRV [27,33] and one study observes normal HF-HRV [30] in CFS patients compared to healthy controls. In response to standing, LF-HRV is higher in CFS patients compared to healthy controls, while there is no difference in HF-HRV[28] . Studies performing other challenge tests measuring only HF-HRV, show lower HF-HRV in CFS patients compared to healthy sedentary controls after treadmill walking [32] and during fixed breathing rates [31]. Furthermore, LF-HRV and HF-HRV responsiveness is decreased during mental stress in CFS patients [34].

Some other studies analyze HR or BP reactivity during a battery of cardiovascular reflex tests to explore autonomic function. Often, there is no difference in HR reactivity to deep breathing [28,35,36], to the Valsalva maneuver [28,35,36] and to isometric exercise [28] in CFS patients compared to healthy controls. One study observes a lower E/I ratio in CFS patients compared to healthy controls [37]. A potential difference in HR reactivity to standing is equivocal: two studies show no differences [28,35], whereas two other studies observe a larger increase in HR to standing [36,37] in CFS patients compared to healthy controls. One study investigates spontaneous BRS with sequential analysis and reports a higher resting BRS in CFS patients compared to healthy controls. During standing, this study demonstrates that CFS patients have a greater decline in BRS than healthy controls [38]. When assessing HR after mental stress, two studies report lower HR reactivity in CFS patients compared to healthy controls [35,39].

Concerning BP responses, studies report no difference in BP reactivity to deep breathing [28,35] or isometric exercise [35] between CFS patients and healthy controls. One study observes more decline in BP during Valsalva phase two [37], and one study observes higher BP reactivity during Valsalva phase four [35], whereas another study found no difference in BP reactivity to the Valsalva maneuver [28] comparing CFS patients and healthy controls. The results on BP reactivity to standing are also ambivalent: two studies report no difference [28,35], two studies report a larger decrease [36,37], and one study reports a smaller decrease [38] in CFS patients compared to controls. Furthermore, a low BP reactivity to mental stress after strenuous exercise in CFS patients compared to healthy controls is found [39]; also no difference in BP reactivity to mental stress has been reported [38]. These differences appear not to be explained by the degree of stressfulness of the mental tasks.

Summarizing the results, we conclude that studies often fail to detect abnormalities in ANS measures during resting conditions. However, alterations become apparent when subjects undergo particular challenge tests. If deviated, parasympathetic activity tends to be decreased, while the pattern is less clear for sympathetic activity. Responsiveness to mental stress in CFS is characterized by lower LF-HRV, HR and BP reactivity, suggesting decreased sympathetic reactivity, and by lower HF-HRV, pointing to decreased parasympathetic reactivity to mental stress.

3.2.2. Fibromyalgia

There are few detailed hypotheses about the role of the ANS in FM. One study supposes that lower capacity of the ANS in response to stress heightens sensitivity for pain [40]. How this link between pain sensitivity and ANS capacity exactly works has not been elaborated on.

There are nine studies examining ANS function in FM, with seven of them using HRV. Five of these studies report higher LF-HRV values in combination with lower HF-HRV values in the resting condition in FM patients compared to healthy controls [41-45]. Only one study reports both low LF-HRV and HF-HRV values [46], and one study reports no difference in LF-HRV and HF-HRV between FM patients and healthy controls [40]. The remarkable finding of low resting LF-HRV has never been replicated. It is noteworthy that in the study reporting this finding, FM patients are significantly older than the healthy controls. Two studies examine the circadian rhythm of HRV in FM. During 24-hour HRV recordings, both report diminished HRV. One study measures normal LF-HRV [46] and the other measures increased LF-HRV values [45], whereas both report lower HF-HRV values in FM patients compared to healthy controls. Two studies use BRS to study autonomic differences in FM, and found no differences between FM patients and healthy controls at rest [40,41].

Six studies examine responsiveness during challenge tests. In response to standing, some studies observe decreased responsiveness both of LF-HRV and HF-HRV [42,43], or only decreased LF-HRV with no difference in HF-HRV [40], or decreased HF-HRV with no measurement of LF-HRV [44]. Responses to the tilt table test in FM patients compared to healthy controls show diminished LF-HRV and HF-HRV reactivity but normal BRS [41] or no difference in both LF-HRV and HF-HRV [46]. The response to mental stress has only been studied once. This study describes diminished LF-HRV reactivity and no difference in HF-HRV or BRS in the response to mental stress between FM patients and healthy controls [40].

Two studies investigate the SSR and analyze HR and BP reactivity during cardiovascular challenge tests. The first study reports an enhanced amplitude of the SSR at rest [47]. HR responsiveness to breathing is diminished in FM patients compared to healthy controls [48]. After mental stress, one study observes no difference in SRR amplitude, but finds a prolonged latency in CFS patients compared to healthy controls [47]. HR responsiveness to mental stress is diminished, while no difference in BP responsiveness is observed comparing FM patients and healthy controls [47].

In summary, there is a clear tendency of decreased baseline parasympathetic activity combined with indications for increased baseline sympathetic activity in FM. Responsiveness of both systems seems to be impaired. Concerning mental stress, some sympathetic parameters indicate impaired responsiveness (diminished HR reactivity and decreased LF-HRV reactivity), while others (BRS and BP reactivity) are normal. No difference in HF-HRV in one study suggests normal parasympathetic activity after mental stress in FM patients compared to healthy controls. There are some reviews discussing a part of the aforementioned studies in FM, mainly focusing on the sympathetic nervous system. Those reviews are in agreement with our conclusion, postulating that there is evidence for increased sympathetic activity in rest, and that there are indications for impaired responsiveness to stressors [49-51].

3.2.3. Irritable Bowel Syndrome

In studies on IBS, it is hypothesized that ANS alterations contribute to abnormal gastrointestinal motility and visceral hypersensitivity. There is no general agreement on the relative roles of these mechanisms in the etiology of IBS, however, both may mediate between stress and IBS [52].

Eight studies determine HRV in IBS, with four studies specifically measuring HRV in rest. The study with the largest patient group (127 IBS patients and 51 healthy controls) observes differences between patients and healthy controls in 10 minutes measurements of HRV. These differences include lower HF-HRV and higher LF-HRV/HF-HRV ratio in patients; separate values of LF-HRV are not shown [52]. The other studies observe no difference in both LF-HRV and HF-HRV at rest in IBS patients compared to healthy controls [53,54] whereas one study finds higher LF-HRV and comparable HF-HRV [55] in IBS patients compared to healthy controls. One study reports normal LF-HRV but lower HF-HRV in IBS patients versus healthy controls [56]. The pattern of normal LF-HRV but lower HF-HRV is in agreement with a study conducted in patients with functional dyspepsia [57], another functional somatic disorder with gastrointestinal symptoms. The fact that the two latter studies conduct 24-hour HRV measurements, thus including resting and active periods, may explain the discrepancies with most other studies.

Measuring HRV after challenge tests, normal LF-HRV with decreased HF-HRV values after standing and isometric exercise become apparent in IBS patients compared to healthy controls [53]. One study reports unchanged LF-HRV and HF-HRV values in IBS patients versus healthy controls after tilt table test [55]. Two studies investigate HRV after deep breathing; both report increased LF-HRV, in one case combined with decreased HF-HRV [54] and in the other case with normal HF-HRV [55].

One study comparing LF-HRV/HF-HRV ratio in IBS patients and healthy controls observes an increased alteration of LF-HRV/HF-HRV in response to a meal, and no differences in response to mental stress [58].

Two studies perform SSR measurements. One study measuring SSR during rest reveals no difference between IBS patients and healthy controls [59]. The other study observes higher SSR amplitude after rectal balloon stimulation in IBS patients compared to healthy controls [52]. Furthermore, two studies examining HR during breathing find a low E/I difference [53,60].

Functional abdominal pain is a functional somatic disorder that is related to IBS in symptomatology. In a study of ANS dysfunction in children with functional abdominal pain, some affected children exhibit excessive HR increase and BP decrease during the tilt test, while HR activity in response to deep breathing and to phase four of the Valsalva maneuver were normal in all cases. Limitations of this study include the small sample size (only eight patients) and the absence of a control group (reference values are based on literature) [61]. Another study in patients with functional abdominal pain assesses HRV using the time domain analysis instead of the commonly used frequency domain analysis. This study reports higher HRV in patients with functional abdominal pain, which is conflicting with results in IBS. An additional control group with patients with explained abdominal pain reveals no differences in HRV compared to healthy controls [62]. Additional studies should investigate if ANS dysfunction in functional abdominal pain differs from IBS.

In summary, a tendency for decreased baseline parasympathetic activity is evidenced by low HF-HRV values, whereas baseline sympathetic activity appears to be normal in IBS patients compared to healthy controls. Clearly, decreased activity of parasympathetic function after stimulation is apparent (notably, lower HF-HRV response to cardiovascular challenge test and decreased E/I difference to deep breathing). Sympathetic function measured by the SSR after rectal balloon stimulation, a challenge test particularly relevant in IBS, is increased in IBS patients compared to healthy controls. The ANS response to mental stress has not been studied in IBS.

4. HPA-Axis

4.1. Physiology and Assessment of the HPA-axis

While the ANS responds almost immediately after stress exposure, the HPA-axis effects become apparent a few minutes later. The physiology of the HPA-axis is extensively outlined elsewhere [20,63]. For the understanding of the research of HPA-axis function in somatization, we will give a brief summary. In response to several afferent stimuli, the hypothalamus releases corticotrophin-releasing hormone (CRH) and arginin vasopressin (AVP). CRH and AVP synergistically induce release of adrenocorticotropic hormone (ACTH) from the pituitary. ACTH is secreted with a diurnal rhythm superimposed upon the pulses of CRH. The lowest serum ACTH concentrations occur in the early night, whereas the highest ACTH concentrations occur between 4:00 and 6:00 a.m. In addition to this diurnal rhythm, ACTH responds to a wide variety of stimuli. ACTH controls the release of cortisol from the adrenal cortex. The diurnal curve of total and free plasma cortisol includes 7 to 13 pulses of cortisol secretion per day. Half of the total daily cortisol is secreted within the major burst before dawn. Cortisol causes negative feedback of its own secretion at several levels, including hippocampus, hypothalamus and pituitary. The main biological effects of cortisol are widespread and include increasing glucose levels and BP, lipolysis, protein catabolism, breakdown of muscle mass, lowering the activity of the immune system and altering mood. These biological effects depend on many factors other than serum blood levels, such as sensitivity of receptors and the presence of other molecules. It is important to realize that the HPA-axis does not function on its own, but interacts with many other bodily systems. For instance, CRH causes increased sympathetic nervous system activity and may also regulate beta-endorphin and its analgesic action. Furthermore, cytokines are also able to interfere with HPA-axis function.

HPA-axis function is being characterized by CRH, ACTH and cortisol levels in serum, saliva and urine at baseline or after challenge tests. Some researchers collect HPA hormones solely at one time point, others collect HPA hormones several times a day or even during consecutive days. ACTH is frequently studied, but the clinical significance of altered levels of this hormone, followed by a normal cortisol response, is unknown. Methods for assessing the HPA-axis are briefly outlined below.

Baseline assessments. CRH, ACTH and cortisol can be measured in blood. Cortisol is also present in saliva; salivary cortisol has been shown to be a reliable indicator of blood cortisol [64]. Non-protein bound cortisol in blood is filtered by the kidney and excreted unchanged in urine. The 24-hour urinary free cortisol (24-h UFC) represents only 2 -3 % of the daily adrenal cortisol output. Measuring 24-hour urinary cortisol metabolites, generated primarily in the liver and accounting for more than 95% of cortisol production, provides an integrated picture of the hormone production over the course of the entire day [65].

Challenge tests. Numerous pharmaceutical agents are able to stimulate the HPA-axis. In the CRH challenge test, the serum ACTH response to exogenous CRH infusion is measured. The ACTH response to exogenous CRH is a good indicator of the existing CRH tonus. In CRH deficiency states, there is an exaggerated ACTH response due to upregulation of CRH receptors in the pituitary. Because CRH and AVP act synergistically on the pituitary to secrete ACTH, exogenous AVP infusion is able to stimulate release of ACTH and cortisol [66]. Serotonin agonists such as buspirone, ipsapirone and d-fenfluramine activate the HPA-axis. The non-selective opioid antagonist naloxone attenuates opiodergic inhibition of the HPA-axis and thus stimulates the HPA-axis. The pro-inflammatory cytokine IL-6 is also capable of stimulating hypothalamic CRH release. Hypoglycaemia due to a controlled insulin infusion is a severe stressor that should elicit a burst of ACTH followed by cortisol secretion. The insulin tolerance test (ITT) is the gold standard for the measurement of HPA-axis function. The ACTH challenge test measures the ability of the adrenal cortex to increase cortisol secretion in response to ACTH administration.

Release of cortisol can be pharmacologically suppressed by dexamethasone and metyrapone. Dexamethasone is a long-acting cortisol analogue which suppresses ACTH secretion. Suppression of ACTH secretion by dexamethasone causes a decrease in adrenal cortisol synthesis and thus a decrease in serum cortisol. The 1 mg overnight dexamethasone suppression test (DST) is commonly used instead of the formal 6-day DST. The suppression test with metyrapone is the gold standard.

In addition to pharmacological provocation, there are also non-pharmacological methods for stimulating HPA-axis function, such as awakening and mental stress. Awakening acts as a mild stressor in a naturalistic setting; the increase of cortisol in at least the first 30 minutes after awakening gives an indication of the responsiveness of the HPA-axis [67]. Mental stress, induced by the Trier Social Stress Test for instance, increases ACTH and subsequent cortisol secretion [68].

4.2. The HPA-Axis in Functional Somatic Disorders

There are several hypotheses about how altered levels of CRH or cortisol can cause functional somatic disorders. Firstly, CRH levels not only modulate the endocrine response, but also influence pain perception. Although acute stress is known to produce analgesia, chronic stress may have the opposite effect, a process mediated by CRH [69,70]. Furthermore, low cortisol levels may cause widespread pain and fatigue [71]. In the opposite, high cortisol levels may also play a role in somatization. It is hypothesized that elevated cortisol levels increase the likelihood of perception and misattribution of bodily signals [72].

4.2.1. Chronic Fatigue Syndrome

Conditions associated with hypocortisolism, such as Addison's disease, lead to fatigue and other symptoms frequently seen in CFS. This has been a reason to assume that a mild form of adrenal insufficiency also exists in CFS.

Since 1991, nearly forty studies have been published on baseline and stimulated HPA-axis function in CFS. The majority of the studies measuring baseline serum ACTH at one time point in the morning or afternoon reports no significant difference between CFS patients and healthy controls [66,73-77]; one study reports higher baseline serum ACTH in CFS patients in the evening [78]. Mean ACTH in serial blood samples of CFS patients is decreased [79,80] or reveals no difference compared to healthy controls [81,82]. Thus, baseline serum ACTH seems unchanged in CFS, although in serial serum samples in two of four studies a lower level is observed.

Besides ACTH, studies also determine baseline serum cortisol levels. In the largest study (62 CFS and 46 healthy controls), the serum morning cortisol appears low [83]. Some studies are in support of this finding of low morning serum cortisol [84-87]. One study finds low serum cortisol in the evening [78]. However, the majority reports no difference [75-77,82,88-92] at one time point in the morning or afternoon; one study observes even high serum cortisol in the evening [73] in CFS patients compared to healthy controls. In serial serum cortisol sampling the findings are also contradictory. One study observes lower peak morning serum cortisol [93], whereas one study observes higher nocturnal peak serum cortisol [94], and four studies observe no difference [66,79,80,82] in CFS patients compared to healthy controls. Another method for baseline cortisol measurement is analyzing salivary cortisol. In serial salivary cortisol sampling, one study reports higher salivary cortisol [95], two studies detect no difference in salivary cortisol [96,97] and one study observes lower salivary cortisol [98] in CFS patients compared to healthy controls. Examining 24-h UFC, the study with the largest patients group (121 CFS and 64 healthy controls) reveals lower levels in CFS patients compared to healthy controls [99]. Some studies support this finding of low 24-h UFC in CFS patients [73,78,100]. Nevertheless, the majority of studies reveals no differences in 24-h UFC [79,81,88,93,97] or 24-h urinary cortisol metabolites [101] comparing CFS patients with healthy controls. In a study which analyzes 3-hourly samples, UFC levels are lower in CFS patients than in healthy controls, but levels of urinary free metabolites are similar in both groups [102]. Some studies look specifically to the circadian rhythm of cortisol in CFS patients. Two studies observe a normal diurnal rhythm [79,102], whereas one study observes a delayed rhythm [81] comparing CFS patients and healthy controls. Another study finds decreased diurnal variation in cortisol levels in CFS patients [82]. Looking at baseline cortisol levels, frequently no difference has been observed between CFS patients and controls, albeit particularly in larger studies cortisol values are lower in CFS patients than controls.

Challenge tests are undertaken on different time points and with different doses. Serum ACTH after the CRH challenge test reveal no difference [73,75], or observed levels appear lower [76,78,84] in CFS patients compared to healthy controls. Serum cortisol levels in CFS patients often reveal no differences compared to healthy controls [73,75,78,88], although two studies report lower serum cortisol after the CRH challenge test in CFS patients [76,86]. Performing the ITT, one study observes no difference in ACTH levels [94], and one study observes lower ACTH in CFS patients compared to healthy controls [103]. Concerning cortisol responses in the ITT, one study observes lower serum cortisol [86], whereas three studies show no difference in serum or salivary cortisol [94,103,104] comparing CFS patients with healthy controls. Other pharmacological challenge tests reveal disturbed ACTH responses but no differences in (subsequent) cortisol responses comparing CFS patients and healthy controls. Although ACTH levels reveal no differences after AVP administration [66], ACTH levels are increased after d-fenfluramine [104] and decreased after ipsapirone [91] and naloxone [77,78] administration comparing CFS patients and healthy controls. Despite of these disturbed ACTH responses, studies observe comparable serum cortisol responses in CFS patients and healthy controls after administration of AVP [66], d-fenfluramine [87,92,104], ipsapirone [91] and naloxone [77,88]. The ACTH challenge test evokes different results. Serum and salivary cortisol levels are equal [90,105,106] or decreased [78,84] after the ACTH challenge test in CFS patients compared to healthy controls. One study investigates salivary cortisol after the DST in CFS patients and in healthy controls, finding lower levels in the former [96].

Sometimes, non-pharmacological tests are utilized to evoke a stress response. In response to awakening, one study reports lower salivary cortisol levels [107] and one study reports no difference in salivary cortisol levels [96] comparing CFS patients and healthy controls. Serum ACTH response is impaired after exercise [74,103] and social stress [103], while subsequent serum and salivary cortisol levels reveal no differences between CFS patients and controls [103]. The meaning of ACTH disturbances without alterations in subsequent cortisol levels is not known.

Other functional somatic disorders with fatigue as the predominant symptom reveal no associations with HPA-axis function. One study examines the relation between salivary cortisol levels after the DST in 41 adults from a community sample with self-perceived fatigue, observing no differences with cortisol levels measured in healthy controls [108]. The time of blood collection, 15:30 hour, is a shortcoming of this study, because the serum cortisol after an overnight DST should be determined at 8:00 hour. The outcome of this study can therefore better be considered as a single baseline blood sample.

In summary, results are far from uniform, with low, normal and high levels of HPA-axis hormones observed, but it appears that of all HPA-axis abnormalities, low levels of basal cortisol are most frequently observed in CFS. There is no remarkable tendency for the deviated cortisol levels to be found at a distinctive time point. In response to challenge tests, ACTH levels are more often disturbed than cortisol levels. The most recent studies mainly reveal normal HPA-axis determinants, although the largest studies reveal baseline hypocortisolism.

In an extensive review published in 2003, Cleare states that evidence is at least firmly against any increases in basal serum cortisol levels or HPA-axis responses to challenge test. In contrast, over half of the studies suggest lowered basal cortisol and/or blunted HPA-axis responses to challenge [109]. Subsequent studies underline this conclusion.

4.2.2. Fibromyalgia

McCain et al. and Ferracioli et al. were the first to investigate the HPA-axis in relation to FM [110,111]. Calculating the mean of serial sampling, the former reports higher serum cortisol in FM patients compared to rheumatoid arthritis patients. Confusingly, lower 24-h UFC values in FM patients are also reported in this study. The latter observes no difference in baseline serum cortisol between FM patients and rheumatoid arthritis patients. Performing the DST, both report more non suppressors in the FM group than in the rheumatoid arthritis group.

Studies which are accomplished afterwards usually include a healthy control group. Many studies examine baseline HPA-axis activity. All studies report comparable baseline serum ACTH in FM patients and healthy controls [81,112-116]. Despite the normal baseline ACTH levels, cortisol levels are sometimes disturbed. Many studies measure serum cortisol at one time point. The three largest studies (68, 63 and 40 patients) all identify lower morning serum cortisol comparing FM patients and healthy controls [83,116,117]. Besides those three large studies, only one other study reports low baseline serum cortisol in the morning [112]. Smaller studies measuring serum cortisol at a single time point find normal levels in FM patients in the morning [113-115,118,119] and evening [115]. Two FM studies and one study with subjects with chronic widespread pain (comparable with FM, albeit less stringent criteria) measure salivary cortisol levels. The study concerning chronic widespread pain detects no difference in the salivary cortisol in the morning but finds low salivary cortisol in the evening in FM patients compared to controls [120]. One FM study detects no difference in salivary cortisol at any of five diurnal time points [121] and the other FM study detects higher mean serial salivary cortisol levels [122] in patients compared to healthy controls. This high baseline cortisol in the latter is an isolated finding when comparing FM patients and healthy controls.

Several studies look at 24-hour cortisol production and excretion. Two studies find no alterations in 24-hour serum cortisol in FM [81,123]. Measuring cortisol in the urine, four studies observe no significant differences in 24-h UFC excretion between FM patients and controls [81,115,124,125], whereas one study measures lower 24-h UFC in the group of FM patients [116]. Studies analyzing differences in the rhythm of cortisol secretion are conflicting: a normal diurnal rhythm is reported [121-123], but also a delay in the rate of decline of cortisol [81] in FM patients compared to healthy controls. Decreased diurnal variation has been observed in FM patients compared to patients with rheumatoid arthritis [110].

Measuring HPA-axis function under stimulated conditions, three studies perform the CRH challenge test. Comparing FM patients to healthy controls, high ACTH responses in two studies [116,126] and no differences in the ACTH response in another study [112] are found. All three studies find no differences in subsequent cortisol responses between FM patients and healthy controls. Performing the ITT, one study observes high ACTH responses [115], whereas another study observes low ACTH responses [126] in FM patients compared to healthy controls. However, both studies show comparable subsequent cortisol levels in FM patients and healthy controls. Two studies conduct the ACTH challenge test. One study fails to detect differences in cortisol response [116], while the other study reports decreased cortisol levels after the ACTH challenge test in FM patients compared to healthy controls

[118]. It is notable that the former performs the ACTH challenge immediately after a DST. Using other pharmacological agents, no differences in HPA-axis reactivity between FM patients and healthy controls appear. Comparable serum ACTH and serum cortisol levels are observed after buspirone [113] and after IL-6 administration [124]. Two FM studies determine serum cortisol (derivates) levels after suppression tests. The first study observes lower 11-deoxycortisol levels after the metyrapone test in FM patients compared to healthy controls [118]. The second study is not able to find differences in cortisol suppression after the DST in FM patients compared to healthy controls [126]. The results in the study in chronic widespread pain underline the conclusion of the latter [120]. Furthermore, HPA-axis responses after physical stress have been measured. One study observes lower serum cortisol values after an exercise test [119], while another study observes no differences in serum ACTH and cortisol between FM patients and healthy controls [114]. The study on chronic widespread pain reveals no differences between FM patients and healthy controls in cortisol response to a pain pressure threshold test [120]. Noteworthy are reported associations between pain experience and cortisol in two studies. In one study, current pain symptoms at awakening and an hour later are positively correlated with actual serum cortisol levels [121]. In another study, the cortisol level is correlated with sensoric pain levels in FM patients but not in healthy controls [113]. Measuring salivary cortisol levels after self perceived mental stress, one study did not find differences in FM patients compared to healthy controls [122].

HPA-axis function has also been determined in other functional somatic disorders with predominantly musculoskeletal symptoms. Studies have been performed in patients with temporomandibular dysfunction [127,128], low back pain [116] and the whiplash syndrome [129]. With respect to baseline serum ACTH levels, musculoskeletal functional somatic disorders do not differ from healthy controls [116]. Studies are not unanimous in baseline cortisol results. Two studies find comparable single time point salivary [128] or serial serum cortisol levels [116] in musculoskeletal functional somatic disorders compared to healthy controls. One study observes higher serial serum cortisol in temporomandibular dysfunction [127]. Another study measures lower 24-UFC in a patients with low back pain compared to healthy controls [128].

HPA-axis challenge tests in musculoskeletal functional somatic disorders reveal heterogeneous results. After the CRH challenge test, one study finds an increased maximal ACTH response in low back pain patients compared to healthy controls. Measuring subsequent cortisol responses, this study reports no difference in serum cortisol [116]. Another study finds enhanced cortisol suppression after DST in the whiplash syndrome compared to healthy controls. In response to awakening, this study detects decreased salivary cortisol in whiplash syndrome patients compared to healthy controls [129]. In response to mental stress, one study observes higher salivary cortisol responses in temporomandibular dysfunction [128].

To recapitulate, studies measuring basal HPA-axis activity frequently failed to find baseline differences between FM patients and controls, however, the two largest studies find reduced basal cortisol levels. Disturbed ACTH responses after dynamic stress tests are frequently seen in FM patients, however, subsequent cortisol levels are often comparable with those of healthy controls. If there are alterations in cortisol response to challenge, frequently a tendency for decreased responsiveness is seen. Studies with mental stress as challenge test are scarce. Musculoskeletal functional somatic disorders reveal a comparable picture. Our conclusion is supported by reviews about the neuroendocrine dysfunction in FM [10,51,130]. They state that basal HPA-axis activity tends to be normal in fibromyalgia. With respect to dynamic responsiveness of the HPA-axis, FM patients show ACTH hypersecretion in response to severe acute stressors, but cortisol responses to various stressors are mostly normal.

4.2.3. Irritable Bowel Syndrome

Hypotheses about specific IBS related HPA-dysfunction concern disturbances of the brain-gut axis. Two studies hypothesize that IBS and HPA-axis function may be related since CRH is known to alter gastric and colonic motility [131,132]. Another study presents an alternative view, stating that increased cortisol secretion due to HPA-axis activation causes suppression of inflammatory processes of the gut and thus increases the risk for gastrointestinal infection [133].

Like in other functional somatic disorders, both baseline and stimulated HPA-axis activity have been measured in IBS studies. All studies on baseline serum ACTH observe no difference between IBS patients and controls [132-135]. Measuring baseline serum cortisol responses, two studies observe no difference in the afternoon [132,135] and one study finds higher serum cortisol in the evening [133] in IBS patients compared to healthy controls. Levels of cortisol in saliva also differ between studies. The largest study reports lower evening salivary cortisol and higher morning salivary cortisol in 55 IBS patients compared to 28 healthy controls [136], while other studies detect no difference between IBS patients and healthy controls in the morning or evening [58,131]. The only study collecting 24-h UFC finds no differences between IBS patients and controls [137].

Results of HPA-axis activity after challenge tests turn out to be heterogeneous. Two studies have performed the CRH challenge test. One study measures high serum ACTH and cortisol [133], whereas another study measures lower serum and salivary cortisol in IBS patients compared to healthy controls [131]. Cortisol levels after the DST have been determined twice. Both studies find no differences between IBS patients and healthy controls in post DST serum cortisol levels [133] or salivary cortisol levels [131]. It is notable that the former study measures cortisol levels in the afternoon after the overnight DST instead of the following morning, which can contribute to the finding of comparable cortisol levels.

Non-pharmacological challenge tests generally evoke comparable results in IBS patients and healthy controls. Salivary cortisol collected after a meal reveals no differences between IBS patients and healthy controls [58]. After mental stress, serum ACTH levels are low [132] or equal [135], but subsequent serum cortisol responses reveal no differences between IBS patients and healthy controls in both studies.

Functional abdominal pain and chronic pelvic pain are two other functional somatic disorders with predominantly gastrointestinal symptoms. One study detects no difference in afternoon serum ACTH and serum cortisol values in 22 patients with functional abdominal pain compared to healthy controls [62]. Another study detects higher salivary cortisol in the morning, but no differences in the afternoon or evening in children with functional abdominal pain compared to healthy controls [138]. In chronic pelvic pain, baseline ACTH and salivary cortisol are unchanged, while salivary cortisol is low after both the CRH challenge test and DST [139].

In summary, there are conflicting findings in HPA-axis activity in IBS, equally to the other functional somatic disorders. Disturbances are mainly apparent in ACTH levels. Baseline high cortisol levels seem slightly more reported in IBS and IBS related disorders compared to other functional somatic disorders, although hypocortisolism is also reported in IBS. Stimulated cortisol responses tend to be comparable with those in healthy controls, also regarding mental stress.

5. Explanations for Discrepancies in Findings

In the previous paragraphs, we discussed results of ANS and HPA-axis function research in functional somatic disorders.

In ANS research, a tendency of increased sympathetic activity and decreased parasympathetic activity is shown in functional somatic disorders. Despite of this tendency, it remains to be elucidated why the severity of ANS alteration varied widely both within and among functional somatic disorders. Some studies report deviations in virtually all ANS tests, while other studies find only one alteration in a battery of ANS tests or report no at all.

Results in HPA-axis function research suggest decreased baseline and stimulated cortisol release in most functional somatic disorders. However, results are highly heterogeneous with high cortisol levels also frequently observed. Moreover, often there is no difference in cortisol levels compared to healthy controls.

We will provide some factors that may explain the discrepancy in findings. Factors that may influence outcomes are in part similar for ANS and HPA-axis function. Therefore, explanations for discrepancy in findings in both stress systems are discussed in the same paragraph.

5.1. Variability in Assessment of Parameters

Large differences in protocols for measuring unstimulated and stimulated stress-axis measurements are clear for both the ANS and HPA-axis.

Remarkable differences in ANS function study methodology include type of cardiovascular reflex test, duration of HR measurements and definition of frequency bands in HRV studies. A wide range of cardiovascular reflex tests has been utilized, with some tests being more specific for sympathetic or parasympathetic disturbances than others. Deviated definition of frequency bands in HRV may form an explanation for inconsistent findings. For example, mid frequency bands (MF-HRV) are sometimes defined at the frequency commonly defined as LF-HRV [55,56] and upper limits of HF-HRV range from 0.35 Hz to 1.0 Hz. Furthermore, in some protocols resting position means supine position and in others the sitting or standing position.

Differences in HPA-axis study methodology include time of sampling, type of stimulation test and standardization of sampling with regard to menstrual cycle. The time of baseline serum or saliva collection varies widely between studies, which limits their comparability owing to the circadian rhythm of HPA-axis hormone secretion. The lack of difference between patients and controls in the evening may be explained by the normal physiology of the diurnal HPA rhythm, in which ACTH and cortisol levels decrease [140], and therefore inter-individual cortisol variability in the evening is very low. Furthermore, the types of stimulation tests comprise a range of chemical, physical and mental stress tests. Studies use different kinds of chemical stress, such as CRH, ACTH, insulin, buspirone, naloxone, d-fenfluramine and ipsapirone administration. Different kinds of physical stress, such as awakening, eating a meal, rectal balloon dilation, exercise and undergoing pain have been used. Mental stress may refer to public speaking, performing cognitive tasks and reporting self-perceived stress. Moreover, even within one type of test, there were methodological differences. Studies may differ in doses of administered chemical agents, for example in the ACTH challenge test or the DST. Furthermore, the way of measuring hormone levels differs between studies. Some researchers measure total hormone release in a time period with serial sampling, while others measure hormone release at one distinctive time point. Furthermore, menstrual cycle may influence HPA responsiveness to psychological stress [141]. The extent to which studies examine this parameter varies widely. Finally, various methods are used to quantify cortisol levels, including competitive binding assays and massaspectometry.

Summarizing, it is already difficult to compare outcomes of the studies solely based on the primary assessment of HPA function, without engaging possible confounders. One can criticise the comparability of these highly differing methods for assessing stress system function in functional somatic disorders.

5.2. Psychiatric Comorbidity

Psychiatric comorbidity is frequently seen in subjects with functional somatic disorders. The question raises if ANS and HPA-axis alterations are specifically due to the aetiology of the functional somatic disorder, or that the ANS and HPA-axis alterations occur mainly in the course of a comorbid psychiatric disorder.

It is possible that major depression is associated with ANS dysfunction, characterized by increased sympathetic and diminished parasympathetic activity [142]. These suggested disturbances are comparable with alterations found in functional somatic disorders. In the majority of studies, psychiatric morbidity has not been determined. Concerning CFS studies, patients with mood disorders are often excluded. In the remainder, exclusion criteria for psychiatric morbidity vary or authors report that none of the patients has a psychiatric disorder. Adjusting for the possible influences of psychiatric diseases on the ANS, one study finds that both concurrent and premorbid depression or anxiety do not correlate with the measures of autonomic function [37].

Indications also exist for disturbances in HPA-axis associated with psychiatric disorders. Examples are enhanced CRH secretion and hypercortisolism in major depression [143,144] and HPA-axis abnormalities in post-traumatic stress disorder and alcoholism [145]. Concerning the HPA-axis, part of the studies have not examined psychiatric disorders in their study participants. In other studies, every history of DSM-IV axis I disorder is considered as an exclusion criterion, while less strict psychiatric exclusion criteria have also been utilized. Some studies state that none of their patients had a psychiatric disorder. Others diagnose psychiatric diagnoses with a structured clinical assessment or questionnaire and keep affected participants in their analyses. Nevertheless, often studies conclude that the influence of psychiatric comorbidity on their results is limited. Several authors report that exclusion of CFS subjects with depression does not alter their results [99,100,103,107]. One study compares the CFS group with a depression group and observes lower salivary cortisol in CFS patients in the evening compared to depressed subjects. However, in the morning there is no difference [98]. Also in temporomandibular dysfunction [127] and gastrointestinal functional disorders [131] authors report that exclusion of subjects with depression does not influence the outcome of the study.

Although psychiatric diseases may be able to disturb ANS and HPA-axis function, it seems that they do not play a major role in discrepancies in findings on functional somatic disorders

5.3. Medication Use

Medication use may interfere with ANS and HPA-axis function. Therefore, differences in medication use could cause differences in findings on stress axes alterations in functional somatic disorders.

Occasionally, medication use has been inquired in ANS research. Some studies report that they exclude participants taking medication such as anticholinergic antidepressants and antihistamines or adrenergic beta blockers. One study reports that 7 out of 103 IBS patients and 2 out of 49 controls were using selective serotonin reuptake inhibitors or anxiolytics, with apparently no influence on HRV [56]. Another study in 12 CFS patients reports that there is no significant difference in parasympathetic activity comparing medicated (desipramine, fluoxetine, bupropion, antihistaminics, NSAIDs) and non-medicated patients [31].

Concerning HPA-axis studies, medication use is more often determined. Some studies ask patients to discontinue their medication, especially when the medication may interfere with HPA-axis function. The kind of discontinued medication and the duration of discontinuation differ between studies. Other studies exclude participants with every current medication use, or only patients using medication that may interfere with HPA-axis function. Examples are glucocorticoids, contraceptives or other hormones, neuroleptics and antidepressants. The different outcomes in HPA-axis findings may be partly explained by the different methods researchers apply to correct for medication use. These differences exist not only among the different functional somatic disorders, but also within one functional somatic disorder.

Although HPA-axis changes in functional somatic disorders may be an epiphenomenon of medication use, after adjusting for medication use usually the same results are obtained. In a large study, both subjects with CFS on and off medication have been shown to have low 24-UFC [99]. Two other studies in CFS and whiplash disorder also find that repeating the analyses after excluding patients with medication use does not alter HPA-axis results [82,129].

Comparable with psychiatric comorbidity, it seems that current medication use does not play a major role on explaining discrepancies in findings in functional somatic disorders

5.4. Life Style

It is conceivable that patients with functional somatic disorders have an altered sleep pattern and undertake fewer activities. These life style features are believed to explain both ANS and HPA-axis alterations found in these disorders to some extent. Since the majority of the studies do not measure activity level of the participants, this is a factor of unknown significance. Some studies select sedentary controls to avoid this confounder to some extent . Others adjust for psychosocial and life style factors. For example, one study finds no correlation between salivary cortisol response to awakening and salivary cortisol day profile with self-reported sleep quality or duration in CFS [96]. In general, the precise role of life style in ANS and HPA-axis function remains uncertain.

5.5. Other Sample Characteristics

As previously mentioned, psychiatric comorbidity and medication use appear to be highly different between study samples. Also other parameters, such as duration of illness, compliance to study protocol or genetic characteristics may vary between patient populations, thus hampering comparison between studies.

It is proposed that alterations in stress systems might be course depending. In ANS function research, sympathetic and parasympathetic disturbances may differ in relation to the phase of the disorder. Decreased responsiveness of the ANS may be a late feature of functional somatic disorders. Likewise, hypercortisolism may be a feature of the acute state of stress, followed by a normocortisolism in the subacute phase, and hypocortisolism on the long term. The consequence may be that stress axes results depend on duration of illness of subjects studied. This is supported by a study concluding that the observed HPA-axis alterations are especially prevalent in subjects having FM for longer than two years [110]. However, duration of illness does not influence ACTH and cortisol levels in a CFS study [82]. Similarly, illness duration is not associated with any endocrine parameter in whiplash syndrome [129].

Variability in study protocol compliance could be another parameter explaining the discrepancies in findings, particularly in case of HPA-axis studies. One study investigated self-report of compliance in a salivary cortisol sampling protocol in female FM patients and healthy controls. Objective compliance among participants unaware of monitoring is 71%, though their self-reported compliance is 93%. Patients were somewhat more compliant than healthy volunteers. Non-compliance with the sampling protocol results in cortisol data that significantly differ from compliant data [146].

It is plausible that specific genes also partially explain variations in ANS and HPA-axis function. Preliminary evidence exists for an association between polymorphisms in genes regulating the HPA-axis and subtypes of unexplained chronic fatigue. It is possible that genetic variations influencing HPA-axis function may contribute to the inconsistencies in HPA-axis findings [147]. Polymorphisms in such genes may form an explanation why only subgroups of patients with functional somatic disorders exhibit alterations in stress systems.

5.6. Referral Bias

Generally, groups investigating the stress axes in somatization select patients from outpatient clinics, and not from primary care or the community. The consequence of this selection method may be that the study population is not representative for the group of patients as a whole. This can be a merit for etiological research, since stress axes changes may be more pronounced in the severe forms of functional disorders found in tertiary care. However, this highly selected type of patient may have higher rates of potential confounders, such as psychiatric comorbidity and medication use. These higher rates of confounders may hinder comparability of studies. Furthermore, one may argue that the utility of the findings is limited for clinical practice, since the general practitioner has to deal with these patients more often than the medical specialist.

In summary, there are numerous disparities in ANS and HPA-axis research methodology. None of these disparities seems to be on its own responsible for the differences in stress system function results in functional somatic disorders. However, it can not be excluded that combinations of small differences may explain the heterogeneous findings.

6. Conclusion

Despite the wide range of methods applied for measuring ANS function in functional somatic disorders, the results suggest the presence of alterations in ANS function. In most of the studies, there is a decreased parasympathetic activity in at least one of the tests in subjects with a functional somatic disorder. The majority of the studies examining sympathetic activity reveal in at least one test an increased activity in patients with functional somatic disorders. Studies that investigate the sympathetic modulation almost unanimously reveal a decreased sympathetic response to stress. This could be explained by receptor saturation by persistent increased sympathetic activity, because of which there is no capacity left for response to further stimulation. Nevertheless, there are some studies which failed to find alterations, or even found alterations in another direction. These ANS alterations are observed across functional somatic disorders.

The results of the reviewed HPA-axis studies in functional somatic disorders did not exclusively support one separate kind of dysfunction. The palette of different findings about the association between somatization and the HPA-axis includes reduced cortisol output, enhanced negative feedback and deviated ACTH and cortisol responses to a variety of challenges. However, locating the anatomical focus of disturbance (hypothalamus, pituitary or adrenals) is not possible. Moreover, the disturbances are subtle.

Since a substantial part of studies found no differences between patients and controls, it can be suggested that alterations in HPA-axis and ANS are not the core of the functional somatic disorders. Nonetheless, they may interfere in perpetuating the experience of physical symptoms. There are several factors which may underlay the inconsistent findings so far. There are disparities in sample characteristics, such as medication use, psychiatric comorbidity, duration of illness and lifestyle factors. There are also disparities in methodology of HPA-axis and ANS function assessment. However, in studies specifically adjusting for these factors, none of the possible confounders seems to excessively interfere in the association between stress axes activity and functional somatic disorders. It can not be excluded that combinations of small disturbances still form a reasonable explanation for the heterogeneous findings. Alternatively, heterogeneous HPA-axis and ANS findings may be an indication for the existence of different pathophysiological subtypes of patients with functional somatic disorders, both within and across disorders.

It is remarkable that the disturbances in the two most important stress systems in the human body, the ANS and HPA-axis, seem contradicting: the tendency of a state of hypocortisolism and overall impaired HPA-axis activity is the opposite of increased sympathetic activity, one of the principal findings in ANS research. However, chronic disturbances in an endocrine system may have a different course than those in a nervous system. This idea is supported by a study in healthy subjects, showing that HPA-axis responses quickly habituate after repeated psychosocial stress, while the sympathetic nervous system shows rather uniform activation patterns with repeated exposure to stress [148].

It is clear that a coherent theory about the role of ANS and HPA-axis function in somatization is needed. Some recommendable directions for future research are discussed in the next paragraph.

7. Recommendations for Future Research

Although there is profound evidence for disturbances in the two main bodily stress systems, none of these disturbances seems to be a condition sine qua non. An integrated etiological model of somatization is needed, including both psychosocial and biomedical factors. Furthermore, researchers should try to avoid frequently mentioned limitations in previous studies.

First of all, measurement of confounding factors (length of illness, perceived life stress, medication use, psychiatric comorbidity, sleep pattern and activity level) must be standardized. Studies should perform pre-defined subgroup analyses for these factors instead of defining them as exclusion criteria, to avoid limited generalizability of studies.

In addition, consensus about a gold standard for ANS and HPA-axis function assessment would mean a major improvement in the field of psychoneuroendocrinology. Up to date, studies differ in sampling methods, time of measurements and type of stimulation Furthermore, studies should standardize methodology for quantifying sympathetic activity. Although the LF-HRV band is frequently considered as a sympathetic measure in the research about the association between the ANS and functional somatic disorders, careful examination of the sympathetic nervous system may require an alternative test. A review with recommendations for measuring HPA-axis and ANS function in psychoneuroendocrinological research is warranted.

The precise nature of the psychoneuroendocrinological findings in somatization has to be further elucidated. Little is known about the entire pathway from benign bodily signals to functional somatic disorders, and the potential role of the ANS and HPA-axis. Additional fundamental studies are needed to gain a more complete understanding of mechanisms of perception of bodily signals in healthy and somatizing individuals. To determine this adequately, it is crucial to investigate the exact sequence of events. An overview of three possible pathways is shown in figure 1. Firstly, chronic stress could induce ANS and HPA-axis alterations, which in turn induce a functional somatic disorder: the etiological pathway. Secondly, chronic stress could induce psychiatric morbidity and concomitant medical treatment, which induce ANS and HPA-axis alterations. In this epiphenomenal pathway, functional somatic complaints develop in parallel as a result of chronic stress. Thirdly, chronic stress could induce a functional somatic disorder, which subsequently induces ANS and HPA-axis alterations, due to lifestyle alterations or pain experience: the consequential pathway. All these mechanisms are likely to contribute to somatization, but it has to be determined which is most weighty.

It should be stressed that studies with a cross sectional design and small clinical samples are not very helpful to solve this problem. Emphasis should be on how ANS and HPA-axis function alterations are connected with somatization. Almost all studies performed so far concerning ANS and HPA-axis function in somatization have a cross sectional design. In this design, it cannot be clarified whether demonstrated psychoneuroendocrinological alterations are a cause or an effect of somatization. Few studies about functional somatic disorders are already longitudinal. One study prospectively observes patients with an acute episode of a high risk infection for developing either IBS or CFS. It is concluded that the nature of the precipitating infection is important, and that premorbid levels of distress are more strongly associated with CFS than with IBS [149]. Unfortunately, this study does not examine ANS or HPA-axis parameters. Another study demonstrates that those at risk of chronic widespread pain also expose abnormalities of HPA-axis function, although less marked than the group with established chronic widespread pain [120]. The actual development of a functional somatic disorder in the high risk group is currently being studied [10]. More equivalent longitudinal studies are needed to elucidate the meaning of the detected psychoneuroendocrinological alterations in somatization. Similar longitudinal investigations could be performed at the other side of the disorder, in those who have recovered from a functional somatic disorder. If disturbances in HPA-axis or ANS function disappear together with the disorder, these psychoneuroendocrinological alterations are more likely to be state than trait markers [11].

Despite the limitations and current uncertainties, preliminary data yield interesting results about the role of the ANS and HPA in somatization. To allow a final conclusion about their relevance, further discussion and careful longitudinal multidisciplinary studies are required.

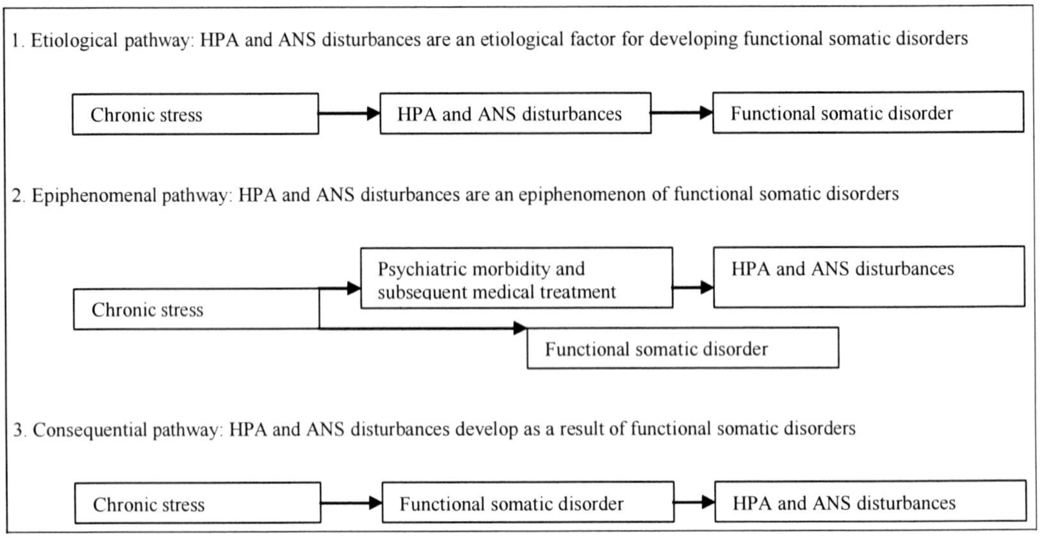

Figure 1. Three simplified representations of theories about the association between functional somatic disorders and psychoneuroendocrinological alterations. Future research should verify the amount of which each pathway contributes to somatization.

REFERENCES

[1] Burton C. Beyond somatisation: a review of the understanding and treatment of medically unexplained physical symptoms (MUPS). *Br J Gen Pract* 2003;53(488):231-9.

[2] Mayou R, Kirmayer LJ, Simon G, Kroenke K, Sharpe M. Somatoform disorders: time for a new approach in DSM-V. *Am J Psychiatry* 2005;162(5):847-55.

[3] Aaron LA, Buchwald D. A review of the evidence for overlap among unexplained clinical conditions. *Ann Intern Med* 2001;134(9 Pt 2):868-81.

[4] Pearlin I.L. Stress and mental health: A conceptual overview. In: Scheid T.L., Horwitz A.V., editors. *A handbook for the study of mental health: Social contexts, theories, and systems.* New York: Cambridge University Press; 1999. p. 161-76.

[5] Chrousos GP, Gold PW. The concepts of stress and stress system disorders. Overview of physical and behavioral homeostasis. *JAMA* 1992;267(9):1244-52.

[6] Prins JB, van der Meer JW, Bleijenberg G. Chronic fatigue syndrome. *Lancet* 2006;367(9507):346-55.

[7] Heim C, Wagner D, Maloney E, Papanicolaou DA, Solomon L, Jones JF, Unger ER, Reeves WC. Early adverse experience and risk for chronic fatigue syndrome: results from a population-based study. *Arch Gen Psychiatry* 2006;63(11):1258-66.

[8] Kato K, Sullivan PF, Evengard B, Pedersen NL. Premorbid predictors of chronic fatigue. *Arch Gen Psychiatry* 2006;63(11):1267-72.

[9] van Houdenhove B, Egle U, Luyten P. The role of life stress in fibromyalgia. *Curr Rheumatol Rep* 2005;7(5):365-70.

[10] Gupta A, Silman AJ. Psychological stress and fibromyalgia: a review of the evidence suggesting a neuroendocrine link. *Arthritis Res Ther* 2004;6(3):98-106.

[11] van Houdenhove B, Egle UT. Fibromyalgia: a stress disorder? Piecing the biopsychosocial puzzle together. *Psychother Psychosom* 2004;73(5):267-75.

[12] Wolfe F, Hawley DJ. Psychosocial factors and the fibromyalgia syndrome. *Z Rheumatol* 1998;57 Suppl 2:88-91.

[13] Folks DG. The interface of psychiatry and irritable bowel syndrome. *Curr Psychiatry Rep* 2004;6(3):210-5.

[14] Blanchard EB, Scharff L. Psychosocial aspects of assessment and treatment of irritable bowel syndrome in adults and recurrent abdominal pain in children. *J Consult Clin* Psychol 2002;70(3):725-38.

[15] Mayer EA. The neurobiology of stress and gastrointestinal disease. *Gut* 2000;47(6):861-9.

[16] Drossman DA. Do psychosocial factors define symptom severity and patient status in irritable bowel syndrome? *Am J Med* 1999;107(5A):41S-50S.

[17] Burke P, Elliott M, Fleissner R. Irritable bowel syndrome and recurrent abdominal pain. A comparative review. *Psychosomatics* 1999;40(4):277-85.

[18] Barry S, Dinan TG. Functional dyspepsia: are psychosocial factors of relevance? *World J Gastroenterol* 2006;12(17):2701-7.

[19] Brett JF, Brief AP, Burke MJ, George JM, Webster J. Negative affectivity and the reporting of stressful life events. *Health Psychol* 1990;9(1):57-68.

[20] Berne RM, Stanton BA. The autonomic nervous system and its central control. In: Berne RM, Levy RL, editors. *Physiology*, Fourth Edition. Missouri: Mosby; 1998. p. 233-48.

[21] Ravits JM. AAEM minimonograph #48: autonomic nervous system testing. *Muscle Nerve* 1997;20(8):919-37.

[22] Vetrugno R, Liguori R, Cortelli P, Montagna P. Sympathetic skin response: basic mechanisms and clinical applications. *Clin Auton Res* 2003;13(4):256-70.

[23] Heart rate variability: standards of measurement, physiological interpretation and clinical use. Task Force of the European Society of Cardiology and the North American Society of Pacing and Electrophysiology. *Circulation* 1996;93(5):1043-65.

[24] Sztajzel J. Heart rate variability: a noninvasive electrocardiographic method to measure the autonomic nervous system. *Swiss Med Wkly* 2004;134(35-36):514-22.

[25] Kobayashi H, Ishibashi K, Noguchi H. Heart rate variability; an index for monitoring and analyzing human autonomic activities. *Appl Human Sci* 1999;18(2):53-9.

[26] Rief W, Barsky AJ. Psychobiological perspectives on somatoform disorders. *Psychoneuroendocrinology* 2005;30(10):996-1002.

[27] Yamamoto Y, LaManca JJ, Natelson BH. A measure of heart rate variability is sensitive to orthostatic challenge in women with chronic fatigue syndrome. *Exp Biol Med (Maywood)* 2003;228(2):167-74.

[28] De Becker P, Dendale P, De MK, Campine I, Vandenborne K, Hagers Y. Autonomic testing in patients with chronic fatigue syndrome. *Am J Med* 1998;105(3A):22S-6S.

[29] Duprez DA, De Buyzere ML, Drieghe B, Vanhaverbeke F, Taes Y, Michielsen W, Clement DL. Long- and short-term blood pressure and RR-interval variability and psychosomatic distress in chronic fatigue syndrome. *Clin Sci (Lond)* 1998;94(1):57-63.

[30] Yataco A, Talo H, Rowe P, Kass DA, Berger RD, Calkins H. Comparison of heart rate variability in patients with chronic fatigue syndrome and controls. *Clin Auton Res* 1997;7(6):293-7.

[31] Sisto SA, Tapp W, Drastal S, Bergen M, DeMasi I, Cordero D, Natelson B. Vagal tone is reduced during paced breathing in patients with the chronic fatigue syndrome. *Clin Auton Res* 1995;5(3):139-43.

[32] Cordero DL, Sisto SA, Tapp WN, LaManca JJ, Pareja JG, Natelson BH. Decreased vagal power during treadmill walking in patients with chronic fatigue syndrome. *Clin Auton Res* 1996;6(6):329-33.

[33] Stewart JM. Autonomic nervous system dysfunction in adolescents with postural orthostatic tachycardia syndrome and chronic fatigue syndrome is characterized by attenuated vagal baroreflex and potentiated sympathetic vasomotion. *Pediatr Res* 2000;48(2):218-26.

[34] Pagani M, Lucini D, Mela GS, Langewitz W, Malliani A. Sympathetic overactivity in subjects complaining of unexplained fatigue. *Clin Sci (Lond)* 1994;87(6):655-61.

[35] Soetekouw PM, Lenders JW, Bleijenberg G, Thien T, van der Meer JW. Autonomic function in patients with chronic fatigue syndrome. *Clin Auton Res* 1999;9(6):334-40.

[36] Winkler AS, Blair D, Marsden JT, Peters TJ, Wessely S, Cleare AJ. Autonomic function and serum erythropoietin levels in chronic fatigue syndrome. *J Psychosom Res* 2004;56(2):179-83.

[37] Freeman R, Komaroff AL. Does the chronic fatigue syndrome involve the autonomic nervous system? *Am J Med* 1997;102(4):357-64.

[38] Peckerman A, LaManca JJ, Qureishi B, Dahl KA, Golfetti R, Yamamoto Y, Natelson BH. Baroreceptor reflex and integrative stress responses in chronic fatigue syndrome. *Psychosom Med* 2003;65(5):889-95.

[39] LaManca JJ, Peckerman A, Sisto SA, DeLuca J, Cook S, Natelson BH. Cardiovascular responses of women with chronic fatigue syndrome to stressful cognitive testing before and after strenuous exercise. *Psychosom Med* 2001;63(5):756-64.

[40] Friederich HC, Schellberg D, Mueller K, Bieber C, Zipfel S, Eich W. Stress and autonomic dysregulation in patients with fibromyalgia syndrome. *Schmerz* 2004.

[41] Furlan R, Colombo S, Perego F, Atzeni F, Diana A, Barbic F, Porta A, Pace F, Malliani A, Sarzi-Puttini P. Abnormalities of cardiovascular neural control and reduced orthostatic tolerance in patients with primary fibromyalgia. *J Rheumatol* 2005;32(9):1787-93.

[42] Cohen H, Neumann L, Shore M, Amir M, Cassuto Y, Buskila D. Autonomic dysfunction in patients with fibromyalgia: application of power spectral analysis of heart rate variability. *Semin Arthritis Rheum* 2000;29(4):217-27.

[43] Cohen H, Neumann L, Alhosshle A, Kotler M, bu-Shakra M, Buskila D. Abnormal sympathovagal balance in men with fibromyalgia. *J Rheumatol* 2001;28(3):581-9.

[44] Martinez-Lavin M, Hermosillo AG, Mendoza C, Ortiz R, Cajigas JC, Pineda C, Nava A, Vallejo M. Orthostatic sympathetic derangement in subjects with fibromyalgia. *J Rheumatol* 1997;24(4):714-8.

[45] Martinez-Lavin M, Hermosillo AG, Rosas M, Soto ME. Circadian studies of autonomic nervous balance in patients with fibromyalgia: a heart rate variability analysis. *Arthritis Rheum* 1998;41(11):1966-71.

[46] Raj SR, Brouillard D, Simpson CS, Hopman WM, Abdollah H. Dysautonomia among patients with fibromyalgia: a noninvasive assessment. *J Rheumatol* 2000;27(11):2660-5.

[47] Thieme K, Rose U, Pinkpank T, Spies C, Turk DC, Flor H. Psychophysiological responses in patients with fibromyalgia syndrome. *J Psychosom Res* 2006;61(5):671-9.

[48] Ulas UH, Unlu E, Hamamcioglu K, Odabasi Z, Cakci A, Vural O. Dysautonomia in fibromyalgia syndrome: sympathetic skin responses and RR Interval analysis. *Rheumatol Int* 2006;26(5):383-7.

[49] Sarzi-Puttini P, Atzeni F, Diana A, Doria A, Furlan R. Increased neural sympathetic activation in fibromyalgia syndrome. *Ann N Y Acad Sci* 2006;1069:109-17.

[50] Petzke F, Clauw DJ. Sympathetic nervous system function in fibromyalgia. *Curr Rheumatol Rep* 2000;2(2):116-23.

[51] Adler GK, Geenen R. Hypothalamic-pituitary-adrenal and autonomic nervous system functioning in fibromyalgia. *Rheum Dis Clin North Am* 2005;31(1):187-202, xi.

[52] Tillisch K, Mayer EA, Labus JS, Stains J, Chang L, Naliboff BD. Sex-specific alterations in autonomic function among patients with irritable bowel syndrome. *Gut* 2005.

[53] Waring WS, Chui M, Japp A, Nicol EF, Ford MJ. Autonomic cardiovascular responses are impaired in women with irritable bowel syndrome. *J Clin Gastroenterol* 2004;38(8):658-63.

[54] Adeyemi EO, Desai KD, Towsey M, Ghista D. Characterization of autonomic dysfunction in patients with irritable bowel syndrome by means of heart rate variability studies. *Am J Gastroenterol* 1999;94(3):816-23.

[55] Karling P, Nyhlin H, Wiklund U, Sjoberg M, Olofsson BO, Bjerle P. Spectral analysis of heart rate variability in patients with irritable bowel syndrome. *Scand J Gastroenterol* 1998;33(6):572-6.

[56] Heitkemper M, Jarrett M, Cain KC, Burr R, Levy RL, Feld A, Hertig V. Autonomic nervous system function in women with irritable bowel syndrome. *Dig Dis Sci* 2001;46(6):1276-84.

[57] Silva L, de O, Souza A, Mesquita MA. Autonomic function in patients with functional dyspepsia assessed by 24-hour heart rate variability. *Dig Dis Sci* 2002;47(1):27-31.

[58] Elsenbruch S, Lovallo WR, Orr WC. Psychological and physiological responses to postprandial mental stress in women with the irritable bowel syndrome. *Psychosom Med* 2001;63(5):805-13.

[59] Lee CT, Chuang TY, Lu CL, Chen CY, Chang FY, Lee SD. Abnormal vagal cholinergic function and psychological behaviors in irritable bowel syndrome patients: a hospital-based Oriental study. *Dig Dis Sci* 1998;43(8):1794-9.

[60] van Orshoven NP, Andriesse GI, van Schelven LJ, Smout AJ, Akkermans LM, Oey PL. Subtle involvement of the parasympathetic nervous system in patients with irritable bowel syndrome. *Clin Auton Res* 2006;16(1):33-9.

[61] Chelimsky G, Boyle JT, Tusing L, Chelimsky TC. Autonomic abnormalities in children with functional abdominal pain: coincidence or etiology? *J Pediatr Gastroenterol Nutr* 2001;33(1):47-53.

[62] Jorgensen LS, Christiansen P, Raundahl U, Ostgaard S, Christensen NJ, Fenger M, Flachs H. Autonomic nervous system function in patients with functional abdominal pain. An experimental study. *Scand J Gastroenterol* 1993;28(1):63-8.

[63] Niewoehner, CB, Bantle J. The adrenal cortex. In: *Endocrine Pathophysiology*. Connecticut: Fence Creek Publishing; 1998. p. 61-76.

[64] Kirschbaum C, Hellhammer DH. Salivary cortisol in psychoneuroendocrine research: recent developments and applications. *Psychoneuroendocrinology* 1994;19(4):313-33.

[65] Raven PW, Taylor NF. Sex differences in the human metabolism of cortisol. *Endocr Res* 1996;22(4):751-5.

[66] Altemus M, Dale JK, Michelson D, Demitrack MA, Gold PW, Straus SE. Abnormalities in response to vasopressin infusion in chronic fatigue syndrome. *Psychoneuroendocrinology* 2001;26(2):175-88.

[67] Clow A, Thorn L, Evans P, Hucklebridge F. The awakening cortisol response: methodological issues and significance. *Stress* 2004;7(1):29-37.

[68] Kudielka BM, Schommer NC, Hellhammer DH, Kirschbaum C. Acute HPA axis responses, heart rate, and mood changes to psychosocial stress (TSST) in humans at different times of day. *Psychoneuroendocrinology* 2004;29(8):983-92.

[69] Lariviere WR, Melzack R. The role of corticotropin-releasing factor in pain and analgesia. *Pain* 2000;84(1):1-12.

[70] Clauw DJ, Chrousos GP. Chronic pain and fatigue syndromes: overlapping clinical and neuroendocrine features and potential pathogenic mechanisms. *Neuroimmunomodulation* 1997;4(3):134-53.

[71] Fries E, Hesse J, Hellhammer J, Hellhammer DH. A new view on hypocortisolism. *Psychoneuroendocrinology* 2005;30(10):1010-6.

[72] Rief W, Shaw R, Fichter MM. Elevated levels of psychophysiological arousal and cortisol in patients with somatization syndrome. *Psychosom Med* 1998;60(2):198-203.

[73] Cleare AJ, Miell J, Heap E, Sookdeo S, Young L, Malhi GS, O'Keane V. Hypothalamo-pituitary-adrenal axis dysfunction in chronic fatigue syndrome, and the effects of low-dose hydrocortisone therapy. *J Clin Endocrinol Metab* 2001;86(8):3545-54.

[74] Ottenweller JE, Sisto SA, McCarty RC, Natelson BH. Hormonal responses to exercise in chronic fatigue syndrome. *Neuropsychobiology* 2001;43(1):34-41.

[75] Kavelaars A, Kuis W, Knook L, Sinnema G, Heijnen CJ. Disturbed neuroendocrine-immune interactions in chronic fatigue syndrome. *J Clin Endocrinol Metab* 2000;85(2):692-6.

[76] Scott LV, Medbak S, Dinan TG. Blunted adrenocorticotropin and cortisol responses to corticotropin-releasing hormone stimulation in chronic fatigue syndrome. *Acta Psychiatr Scand* 1998;97(6):450-7.

[77] Scott LV, Burnett F, Medbak S, Dinan TG. Naloxone-mediated activation of the hypothalamic-pituitary-adrenal axis in chronic fatigue syndrome. *Psychol Med* 1998;28(2):285-93.

[78] Demitrack MA, Dale JK, Straus SE, Laue L, Listwak SJ, Kruesi MJ, Chrousos GP, Gold PW. Evidence for impaired activation of the hypothalamic-pituitary-adrenal axis in patients with chronic fatigue syndrome. *J Clin Endocrinol Metab* 1991;73(6):1224-34.

[79] Di Giorgio A, Hudson M, Jerjes W, Cleare AJ. 24-hour pituitary and adrenal hormone profiles in chronic fatigue syndrome. *Psychosom Med* 2005;67(3):433-40.

[80] Racciatti D, Guagnano MT, Vecchiet J, De Remigis PL, Pizzigallo E, Della VR, Di ST, Merlitti D, Sensi S. Chronic fatigue syndrome: circadian rhythm and hypothalamic-pituitary-adrenal (HPA) axis impairment. *Int J Immunopathol Pharmacol* 2001;14(1):11-5.

[81] Crofford LJ, Young EA, Engleberg NC, Korszun A, Brucksch CB, McClure LA, Brown MB, Demitrack MA. Basal circadian and pulsatile ACTH and cortisol secretion in patients with fibromyalgia and/or chronic fatigue syndrome. *Brain Behav Immun* 2004;18(4):314-25.

[82] MacHale SM, Cavanagh JT, Bennie J, Carroll S, Goodwin GM, Lawrie SM. Diurnal variation of adrenocortical activity in chronic fatigue syndrome. *Neuropsychobiology* 1998;38(4):213-7.

[83] Gur A, Cevik R, Nas K, Colpan L, Sarac S. Cortisol and hypothalamic-pituitary-gonadal axis hormones in follicular-phase women with fibromyalgia and chronic fatigue syndrome and effect of depressive symptoms on these hormones. *Arthritis Res Ther* 2004;6(3):R232-R238.

[84] Segal TY, Hindmarsh PC, Viner RM. Disturbed adrenal function in adolescents with chronic fatigue syndrome. *J Pediatr Endocrinol Metab* 2005;18(3):295-301.

[85] Cevik R, Gur A, Acar S, Nas K, Sarac AJ. Hypothalamic-pituitary-gonadal axis hormones and cortisol in both menstrual phases of women with chronic fatigue syndrome and effect of depressive mood on these hormones. *BMC Musculoskelet Disord* 2004;5:47.

[86] Kirnap M, Colak R, Eser C, Ozsoy O, Tutus A, Kelestimur F. A comparison between low-dose (1 microg), standard-dose (250 microg) ACTH stimulation tests and insulin tolerance test in the evaluation of hypothalamo-pituitary-adrenal axis in primary fibromyalgia syndrome. *Clin Endocrinol (Oxf)* 2001;55(4):455-9.

[87] Cleare AJ, Bearn J, Allain T, McGregor A, Wessely S, Murray RM, O'Keane V. Contrasting neuroendocrine responses in depression and chronic fatigue syndrome. *J Affect Disord* 1995;34(4):283-9.

[88] Inder WJ, Prickett TC, Mulder RT. Normal opioid tone and hypothalamic-pituitary-adrenal axis function in chronic fatigue syndrome despite marked functional impairment. *Clin Endocrinol (Oxf)* 2005;62(3):343-8.

[89] van Rensburg SJ, Potocnik FC, Kiss T, Hugo F, van ZP, Mansvelt E, Carstens ME, Theodorou P, Hurly PR, Emsley RA, Taljaard JJ. Serum concentrations of some metals and steroids in patients with chronic fatigue syndrome with reference to neurological and cognitive abnormalities. *Brain Res Bull* 2001;55(2):319-25.

[90] Hudson M, Cleare AJ. The 1microg short Synacthen test in chronic fatigue syndrome. *Clin Endocrinol (Oxf)* 1999;51(5):625-30.

[91] Dinan TG, Majeed T, Lavelle E, Scott LV, Berti C, Behan P. Blunted serotonin-mediated activation of the hypothalamic-pituitary-adrenal axis in chronic fatigue syndrome. *Psychoneuroendocrinology* 1997;22(4):261-7.

[92] Yatham LN, Morehouse RL, Chisholm BT, Haase DA, MacDonald DD, Marrie TJ. Neuroendocrine assessment of serotonin (5-HT) function in chronic fatigue syndrome. *Can J Psychiatry* 1995;40(2):93-6.

[93] Hamilos DL, Nutter D, Gershtenson J, Redmond DP, Clementi JD, Schmaling KB, Make BJ, Jones JF. Core body temperature is normal in chronic fatigue syndrome. *Biol Psychiatry* 1998;43(4):293-302.

[94] Moorkens G, Berwaerts J, Wynants H, Abs R. Characterization of pituitary function with emphasis on GH secretion in the chronic fatigue syndrome. *Clin Endocrinol (Oxf)* 2000;53(1):99-106.

[95] Wood B, Wessely S, Papadopoulos A, Poon L, Checkley S. Salivary cortisol profiles in chronic fatigue syndrome. *Neuropsychobiology* 1998;37(1):1-4.

[96] Gaab J, Huster D, Peisen R, Engert V, Schad T, Schurmeyer TH, Ehlert U. Low-dose dexamethasone suppression test in chronic fatigue syndrome and health. *Psychosom Med* 2002;64(2):311-8.

[97] Young AH, Sharpe M, Clements A, Dowling B, Hawton KE, Cowen PJ. Basal activity of the hypothalamic-pituitary-adrenal axis in patients with the chronic fatigue syndrome (neurasthenia). *Biol Psychiatry* 1998;43(3):236-7.

[98] Strickland P, Morriss R, Wearden A, Deakin B. A comparison of salivary cortisol in chronic fatigue syndrome, community depression and healthy controls. *J Affect Disord* 1998;47(1-3):191-4.

[99] Cleare AJ, Blair D, Chambers S, Wessely S. Urinary free cortisol in chronic fatigue syndrome. *Am J Psychiatry* 2001;158(4):641-3.

[100] Scott LV, Dinan TG. Urinary free cortisol excretion in chronic fatigue syndrome, major depression and in healthy volunteers. *J Affect Disord* 1998;47(1-3):49-54.

[101] Jerjes WK, Taylor NF, Peters TJ, Wessely S, Cleare AJ. Urinary cortisol and cortisol metabolite excretion in chronic fatigue syndrome. *Psychosom Med* 2006;68(4):578-82.

[102] Jerjes WK, Peters TJ, Taylor NF, Wood PJ, Wessely S, Cleare AJ. Diurnal excretion of urinary cortisol, cortisone, and cortisol metabolites in chronic fatigue syndrome. *J Psychosom Res* 2006;60(2):145-53.

[103] Gaab J, Huster D, Peisen R, Engert V, Heitz V, Schad T, Schurmeyer TH, Ehlert U. Hypothalamic-pituitary-adrenal axis reactivity in chronic fatigue syndrome and health under psychological, physiological, and pharmacological stimulation. *Psychosom Med* 2002;64(6):951-62.

[104] Bearn J, Allain T, Coskeran P, Munro N, Butler J, McGregor A, Wessely S. Neuroendocrine responses to d-fenfluramine and insulin-induced hypoglycemia in chronic fatigue syndrome. *Biol Psychiatry* 1995;37(4):245-52.

[105] Gaab J, Huster D, Peisen R, Engert V, Heitz V, Schad T, Schurmeyer T, Ehlert U. Assessment of cortisol response with low-dose and high-dose ACTH in patients with

chronic fatigue syndrome and healthy comparison subjects. *Psychosomatics* 2003;44(2):113-9.

[106] Scott LV, Medbak S, Dinan TG. The low dose ACTH test in chronic fatigue syndrome and in health. *Clin Endocrinol (Oxf)* 1998;48(6):733-7.

[107] Roberts AD, Wessely S, Chalder T, Papadopoulos A, Cleare AJ. Salivary cortisol response to awakening in chronic fatigue syndrome. *Br J Psychiatry* 2004;184:136-41.

[108] Hyyppa MT, Lindholm T, Lehtinen V, Puukka P. Self-perceived fatigue and cortisol secretion in a community sample. *J Psychosom Res* 1993;37(6):589-94.

[109] Cleare AJ. The neuroendocrinology of chronic fatigue syndrome. *Endocr Rev* 2003;24(2):236-52.

[110] McCain GA, Tilbe KS. Diurnal hormone variation in fibromyalgia syndrome: a comparison with rheumatoid arthritis. *J Rheumatol Suppl* 1989;19:154-7.

[111] Ferraccioli G, Cavalieri F, Salaffi F, Fontana S, Scita F, Nolli M, Maestri D. Neuroendocrinologic findings in primary fibromyalgia (soft tissue chronic pain syndrome) and in other chronic rheumatic conditions (rheumatoid arthritis, low back pain). *J Rheumatol* 1990;17(7):869-73.

[112] Riedel W, Schlapp U, Leck S, Netter P, Neeck G. Blunted ACTH and cortisol responses to systemic injection of corticotropin-releasing hormone (CRH) in fibromyalgia: role of somatostatin and CRH-binding protein. *Ann N Y Acad Sci* 2002;966:483-90.

[113] Malt EA, Olafsson S, Lund A, Ursin H. Factors explaining variance in perceived pain in women with fibromyalgia. *BMC Musculoskelet Disord* 2002;3:12.

[114] Gursel Y, Ergin S, Ulus Y, Erdogan MF, Yalcin P, Evcik D. Hormonal responses to exercise stress test in patients with fibromyalgia syndrome. *Clin Rheumatol* 2001;20(6):401-5.

[115] Adler GK, Kinsley BT, Hurwitz S, Mossey CJ, Goldenberg DL. Reduced hypothalamic-pituitary and sympathoadrenal responses to hypoglycemia in women with fibromyalgia syndrome. *Am J Med* 1999;106(5):534-43.

[116] Griep EN, Boersma JW, Lentjes EG, Prins AP, van der Korst JK, de Kloet ER. Function of the hypothalamic-pituitary-adrenal axis in patients with fibromyalgia and low back pain. *J Rheumatol* 1998;25(7):1374-81.

[117] Gur A, Cevik R, Sarac AJ, Colpan L, Em S. Hypothalamic-pituitary-gonadal axis and cortisol in young women with primary fibromyalgia: the potential roles of depression, fatigue, and sleep disturbance in the occurrence of hypocortisolism. *Ann Rheum Dis* 2004;63(11):1504-6.

[118] Calis M, Gokce C, Ates F, Ulker S, Izgi HB, Demir H, Kirnap M, Sofuoglu S, Durak AC, Tutus A, Kelestimur F. Investigation of the hypothalamo-pituitary-adrenal axis (HPA) by 1 microg ACTH test and metyrapone test in patients with primary fibromyalgia syndrome. *J Endocrinol Invest* 2004;27(1):42-6.

[119] van Denderen JC, Boersma JW, Zeinstra P, Hollander AP, van Neerbos BR. Physiological effects of exhaustive physical exercise in primary fibromyalgia syndrome (PFS): is PFS a disorder of neuroendocrine reactivity? *Scand J Rheumatol* 1992;21(1):35-7.

[120] McBeth J, Chiu YH, Silman AJ, Ray D, Morriss R, Dickens C, Gupta A, Macfarlane GJ. Hypothalamic-pituitary-adrenal stress axis function and the relationship with chronic widespread pain and its antecedents. *Arthritis Res Ther* 2005;7(5):R992-R1000.

[121] McLean SA, Williams DA, Harris RE, Kop WJ, Groner KH, Ambrose K, Lyden AK, Gracely RH, Crofford LJ, Geisser ME, Sen A, Biswas P, Clauw DJ. Momentary relationship between cortisol secretion and symptoms in patients with fibromyalgia. *Arthritis Rheum* 2005;52(11):3660-9.

[122] Catley D, Kaell AT, Kirschbaum C, Stone AA. A naturalistic evaluation of cortisol secretion in persons with fibromyalgia and rheumatoid arthritis. *Arthritis Care Res* 2000;13(1):51-61.

[123] Klerman EB, Goldenberg DL, Brown EN, Maliszewski AM, Adler GK. Circadian rhythms of women with fibromyalgia. *J Clin Endocrinol Metab* 2001;86(3):1034-9.

[124] Torpy DJ, Papanicolaou DA, Lotsikas AJ, Wilder RL, Chrousos GP, Pillemer SR. Responses of the sympathetic nervous system and the hypothalamic-pituitary-adrenal axis to interleukin-6: a pilot study in fibromyalgia. *Arthritis Rheum* 2000;43(4):872-80.

[125] Maes M, Lin A, Bonaccorso S, van HF, Van GA, Delmeire L, Biondi M, Bosmans E, Kenis G, Scharpe S. Increased 24-hour urinary cortisol excretion in patients with post-traumatic stress disorder and patients with major depression, but not in patients with fibromyalgia. *Acta Psychiatr Scand* 1998;98(4):328-35.

[126] Griep EN, Boersma JW, de Kloet ER. Altered reactivity of the hypothalamic-pituitary-adrenal axis in the primary fibromyalgia syndrome. *J Rheumatol* 1993;20(3):469-74.

[127] Korszun A, Young EA, Singer K, Carlson NE, Brown MB, Crofford L. Basal circadian cortisol secretion in women with temporomandibular disorders. *J Dent Res* 2002;81(4):279-83.

[128] Jones DA, Rollman GB, Brooke RI. The cortisol response to psychological stress in temporomandibular dysfunction. *Pain* 1997;72(1-2):171-82.

[129] Gaab J, Baumann S, Budnoik A, Gmunder H, Hottinger N, Ehlert U. Reduced reactivity and enhanced negative feedback sensitivity of the hypothalamus-pituitary-adrenal axis in chronic whiplash-associated disorder. *Pain* 2005;119(1-3):219-24.

[130] Geenen R, Jacobs JW, Bijlsma JW. Evaluation and management of endocrine dysfunction in fibromyalgia. *Rheum Dis Clin North Am* 2002;28(2):389-404.

[131] Bohmelt AH, Nater UM, Franke S, Hellhammer DH, Ehlert U. Basal and stimulated hypothalamic-pituitary-adrenal axis activity in patients with functional gastrointestinal disorders and healthy controls. *Psychosom Med* 2005;67(2):288-94.

[132] Posserud I, Agerforz P, Ekman R, Bjornsson ES, Abrahamsson H, Simren M. Altered visceral perceptual and neuroendocrine response in patients with irritable bowel syndrome during mental stress. *Gut* 2004;53(8):1102-8.

[133] Dinan TG, Quigley EM, Ahmed SM, Scully P, O'Brien S, O'Mahony L, O'Mahony S, Shanahan F, Keeling PW. Hypothalamic-pituitary-gut axis dysregulation in irritable bowel syndrome: plasma cytokines as a potential biomarker? *Gastroenterology* 2006;130(2):304-11.

[134] Elsenbruch S, Lucas A, Holtmann G, Haag S, Gerken G, Riemenschneider N, Langhorst J, Kavelaars A, Heijnen CJ, Schedlowski M. Public speaking stress-induced

neuroendocrine responses and circulating immune cell redistribution in irritable bowel syndrome. *Am J Gastroenterol* 2006;101(10):2300-7.

[135] Dickhaus B, Mayer EA, Firooz N, Stains J, Conde F, Olivas TI, Fass R, Chang L, Mayer M, Naliboff BD. Irritable bowel syndrome patients show enhanced modulation of visceral perception by auditory stress. *Am J Gastroenterol* 2003;98(1):135-43.

[136] Patacchioli FR, Angelucci L, Dellerba G, Monnazzi P, Leri O. Actual stress, psychopathology and salivary cortisol levels in the irritable bowel syndrome (IBS). *J Endocrinol Invest* 2001;24(3):173-7.

[137] Heitkemper M, Jarrett M, Cain K, Shaver J, Bond E, Woods NF, Walker E. Increased urine catecholamines and cortisol in women with irritable bowel syndrome. *Am J Gastroenterol* 1996;91(5):906-13.

[138] Tornhage CJ, Alfven G. Diurnal salivary cortisol concentration in school-aged children: increased morning cortisol concentration and total cortisol concentration negatively correlated to body mass index in children with recurrent abdominal pain of psychosomatic origin. *J Pediatr Endocrinol Metab* 2006;19(6):843-54.

[139] Heim C, Ehlert U, Hanker JP, Hellhammer DH. Abuse-related posttraumatic stress disorder and alterations of the hypothalamic-pituitary-adrenal axis in women with chronic pelvic pain. *Psychosom Med* 1998;60(3):309-18.

[140] Berne RM, Stanton BA. The hypothalamus and pituitary gland. In: *Physiology, Fourth Edition.* Missouri: Mosby; 1998. p. 872-909.

[141] Kirschbaum C, Kudielka BM, Gaab J, Schommer NC, Hellhammer DH. Impact of gender, menstrual cycle phase, and oral contraceptives on the activity of the hypothalamus-pituitary-adrenal axis. *Psychosom Med* 1999;61(2):154-62.

[142] Guinjoan SM, Bernabo JL, Cardinali DP. Cardiovascular tests of autonomic function and sympathetic skin responses in patients with major depression. *J Neurol Neurosurg Psychiatry* 1995;59(3):299-302.

[143] Parker KJ, Schatzberg AF, Lyons DM. Neuroendocrine aspects of hypercortisolism in major depression. *Horm Behav* 2003;43(1):60-6.

[144] Antonijevic IA. Depressive disorders -- is it time to endorse different pathophysiologies? *Psychoneuroendocrinology* 2006;31(1):1-15.

[145] Ehlert U, Gaab J, Heinrichs M. Psychoneuroendocrinological contributions to the etiology of depression, posttraumatic stress disorder, and stress-related bodily disorders: the role of the hypothalamus-pituitary-adrenal axis. *Biol Psychol* 2001;57(1-3):141-52.

[146] Broderick JE, Arnold D, Kudielka BM, Kirschbaum C. Salivary cortisol sampling compliance: comparison of patients and healthy volunteers. *Psychoneuroendocrinology* 2004;29(5):636-50.

[147] Smith AK, White PD, Aslakson E, Vollmer-Conna U, Rajeevan MS. Polymorphisms in genes regulating the HPA axis associated with empirically delineated classes of unexplained chronic fatigue. *Pharmacogenomics* 2006;7(3):387-94.

[148] Schommer NC, Hellhammer DH, Kirschbaum C. Dissociation between reactivity of the hypothalamus-pituitary-adrenal axis and the sympathetic-adrenal-medullary system to repeated psychosocial stress. *Psychosom Med* 2003;65(3):450-60.

[149] Moss-Morris R, Spence M. To "lump" or to "split" the functional somatic syndromes: can infectious and emotional risk factors differentiate between the onset of chronic fatigue syndrome and irritable bowel syndrome? *Psychosom Med* 2006;68(3):463-9.

INDEX

#

A

B

C

D

E

F

G

H

I

J

K

L

M

N

O

P

S

T

U

V

W

X

Y